THE
SOLAR SYSTEM

THE SOLAR SYSTEM

Volume 3

Saturn's Satellites—X-Ray and
Gamma-Ray Astronomy
Appendixes
Index

Editors

David G. Fisher and **Richard R. Erickson**
Lycoming College

SALEM PRESS
Pasadena, California   Hackensack, New Jersey

Editor in Chief: Dawn P. Dawson

Editorial Director: Christina J. Moose *Production Editor:* Joyce I. Buchea
Acquisitions Editor: Mark Rehn *Page Design:* James Hutson
Manuscript Editor: Jennifer L. Campbell *Layout:* William Zimmerman
Photo Editor: Cynthia Breslin Beres *Editorial Assistant:* Dana Garey

Cover photo: (©Andrea Danti/Dreamstime.com)

Library of Congress Cataloging-in-Publication Data

The solar system / editors, David G. Fisher, Richard R. Erickson.
  p. cm.
Includes bibliographical references and index.
  ISBN 978-1-58765-530-2 (set : alk. paper) — ISBN 978-1-58765-531-9 (v. 1 : alk. paper) —
ISBN 978-1-58765-532-6 (v. 2 : alk. paper) — ISBN 978-1-58765-533-3 (v. 3 : alk. paper) —
1. Solar system. 2. Astronomy. I. Fisher, David G. II. Erickson, Richard R., 1945-
  QB501.S627 2010
  523.2—dc22
                                                                            2009013008

First Printing

Contents

Units of Measure

Common prefixes for metric units—which may apply in more cases than shown above—include *giga-* (1 billion times the unit), *mega-* (one million times), *kilo-* (1,000 times), *hecto-* (100 times), *deka-* (10 times), *deci-* (0.1 times, or one tenth the unit), *centi-* (0.01, or one hundredth), *milli-* (0.001, or one thousandth), and *micro-* (0.0001, or one millionth).

Unit	Quantity	Symbol	Equivalents
Acre	Area	ac	43,560 square feet 4,840 square yards 0.405 hectare
Ampere	Electric current	A *or* amp	1.00016502722949 international ampere 0.1 biot *or* abampere
Angstrom	Length	Å	0.1 nanometer 0.0000001 millimeter 0.000000004 inch
Astronomical unit	Length	AU	92,955,807 miles 149,597,871 kilometers (mean Earth-Sun distance)
Barn	Area	b	10^{-28} meters squared (approx. cross-sectional area of 1 uranium nucleus)
Barrel (dry, for most produce)	Volume/capacity	bbl	7,056 cubic inches; 105 dry quarts; 3.281 bushels, struck measure
Barrel (liquid)	Volume/capacity	bbl	31 to 42 gallons
British thermal unit	Energy	Btu	1055.05585262 joule
Bushel (U.S., heaped)	Volume/capacity	bbl	2,747.715 cubic inches 1.278 bushels, struck measure
Bushel (U.S., struck measure)	Volume/capacity	bsh *or* bu	2,150.42 cubic inches 35.238 liters
Candela	Luminous intensity	cd	1.09 hefner candle
Celsius	Temperature	C	1° centigrade
Centigram	Mass/weight	cg	0.15 grain
Centimeter	Length	cm	0.3937 inch
Centimeter, cubic	Volume/capacity	cm^3	0.061 cubic inch
Centimeter, square	Area	cm^2	0.155 square inch
Coulomb	Electric charge	C	1 ampere second

Unit	Quantity	Symbol	Equivalents
Cup	Volume/capacity	C	250 milliliters 8 fluid ounces 0.5 liquid pint
Deciliter	Volume/capacity	dl	0.21 pint
Decimeter	Length	dm	3.937 inches
Decimeter, cubic	Volume/capacity	dm^3	61.024 cubic inches
Decimeter, square	Area	dm^2	15.5 square inches
Dekaliter	Volume/capacity	dal	2.642 gallons 1.135 pecks
Dekameter	Length	dam	32.808 feet
Dram	Mass/weight	dr *or* dr avdp	0.0625 ounce 27.344 grains 1.772 grams
Electron volt	Energy	eV	$1.5185847232839 \times 10^{-22}$ Btus $1.6021917 \times 10^{-19}$ joules
Fermi	Length	fm	1 femtometer 1.0×10^{-15} meters
Foot	Length	ft *or* '	12 inches 0.3048 meter 30.48 centimeters
Foot, cubic	Volume/capacity	ft^3	0.028 cubic meter 0.0370 cubic yard 1,728 cubic inches
Foot, square	Area	ft^2	929.030 square centimeters
Gallon (British Imperial)	Volume/capacity	gal	277.42 cubic inches 1.201 U.S. gallons 4.546 liters 160 British fluid ounces
Gallon (U.S.)	Volume/capacity	gal	231 cubic inches 3.785 liters 0.833 British gallon 128 U.S. fluid ounces
Giga-electron volt	Energy	GeV	$1.6021917 \times 10^{-10}$ joule
Gigahertz	Frequency	GHz	—
Gill	Volume/capacity	gi	7.219 cubic inches 4 fluid ounces 0.118 liter
Grain	Mass/weight	gr	0.037 dram 0.002083 ounce 0.0648 gram

Units of Measure

Unit	Quantity	Symbol	Equivalents
Gram	Mass/weight	g	15.432 grains 0.035 avoirdupois ounce
Hectare	Area	ha	2.471 acres
Hectoliter	Volume/capacity	hl	26.418 gallons 2.838 bushels
Hertz	Frequency	Hz	$1.08782775707767 \times 10^{-10}$ cesium atom frequency
Hour	Time	h	60 minutes 3,600 seconds
Inch	Length	in *or* ″	2.54 centimeters
Inch, cubic	Volume/capacity	in^3	0.554 fluid ounce 4.433 fluid drams 16.387 cubic centimeters
Inch, square	Area	in^2	6.4516 square centimeters
Joule	Energy	J	$6.2414503832469 \times 10^{18}$ electron volt
Joule per kelvin	Heat capacity	J/K	$7.24311216248908 \times 10^{22}$ Boltzmann constant
Joule per second	Power	J/s	1 watt
Kelvin	Temperature	K	–272.15 Celsius
Kilo-electron volt	Energy	keV	$1.5185847232839 \times 10^{-19}$ joule
Kilogram	Mass/weight	kg	2.205 pounds
Kilogram per cubic meter	Mass/weight density	kg/m^3	$5.78036672001339 \times 10^{-4}$ ounces per cubic inch
Kilohertz	Frequency	kHz	—
Kiloliter	Volume/capacity	kl	—
Kilometer	Length	km	0.621 mile
Kilometer, square	Area	km^2	0.386 square mile 247.105 acres
Light-year (distance traveled by light in one Earth year)	Length/distance	lt-yr	5,878,499,814,275.88 miles 9.46×10^{12} kilometers
Liter	Volume/capacity	L	1.057 liquid quarts 0.908 dry quart 61.024 cubic inches
Mega-electron volt	Energy	MeV	—
Megahertz	Frequency	MHz	—

Unit	Quantity	Symbol	Equivalents
Meter	Length	m	39.37 inches
Meter, cubic	Volume/capacity	m^3	1.308 cubic yards
Meter per second	Velocity	m/s	2.24 miles per hour 3.60 kilometers per hour
Meter per second per second	Acceleration	m/s^2	12,960.00 kilometers per hour per hour 8,052.97 miles per hour per hour
Meter, square	Area	m^2	1.196 square yards 10.764 square feet
Metric. *See* unit name			
Microgram	Mass/weight	mcg *or* µg	0.000001 gram
Microliter	Volume/capacity	µl	0.00027 fluid ounce
Micrometer	Length	µm	0.001 millimeter 0.00003937 inch
Mile (nautical international)	Length	mi	1.852 kilometers 1.151 statute miles 0.999 U.S. nautical miles
Mile (statute or land)	Length	mi	5,280 feet 1.609 kilometers
Mile, square	Area	mi^2	258.999 hectares
Milligram	Mass/weight	mg	0.015 grain
Milliliter	Volume/capacity	ml	0.271 fluid dram 16.231 minims 0.061 cubic inch
Millimeter	Length	mm	0.03937 inch
Millimeter, square	Area	mm^2	0.002 square inch
Minute	Time	m	60 seconds
Mole	Amount of substance	mol	6.02×10^{23} atoms or molecules of a given substance
Nanometer	Length	nm	1,000,000 fermis 10 angstroms 0.001 micrometer 0.00000003937 inch
Newton	Force	N	$6.14124095407198 \times 10^{25}$ atomic weight 0.224808943099711 pound force 0.101971621297793 kilogram force 100,000 dynes

Unit	Quantity	Symbol	Equivalents
Newton meter	Torque	N·m	0.7375621 foot-pound
Ounce (avoirdupois)	Mass/weight	oz	28.350 grams 437.5 grains 0.911 troy or apothecaries' ounce
Ounce (troy)	Mass/weight	oz	31.103 grams 480 grains 1.097 avoirdupois ounces
Ounce (U.S., fluid or liquid)	Mass/weight	oz	1.805 cubic inch 29.574 milliliters 1.041 British fluid ounces
Parsec	Length	pc	30,856,775,876,793 kilometers 19,173,511,615,163 miles
Peck	Volume/capacity	pk	8.810 liters
Pint (dry)	Volume/capacity	pt	33.600 cubic inches 0.551 liter
Pint (liquid)	Volume/capacity	pt	28.875 cubic inches 0.473 liter
Pound (avoirdupois)	Mass/weight	lb	7,000 grains 1.215 troy or apothecaries' pounds 453.59237 grams
Pound (troy)	Mass/weight	lb	5,760 grains 0.823 avoirdupois pound 373.242 grams
Quart (British)	Volume/capacity	qt	69.354 cubic inches 1.032 U.S. dry quarts 1.201 U.S. liquid quarts
Quart (U.S., dry)	Volume/capacity	qt	67.201 cubic inches 1.101 liters 0.969 British quart
Quart (U.S., liquid)	Volume/capacity	qt	57.75 cubic inches 0.946 liter 0.833 British quart
Rod	Length	rd	5.029 meters 5.50 yards
Rod, square	Area	rd^2	25.293 square meters 30.25 square yards 0.00625 acre
Second	Time	s *or* sec	$\frac{1}{60}$ minute $\frac{1}{3600}$ hour

Unit	Quantity	Symbol	Equivalents
Tablespoon	Volume/capacity	T *or* tb	3 teaspoons 4 fluid drams
Teaspoon	Volume/capacity	t *or* tsp	0.33 tablespoon 1.33 fluid drams
Ton (gross or long)	Mass/weight	t	2,240 pounds 1.12 net tons 1.016 metric tons
Ton (metric)	Mass/weight	t	1,000 kilograms 2,204.62 pounds 0.984 gross ton 1.102 net tons
Ton (net or short)	Mass/weight	t	2,000 pounds 0.893 gross ton 0.907 metric ton
Volt	Electric potential	V	1 joule per coulomb
Watt	Power	W	1 joule per second 0.001 kilowatt $2.84345136093995 \times 10^{-4}$ ton of refrigeration
Yard	Length	yd	0.9144 meter
Yard, cubic	Volume/capacity	yd^3	0.765 cubic meter
Yard, square	Area	yd^2	0.836 square meter

Alphabetical List of Contents

Volume 1

Volume 2

Volume 3

Category List of Contents

LIST OF CATEGORIES

Asteroids. *See* **Small Bodies**

Comets. *See* **Small Bodies**

The Cosmological Context

Dwarf planets. *See* **Small Bodies**

Earth

The Jovian System

The Saturnian System

Scientific Methods

Small Bodies

The Solar System as a Whole

The Stellar Context

THE
SOLAR SYSTEM

Saturn's Satellites

Categories: Natural Planetary Satellites; Planets and Planetology; The Saturnian System

Saturn has a remarkably diverse set of satellites. They include gigantic Titan, which retains a thick atmosphere; Enceladus, possessing a vastly reworked surface that includes active geysers; Hyperion, a disk-shaped satellite whose rotation is erratic; Phoebe, moving in a retrograde orbit; and a coorbiting pair called Janus and Epimetheus, to name some of the most interesting of the sixty confirmed satellites.

OVERVIEW

Prior to the space age, Saturn was known as the beautiful ringed world of the solar system. Many of its numerous larger satellites were discovered prior to the time of interplanetary spacecraft, the most notable being Titan, Saturn's largest satellite and the only one known from telescopic observation to maintain a thick atmosphere. Prior to the Voyager flybys, planetary scientists expected all of the other Saturn satellites to be relatively uninteresting ice inactive worlds. Only Iapetus was a curiosity, since it displayed a very reflective side and an extremely dark side as well. The Voyager flyby results and the Cassini orbiter images and observations revealed Saturn's system to be a miniature solar system in its own right with a variety of extremely interesting and diverse satellites.

When Voyager 1 passed by Saturn's largest satellite Titan in November, 1980, scientists were somewhat disappointed with imagery transmitted back to Earth. Titan appeared as a uniform orange sphere whose outline was blurred by a dense cloud cover. Closer examination found a higher layer of ultraviolet haze. The southern hemisphere has a slightly darker cast than the northern hemisphere. A clear equatorial boundary was noted, and a darker polar ring is evident in some photographs from Voyager 2. Beneath those clouds, Titan proved more interesting. Voyager 1's close passage behind the disk of Titan allowed the use of its radio transmissions to probe the satellite's atmosphere. The pressure at ground level is 1.5 times that of Earth. If Titan's lower surface gravity is taken into account, the implication is that every square meter of Titan has ten times as much gas above its surface as Earth does.

Methane was spectroscopically detected from Earth, but the prime component of Titan's atmosphere proved to be nitrogen. It was suspected that as much as 10 percent of the atmosphere is argon, and methane makes up between 1 and 6 percent of the rest of the atmosphere, increasing in concentration near Titan's surface.

At higher altitudes, solar ultraviolet rays break methane down, and new molecules form as some hydrogen is lost. Spectroscopic observations show traces of hydrogen, ethane, propane, ethylene, diacetylene, hydrogen cyanide, carbon monoxide, and carbon dioxide. Together, these components form the petrochemical smog that so frustrated the Voyager imaging team.

The upper optical haze layer lies about 280 kilometers above the surface. The main cloud deck is about 200 kilometers from the surface. Titan's solid surface is 400 kilometers smaller in diameter than previously thought, smaller than both Ganymede and Callisto in the Jupiter system. Why do these Jovian satellites not have atmospheres? Titan orbits at a greater distance from Saturn than either of these satellites do from Jupiter, so its tidal stress is less. Furthermore, Saturn is twice as far from the Sun as is Jupiter, so solar radiation intensity at Titan is four times weaker than in the Jovian system.

Beneath those tantalizing orange clouds, the surface temperature is only 94 kelvins; combined with the fact that Titan's atmospheric pressure is 1.5 bars, this temperature suggested the possibility of an ethane and/or methane sea on Titan's surface. If tidal stresses heat the interior enough, there may even be icy geysers. The possibility of life arising at such low temperatures appears unlikely, but certainly the carbon chemistry on Titan must be very interesting.

The European Space Agency (ESA) provided the Huygens probe, a combinations atmospheric entry probe and soft lander, for National Aeronautics and Space Administration's (NASA's)

Cassini program. The Orbiter carried Huygens from launch in October, 1997, to release on Christmas, 2004. For the next two weeks the probe flew independent from the Cassini orbiter, and then it entered Titan's atmosphere on January 14, 2005, and dropped down under a large parachute to a safe touchdown near Titan's Xanadu region. Some researchers expected Huygens to splash down in a cryogenic sea of liquid hydrocarbons. It became clear rather quickly that Huygens had "plopped" down in what some referred to as Titanian mud. Evidence of liquid action on the surface was found, but the original idea of liquid hydrocarbon seas were dashed. Analysis of data sent from the probe on the way down to impact revealed several layers in the atmosphere, most notably a thick haze between 18 and 20 kilometers above the surface. An Aerosol Collector and Pyrolyzer collected samples at different altitudes to determine the pressure of volatiles and organic materials. The probe was also outfitted with a gas chromatograph mass spectrometer to determine atmospheric composition. Titan's atmosphere proved to be hazier than expected, as dust particle concentration was greater than previously believed. Wind data suggested that Titan's atmosphere circulated gas from the south to north pole and back again in periodic fashion.

Winds would play a large role in planetary dynamics for this complex world. Indeed, two years after the probe's several hours of data were collected, Cassini scientists came to the conclusion that Titan's crust moves on a subsurface ocean with crust movements in part driven by wind actions. That movement was noted by comparing radar data from the Orbiter taken at different times during the mission in concert with available Huygens data. The proposed liquid subsurface layer would be located 50 to 100 kilometers beneath the crust and include liquid ammonia in a water ice slush. Floating on this layer, the crust was seen to move as much as 30 kilometers over the course of several years of Cassini observations.

Further examples of Huygens and Cassini images eventually found the evidence proving the existence of ancient shorelines and the presence of liquid on the surface of Titan. Computer models of this dynamic world had to be greatly altered due to Huygens and Cassini data, and many new questions were raised to give Titan an even more mysterious nature. However, it still was believed to be a world rather similar in some ways to an early Earth, just a world frozen at an early point of physical and chemical evolution prior to the development of life. Down in the subsurface liquid layer higher temperatures could permit complex biochemistry, but there was no information produced by Cassini to investigate that supposition.

Mimas, with its huge crater, Herschel. (NASA/JPL/Space Science Institute)

By measuring gravitational perturbations on the Voyager spacecraft as they flew through the Saturnian system, scientists at the Jet Propulsion Laboratory (JPL) could determine the masses and densities of Saturn's middle-sized satellites. Rhea's bulk density of 1.3 grams per cubic centimeter suggests that it contains more ice and fewer silicates than Titan. It is worth noting that the density of bodies depends not only on their composition but also on how tightly packed they are. The greater the mass, the higher the gravity, and thus the greater the density. Thus Titan, Ganymede, and Callisto, which all approximate the size of the planet Mercury, have very similar densities, about 1.9 grams per cubic centimeter, and the smaller, icy satellites are less dense, even though the composition may be quite similar to that of larger satellites.

Rhea and all the smaller Saturnian satellites lack atmospheres and show some signs of older, cratered surfaces. The trailing side of Rhea is covered with pale, wispy streaks, a type of feature it shares with Dione. These streaks may be evidence of venting of water vapor from these satellites' interiors, perhaps from tidally induced volcanism in the past. On Earth's moon, such activity was on the side facing Earth rather than on the trailing side. Perhaps these wisps were once found on Rhea's leading side but were eventually eroded, much as meteoric dust erases all but the youngest ray patterns around lunar craters.

During a Cassini flyby of Rhea in November, 2005, some surprising results were obtained. While the spacecraft's magnetometer did not pick up any interactions with Saturn's magnetosphere that would have indicated even a meager atmosphere about Rhea, there was evidence of a broad debris disk and one structured ring about this satellite. The debris disk extended several thousand kilometers out from Rhea, hence it was several Rhea radii in expanse. Computer simulations of the gravitational interactions of Saturn and Rhea indicated that this ring could exist for a considerable time. The source for the ring particles about the small satellite could have been a large impact event with Rhea.

Slightly smaller than Rhea, Dione is bit more dense (1.4 grams per cubic centimeter). Its wispy patterns are more marked than those of Rhea, and it also has some long cracks and large areas of fairly fresh ice, without large craters in evidence. Ice may have flowed through cracks during the cooling phase of the satellite. Unlike most substances, water expands when it freezes. Therefore, satellites made mainly of ice, or differentiated with rocky cores and mantles composed largely of water, might show such expansion cracks. Such cracks are evident on Tethys, and were noted by Voyager 2 on the Uranian moons Ariel and Titania in January, 1986.

Tethys is similar in size to Dione, but it features one huge impact crater. The crater's floor is quite flat, suggesting internal flooding resulting from impact heating. Running from the crater three-quarters of the way around Tethys is a single gigantic valley system, Ithaca Chasma. Tethys's craters are lower in relief than lunar craters. The icy crust of these Saturnian satellites is more plastic than is lunar crust. Older terrain cratering appears to have been just as heavy as on Earth's moon, resulting craters appear less rugged than the lunar highlands.

The small satellite Mimas features a notable exception to the above rule. One huge crater, Herschel, is one-third as large as Mimas itself. This crater is very deep, about nine kilometers, with a central peak about six kilometers high. It constitutes one of the most striking geological features in the solar system. This crater on such a small roughly spherical body gives it the appearance of the Death Star station in the movie *Star Wars*. Many of the planetary scientists on the Voyager and Cassini teams, and astronomy professors world wide, fondly refer to Mimas as the "Death Star Moon."

It is likely that an impact of any greater force would have broken Mimas apart. Such huge impacts might also explain the very jumbled appearance of Uranus's Miranda, which was apparently broken into several large pieces, then haphazardly reassembled by gravity later.

One highly speculative hypothesis may account for such massive impacts and the intense cratering that is evident throughout the solar system. Perhaps a terrestrial planet in an un-

Saturn's second largest moon, Rhea, may have a debris ring of its own, as illustrated in this artist's rendition. (NASA/JPL/JHUAPL)

stable orbit beyond Mars and an outer icy Jovian satellite were totally fragmented in a high-energy head-on collision. The lighter, icy debris might account for some of the comets. Heavier chunks may have found a relatively stable orbit and formed the asteroid belt. Many chunks and particles, however, would have been scattered in all directions to impact other worlds. In some cases, the impact would have been forceful enough to send up debris which, in turn, would bombard neighboring worlds. This scenario might explain ices found on certain asteroids, the existence of captured satellites such as Phoebe and Phobos, and the fact that meteoritic material seems to have originated on a differentiated planet. It might also explain how fragments which scientists agree came from lunar basalts and from Mars are found on Earth.

While almost a twin of Mimas in size, and orbiting just beyond it in the satellite system, Enceladus is a very different world up close. Even the Pioneer 11 data indicated that it has an albedo near 100 percent. It seems to be made of fresher ice, reflecting far more sunlight than most Saturnian satellites. Had its material

been older, dark meteoritic and cometary dust would have darkened it. Voyager's cameras revealed that one of its hemispheres is heavily cratered and fairly old, but the opposite side features smooth plains cut by grooved terrain, similar to Ganymede. This evidence of much rifting and recent internal activity appears on a satellite about one-tenth the size of Ganymede. A count of ring particles in Saturn's extended E ring also found that they peaked near Enceladus. Just as the dust ring of Jupiter is supplied by Io's volcanoes, icy geysers on Enceladus periodically shoot debris above this active world. Why is this small world active at all? Mimas lies closer to Saturn and thus is more tidally stressed, yet it shows no such activity. Nor does the proximity of any other large satellite seem to account for the heating required to generate such activity. Such activity on smaller satellites is not unique; Uranus's moon, Ariel, which is similar in size to Mimas, shows obvious broad rift valleys. The source of heating for this extensive and possibly continuing crustal activity is a mystery.

Cassini found geysers on Enceladus near its

south pole along long cracks which essentially act like vents. Fresh crystalline ice forms at the site of these cracks and colors the features distinctively. Cassini scientists dubbed these nearly parallel cracks found at the south polar region "Tiger Stripes." The Tiger Stripes were found to be 124 kilometers long and 40 kilometers apart. This activity was not new, so why did the Voyagers fail to see these geysers? Voyager 2 flew over Enceladus's north pole and missed them. Cassini's near-infrared mapping spectrometer and solid state imager both examined the ice around the Tiger Stripes. Freshly formed ice was crystalline. As time progresses that pristine ice becomes radiation-damaged amorphous ice.

Data and the geyser actions strongly suggested the presence of a subsurface ocean on Enceladus. However, calculations about heat transport inside Enceladus led researchers to believe that the satellite's subsurface ocean would not be able to exist for more than 30 million years if it were warmed only by heat escaping from the core toward the crust. Since that ocean most likely has been in existence for more than 30 million years, the heating mechanism for both the subsurface and the cryovolcanic activity at the satellite's south pole must be from tidal flexing. Only that could provide the 5.8 gigawatts of heat Cassini saw emerging from the Tiger Stripes over which it flew on a close fly. Since internal heat sources apart from tidal flexing produce only 0.32 gigawatts, without tidal heating Enceladus's subsurface ocean would have frozen. But the story here is more complex than that. Without the subsurface ocean, tidal flexing of the magnitude necessary to produce the observed heat would not be possible, and without the heat for the tidal flexing the ocean would freeze.

Iapetus confronts scientists with another mystery. A portion of this satellite is extremely dark, whereas the rest of Iapetus has an albedo typical of an icy surface. Iapetus's dark side is six times lower in albedo than that icy portion. Within the darker portion is an irregular dark spot. This pattern of darker leading hemispheres is also seen on Rhea and Dione, but to a far lesser degree. The brighter, icy side does have an albedo of 50 percent. This is typical of

older water-ice crusts, and it shows heavy cratering typical of other similar satellites. Some larger craters near the boundary between the hemispheres have light-colored walls, with darker flat floors, like some larger craters on Earth's moon. What is the reddish-black material that gives the darker side an albedo of only 5 percent? It is probably an organic tar, and it appears to be a good match with carbonaceous chondrite meteorites, the dark rings of Uranus, and the black crust of Halley's comet. Carbon almost always appears in an oxidized form (carbon dioxide, carbonic acid, carbonate rocks, carbohydrates) in the inner solar system. Dark neutral carbon is a major solid material found in the outer solar system. It was impossible to judge the age of Iapetus' dark spot, as Voyager's cameras could not pick up any details on the dark side. In fact, some photographs even make the dark side disappear into the blackness of space. The concentration of the dark material on the leading side suggested an external source for this coating. It had been hypothesized that dark Phoebe was responsible. Phoebe's color, however, does not match the black side of Iapetus. The dark floor of some of Iapetus's craters constitutes evidence for an internal origin. New data and insights would have to wait for Cassini to pass by Iapetus at a much closer distance than had the Voyagers.

Cassini flew within 1640 kilometers of Iapetus on September 10, 2007. Unfortunately during the encounter, Cassini entered a safe mode after on-board delicate solid state electronics suffered a cosmic ray hit. Fortunately, most of the science harvest was recovered after a short delay in playback. Among the findings was a raised area around the satellite's midsection that gave Iapetus the appearance of a walnut. Why the equatorial bulge on this unusual satellite? Julie Castillo of the Jet Propulsion Laboratory advanced an innovative explanation for the unique feature. Castillo invoked a high rotational speed early in the satellite's history coupled with heat from internal radioactivity, perhaps from aluminum 26 and iron 60 isotopes, that softened the satellite to form the equatorial bulge. Consideration of the time frame in which tidal forces forced the spin rate to diminish led to the conclusion that this particular pair of isotopes would be required, since they would be abundant and would generate heat quickly due to rapid radioactive decay. Then, from a softened and malleable state, the satellite's bulge was frozen in place before Iapetus's spin rate slowed down. More investigation would be needed to confirm or refute this theory. Unless extended mission priorities are changed and trajectories reevaluated, this could be the closest Cassini would ever get to this highly unusual satellite.

Saturn's outermost known satellite Phoebe may be a captured asteroid from the far reaches of the main belt, and therefore similar to Chiron. Phoebe's orbit is retrograde, like those of Jupiter's four outermost moonlets. All these small worlds are quite distant from the gas giants whose gravity trapped them. Cassini encountered Phoebe on the way into Saturn orbit insertion. Photographs revealed Phoebe's surface to look almost spongelike, not at all like the other Saturnian satellites.

Like the dark side of Iapetus, Phoebe has an albedo of about 5 percent, which is similar to those of two other captured asteroids, Deimos and Phobos, which orbit Mars. At 200 kilometers in diameter, Phoebe is much rounder than the odd "hamburger moon," Hyperion, which orbits between Tethys and Iapetus. Puck, a satellite of Uranus, is similarly round and dark, and about the same size as Phoebe. This fact suggests that a round shape is the norm for dark, primitive bodies such as these, and that something unusual happened with Hyperion.

Hyperion's shape is quite striking. It is a huge disk, about 250 kilometers across but only 150 kilometers thick. Like Phoebe, its surface is dark, old, and heavily cratered. Stranger still is Hyperion's rotation period. It has not yet been well defined. Like Earth's moon, Saturn's other satellites are tidally locked, with one side permanently facing the planet. As the Voyager 2 team tried to orient photographs of Hyperion to map it, however, they found that it was rotating chaotically. It appears to have no regular rotation period; it tumbles irregularly. Close coupling between Hyperion's eccentric orbit and that of Titan may cause this unique effect.

Phoebe, outermost known satellite, with its heavily cratered surface, may be a captured asteroid. Two images of the satellite shown here were imaged from the Sun-Phoebe spacecraft. (NASA/JPL/Space Science Institute)

The nine satellites discussed to this point were known well before the Voyager missions, but Voyager photographs found or confirmed eight more satellites, making Saturn's the most numerous satellite system. Several photographs suggested the existence of even more Saturnian satellites, but their periods of revolution and orbits had to be determined before they were formally recognized. All these new satellites are much closer to Saturn than are Phoebe and Iapetus, and they show a much more reflective, icy surface like Enceladus. None of these satellites is large enough to be nearly spherical or to have become differentiated. All of them have quite interesting orbits.

Just as the Trojan asteroids share Jupiter's orbit, so two of Saturn's middle-sized satellites have smaller companions in their orbits. Dione has two; Helene, the leading one, appears quite elongated, while the following one appears rounder. Lagrangian, Tethys's companion, is a smaller version of Mimas, with a huge crater from an impact that almost destroyed it.

Saturn's coorbital satellites were first spotted in 1966, but for more than a decade thereafter they were mistaken for a single satellite with an orbit under that of Mimas. Even prior to the Voyager flights, however, observers repeatedly noticed inconsistencies that led some to argue that there must be two satellites sharing the same orbit. Janus, the larger, is about 200 kilometers across, and Epimetheus is about 150 kilometers across. Actually, their orbits are not quite identical. The inner satellite has a period of 16.664 hours; the outer one has a period of 16.672 hours, or a difference of 29 seconds per orbit. Every four years, the inner satellite overtakes the outer at the speed of nine meters per second, and they exchange orbits. This close relationship and the irregular, elongated appearances of these satellites suggest they were once part of a single larger one split apart by a collision into the two pieces now sharing the same orbit.

The inner three satellites discovered by Voyagers 1 and 2 are all closely associated with Saturn's rings. Atlas, a tiny, football-shaped body, orbits just outside the bright A ring of Saturn. Prometheus is a shepherding moon, keeping the particles in Saturn's F ring in place from the in-side of that ring. Pandora plays a similar role on the outside of the F ring. Their close relationship to this thin set of ringlets may explain why the F ring sometimes appears braided. Additional satellites were identified in subsequent reviews of Voyager and other available data. Then, with the arrival of the Cassini probe in the Saturn system, the number of recognized satellites again increased significantly, reaching 60 by 2008.

KNOWLEDGE GAINED

While Jupiter possesses four satellites comparable in size to Earth's moon or even to Mercury, Saturn has only one, Titan. Like Jupiter's Ganymede and Callisto, Titan is comparable to Mercury in size but only about one-third as massive and dense. Its exact dimensions were still in debate prior to the Voyager missions. Its visible orange disk made it appear to be the largest known natural satellite; out-of-date astronomy textbooks will list Titan as the largest satellite in the solar system. Spectroscopic observations plainly revealed an atmosphere with gaseous methane and other hydrocarbons. Just how deep was the atmosphere, and what was it made of? These questions led the Voyager 1 team to target Titan as a main mission objective and to guide one probe closer to this satellite than to any other body on its mission.

The chief discoveries that resulted concerned Titan's atmosphere. It is thick, twice as dense as Earth's, but like Earth's, Titan's atmosphere is made primarily of nitrogen. Orange clouds appear to be a hydrocarbon smog, with complex organic chemistry taking place there. Surface temperatures and pressures lie close to the triple point of methane, so the surface might experience methane rains that would build up into lakes of liquid methane and freeze into methane ice at the poles. Confirmation of that would have to wait for a probe outfitted with imaging radar and/or a lander. Thus the origin of the Cassini mission. Combination of radar images taken from orbit with data from the Huygens lander eventually confirmed the presence of cryogenic lakes and found ancient shorelines of lakes no longer existent. Huygens appeared to have landed in a wet slushlike material at cryogenic temperatures rather than floating on a lake or sea or even having hit a hard icy surface.

By the time its primary mission was completed, Cassini had flown past Titan several dozen times at varying distances. Perhaps one of Cassini's biggest surprises was the detection that the surface of Titan moved as much as 30 kilometers between the earliest flyby of the Cassini primary mission (2004) and some near the time that the extended mission was approved (2008). This suggested that the crust floated on a layer of fluid, meaning the large satellite likely has an underground ocean, presumably a mixture of water and ammonia.

All of Saturn's remaining satellites are smaller than Earth's moon. Rhea is next in size, about half as large as the Moon at 1,500 kilometers in diameter; Voyager 1 showed its icy surface to be cratered, but with fresher ice creating wispy terrain. Dione is next in size, at 1,100 kilometers in diameter, and has even more wispy terrain than Rhea. Tethys featured a huge, flattened crater on one side, with a great crack or rift running to the other side.

The innermost of the satellites well-known prior to the Voyager flybys are Mimas and Enceladus, both about 500 kilometers in diameter. Mimas was found by Voyager 1 to have a dramatic impact crater one-third as large as the satellite. Enceladus is one of the most puzzling satellites, with tidal stresses producing plate activity, according to Voyager 2 data. These satellites are, in order from Titan inward: Dione, Tethys, Rhea, Enceladus, and Mimas. Prior to the Voyagers, only their orbital periods and approximate diameters were known, based on their brightness. No one had actually seen their disks. Cassini provided data that made Enceladus much more interesting to planetary scientists.

The brightness of Iapetus presented a major problem. When Gian Domenico Cassini found it in 1671, he realized that this odd satellite must be far brighter on one side (the leading hemisphere as it orbits Saturn) than on the other. Diameter measurements were impossible until Voyager photographed the disk. It proved to be about half as big as the Moon, with one side bright and icy. The other side is mostly covered with a layer of tarlike black material that hid any surface features from the cameras on Voyager 2. Similarly, little was known about Hyperion, another dark satellite orbiting between Titan and Iapetus; it was found by Voyager 2 to be irregularly shaped and tumbling without any rotational period.

Phoebe, the outermost known satellite, is distinguished by its retrograde orbit, like four of the outermost satellites of Jupiter. Like the dark side of Iapetus, Phoebe may be covered with carbon-rich material. More puzzling, the existence of Janus, a tenth moon, had been suspected, but before Voyager, photographs showed it in the wrong place. Voyager 1 detected two satellites sharing the same orbit. The other seven of Saturn's major satellites were not known prior to the Voyager missions. Cassini added considerably to the total list of Saturn's family of satellites.

CONTEXT

Practically nothing was known about Saturn's satellites prior to the Voyager flybys. Titan, Dione, Mimas, and Rhea were examined most fully by Voyager 1, in November, 1980. Until the arrival of Cassini in Saturn orbit, most information about Enceladus, Iapetus, Hyperion, and Tethys had come from Voyager 2 in August, 1981. Much about these satellites were discovered or confirmed by Voyager 1, but thanks to improved orbital data, they were best photographed by Voyager 2. Clearly, a strong argument can be made for using two spacecraft in flyby missions.

In brief, the Voyager missions found Saturn's satellite family to be a very diverse lot. Even satellites similar in size and mass, such as Mimas and Enceladus, appeared very different up close, and obviously were shaped by different processes. Each satellite has its own history of impacts. Tidal stress has played an important role in the evolution of many of these bodies, as it has in the Jovian satellite system. Each satellite has its own fascinating evolutionary story to be interpreted by geologists.

With Cassini repeatedly orbiting Saturn and conducting numerous flybys of many of the satellites, planetary scientists were able to make comparisons over time. Just as the Voyagers had piqued interest in satellites that had once been thought to be merely crater-pocked ice balls, Cassini images revealed many of the satellites not well studied by the Voyagers to also be rather intriguing in totally unexpected ways. Interest in Endeladus, for example, increased greatly due to Cassini observations.

J. Wayne Wooten and David G. Fisher

FURTHER READING

Consolmagno, Guy. *Worlds Apart: A Textbook in Planetary Sciences*. Englewood Cliffs, N.J.: Prentice Hall, 1994. A text accessible to college-level science and nonscience readers alike. Presents subjects at low-level mathematics and also involves integral calculus where required. Demonstrates how the area of planetary science progresses by questioning previous understanding in light of new observations.

Encrenaz, Thérèse, et al. *The Solar System*. New York: Springer, 2004. A thorough exploration of the solar system from early telescopic observations through the space missions that have investigated all planets with the exception of Pluto by the publication date. Takes an astrophysical approach to give our solar system a wider context as just one member of similar systems throughout the universe.

Harland, David M. *Cassini at Saturn: Huygens Results*. New York: Springer, 2007. This text provides a thorough explanation of the entire Cassini program, including the Huygens landing on Saturn's largest satellite. Essentially a complete collection of NASA releases from the start of Cassini flight operations through the majority of Cassini's seventy orbits of its primary mission. Cassini's primary mission concluded a year after this book entered print. Technical writing style but accessible to a wide audience.

_____. *Mission to Saturn: Cassini and the Huygens Probe*. New York: Springer Praxis, 2002. Another book in Springer's Space Exploration Series, this is a technical description of the Cassini program, its science goals and the instruments used to accomplishment those goals. Written before Cassini arrived at Saturn. Provides a historical review of pre-Cassini knowledge of the Saturn system.

Hartmann, William K. *Moons and Planets*. 5th ed. Belmont, Calif.: Thomson Brooks/Cole, 2005. An updated version of a classic text on planetary science. The chapter on Saturn covers all aspects of ground-based and spacecraft observations of Saturn.

Irwin, Patrick G. J. *Giant Planets of Our Solar System: An Introduction*. 2d ed. New York: Springer, 2006. Suitable as a textbook for upper-level college courses in planetary science. Focuses on Jupiter, Saturn, Uranus, and Neptune and their satellites, rings, and magnetic fields. Filled with figures and photographs. Available to the serious general audience.

Leverington, David. *Babylon to Voyager and Beyond: A History of Planetary Astronomy*. New York: Cambridge University Press, 2003. An historical approach to planetary science. Heavily illustrated, concludes with a summary of spacecraft discoveries. Suitable for general readers and the astronomy community.

Lorenz, Ralph, and Jacqueline Mitton. *Lifting Titan's Veil: Exploring the Giant Moon of Saturn*. Cambridge, England: Cambridge University Press, 2002. An in-depth examination of all that was known about Titan prior to the Cassini-Huygens mission written by an engineer who worked for the European Space Agency and an astrophysicist who was Press Officer for the Royal Astronomical Society. Describes the mission of Cassini-Huygens, but the book was published before the spacecraft arrived at Saturn.

Morrison, David, and Tobias Owen. *The Planetary System*. 3d ed. San Francisco: Pearson/Addison-Wesley, 2003. A fine survey of the solar system, and very current for the editing date. Intended for use as a introductory text, it is very well organized. Highly recommended for the general reader.

Van Pelt, Michel. *Space Invaders: How Robotic Spacecraft Explore the Solar System*. New York: Springer, 2006. An historical account of robotic planetary science missions attempted by all spacefaring nations written by an European Space Agency cost and systems engineer. As such the narrative not only explains the science but also provides a behind-the-scenes description of the development of a space exploration mission from concept proposal to flight operation.

See also: Enceladus; Iapetus; Io; Jovian Planets; Jupiter's Ring System; Jupiter's Satellites; Neptune's Ring System; Neptune's Satellites; Planetary Ring Systems; Planetary Satellites; Saturn's Ring System; Titan; Uranus's Rings; Uranus's Satellites.

Search for Extraterrestrial Intelligence

Category: Life in the Solar System

The search for intelligent life in the universe is perhaps the most profound of human endeavors. If extraterrestrial intelligence were discovered and communication established, it would irrevocably alter the conception of humanity's place in the universe. Contemporary science suggests that given a suitable planetary environment and sufficient time, life will evolve, and since intelligence and technology have high survival value, they will inevitably follow. Given that there are a myriad of suitable planets in the galaxy, intelligent life-forms, willing and able to communicate, should be abundant.

OVERVIEW

The search for extraterrestrial intelligence is a relatively recent exploratory science which assumes that if intelligent life, more technologically advanced than us, exists elsewhere in the universe, they will attempt to communicate by broadcasting messages using radio waves. Detecting and decoding such messages, however, requires considerable effort. If there is only a vanishingly small probability that an alien civilization has evolved, it would be a futile waste of time and money to search for a message. In order to arrive at an estimate of the number of technological civilizations with the ability and desire to communicate, Dr. Frank Drake con-

vened a small group of scientific experts from various fields at the National Radio Astronomy Observatory (Green Bank, West Virginia) in 1961. This group formulated a simple equation to estimate the possible number of communicative civilizations. The equation is:

$$N = R^* \, f_p \, n_e \, f_l \, f_i \, f_c \, L$$

where N is the number of advanced technological civilizations, R^* represents the rate of formation of suitable stars, f_p stands for the fraction of stars with planetary systems, n_e symbolizes the number of planets in a solar system which are suitable for life, f_l corresponds to the fraction of suitable planets on which life evolves, f_i signifies the fraction of inhabited planets on which intelligent life evolves, f_c characterizes the fraction of planets with intelligence on which a technological civilization can emerge, and L embodies the length of time the technological civilization would broadcast signals. When these factors are multiplied, the equation will yield an estimate of N.

After considerable discussion, the group narrowed the possible range of values for each factor to the following values. The average rate of star formation is the 100 billion stars in our galaxy divided by its 10 billion year age, yielding about 10 stars per year. The fraction of stars with planetary systems was estimated to be about one, but the number of habitable planets was thought to be only about one in ten (0.1). Since it is believed that life evolves wherever the conditions are conducive, the fraction of habitable planets on which life evolves was taken to be one. Because intelligence has high survival value, it was assumed that the fraction of life-bearing planets on which intelligence evolves is also one. The fraction of planets on which intelligent life develops technology is taken to be one because of technology's high survival value, and from the fact that tool-using cultures arose independently at multiple locations on Earth.

The final factor, the average lifetime of a technological civilization, is the most difficult to estimate. The Earth has had the ability to communicate by radio waves for less than a hundred years, and this state may not last another hundred years if the environmental and social problems which plague Earth are not addressed and solved. If an advanced civilization can continue in its communicative state for at least a thousand years, it could probably last for a million years. When the foregoing factors, momentarily excluding L, are multiplied, the result is a value of about one. Thus, if a technological civilization exists for only 200 years, the number of communicators in our galaxy will be only about 200 out of the 100 billion stars present; they would be few and far between. If, on the other hand, advanced civilizations can exist for at least a million years, there will be enough potential communicators to make the search worthwhile. Scientists working on the Search for Extra-terrestrial Intelligence (SETI) project have chosen the optimistic view.

The SETI Institute (Mountain View, California), was founded in 1984 as a private, nonprofit organization dedicated to searching the skies for technological indicators of extraterrestrial intelligence. Because of the vast distance between stars, the fact that no material object can travel faster than the speed of light, and the prohibitive cost of interstellar space travel, it is assumed that an alien civilization would attempt to locate other intelligent species by using radio waves to carry nonrandom messages. Two problems then present themselves. Which stars are likely candidates for intelligent life to evolve, and what radio frequencies should be scanned for possible messages? The stars most likely to have habitable zones where planets would be neither too hot nor too cold for life are main sequence G- and K-type stars. G-type stars, such as our Sun, have a surface temperature of about 6,400 kelvins and a habitable zone that extends from Venus through Mars. K-type stars have a surface temperature of approximately 4,250 kelvins and a somewhat narrower habitable zone.

Stars hotter than G probably evolve too quickly for life to develop intelligence, while stars cooler than K would have such a narrow habitable zone that a planet is unlikely to occupy it.

Only certain radio frequencies are useful for communication. From Earth's surface, the available window of frequencies ranges from 1 gigahertz (GHz) to about 30 GHz. Frequencies lower than 1 GHz would be lost in the background radio noise of the Galaxy; frequencies higher than 30 GHz are absorbed by Earth's atmosphere. There are still far too many possible frequencies available in this window to make a search practical. There are, however, two important frequencies within this window that would be likely choices for an advanced civilization, wishing to establish contact with an alien civilization, to broadcast a message. These are the 14.3 GHz frequency, radiated by neutral hydrogen (H), and the 16.7 GHz frequency, emitted by the

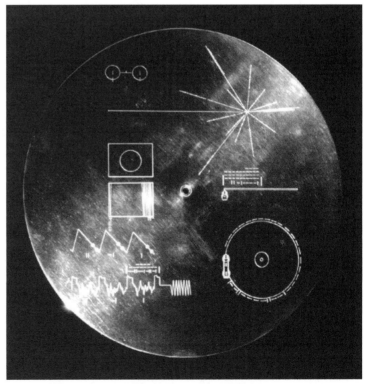

The two Voyager spacecraft carried identical copies of this gold-plated copper record, which was designed to tell the story of Earth to other intelligent civilizations that might find the spacecraft. Data included music, greetings in sixty languages, sounds from Earth, diagrams, and photographs. (NASA)

hydroxyl ion (OH). Because hydrogen is the most abundant element in the universe and because H and OH combine to form water (HOH, or H_2O), any other intelligent species based on water would be reasonably likely to choose one of these frequencies on which to broadcast. Although it is possible that an alien life-form could be formed from other molecules, there are valid chemical reasons to assume that this is unlikely.

KNOWLEDGE GAINED

No direct evidence of extraterrestrial life has ever been observed, but there exists some significant data related to the origin of life. Complex carbon molecules, the precursors of life, have been found in interstellar clouds and in certain meteorites from our solar system. Based on the not unreasonable assumption that intelligent life is abundant in our galaxy, the National Aeronautics and Space Administration (NASA) funded a search for alien radio messages during the late 1980's. This project was canceled in the early 1990's because politicians feared public ridicule for funding a search for "little green men." Although the annual cost was small (about the same as one Air Force attack helicopter) it was felt that the money would be better spent elsewhere. The SETI Institute continued the search using the radio telescopes at major radio observatories. Stars deemed likely to have planets on which intelligent life could evolve are monitored for nonrandom radio emissions; to date none has been identified.

A joint effort between the SETI Institute and the Radio Astronomy Laboratory at the University of California, Berkeley, is the Allen Telescope Array (ATA), a composite of 350 separate radio dish antennas at one location in northeast California. Designed for innovative astronomical research, the array is also being used to search for nonrandom extraterrestrial signals. Because the ATA is equivalent to one huge radio telescope, it can collect enormous amounts of data every hour of the day, every day of the year. It can also scan a wider portion of the sky and do so more quickly than existing telescopes. The ATA is programmed to survey one million stars within 1,000 light-years for extraterrestrial signals in the frequency range of 1 to 10 GHz. It will also survey the billions of stars of the inner galactic plane in the frequency range from 14.3 GHz through 16.7 GHz.

Another innovative experiment searches the sky for intelligent signals in the form of powerful pulsating flashes of laser light beamed from solar systems many light-years distant. The new pulse-detection system, coupled to the Lick Observatory's 40-inch telescope, is able to detect beacons pulsating at less than a billionth of a second with an error rate of only about one per year. This system can be automated and promises results less ambiguous than previous optical systems in which false alarms occurred daily.

CONTEXT

From the ancient Greek concept that Earth was the center of the universe, humans have always assumed their place in the cosmos to be pivotal and unique. As science advanced in the sixteenth and seventeenth centuries, Nicolaus Copernicus and Johannes Kepler showed that the Sun was the center of the known universe and Earth was merely another planet. In the early twentieth century, Edwin Hubble discovered that the Sun is a very common G-type star in a galaxy of a hundred billion stars and that the Milky Way galaxy is only one among billions in the universe.

There is nothing particularly unique about the Sun. Many humans, however, clung to the notion that few other stars were likely to have Earth-like planets. By the mid-twentieth century, however, astronomers had concluded that such planetary systems would be quite common. In order to preserve the illusion of uniqueness, it was then assumed that life on Earth was unique. Discoveries have shown, however, that the molecular precursors of life, such as amino acids, have formed elsewhere in the solar system, leading biologists to conclude that life is probably abundant in the universe. Since intelligence has a high survival value, it follows that intelligent life would not be uncommon either. Because a civilization would necessarily be more advanced to send an intense signal through interstellar space, humans would also be denied their unique rank as the supreme intelligence. Confirmation of a superior extraterrestrial intelligence would be the kiss of death to the lingering remnants of the unjustifiably self-centered sense of our uniqueness and importance in the universe.

George R. Plitnik

FURTHER READING

Cameron, A. G. W., ed. *Interstellar Communication*. New York: W. A. Benjamin, 1963. One of the earliest books to present a serious consideration of the possibilities and ramifications of communicating with extraterrestrials, this collection of reprints and original articles covers all the extant knowledge available at the time of publication.

Chaisson, E., and S. McMillan. *Astronomy: A Beginner's Guide to the Universe.* 5th ed. Upper Saddle River, N.J.: Pearson/Prentice Hall, 2007. This accessible work, written for laypersons with inquisitive minds, has an entire chapter devoted to life in the universe. Commencing with the nature and origin of life on Earth, it first considers the possibility of life elsewhere in our solar system and then details the best estimates of how many communicative civilizations may exist in our galaxy. Finally, the various searches being conducted for extraterrestrial intelligence are discussed.

Christian, James L. *Extraterrestrial Intelligence: The First Encounter.* Buffalo, N.Y.: Prometheus, 1976. A fascinating speculative account based on the not unreasonable assumption that extraterrestrial life is abundant in our galaxy and able and willing to communicate. Fourteen authors consider what might be humanity's response, in terms of our view of our place in the universe and how our theologies may respond, when contact is finally made.

Ponnamperuma, C., and A. G. W. Cameron. *Interstellar Communication: Scientific Perspectives.* Boston: Houghton Mifflin, 1974. Covers every aspect of interstellar communication from a scientific perspective. Various experts discuss the prevalence of planetary systems in our galaxy, the likelihood of communicative intelligences evolving on other planets, methods of communicating, and interstellar probes.

Seeds, Michael A. *Foundations of Astronomy.* 9th ed. Belmont, Calif.: Thomson Brooks-Cole, 2007. This lavishly illustrated text commingles experimental evidence and theory to provide deep but well-explained elucidations of many fascinating facets of the universe. The final chapter, "Life on Other Worlds," is a comprehensive survey of the nature of life, its origin on Earth, life in our solar system, life on other planetary systems, communicating with distant civilizations, and estimating the number of technological civilizations in our galaxy.

Shklovskii, I. S., and Carl Sagan. *Intelligent Life in the Universe.* San Francisco: Holden-Day, 1966. The definitive tome covering the entire panorama of the natural evolution of the universe, including the development of intelligence and technical civilizations in our galaxy. Of the thirty-five chapters, eleven deal with life in the universe; the final chapter discusses intelligent life in the universe in great detail.

See also: Extraterrestrial Life in the Solar System; Habitable Zones; Life's Origins; Main Sequence Stars; Mars: Possible Life; Radio Astronomy.

Solar Chromosphere

Category: The Sun

The solar chromosphere is the layer of the Sun's atmosphere, a few thousand kilometers thick, immediately above the photosphere. The gas is warmer, thinner, and more transparent than in the photosphere. The chromosphere is most directly visible from Earth as a layer of color during a total solar eclipse. Its spectrum shows bright emission lines, unlike the dark absorption lines of the photosphere. The element helium was first discovered in the Sun's chromosphere during a total solar eclipse.

OVERVIEW

The Sun is a ball of gas without a solid surface. The Sun's photosphere, however, is usually considered to mark the surface of the Sun. Relatively opaque, the photosphere blocks our view of the solar interior. It is also the coolest layer of the Sun. When astronomers observe the solar spectrum, they see dark absorption lines from the photosphere because the photosphere is cooler than the hot solar interior.

The chromosphere is the layer of the Sun's atmosphere directly above the photosphere. The chromosphere is only a few thousand kilometers thick. The temperature of the gas in the chromosphere is slightly higher than that in the photosphere. The bottom layer of the photosphere is at 6,600 kelvins, but the coolest level at the top of the photosphere is at 4,400 kelvins. For reasons that are not completely understood, the temperature in the chromosphere climbs from this low to about 6,000 kelvins as the distance from the photosphere increases. The chromosphere has a lower density than the photosphere; its density decreases with height above the photosphere from nearly 10^{-5} kilogram/meter^3 at the photosphere-chromosphere boundary to about 10^{-10} kilograms/meter^3 in the transition region between the chromosphere and the corona.

The layer of the Sun's atmosphere above the chromosphere is the corona. In the approximately 100-kilometer-thick transition region between the chromosphere and corona, the temperature rapidly increases with height above the photosphere, from about 6,000 kelvins to a few hundred thousand kelvins. The temperature in the corona can be millions of kelvins, but the gas is very thin. In the transition region, the density drops rapidly to about 10^{-12} kilograms/meter^3.

Because the chromosphere is so thin and transparent compared to the photosphere, it cannot be seen when the photosphere is visible. The much brighter photosphere overwhelms the fainter chromosphere. The chromosphere is briefly visible during total solar eclipses. When the Moon blocks light from the photosphere, the chromosphere shines briefly until the Moon also blocks the chromosphere. The chromosphere is visible only briefly, just before and just after totality. It is also possible to

Facts About the Sun

	Sun	Earth
Mass (10^{24} kg)	1,989,100	5.9742
Volume (10^{12} km^3)	1,412,000	1.083
Volumetric mean radius (km)	696,000	6,371
Mean density (kg/m^3)	1,408	5,515
Surface gravity at equator (m/s^2)	274.0	9.80
Escape velocity (km/s)	617.7	11.2
Ellipticity (oblateness)	0.00005	0.00335
Absolute magnitude	+4.83	—
Luminosity (10^{24} J/s)	384.6	—
Mean distance from Earth (10^6 km)	149.6	—

The Sun's Atmosphere

Surface gas pressure (top of photosphere)	0.868 millibar
Effective temperature	5,778 kelvins
Temperature at bottom of photosphere	6,600 kelvins
Temperature at top of photosphere	4,400 kelvins
Temperature at top of chromosphere	~30,000 kelvins
Photosphere thickness	~400 km
Chromosphere thickness	~2,500 km
Sunspot cycle	11.4 yrs.

Photosphere composition:

Hydrogen	90.965%
Helium	8.889%
Oxygen	774 ppm
Carbon	330 ppm
Neon	112 ppm
Nitrogen	102 ppm
Iron	43 ppm
Magnesium	35 ppm
Silicon	32 ppm
Sulfur	15 ppm

Source: Data are from the National Aeronautics and Space Administration/Goddard Space Flight Center, National Space Science Data Center.

see the chromosphere using a coronograph, which is a disk in the focal plane of the telescope that blocks the photosphere to reveal the chromosphere and corona.

A dark, or absorption line, spectrum is produced when light from a hot, compressed gas passes through a cooler, thin gas. Therefore, the Sun's photosphere produces an absorption line spectrum. A bright, or emission line, spectrum results from a hot, thin gas. When astronomers observe the Sun's chromospheric spectrum, they are observing the chromosphere at the edge of the Sun, so the hotter interior is not directly behind the chromosphere in their line of sight. Hence, they observe emission lines. These emission lines flash into view for a short time just before and after the total portion of the eclipse. The chromospheric spectrum thus revealed is therefore called the flash spectrum.

The Sun is mostly hydrogen, and the brightest emission line in the chromospheric spectrum is the red hydrogen H-alpha line. This red emission line gives the chromosphere its reddish-pinkish color. The color also gives the chromosphere its name; *chromo* comes from the Greek word for color. Two spectral lines from the element calcium, the H and K lines, are also prominent in the chromospheric spectrum.

Astronomers use the H-alpha line and the H and K lines of calcium to observe the surface of the chromosphere. The chromosphere is relatively transparent at most wavelengths, but it is fairly opaque at the wavelengths of these spectral lines. Hence, observing the Sun at these wavelengths allows astronomers to image the chromosphere rather than the photosphere. The light at these wavelengths originates from different depths in the chromosphere, so comparing images made at the different wavelengths gives astronomers a three-dimensional image of the Sun's surface. Lyot filters made of calcite crystals are best for this purpose, but they are very expensive. Interference filters are less expensive, and H-alpha interference filters are available at a reasonable cost to allow amateur astronomers to observe the Sun safely with small telescopes.

Just below the photosphere, the Sun transfers energy from the interior via convection currents. These convection currents produce structures on the surface of the photosphere, called granules, that are the tops of the convection current cells. Faculae are bright areas on the solar photosphere that are often found in the lower areas marking the boundaries between the photospheric granules. When faculae extend upward from the photosphere into the chromosphere, they are called plages. Plages are therefore brighter than normal regions of the Sun's chromosphere. Plage regions can have opposite magnetic polarities, so magnetic field lines flow outward from one plage and connect to another plage where they flow back into the Sun's interior. Solar material often streams along these magnetic field lines to produce prominences. Prominences begin and end in the chromosphere, but extend well out into the corona.

Perhaps the most important feature of the chromosphere is the spicule. The chromosphere is covered with spicules. They are vertical streams of chromospheric material moving upward into the corona. The total mass of material moving upward into the corona from spicules is approximately the mass of the entire corona every few minutes. Because the mass of the corona is not rapidly increasing and the solar wind is not that much mass, astronomers know that most of this material must also be falling back into the chromosphere. However, astronomers are not able to observe the falling material. These spicules form a chromospheric bright network, along the boundaries of large supergranules on the Sun's photosphere, that is related to the Sun's magnetic activity.

The Sun's bright magnetic network may play a role in climate changes on Earth. When the Sun is at the maximum of its sunspot cycle and has the largest portion of its surface covered by sunspots, it has a very slightly higher energy output than when it is at sunspot minimum. Even though sunspots tend to reduce the Sun's energy output, brighter areas such as faculae (which become plages when they extend to the chromosphere) and the bright chromospheric magnetic network more than compensate for the energy loss caused by the darker sunspots. During the prolonged Maunder minimum of sunspot activity in the late seventeenth century, Earth's climate experienced its coldest period of the last millennium. From about 1100 to 1250 there was a medieval grand maximum of sunspot activity, corresponding to a warmer than normal period on Earth. Historical climate evidence suggests that solar luminosity variations caused by the Sun's magnetic activity affect Earth's climate. Sunspots and faculae on the photosphere and plages and the bright magnetic spicule network on the photosphere all play a role in changing the Sun's energy output.

KNOWLEDGE GAINED

During a solar eclipse in 1686, not long after the invention of spectroscopy, Jules Janssen, working in France, observed bright emission lines from the chromosphere. One line was very close to the wavelength of a well-known pair of yellow emission lines from the element sodium. This line was so bright that Janssen was able to study it in detail even when there was no eclipse. About the same time, Norman Lockyer, from Britain, also observed this bright line outside eclipses. Further detailed study revealed that this newly discovered spectral line was not quite the right wavelength to be sodium. Furthermore, it did not match the wavelengths of lines from any known element at the time. This newly discovered element was named helium after the word *helios*, for Sun. In 1895, helium was finally isolated on Earth.

As the closest star to us, our Sun is the best studied star and the only star for which it is possible to study surface details. Studying the solar chromosphere as well as the Sun's other regions in detail reveals much about atmospheres of other stars. A more complete understanding of stellar atmospheres, in turn, helps astronomers fine-tune their stellar models and improve their understanding of all aspects of stars.

One of the enduring mysteries about the Sun is why the temperature increases outward in the chromosphere and the corona. The Solar and Heliospheric Observatory (SOHO), launched in 1995, shed light on that question when it observed a magnetic carpet on the Sun's surface. About four thousand magnetic field lines loop up daily from the Sun's interior and then back down into the interior. Because these loops resemble the loops in a carpet, this phenomenon has been called the Sun's magnetic carpet. When these loops burst from the Sun's turbulence, they release energy to heat the outer layers of the Sun's atmosphere.

CONTEXT

Although the chromosphere is a relatively thin layer of the Sun's atmosphere, it is quite important to us. It is not possible to understand the photosphere below the chromosphere, or the corona above it, without understanding the chromosphere. For example, plages in the chromosphere are directly related to faculae in the photosphere. Spicules in the chromosphere extend vertically upward into the lower portions of the corona. Hence photospheric, chromospheric, and coronal phenomena are all interconnected.

The chromosphere plays a role in solar magnetic activity. Interactions between solar magnetic storms and Earth's magnetosphere can affect auroral activity on Earth and, in the case of strong magnetic storms, can affect radio communications on Earth.

The chromosphere and chromospheric phenomena play an important role in the Sun's magnetic activity cycle. If the Sun's luminosity changes with its activity cycle, the chromosphere likely played a role in past climate changes on Earth. It has not been proven, but some scientists think that solar variations may also play a contributing role in Earth's warming. In order to understand this possible role in Earth's climate fully, scientists must have a fuller understanding of the Sun's chromosphere.

Paul A. Heckert

FURTHER READING

Chaisson, Eric, and Steve McMillan. *Astronomy Today*. 6th ed. New York: Addison-Wesley, 2008. Chapter 16 of this very readable introductory astronomy textbook covers the Sun.

Frazier, Kendrick. *Our Turbulent Sun*. Englewood Cliffs, N.J.: Prentice-Hall, 1980. A good account of the basic knowledge about the Sun through the publication date.

Freedman, Roger A., and William J. Kaufmann III. *Universe*. 8th ed. New York: W. H. Freeman, 2008. Chapter 16 of this introductory astronomy textbook is a complete and up-to-date overview of our knowledge of the Sun.

Golub, Leon, and Jay M. Pasachoff. *Nearest Star: The Surprising Science of Our Sun.* Cambridge, Mass.: Harvard University Press, 2001. A detailed summary of solar science, including a section on the chromosphere that is more detailed than in most.

Heckert, Paul A. "Solar and Heliospheric Observatory." In *USA in Space.* 3d ed. Edited by Russell Tobias and David G. Fisher. Pasadena, Calif.: Salem Press, 2006. Describes the SOHO solar observatory mission, which was used to study the Sun's outer layers, including the chromosphere. This mission revealed much about the Sun's magnetic activity and why the temperature increases in the chromosphere and corona.

Hester, Jeff, et al. *Twenty-First Century Astronomy.* New York: W. W. Norton, 2007. Chapter 13 of this astronomy textbook is about the Sun.

Morrison, David, Sidney Wolf, and Andrew Fraknoi. *Abell's Exploration of the Universe.* 7th ed. Philadelphia: Saunders College Publishing, 1995. The Sun is covered in chapter 26 of this classic astronomy textbook.

Zeilik, Michael. *Astronomy: The Evolving Universe.* 9th ed. Cambridge, England: Cambridge University Press, 2002. An extremely well written introductory astronomy textbook. Chapter 12 is an overview of the Sun.

Zeilik, Michael, and Stephen A. Gregory. *Introductory Astronomy and Astrophysics.* 4th ed. Fort Worth, Tex.: Saunders College Publishing, 1998. Designed for undergraduate physics or astronomy majors, this textbook goes into more mathematical depth than most introductory astronomy textbooks. Chapter 10 covers the Sun, including the chromosphere.

See also: Coronal Holes and Coronal Mass Ejections; Electromagnetic Radiation: Nonthermal Emissions; Electromagnetic Radiation: Thermal Emissions; Infrared Astronomy; Red Giant Stars; Solar Corona; Solar Evolution; Solar Flares; Solar Geodesy; Solar Infrared Emissions; Solar Interior; Solar Magnetic Field; Solar Photosphere; Solar Radiation; Solar Radio Emissions; Solar Seismology; Solar Structure and Energy; Solar System: Origins; Solar Ultraviolet Emissions; Solar Variability; Solar Wind; Solar X-Ray Emissions; Sunspots; Thermonuclear Reactions in Stars; Ultraviolet Astronomy.

Solar Corona

Category: The Sun

The solar corona is the outermost, high-temperature portion of the solar atmosphere. This region of the solar atmosphere is the originating site of the solar wind, a flux of charged particles emitted by the Sun, and is considered to be the inner boundary of the interplanetary medium.

OVERVIEW

The solar corona is the outer atmosphere of the Sun. Well known from antiquity, it can be seen with the unaided eye during a total solar eclipse as a glowing white halo around the silhouette of the Moon. The rest of the time, when the Moon is not blocking the bright light from the Sun's photosphere, the corona is too faint at visible wavelengths to be observed without special instruments. The corona was not positively confirmed as a solar feature, rather than an artifact of Earth's atmosphere, until photographs of it were noted to look the same from widely separated sites on Earth.

Spectroscopic studies of the corona show that its visible light emission is a continuous background of all visible wavelengths together with a few superimposed broad emission lines. Thus, its spectrum is quite different from that of the photosphere, the visible solar disk, which has a visible light spectrum consisting of a continuous background with many dark absorption lines, produced by some of the elements found in the Sun. The coronal emission lines did not have wavelengths that matched those identified from any known chemical element, and some astronomers suggested that they were produced by a new element, dubbed "coronium." In 1925, similar lines were observed in the spectrum of a nova (an exploding star), RR Pectoris. In 1939, Walter Grotrian identified a red coronal line as being emitted by iron ions with nine electrons removed (called Fe X, X being the Roman numeral 10), and in 1942, Bengt Edlen identified a green coronal line as being emitted from iron ions with thirteen electrons removed (Fe XIV). It turned out that highly ionized atoms of familiar elements such as iron, calcium, and argon are the source of the coronal emission lines.

The high degree of ionization requires high temperatures in excess of one million kelvins. It is now thought that this high temperature is produced by the Sun's strong magnetic field. Magnetic waves from the turbulent convective photosphere follow magnetic field lines upward into the lower density corona (about 10^{15} particles per cubic meter or less, compared to 10^{22} particles per cubic meter in the upper photosphere), becoming supersonic shock waves that transfer their energy to the coronal gas. In addition, the complex pattern of magnetic field lines may twist and reconnect, releasing energy. This heating causes the temperature to rise steeply from a minimum of about 4,400 kelvins in the upper part of the photosphere to about 1.5 to 2 million kelvins in the corona. In spite of its high temperature, because of its low density the corona is only about one-millionth as bright as the photosphere at visible wavelengths (which is why the corona usually cannot be seen with the human eye except when the Moon covers the much brighter photosphere during total solar eclipses).

In the early 1930's, a French scientist, Bernard Lyot, perfected the coronagraph, a telescope that artificially created an eclipse of the solar disk. The invention of this type of telescope allowed observation of the corona at times other than total solar eclipses, and for the first time it was possible to follow the evolution of the outer atmosphere of the Sun. The coronagraph is now standard equipment at many ground-based solar observatories, and similar instruments have also been used for studies of the corona performed from orbiting spacecraft. The development of this instrument has been vital to constantly monitoring the corona and studying how it evolves with time. Studies of the corona involving determination of its temperature, density, and chemical composition occupied researchers for the next three decades.

A major advance in solar physics was the publication in 1958 of a classic paper on the dynamics of interplanetary gas, by Eugene Parker of the University of Chicago. He deduced that, as a consequence of the high temperature and low density of the corona, there should be a continual outflow of ionized gas, which he called the solar wind, from its outer edges. Confirmation of the existence of the solar wind was quickly forthcoming from early satellite experiments and, with it, the recognition that Earth is actually surrounded by this flow of charged particles. Earth is exposed to a constant flux of charged particles—mostly electrons, protons (hydrogen nuclei), and alpha particles (helium nuclei)—from the Sun, moving at speeds of a few hundred kilometers per second. However, only a tiny fraction of the Sun's mass is lost via the solar wind, about 10^{-14} of its mass per year.

The corona may be observed not only at visible wavelengths but also in the ultraviolet and X-ray bands of the electromagnetic spectrum. A major advantage of observing at ultraviolet and X-ray wavelengths is that the corona is much brighter in these parts of the spectrum than the photosphere is (just the opposite of the situation at visible wavelengths), so the corona is easily observed against the disk of the Sun. However, Earth's atmosphere effectively blocks ultraviolet and X-ray radiation from reaching Earth's surface, so these wavelength regions must be observed above our atmosphere. Thus it was natural to turn to the brightest source in the sky, the Sun, as the first astronomical object to study with instruments carried above the absorbing atmosphere. The first ultraviolet observations of the Sun were obtained with a spectrograph attached to a fin of a V-2 rocket, which was launched from the White Sands test facility on October 10, 1946. The results demonstrated conclusively the value of sending observational equipment above the absorbing blanket of air and opened a new field of ultraviolet astronomy.

As a step toward the exploration of the solar spectrum in these two wavelength regions, a series of sounding rocket experiments was conducted by the University of Colorado, the Air Force Cambridge Research Laboratory (AFCRL), and the Naval Research Laboratory (NRL) between 1958 and 1965. These experiments succeeded in characterizing the intensity and the wavelength distribution of the ultraviolet and soft X radiation from the solar corona. Many of these pioneering efforts were undertaken by Richard Tousey of NRL, a leader in this expanding field in the 1960's. The problem with using sounding rockets is that they are above Earth's atmosphere for no more than a few minutes, so long-duration observations are impossible. The solution is to launch spacecraft into orbits around the Earth and in some cases around the Sun, so the Sun may be observed for prolonged periods of time. Some of the first of these were the Orbiting Solar Observatories (OSOs), launched by the National Aeronautics and Space Administration (NASA). Nine of these spin-stabilized satellites were launched over a thirteen-year period, from March, 1962, through June, 1975. The major research gains made through the use of these satellites included a definition of the differences between active solar regions and quiet regions as seen in the ultraviolet, evidence that material found in the corona is cooler over the solar poles than at the equator, and detection of coronal "hole" features, large low-density regions of the corona that typically last for several of the Sun's twenty-seven-day rotation periods. A white-light coronagraph flown on OSO 7 detected the ejection of a huge mass of material upward from the Sun's disk, a mass that evidently had sufficient speed to escape the solar gravitational field.

Initially, NASA intended to conduct a second series of solar investigations from satellites, the Advanced Orbiting Solar Observatory (AOSO) series. These advanced spacecraft were to have improved pointing capabilities and were intended to employ higher spatial resolution for studies of the solar atmosphere. This program was integrated into the crewed Skylab program,

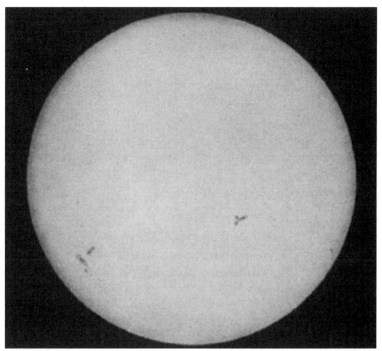

This image of the Sun was made in 1845 as a daguerreotype by physicists Louis Fizeau and Jean-Bernard-Léon Foucault. Sunspots are visible on the surface. (National Science Foundation, High Altitude Observatory)

and a set of high-resolution solar instruments, including a white-light coronagraph and several ultraviolet and X-ray imaging telescopes, was placed into Earth orbit with the Apollo Telescope Mount (ATM) system on Skylab. Of the many factors that made the ATM-Skylab mission uniquely valuable, two had special bearing on studies of the solar corona. First, the nine-month duration of the mission permitted uninterrupted observation of the evolution and activity of the Sun's atmosphere over a very wide range of wavelengths that are not observable from Earth's surface. Second, since this was a crewed mission, it was possible for the science teams to exploit interactive observing modes, where the data obtained suggested new observations, with the astronauts controlling the instrumentation to optimize the collection of scientific data.

The ATM-Skylab experiments, along with the data and knowledge gained from them about the solar corona, form a cornerstone for modern solar research concerning the outer portions of the Sun's atmosphere. The ultraviolet and X-ray telescopes carried in this orbiting observatory were able to return new images of how the material in the corona is distributed over the disk of the Sun. As never before, astronomers were able to specify the morphology and evolution of the distribution of mass in the outer atmosphere. Coronal holes and their relationship to the solar wind were investigated in detail with ATM data. It also became clear, using the images from the white-light coronagraph, that there is considerable transient activity in the corona, and numerous mass ejection events were detected.

Based on the ATM-Skylab experience, a second white-light coronagraph was launched on the NASA Solar Maximum Mission (SMM) spacecraft in February, 1980. This instrument operated successfully for nine months until it was subject to an electronics failure. The equipment was later repaired in orbit by astronauts who were transported to the satellite by the space shuttle (STS-41C). The electronics package was successfully replaced, and the coronagraph experiment was subsequently operated for a number of years. Thus, the SMM coronagraph has been used to define the variations of the solar corona over a rather long period of time.

KNOWLEDGE GAINED

The physical characteristics of the solar corona are now reasonably well known, as are its basic evolutionary characteristics. The corona is made up of a fully ionized mixture of the solar elements, mostly hydrogen and helium. Metals and other elements constitute a minor fraction of the total mass. The solar mixture is ionized by the high temperature of the outer atmosphere. The high ionization states of the elements identified by their emission lines in the spectrum of the corona require temperatures on the order of 1.5 to 2 million kelvins.

The solar surface exhibits numerous regions in which there is a concentration of magnetic flux. Magnetic forces are frequently strong enough to dominate the motion of the coronal plasma (ionized gas). Unlike Earth's atmosphere, which is controlled by the interaction of pressure and gravitational force, the Sun's outer atmosphere reacts to the interplay of three forces: pressure, gravity, and magnetism. Active regions, associated with sunspots, often show coronal material to be configured into loop or arch patterns, as if the material is confined to specific magnetic field structures. Coronal holes tend to form where the magnetic field lines are open and oriented radially outward into interplanetary space.

The white glow of the corona seen during total solar eclipses is produced by the scattering of photospheric light off the free electrons in the coronal plasma. On average, the electron density near the base of the corona is about 10^{14} to 10^{15} per cubic meter; it is greater near active regions and sunspots and less in coronal hole regions. The major white-light structures tend to be long streamers, which extend from the limb (edge) of the Sun more or less radially, and loops, which are almost always associated with concentrations of surface magnetic fields. Streamers viewed at eclipse are often found to extend outward 4 to 6 solar radii. (The nominal value of the solar radius is 700,000 kilometers, a length approximately equal to twice the distance from Earth to the Moon.) Streamers occur over two kinds of solar disk features: Long, linear streamers occur frequently over magnetically active regions and sunspots, and helmet streamers, which are shaped somewhat like bowling pins, are often seen over magnetic neutral lines, areas in which the magnetic polarity switches from one sign to the other.

The Sun has a twenty-two-year cycle of magnetic activity. Sunspots rise and fall in number with a period which is half of the magnetic cycle, so that concentrations of magnetic flux have been observed to be at maxima in 1958, 1969, and 1980, for example. The total amount of material in the corona is modulated by this cyclic variation of magnetic flux, and the total mass of the corona varies by about a factor of 2 over the sunspot cycle. Also over the sunspot cycle, the number of coronal streamers occurring at any given time varies by a factor of 2.

By observing sunspots and other solar features, it is possible to establish the rotation period of the Sun. Near the equator, it is found to be about twenty-seven days. The basic rotation rate of the corona is approximately equal to the equatorial rotation rate, and the bright features used to follow coronal rotation have lifetimes that range from one to five months. By the late 1930's, it was known that some effect of the Sun dominated Earth's upper atmosphere. Ionospheric disturbances and perturbations of Earth's magnetic field were observed to follow a twenty-seven-day period (like the Sun's rotation). These so-called M-regions could not, however, be correlated with distinct solar structures, such as an enhanced concentration of magnetic field. Following the discovery of the solar wind and its variation in space and time, the problem of locating the solar origins for Earth-perturbing wind streams began to receive much attention from scientists. During the 1970's, the origins of the high-speed solar wind streams were eventually identified with open magnetic field configurations and solar coronal holes, based on the OSO 7 and ATM-Skylab data sets. This correlation constituted a major breakthrough in the task of associating the interplanetary magnetic field and flux configuration at Earth with the physical conditions of the corona at the base of the solar wind.

Along with the solar wind, there are other sporadic ejections of solar material which escape the pull of solar gravity. First detected in the early 1970's, coronal mass ejections (CMEs) were finally explained using data from the white-light coronagraph carried in the ATM-Skylab observatory. Often appearing as huge loops or bubblelike structures having a dimension of half a solar radius in the lower corona, the ejections expand as they rise through the corona and may exceed the size of the Sun at heights above 5 solar radii. The average amount of mass ejected is on the order of 10^{12} kilograms, and their mean outward velocity from the Sun's surface is typically 300 to 400 kilometers per second.

Some CMEs are associated with flare activity, the catastrophic conversion of magnetic energy into thermal energy in or near sunspot regions. Occasionally, CME events occur in areas that have no obvious sunspots. CME events occur about once every three to five days during solar minimum, and the production of events can reach one to two per day during times near the maximum of the activity cycle. CME events

constitute one of the most energetic phenomena detected on the Sun; typically, the kinetic energy of such an event can exceed 10^{25} joules.

CONTEXT

The corona provides scientists with a laboratory for studies of how low-density, high-temperature plasmas interact with magnetic fields, and it affords investigators a view of a situation in which pressure, gravity, and electromagnetic forces operate simultaneously. The study of such interactions is known as magnetohydrodynamics (MHD), and the solar corona offers investigators an example at close range where complete MHD processes, such as coronal mass ejections, may be observed over a wide variety of both temporal and spatial scales.

Recognition that Earth is, in fact, subjected to constant bombardment by the solar wind flux has given new impetus to studies of how this wind is generated and how it is controlled by MHD processes. At the center of the solar system, there is a magnetic star (the Sun) that modulates the interplanetary space beyond it. The corona, reflecting the magnetic organization of the Sun over a great variety of spatial scales, is astronomers' best clue to the initial organization of the interplanetary magnetic field and the structure of solar wind flow. Once the locations of coronal hole structures are identified, either by observing in the X-ray or ultraviolet regions or by using limb observations from coronagraphs, it is possible to predict when the sub-Earth point on the Sun is occupied by a coronal hole. This knowledge allows the prediction, with fair accuracy, of when Earth will be subjected to a high-speed stream of solar wind.

The geophysical significance of coronal observations has led to international collaboration in satellite investigations of the Sun. The Solar and Heliospheric Observatory (SOHO) satellite, launched in 1995 and operated cooperatively by the National Aeronautics and Space Administration (NASA) and the European Space Agency (ESA), provides continuous monitoring of the Sun from a stable orbit around the Sun at the Earth-Sun L1 Lagrangian point. (This is the point, located 1.5 million kilometers from Earth along the line between the Earth and Sun, where the gravitational pull of the Sun and Earth are equal.) Among SOHO's many devices for studying various aspects of the Sun, it carries plasma diagnostic instruments for investigation of the solar wind along with several telescopes for solar coronal studies.

Studies of the Sun's corona have provided insights into observed features of other stars. In the late 1970's and early 1980's, X-ray observations from the High-Energy Astronomical Observatories (HEAOs) were interpreted to show, to the surprise of many investigators, that almost all types of stars have outer coronas that radiate at the short wavelengths characteristic of high temperatures. It is now accepted that coronas are a standard feature of stellar atmospheric structure.

Similarly, spectroscopic studies performed in the 1980's with the International Ultraviolet Explorer (IUE) satellite demonstrated that stellar winds, outflows of ionized gas like the solar wind, are common from other stars. However, unlike the Sun, which is a star of moderate luminosity and surface temperature, luminous hot stars have stellar winds driven by the radiation pressure of the intense ultraviolet radiation they emit. Again, insight gained from Earth's own star has aided in the identification and interpretation of a common stellar process.

Richard R. Fisher

FURTHER READING

Chaisson, Eric, and Steve McMillan. *Astronomy Today*. 6th ed. New York: Addison-Wesley, 2008. A well-written college-level textbook for introductory astronomy courses. Contains a good description of the solar corona and the SOHO mission.

Eddy, John A. *A New Sun: The Solar Results from Skylab*. NASA SP-402. Washington, D.C.: Government Printing Office, 1979. A summary of the operations of the ATM-Skylab solar observatory. The text is clear and nontechnical, and the book is extensively illustrated with reproductions of many of the most important images returned from this mission.

Foukal, Peter. *Solar Astrophysics*. 2d rev. ed. Weinheim, Germany: Wiley-VCH, 2004. A detailed look at the Sun and our understanding of it. Also covers the history of solar astrophysics. Suitable for undergraduates.

Fraknoi, Andrew, David Morrison, and Sidney Wolff. *Voyages to the Stars and Galaxies*. Belmont, Calif.: Brooks/Cole-Thomson Learning, 2006. A well-written, thorough college textbook for introductory astronomy courses. Offers a good description of the solar corona.

Frazier, Kendrick. *Our Turbulent Sun*. Englewood Cliffs, N.J.: Prentice-Hall, 1980. An accessible review of solar physics as of its publication date, this work includes basic discussions of solar-terrestrial interactions and the impact of solar activity on Earth. Illustrations show coronal holes and the interplanetary magnetic configuration. An excellent reference for a reader seeking information on basic problems in solar physics.

Freedman, Roger A., and William J. Kaufmann III. *Universe*. 8th ed. New York: W. H. Freeman, 2008. College-level introductory astronomy textbook, thorough and well-written. Provides a good description of the solar corona.

Hirsh, Richard F. *Glimpsing an Invisible Universe*. Cambridge, England: Cambridge University Press, 1985. A historical review of the development of X-ray astronomy from the beginnings through the mid-1980's. NASA sources and interviews with experimenters are used as background for this review.

Schneider, Stephen E., and Thomas T. Arny. *Pathways to Astronomy*. 2d ed. New York: McGraw-Hill, 2008. Very thorough college textbook for introductory astronomy courses, divided into short sections on specific topics. Includes a section on the Sun's atmosphere.

Tucker, Wallace, and Riccardo Giacconi. *The X-ray Universe*. Cambridge, Mass.: Harvard University Press, 1985. An interesting history of the development of the discipline of X-ray astronomy by scientists who participated in many of the more significant events of the mid-twentieth century.

See also: Auroras; Coronal Holes and Coronal Mass Ejections; Electromagnetic Radiation: Nonthermal Emissions; Electromagnetic Radiation: Thermal Emissions; Infrared Astronomy; Red Giant Stars; Solar Chromosphere; Solar Evolution; Solar Flares; Solar Geodesy; Solar Infrared Emissions; Solar Interior; Solar Magnetic Field; Solar Photosphere; Solar Radiation; Solar Radio Emissions; Solar Seismology; Solar Structure and Energy; Solar System: Origins; Solar Ultraviolet Emissions; Solar Variability; Solar Wind; Solar X-Ray Emissions; Sunspots; Thermonuclear Reactions in Stars; Ultraviolet Astronomy.

Solar Evolution

Category: The Sun

The Sun, the closest representative of the stars that populate the universe, has been a reliable benchmark for testing theories of stellar astrophysics for centuries. Much has been learned about the Sun's formation, present status, and future evolution from the developments of modern physics, their powerful theoretical models, and a wealth of observations of both this star and others like it.

OVERVIEW

The Sun was born a little over 4.5 billion years ago from a 10-kelvin, interstellar cloud, roughly 10^{14} kilometers across, of cold atomic and molecular gas. Triggered by an event like a nearby supernova, supersonic turbulence in the cloud caused a compressed region to collapse under its own gravity, fragmenting into smaller (on the order of 10^{12} kilometers) pieces on a timescale of about 2 million years. These clumps flattened into disks from their angular momenta, feeding the disk centers as they collapsed. In the disk that would become our solar system, the center accumulated 1-2 solar masses after an additional 10^{4} years and became opaque to its own radiation. Consequently, the central temperature rose to about 10,000 kelvins, and the collapsing mass became a protostar within another 10^{5} years.

This protostar, about the size of Mercury's orbit and several times the Sun's current luminosity, contracted further over the next million years. In the process, the early Sun entered the T-Tauri stage of evolution, exhibiting strong protostellar winds and bipolar outflows of jets that became less and less collimated as the surrounding disk flattened and dissipated. These jets' compositions concordantly evolved from being primarily molecular to atomic as the young Sun's temperature rose. This activity subsided as the Sun's protostellar evolution slowed over the next 10^{7} years, and gravity struggled to compress the hot, ionized stellar material further. When the core reached a temperature of 10 million kelvins, fusion of hydrogen into helium through the proton-proton chain was possible.

Following a 30-million-year period of slight contraction, the Sun settled into its current state on the main sequence of stellar evolution, with a radius of 6.96×10^{5} kilometers, a luminosity of 3.83×10^{36} watts, and surface and central temperatures of 5,780 kelvins and 15×10^{6} kelvins, respectively. According to computer models and meteoritic evidence, 4.5 billion years have elapsed since that time. Early in its main sequence history, the Sun increased its luminosity by about 30 percent, as the core temperature and fusion reaction rates rose with the increasing mean atomic weight of the nuclear end products deposited in the interior.

The Sun is expected to remain on the main sequence for another 5.5 billion years, maintaining its normal solar cycles and associated magnetic activity. The Sun's luminosity will also continue to increase slowly as more helium "ash" settles in the core. After a total lifetime of 10 billion years on the main sequence, the Sun's core will be composed of enough helium to shut down hydrogen fusion at its center. Though fusion will still occur in a shell surrounding the depleted core center, hydrodynamic equilibrium will no longer be maintained, and the core will begin to collapse under its own gravity. The subsequent release of gravitational energy will heat the hydrogen-burning shell, increasing the nuclear reaction rates, which in turn will increase the gas pressure on the surrounding solar layers.

Arriving at the subgiant stage in its evolution, the Sun's surface temperature will fall to about 4,000 kelvins, and the solar envelope will expand to a radius three times its current size. The Sun will spend 10^{8} years in this stage before becoming a red giant star, 100 solar radii in size—about the size of Mercury's orbit—and several hundred times its present luminosity. The surface temperature will not change appreciably during this time, however.

About 10^5 years afterward, the core will have contracted to the point where the central temperature will reach 100 million kelvins, and helium fusion can occur through the triple-alpha process. The central density of this compressed core will be an extremely high 10^8 kilograms/meter^3, and electron degeneracy pressure will stabilize the core against further collapse. Since a degenerate core is largely insensitive to the additional energy generated by helium fusion, the core will not expand as it is heated, and a runaway "helium flash" will ensue for several hours. The tremendous amount of energy dumped into the core will then heat it to the point where thermal pressure can take over, expanding the core and establishing a new balance with gravity for the next 10^5 years.

The triple alpha process will then proceed at a steady rate in the core, surrounded by a lower-temperature hydrogen-fusing shell. The Sun will reach the horizontal branch of stellar evolution, with a slightly higher surface temperature. The solar envelope will also shrink back to 10 main sequence solar radii. When the core's helium fuel is exhausted, it will be left as carbon ash, surrounded by concentric shells of helium-fusing, then hydrogen-fusing, layers. As before, the core will contract until electron degeneracy pressure dominates over thermal pressure, reaching a central density of 10^8 kilograms/meter^3 and a temperature of 250 million kelvins.

At this red supergiant stage, the central temperature is insufficient to fuse carbon, but the compression will drive the surrounding helium- and hydrogen-burning shells to higher temperatures and luminosities. This will cause the outer layers to expand to 500 solar radii, or about the size of Earth's orbit, cooling to a surface temperature of 4,000 kelvins. The exact size of the Sun at this point is unknown, depending on the severity of mass loss from winds ejected in the red giant phase. The Sun will last a relatively short 10^4 years in this stage. Shell helium burning will happen in a series of violent spurts, causing the solar envelope to fluctuate in size. Additionally, photons produced by electron-nuclei recombinations in the envelope will push the layers out farther with each expansion phase. Eventually, the outer layers will be ejected as a planetary nebula, enriching the surrounding interstellar medium and leaving behind the Sun's compact carbon core. As a white dwarf, this Earth-sized object will cool in a leisurely manner, over many billions of years, to become a black dwarf at a temperature very nearly that of absolute zero.

KNOWLEDGE GAINED

Our understanding of the Sun's history is gleaned mostly from theoretical models. In 1644, René Descartes proposed the theory of vortices, roughly outlining solar genesis from infalling swirling gas. Later, Emanuel Swedenborg's 1734 nebular hypothesis postulated that the Sun was formed by a rotating nebula, an idea further explained in Immanuel Kant and Pierre-Simon Laplace's independently formulated nebular hypotheses. In 1755 and 1796, respectively, they invoked the conservation of angular momentum to picture a collapsing cloud rotating and contracting into a protostellar disk.

In the early twentieth century, James Jeans established the physical criteria governing hydrodynamic equilibrium and the conditions necessary for a cloud to collapse. Current models of star formation favor supersonic turbulence in the parent interstellar cloud, possibly from the shocks of a nearby supernova, as the spark necessary to trigger gravitational collapse. This idea was originally posited by Carl von Weizsäcker in 1944 and Dirk ter Haar in 1950, and it resurfaced in the 1990's with the advent of modern computational power. While still a vaguely understood subject, it is theorized that turbulence was responsible for defining the structure and evolution of the presolar molecular clouds, providing the high compression and transport of angular momentum required for gravity to induce further collapse. The actual isothermal collapse was investigated as a simple case by Richard Larson and Michael Penston in 1969 and also by Frank Shu, who explored the inside-out collapse model that produces protostars. More rigorous investigations of star formation have since been conducted, with thought given to the roles of complex magnetic effects, turbulent viscosities, chemical compositions, and rotation.

Observationally, astronomers have compared these predictions with sunlike stars in various stages of development. Radio observations of the M20 nebula provide images of many stages of stellar evolution, from the parent cloud, fragmentation, and collapse to emission nebulae lit by the first generation of high-mass stars. The 1970's and 1980's saw the discovery of successively lower mass protostars closer and closer to the solar system. For example, observations by the Infrared Astronomical Satellite (IRAS) identified Barnard 5, a currently forming solar-type star. Radio and infrared observations of hydrogen and carbon monoxide have found winds of 100 kilometers per second, as well as expanding knots of water and bipolar radio jets characteristic of protostars.

At higher energies, Chandra, XMM-Newton, and Einstein Observatory X-ray satellites have also observed nascent solar-type stars and star-forming regions for clues to our Sun's past. Although the evolution of the Sun's X-ray luminosity depends on its poorly known initial rotation, astronomers know that it declined gradually from the outset of the main sequence for 100 million years and then dropped by a factor of 1,000 until the present day. This decline is connected to the decline of the solar corona's temperature with time. As for the emerging Sun's immediate environment, it has been suggested that the abundance of neutron-rich iron 60 (Fe^{60}) in some meteorites implies that supernovae were nearby. This would indicate that the Sun was born in a fairly crowded environment similar to the active star-forming regions in the Orion nebula.

The current isolation of the solar system is probably a result of a series of gravitational interactions with other protostars that ultimately ejected the emerging solar system from its crowded neighborhood. Similar isotope-decay analyses have been applied to primordial gas-rich meteorites, deducing the composition and strength of the solar wind within one billion years of the Sun's formation. Observations of current T-Tauri stars in both the X and ultraviolet ranges indicate that the Sun emitted energetic particles and winds as flares at this stage,

producing precompacted, irradiated grains with peculiar isotope ratios in the circumstellar disk. Excess neon 21 (Ne^{21}) in meteoritic grains is often seen as evidence for this process. After reaching the main sequence, ancient meteoritic evidence further shows that the solar wind flux gradually declined to its present value. Furthermore, radionuclides in lunar rock samples show that the Sun's proton emission has remained relatively constant over the past five million years, aside from variations from the eleven-year solar cycle. This is also true for heavy ions ejected in flares over this timescale, with the exception that the most ancient flares had an overall enrichment in the trans-iron group of nuclei.

Verification of the Sun's lifetimes in its stages of evolution comes from confirming the computed standard model of solar structure, specifically the nuclear reactions occurring in the core and their observable properties. For instance, the Solar and Heliospheric Observatory (SOHO), launched in 1995, probed the interior composition and temperature structure of the Sun by "listening" to internal pressure waves reflecting off the photosphere. This application of helioseismology has indirectly validated the lifetimes of its various evolutionary stages by substantiating the standard model's predictions of solar composition with depth. Further insight into the Sun's future is gained through observing the evolution of other stars. The Ring and Helix nebulae, for example, are photogenic examples of planetary nebulae ejection and the death throes of Sun-like stars.

CONTEXT

Interestingly, theories of solar formation and evolution seem historically motivated by coincidental developments in physics, taking advantage of the increasing availability and sophistication of quantitative measurements and computations. Early speculations of star formation prior to the nineteenth century, without the advanced physics needed to support them, were easily discarded. In the 1840's, J. Robert Mayer and John James Waterson realized that recently studied chemical and electrical energy sources would be unable to provide the Sun's luminosity for any reasonable timescale. Bolstered by the triumph of thermodynamic principles like energy conservation, William Thomson and Hermann von Helmholtz advocated gravitational contraction as the Sun's energy source. In the twentieth century, solving this problem required the synthesis of two separate fields of science—astronomy and atomic physics—into the new field of astrophysics.

Arthur Eddington and Henry Russell analyzed the interplay of pressure and gravity to understand high-temperature stellar interiors and radiative equilibrium, while Ernest Rutherford and Niels Bohr laid the foundation for solar physics by establishing quantum mechanics. In 1917 and 1920, Eddington and Harlow Shapley concluded that stars must have ages greater than the tens of millions of years allotted by the gravitational collapse scenario.

From 1920 onward, favor shifted from electron-proton annihilation to fusion reactions; these would offer lifetimes measured in trillions and billions of years, respectively. The latter solution won with arguments for a multibillion-year universe from Edwin Hubble's research in receding galaxies, quantum mechanical arguments for the possibility of fusion in hot stellar cores, and Hans Bethe's robust proposal of the proton-proton chain and carbon-nitrogen-oxygen (CNO) cycle (a series of thermonuclear reactions) in 1938. This energy-generation mechanism, along with the associated main sequence lifetime and later evolution, has since been supported by a plethora of increasingly sensitive observational data across the electromagnetic spectrum.

Further progress is expected, especially for uncovering the Sun's early evolution, with the launch and operation of the James E. Webb Space Telescope. This satellite, along with other post-Hubble telescopes, will probe dust-obscured interstellar clouds at infrared wavelengths and at high resolution to help elucidate the intricate picture of low-mass star formation.

Brendan Mullan

FURTHER READING

Green, Simon F., Mark H. Jones, and S. Jocelyn Burnell. *An Introduction to the Sun and Stars*. New York: Cambridge University Press, 2004. A text for introductory-level university astronomy courses or the self-motivated amateur astronomer, discussing the basic physical characteristics of the Sun and other stars. Complex mathematics is avoided.

Lankford, John, ed. *History of Astronomy: An Encyclopedia*. New York: Garland, 1997. A series of very easy-to-digest essays on several key topics in astronomy, within a historical context. Concentrates on the people and social settings behind their science.

Montesinos, Benjamín, Alvaro Giménez, and Edward F. Guinan, eds. *The Evolving Sun and Its Influence on Planetary Environments*. San Francisco: Astronomical Society of the Pacific, 2001. The proceedings of a professional workshop on a range of contemporary topics in solar astronomy, these articles are challenging and technical.

Smith, Michael D. *The Origin of Stars*. London: Imperial College Press, 2004. An excellent text on star formation for undergraduates. The book also doubles as a highly readable and transparent guide for the educated nonspecialist who can loosely follow the meanings of the equations.

Sonett, C. P., M. S. Giampapa, and M. S. Matthews, eds. *The Sun in Time*. Tucson: University of Arizona Press, 1991. This volume is an interdisciplinary collection of essays on the Sun and its impact on the solar system, on a variety of timescales and through many physical mechanisms. Advanced and specialized.

Stix, Michael. *The Sun*. 2d ed. New York: Springer, 2002. A brief introduction to current knowledge of the Sun, meant for the unspecialized scientist-in-training who has a basic conceptual grasp of topics such as thermodynamics and hydrodynamics.

Unsöld, Albrecht, and Bodo Baschek. *The New Cosmos: An Introduction to Astronomy and Astrophysics*. 5th ed. New York: Springer, 2001. A guide to developments in astronomy and astrophysics, geared toward students and researchers. Some prior knowledge of mathematics and physics is recommended.

See also: Auroras; Coronal Holes and Coronal Mass Ejections; Infrared Astronomy; Interplanetary Environment; Nuclear Synthesis in Stars; Red Giant Stars; Solar Chromosphere; Solar

Corona; Solar Flares; Solar Geodesy; Solar Infrared Emissions; Solar Interior; Solar Magnetic Field; Solar Photosphere; Solar Radiation; Solar Radio Emissions; Solar Seismology; Solar Structure and Energy; Solar System: Origins; Solar Ultraviolet Emissions; Solar Variability; Solar Wind; Solar X-Ray Emissions; Sunspots; Thermonuclear Reactions in Stars; Ultraviolet Astronomy.

Solar Flares

Category: The Sun

A solar flare is a high-energy outburst in the chromosphere of the Sun that emits a variety of electromagnetic radiation ranging from energetic gamma rays to long-wavelength radio waves, along with high-energy charged-particle radiation. If the charged particles reach Earth's atmosphere, they produce various phenomena such as auroras and long-distance communication disruptions.

OVERVIEW

Solar flares are sudden outbursts of electromagnetic and particle radiation in the Sun's chromosphere, releasing from 10^{22} up to 10^{30} joules of energy in a matter of minutes to hours. The intensity of emission rises in a few minutes in catastrophic eruptions, increasing more than ten times in brightness in the visible range alone, while radio, ultraviolet, and X-ray emissions may increase a thousand times. In those few minutes, the brightened area may expand to include a billion square kilometers of the Sun's surface (up to a thousandth of the entire solar disk) with temperatures from ten million up to a hundred million kelvins at the center.

This explosive development usually begins in the upper portion of the chromosphere and then moves upward at a rate up to 100 kilometers per second, often reaching heights of 7,000 to 16,000 kilometers above the photosphere. The brightest flares tend to be the most explosive, reaching a peak in five to ten minutes and then fading over a period of up to two hours.

Solar flares occur where there are strong magnetic fields in the Sun's chromosphere. Large flares are magnetically complex and may have a visible filamented structure. The more magnetically complex the solar environment, the more likely flares are to occur, since colliding and reconnecting magnetic field lines are the most common cause of flares.

The strongest magnetic fields undergoing the greatest changes are associated with and located near the centers of sunspot groups. The intense magnetic fields impede radiation from below the photosphere, creating areas called sunspots, up to tens of thousands of kilometers in diameter, that are cooler and therefore not as bright as most of the photosphere. Sunspot groups usually consist of two main sunspots with opposite magnetic polarities, with many smaller spots clustered around them. Individual sunspots appear and disappear with varying frequency, but their numbers are cyclical, reaching a maximum every eleven years.

Flares occur most frequently above rapidly developing sunspot groups, usually within the first ten to fifteen days of the life of the group. This timing takes place because the most complex magnetic phase of a sunspot group occurs during this period. The flares tend to appear and reappear in association with the same active sunspot regions. Although flares alter the magnetic field, multiple reappearances are possible, which implies that the complex magnetic configuration reestablishes itself between flares.

Another solar feature associated with flares are prominences, glowing plumes or arches of ionized gases that rise above the chromosphere into the corona. They frequently appear as loops of glowing gas that reach from 20,000 to 50,000 kilometers above the Sun's photosphere along magnetic field lines. Other prominences appear as filaments of gas that stream away from sunspots. Prominences, like flares, usually occur in the same latitude belts as sunspots. When they are near spots, they tend to vary rapidly. When away from spots, prominences may be quite stable and last up to three hundred days. Prominences are usually not strong enough to eject matter from the Sun, but eruptive prominences may be so energetic as to propel matter into

Dramatic solar flares and coronal mass ejections in January, 2002. (NASA/ESA/SOHO)

accelerate charged particles to speeds sufficient to escape from the Sun. Flares eject billions of tons of ionized gas into space at speeds of up to 500 kilometers per second. The ejection of charged particles causes disturbances in the corona called flare surges, which frequently follow the magnetic field lines in the region. Surges increase with the strength of the flare. Flares are one of the sources of shockwave phenomena that sometimes are observed upon the Sun's surface. At times, the effect of a flare can be traced out to 600,000 kilometers by the effect of the shock wave on thin, gaseous filaments in the Sun's atmosphere. The magnetic field is often significantly altered or dissipated after the flare, although there is a gradual recovery of the magnetic field and the potential for further flares in the same area.

Waves of plasma (charged particles) ejected from solar flares travel outward through interplanetary space. Once they reach Earth, they produce various effects in Earth's atmosphere, such as auroras and long-distance communication disruptions. The fastest particles are high-energy protons moving at an appreciable fraction of the speed of light. A sufficient number of protons may be emitted during a flare to raise the cosmic-ray background on Earth to 180 percent of its normal level.

APPLICATIONS

Scientists are interested in flares because of the fascinating physics associated with their behavior, the grandeur of the size of the phenomena, and the intense dynamism they represent. Their main concern with flares, however, results from the fact that these outbursts can have a significant effect on our planet. Two basic kinds of effects occur: those that are nearly simultaneous with the flare and those that are delayed a day or more.

The simultaneous effects begin at the same

space. Prominences provide some of the charged particles that cause magnetic storms and auroras on Earth, and they are a more modest source of X rays than flares.

Flares emit a wide variety of electromagnetic radiation, generally created by nonthermal mechanisms. Their emissions range from gamma rays and X rays at wavelengths as short as 0.1 nanometer through the spectrum of ultraviolet, visible, and infrared, out to radio wavelengths of up to 10 kilometers. The different kinds of radiation come from different portions of the flare and from different altitudes within it. Thousands of flares occur during each eleven-year sunspot cycle, but flares visible in the white or integrated light portion of the spectrum, are quite rare. The lower in the chromosphere and thus the closer to the photosphere that the flare develops, the more likely that it will be a white-light flare. Only about fifty white-light flares have been observed in the past 150 years.

In addition to emitting electromagnetic radiation, flares emit high-energy charged particles. Magnetic fields in the Sun's atmosphere

time or shortly after the flare is first observed—they are caused by electromagnetic radiation (all forms of which travel at the speed of light) or by high-energy particles traveling almost as fast as light. The most significant simultaneous effects occur in the upper portion of the Earth's atmosphere known as the ionosphere. In this very tenuous gaseous region, the number of free electrons increases, and so the electrical charge increases in an abrupt fashion, called either a "kick" or a "crochet." There is a rapid shortwave radio fade-out that usually begins with the peak of the flare. The fade-out lasts for about twenty-five minutes and then the signal begins to recover. During the fade-out, shortwave radio signals drop to as low as one-tenth of their normal levels. By contrast, the increased reflectivity of the ionosphere for long radio wavelengths of up to 10 kilometers is enhanced. This increased reflectivity also occurs at a lower level in the ionosphere and produces an anomaly in the phasing of the long-wavelength radio waves received directly, compared to those reflected by the ionosphere. Also, the changes in the ionosphere tend to suppress the normal background radio noise.

The delayed effects are caused by the arrival of the particles that travel more slowly and reach the Earth a day or so later. Prominent among the delayed effects are the aurora borealis and aurora australis (the northern and southern lights), which can appear as glowing rays and billowing draperies in the sky. This shimmering display of light is caused when an increased flux of charged particles from the Sun (such as a flare can produce) overloads the Van Allen radiation belts surrounding Earth. The charged particles leak through Earth's magnetic field near the magnetic poles and cascade down into the upper atmosphere. There the high-speed charged particles smash into atmospheric atoms and molecules, exciting them to higher energy states. Then they return to lower energy states by emitting photons of various colors of visible light. Auroras most commonly are seen near Earth's poles, inside the Arctic and Antarctic circles. However, larger flares can produce auroras seen over wider areas; in North America, for example, occasionally they can be seen as far south as the U.S./Mexican border.

Not all flares cause auroras, and not all auroras are caused by flares, since the Sun emits charged particles more or less continuously as the solar wind.

Flares send particles in all directions, but Earth receives the most particles when it is directly above the flare—that is, when the flare is near the center of the solar disk as seen from Earth. The most intense magnetic storms will follow such an event.

The magnetic field deflects high-energy charged particles, and the atmosphere blocks high-energy X rays and gamma rays. Together they provide a shield protecting life on Earth. Flares pose a potentially lethal hazard for both space travelers and uncrewed probes that venture beyond the Earth's atmosphere and magnetic field, and better prediction of solar flares as well as better shielding of spacecraft are needed. In addition, the blast of charged particles from a solar flare can increase the density of Earth's thin upper atmosphere, creating extra drag on satellites in orbit. For example, a particularly strong flare on March 6, 1989, caused a drop of a kilometer in the orbit of the Solar Maximum Mission satellite.

The development of ever smaller and denser microcircuitry has made devices using them more susceptible to the effects of high-energy particles. These single-event effects may damage semiconductor circuits. High-energy particles can alter random access memory (RAM) in computers, depositing charge and actually changing stored information and instructions. This has become an increasing problem for computerized control systems, which may eventually be forced to use protective shielding, as is employed with some satellites. A similar but macroscopic problem is that enough charge can be deposited on long metal cables and pipelines to be hazardous, necessitating electrical grounding.

CONTEXT

For a period of five minutes, a little after 11:00 A.M. on September 1, 1859, while mapping sunspots, Richard C. Carrington saw two bright patches of light on the Sun. They moved over the surface of one of the spots he was mapping, with the spot remaining unchanged. Carrington reasoned that he had observed a solar atmospheric phenomenon that had occurred above the sunspot. Independently, R. Hodgson observed the same two spots and also reported them. Their sightings marked the first recorded visual observation of a solar flare in the astronomical literature. Carrington noted that the flare occurred during an intense magnetic storm that lasted from August 28 to September 4, 1859. From then on, flares were occasionally spotted visually. When flares were observed visually through a spectroscope, momentary bright reversals of the dark absorption lines in the solar spectrum were seen. The connection between flares and magnetic storms continued to be noted, and a sense of causation grew among most astronomers, although as late as 1892, the physicist Lord Kelvin disputed the connection. Prior to the 1890's, the distinction between a prominence and a flare was not made.

The systematic study of flares began with the invention of the spectrohelioscope in 1891 by George Ellery Hale. The sunspot cycle peaks of 1936, 1947, and 1958 were studied from ground-based observatories, with the most intense effort made during the International Geophysical Year from 1957 to 1958.

Although much was learned about flares, there was an increasing awareness that there probably were emissions in other parts of the electromagnetic spectrum that could not be detected from Earth's surface because those wavelengths were blocked by the Earth's atmosphere. Instruments borne above the bulk of the Earth's atmosphere by balloons made the first observations of some of these wavelength regions, but the most significant advances came with the advent of the space age. From 1949 to 1960, Herbert Friedman and a team of scientists observed X rays from the Sun through a complete sunspot cycle using sounding rockets to carry detectors above the atmosphere. However, the flights of sounding rockets are suborbital, so the time above the atmosphere was limited to a few minutes.

Much of what is known about flares, their structure, and the mechanisms that cause them has come from satellite-based research, since satellites allow continuous observation. The first satellite devoted to solar studies was Solar

Radiation I (1960 Eta 2), launched in August of 1960. The most significant early space-based studies were of X-ray emissions from the Sun. The early results pointed to the areas around sunspots as strong X-ray-emitting regions, with X-ray emission varying dramatically over the eleven-year sunspot cycle.

Another milestone in flare research was the launching of the Solar Maximum Mission (SMM) in 1980. Eight different collectors provided a flow of information from a variety of overlapping wavelengths through the 1991-1992 sunspot peak. The great flare of March 6, 1989, was the largest since satellite observations began. This particular flare released as much energy as 10 trillion 1-megaton hydrogen bombs. The great flare was followed by ten more successive, extremely strong flares. New technology made the thorough study of this series of events possible. Knowledge of the Sun, its features and emissions, continues to expand rapidly with satellite-based observing devices.

A long-standing question about flares has concerned their classification. In the past, some astronomers argued that there was no significant difference among flares; there were simply different power levels of the same basic event. Strong flares allowed observation of all the associated phenomena, while with weaker flares, some phenomena were beyond the ability of existing equipment to detect. Early classification systems simply arranged flares on a scale of one to four, depending on brightness (usually in the visible spectrum). Some scientists added a category of −1 to represent "subflares" (microflares) or exceedingly dim ones. Increased knowledge of the radiation at various wavelengths, especially gamma rays and hard X rays, demonstrated that the brightest visible flares were not the brightest emitters at other wavelengths, and vice versa. This knowledge has led to a new classification system based on several criteria. The letters C, M, and X are used as classes followed by a number from 1 to 15. An X15 flare is the strongest measurable.

Scientists continue to be surprised by the diversity of phenomena that ongoing satellite-based research has revealed. More attention is being devoted to the particle emissions from flares that have an impact on communications.

The expansion of space-based flare-sensing networks will warn of forthcoming communication problems, allowing us to switch to alternative satellite links in order to improve continuity in communications. More attention will also be directed to ultrashort-wavelength electromagnetic radiation, and more study of solar magnetic fields is needed as well.

Ivan L. Zabilka

FURTHER READING

Bai, T., and P. A. Sturrock. "Classification of Solar Flares." *Annual Review of Astronomy and Astrophysics* 27 (September, 1989): 421-467. While highly technical and rather narrowly defined topically, this article remains significant for the more than two hundred bibliographic entries it contains. There is enough description to give the general reader a sense of how complicated the problem of classifying flares remains.

Chaisson, Eric, and Steve McMillan. *Astronomy Today*. 6th ed. New York: Addison-Wesley, 2008. A well-written college-level textbook for introductory astronomy courses containing a good description of solar flares.

Ellison, Mervyn Archdall. *The Sun and Its Influence*. London: Routledge & Kegan Paul, 1955. Primarily of historical interest, since much of the book is outdated, chapter 5 discusses what was known of the physics of flares in a narrative form.

Emslie, A. Gordon. "Explosions in the Solar Atmosphere." *Astronomy* 15 (November, 1987): 18-23. An excellent popular survey of the kinds of research under way in the late 1980's. Provides clearly written background information on the nature of flares themselves.

Foukal, Peter. *Solar Astrophysics*. 2d rev. ed. Weinheim, Germany: Wiley-VCH, 2004. A detailed look at the Sun and our understanding of it. Also covers the history of solar astrophysics. Suitable for undergraduates.

Fraknoi, Andrew, David Morrison, and Sidney Wolff. *Voyages to the Stars and Galaxies*. Belmont, Calif.: Brooks/Cole-Thomson Learning, 2006. A well-written, thorough college textbook for introductory astronomy courses, containing a good description of solar flares.

Freedman, Roger A., and William J. Kaufmann III. *Universe*. 8th ed. New York: W. H. Freeman, 2008. College-level introductory astronomy textbook with a good description of solar flares.

Kundu, M. R., B. Woodgate, and E. J. Schmahl, eds. *Energetic Phenomena of the Sun*. Boston: Kluwer, 1989. A volume in the Astrophysics and Space Science Library that was the result of three meetings held at the Goddard Space Flight Center in January and June of 1983 and February of 1984. While highly technical, it offers some descriptive passages of interest to the general reader. Especially valuable for its bibliography.

Maxwell, Alan. "Solar Flares and Shock Waves." *Sky and Telescope* 66 (October, 1983): 285-288. Includes a summary of information about flare-generated shock waves as a source of acceleration for particles emitted by the Sun. Considers some difficult concepts and presents them clearly.

Meadows, A. J. *Early Solar Physics*. Elmsford, N.Y.: Pergamon Press, 1970. A brief historical account containing a fine summary and significant information about related topics. The original reports of Carrington and Hodgson are reprinted in appendixes.

Rust, David M. "Solar Flares, Proton Showers, and the Space Shuttle." *Science* 216 (May 28, 1982): 939-946. Surveys the dangers of radiation from flares to occupants of the space shuttle, suggesting that no adequate level of shielding is possible. Excellent diagrams; only semitechnical.

Ryan, James M. "The Solar Maximum Mission." *Astronomy* 9 (May, 1981): 6-16. An excellent overview of the state of knowledge about solar flares in 1980. Surveys the results from the Solar Maximum Mission during the ten months before the spacecraft failed. Filled with readable information.

Schneider, Stephen E., and Thomas T. Arny. *Pathways to Astronomy*. 2d ed. New York: McGraw-Hill, 2008. A thorough college textbook for introductory astronomy courses, with a section on solar activity and flares.

Verschuur, Gerrit. "The Day the Sun Cut Loose." *Astronomy* 17 (August, 1989): 48-51. An excellent popular summary of the magnitude and effects of the great flare of March 6, 1989. Contains a reliable summary of information about the general nature of flares.

See also: Auroras; Coronal Holes and Coronal Mass Ejections; Earth's Magnetic Field at Present; Earth's Magnetosphere; Earth-Sun Relations; Interplanetary Environment; Solar Chromosphere; Solar Corona; Solar Evolution; Solar Geodesy; Solar Infrared Emissions; Solar Interior; Solar Magnetic Field; Solar Photosphere; Solar Radiation; Solar Radio Emissions; Solar Seismology; Solar Structure and Energy; Solar Ultraviolet Emissions; Solar Variability; Solar Wind; Solar X-Ray Emissions; Sunspots; Van Allen Radiation Belts.

Solar Geodesy

Category: The Sun

Solar geodesy is the study of the size and shape of the Sun. In attempts to measure the shape of the Sun precisely, astronomers accidentally discovered solar oscillations, complex rhythmic pulsations involving both the deep interior and the atmosphere of the Sun.

OVERVIEW

Geodesy is the mathematical study of the size and shape of the Earth, and how these affect the precise location of points on the Earth's surface. Solar geodesy is the application of this discipline to the Sun, especially the study of the size and shape of the Sun. The main impetus for such studies up to the mid-1900's was twofold: (1) to see if the Sun is oblate due to its rotation, and (2) to find out if the Sun is slowly shrinking. Although not conclusively settled, it appears neither of these actually is the case. However, detailed measurements revealed something unexpected: The Sun undergoes complex oscillations, or as one researcher put it, "The Sun rings like a bell." This discovery led to a new branch of solar studies called solar seismology, or helioseismology, which uses these oscillations as probes of the Sun's interior, analogous to the

way geologists use seismic waves as probes of the Earth's interior. The Sun's vibrations are geometrically complex, and solar physicists can infer the physical nature of the solar interior by analyzing the timing and amplitudes of the many vibration patterns observed on the Sun's surface. Hence, solar geodesy today is focused largely on helioseismology.

In 1960, Robert Leighton of the California Institute of Technology observed that small regions on the Sun's surface were oscillating or pulsating with a period of approximately five minutes. He detected this oscillation by using the Doppler effect, a shifting of the wavelengths or frequencies of electromagnetic radiation. Motion of the source toward the observer causes a "blueshift," a shift to shorter wavelengths. Motion of the source away from the observer causes a "redshift," a shift to longer wavelengths. The amount of the shift is in proportion to the speed of the source toward or away from the observer. Leighton wondered whether the pulsations he observed were occurring only in small regions on the Sun, perhaps a few thousand kilometers in extent, or over a more extended region, possibly even the entire Sun. Although his discovery was noted and subsequently confirmed by many other observers at many observatories, these confirming observations could indicate only that the oscillations were localized. The data would not permit any conclusion regarding coherent motion (connected or related motion over a wide region) on the Sun.

In the early 1960's, a group of researchers at Princeton University led by Robert H. Dicke claimed that they had measured an oblateness (a distortion of a sphere resulting from compression along the polar axis and stretching around the equator) in the shape of the Sun. Using a highly specialized telescope of their own construction, they made measurements of the solar equatorial and polar diameters. They claimed the equatorial diameter to be slightly greater than the polar diameter.

They made these measurements to test Albert Einstein's general theory of relativity against an alternative, the Brans-Dicke scalar-tensor theory, developed by Dicke and Carl Brans. Both theories predicted, among other things, that the orbit of the planet Mercury around the Sun should precess, meaning that Mercury's elliptical orbit itself should slowly rotate around the focus occupied by the Sun. According to the theory of general relativity, this would be due to the warpage of space-time by the Sun's mass. According to the Brans-Dicke theory, however, a small oblateness in the shape of the Sun would cause the same thing. Thus, the oblateness measurements were crucial in distinguishing between general relativity and the Brans-Dicke scalar-tensor theory.

In the mid-1960's, Henry Hill at the University of Arizona designed and built another specialized telescope for detecting distortions in the Sun's surface. Hill's preliminary measurements, however, could not confirm the measurements of the Princeton group. By the late 1960's, other astronomers had shown that the Princeton results probably were due to solar activity that was producing increases in brightness in the Sun's equatorial regions. The increased brightness was caused by plages— bright, patchy regions on the Sun produced by magnetic activity. The plages are usually found near sunspots or centers of magnetic activity, which tend to concentrate within about 30° to 40° north and south of the Sun's equator. Measurements of the solar equatorial diameter are highly influenced by plages. There is a tendency to overestimate the edge of the Sun, or its limb, because of the bright glow of these plages near the solar equator.

In 1975, Hill and Robin Stebbins concluded that the Sun was not oblate but apparently fluctuated or oscillated rhythmically over a large region, possibly its entire surface. Further research verified that the observed phenomena were genuinely solar, not introduced by Earth's atmosphere or telescopic effects. Delicate instruments always have random fluctuations, or "noise," associated with their measurements, but researchers showed convincingly that the observed oscillations were real and not simply the misinterpretation of observational noise.

It came to be accepted that the Sun shakes or vibrates in a range of spectacular ways like a ringing bell. In this process, the Sun's shape undergoes tiny, patterned distortions of a rhythmic nature. These distortions are in effect three-dimensional waves that pass from the deep interior of the Sun to the surface and also move about the Sun's circumference. The periods of these measured oscillations range from about 3 minutes to 160 minutes and perhaps longer.

These oscillation patterns carry information about the deep solar interior, which cannot be observed directly since the Sun's interior is opaque. The waves or disturbances producing these oscillations originate at different levels within the Sun, some just beneath the surface and others farther within the interior. The observed properties of the oscillations—their timing and the extent of their displacement— depend upon the environment through which they pass on their way to the surface, where solar physicists detect them spectroscopically using the Doppler effect. Just as geophysicists study seismic waves traveling through Earth as a result of earthquakes, solar physicists study solar oscillations traveling through the Sun in order to study the solar interior.

Why does the Sun shake? What in the solar interior sets the oscillations in motion, and how do they move through the Sun? All physical bodies, whether they be solid or fluid (fluids include gases and liquids), can oscillate or shake with a variety of frequencies or periods. Virtually any disturbance or natural internal motion can start the oscillations. Convection (the process whereby heat is transferred in the outer regions of the solar interior) is one such stimulus. Solar convection involves the ascent of hot, lower-density bubbles of gas. These bubbles are heated in the deeper, hotter interior; with lower density, they rise buoyantly upward and convey the heat to higher, cooler layers. The motion of the bubbles disturbs the surrounding gases and starts them oscillating, and the oscillations move throughout the Sun. Just as the length of an organ pipe determines the note played, certain fractions of that length produce overtones or harmonics. The Sun's spherical shape, interior density, and temperature determine both the frequency and the length of the waves that are set in motion, regardless of the process that caused those waves.

In a structure such as the Sun, the oscillations can be of two types, depending on the nature of the force that maintains the oscillation. Either gravity or pressure can supply the restoring force (the mechanism that brings the displaced fluid back to its original position, thus maintaining the motion necessary to produce the pulsations or waves). Small pressure fluctuations give rise to acoustic or soundlike waves, alternate compressions and rarefactions that move through the fluid at the speed of sound. Waves of this type are referred to as "p modes." Just below the visible solar surface, the convection produces a deafening roar, similar to that produced by a jet or rocket engine.

Gravity waves are created when an element or small volume of fluid is displaced and subsequently returned to its position by gravity. This type of wave can occur only when the density or compactness of the material varies with depth. On Earth, water waves larger than small ripples are of this type. The wave moves toward the shore, and its vertical displacement is restored by gravity, causing the wave to move along the surface of the water. (These gravity waves are not to be confused with the gravitational waves predicted by Einstein's theory of relativity, which are of a completely different nature.)

Any movement or perturbation in the interior of the Sun can start the quivering process and produce an oscillation. Individual oscillations can be manifested in many ways, with a variety of wavelengths and nodes (regions or points free of oscillations; in a vibrating string, the node would be the tied-down point not undergoing vibration). Because of the three-dimensional, spherical structure of the Sun, a virtually endless variety of modes or patterns can occur in it. Some modes encompass the Sun's entire structure, while others take place in localized regions.

The quivering waves generated by these oscillations reflect from the surface layers of the Sun and speed back toward the deep interior, where in turn they are refracted or bent back toward the surface. The layer of refraction depends on the speed of the waves. In many cases, the speed is the local speed of sound for the solar interior gas, which in turn depends on the temperature of the gas at that location. The reflection from the surface is caused by a rapid de-

crease in density, since the surface represents an abrupt boundary between the solar gas and space. Since sound cannot travel through a vacuum, the waves are reflected at the boundary back into the Sun. Such a process can occur again and again, resulting in waves moving along curved paths and bouncing and refracting their way completely around the interior of the Sun.

KNOWLEDGE GAINED

The Sun oscillates like a giant, spherical bell. This discovery was first made by Robert Leighton in 1960. The observed oscillations were about five minutes in duration. It was not known at that time whether the oscillatory motion involved small, localized regions on the Sun or much larger regions. Hill's 1975 observations found oscillations of relatively short duration, with periods of a few minutes. A considerable effort went into showing that these oscillations were real and not the result of distortions caused by Earth's atmosphere or by instrumental effects. By 1980, however, similar oscillations had been observed at other observatories, and some had been shown to have periods between two and three hours long.

These oscillations were found to be of two types: (1) pressure waves and (2) gravity waves, based on the natural force that maintains the oscillations. In the case of short-period waves, those lasting a few to several minutes, the wave is essentially a sound wave traveling through the solar interior. Such waves are known as pressure waves, since alternate compression and rarefaction of local gas produces the oscillations. Long-period waves, with oscillation periods as long as two or three hours, have gravity as the driving, restoring force. A single oscillation can be set up in an almost infinite number of geometric shapes, or modes. Many of these patterns have striking geometrical beauty and symmetry.

Observations of solar oscillations have revealed many new things about the nature of the Sun's interior and have confirmed other things. For example, at the surface, the Sun's equatorial zone is observed to rotate somewhat faster than do regions north or south of the equator; this phenomenon is known as differential rotation. The rotation of the Sun's interior cannot be directly observed, but observations of surface waves generated in the deep and shallow solar interior indicate a complex rotation pattern, with "flowing rivers" and "zones" and layers rotating at different and sometimes varying rates. Other studies find oscillation patterns that are consistent with the Sun's deep interior temperature being about 15 million kelvins, a value derived from computer models of the Sun's interior.

Early theories of stellar evolution and energy generation held that stars should slowly contract gravitationally as they age, so solar geodesy once was concerned especially with measuring such shrinkage in the Sun. Solar eclipses provided much of the raw data. If the Sun were shrinking, annular eclipses would have been more common and the duration of totality during total eclipses would have been shorter in the past. Although some early studies claimed to have found these effects, they have not been substantiated by more recent work that better accounts for instrumental errors.

CONTEXT

Later studies concentrated on measuring solar oblateness, but claims that the Sun is slightly oblate have not been confirmed. Instead, solar observers have discovered that the Sun oscillates in complex patterns. By observing oscillations visible on the Sun's surface, astronomers are assembling a detailed picture of the solar interior, in much the same way that geologists and geophysicists gain knowledge of Earth's interior by the study of waves produced in earthquakes. The most extensive series of observations of solar oscillations have been made by the Global Oscillations Network Group (GONG), an array of small telescopes around the world that provide continuous monitoring of solar oscillations over many years. The data gathered by GONG are enabling astronomers to differentiate real solar oscillations (together with their amplitudes and periods) from noise produced by instrumentation or Earth's rotation. An average value for the diameter of the Sun is 1,392,000 kilometers, or the equivalent of 110 Earths lined up side by side.

James C. LoPresto

FURTHER READING

Chaisson, Eric, and Steve McMillan. *Astronomy Today*. 6th ed. New York: Addison-Wesley, 2008. Very well-written college-level textbook for introductory astronomy courses. Good description of solar oscillations.

Foukal, Peter. *Solar Astrophysics*. 2d rev. ed. Weinheim, Germany: Wiley-VCH, 2004. A detailed look at the Sun and our understanding of it. Also covers the history of solar astrophysics. Suitable for undergraduates.

Fraknoi, Andrew, David Morrison, and Sidney Wolff. *Voyages to the Stars and Galaxies*. Belmont, Calif.: Brooks/Cole-Thomson Learning, 2006. A well-written, thorough college textbook for introductory astronomy courses. Includes a good description of the Sun's atmosphere and energy output.

Frazier, Kendrick. *Our Turbulent Sun*. Englewood Cliffs, N.J.: Prentice-Hall, 1980. This book summarizes the research and observational discoveries made about the Sun from about 1950 to 1981. When possible, the author relates these developments to the impact the various phenomena have on Earth, considering questions such as whether the solar activity cycle affects Earth's climate.

Freedman, Roger A., and William J. Kaufmann III. *Universe*. 8th ed. New York: W. H. Freeman, 2008. College-level introductory astronomy textbook with a good description of solar oscillations.

Giovanelli, Ronald. *Secrets of the Sun*. Cambridge, England: Cambridge University Press, 1984. This book outlines modern solar physics with the aid of spectacular black-and-white images from numerous observatories. Giovanelli focuses on the physical nature of the various atmospheric components of the Sun: the photosphere, the chromosphere, and the corona.

Harvey, John W., James R. Kennedy, and John W. Leibacher. "GONG: To See Inside Our Sun." *Sky and Telescope* 74 (November, 1987): 470-476. This article examines the field of helioseismology, using detailed drawings and diagrams. The article outlines an ambitious project to put into place small automated telescopes around the world to study solar oscillations over a five-to-ten-year period.

Leibacher, John W., Robert W. Noyes, Juri Toomre, and Roger K. Ulrich. "Helioseismology." *Scientific American* 253 (September, 1985): 48-57. Colorful diagrams illustrate a detailed description of how acoustic waves within the Sun cause the surface to heave up and down. The waves' timing and geometrical patterns are described, with emphasis on how study of them leads to knowledge of the structure, composition, and dynamics of the invisible interior of the Sun. Somewhat challenging reading.

See also: Red Giant Stars; Solar Chromosphere; Solar Corona; Solar Evolution; Solar Flares; Solar Infrared Emissions; Solar Interior; Solar Magnetic Field; Solar Photosphere; Solar Radiation; Solar Radio Emissions; Solar Seismology; Solar Structure and Energy; Solar System: Origins; Solar Ultraviolet Emissions; Solar Variability; Solar Wind; Solar X-Ray Emissions; Sunspots; Thermonuclear Reactions in Stars; Ultraviolet Astronomy.

Solar Infrared Emissions

Category: The Sun

Most solar infrared and far-infrared radiation is emitted from the coolest layers in the solar atmosphere, which are found in the upper portion of the photosphere. Analysis of this radiation not only allows scientists to understand these important layers of the solar atmosphere but also provides observational confirmation of the simplest known interaction between matter and radiation: local thermodynamic equilibrium.

OVERVIEW

Infrared (or IR) radiation is a form of electromagnetic radiation between visible light and microwaves (short-wavelength radio waves). Gases in Earth's atmosphere (especially water vapor) absorb much of the IR radiation—particularly the "far infrared," the longer-wavelength IR—before it reaches the ground. Effective study of IR radiation from astronomical sources must be done using telescopes at dry, high-altitude, mountaintop observatories above the densest part of Earth's atmosphere, or preferably using space telescopes completely outside our atmosphere.

The IR radiation emitted by the Sun represents only a tiny fraction of the total solar radiative energy. The rate at which the Sun emits IR energy is only about 0.058 percent of the total solar luminosity, which is the rate at which the Sun radiates energy over the entire electromagnetic spectrum. The solar constant is the average total solar irradiance (electromagnetic energy per unit time per unit area) impinging on the top of Earth's atmosphere. Its value is 1,368 ± 7 watts per square meter, as measured by the Solar Maximum Mission (SMM). Of this, only 0.802 ± 0.026 watts per square meter is in the IR.

The IR is the simplest part of the solar spectrum. The radiation is almost continuous over the entire IR spectral range, with very few dark absorption lines. The continuous background spectrum is thermal radiation, such as a blackbody (a perfect thermal radiator) emits. Absorption lines (called Fraunhofer lines in the solar spectrum) are formed when photons with specific energies are absorbed by electrons jumping to higher energy levels in atoms or molecules, or by molecules going to higher-energy rotational or vibrational states. Consequently, less energy remains in the spectrum at the wavelengths corresponding to the absorbed photon energies, so the spectrum looks darker at those wavelengths. The absorption lines that are present serve as a chemical fingerprint revealing that specific elements and compounds are present in the source. However, the absence of absorption lines in some part of the spectrum—for example, in the IR—does not necessarily indicate certain elements or compounds are not present, just that the physical conditions are not right for them to absorb in that spectral range if they are present in the source.

IR radiation arises primarily in the upper photosphere (the visible surface of the Sun) and the lower chromosphere (the layer of the Sun's atmosphere, a few thousand kilometers thick, immediately above the photosphere). These levels of the solar atmosphere consist of homogeneous strata. From place to place and layer to layer, the gas and its behavior show a remarkable similarity. This is where the solar atmospheric temperature falls to its minimum value and begins to rise to the higher temperatures found in the chromosphere and corona, the extensive outer layers of solar atmosphere above the photosphere.

The photosphere consists of a series of layers from which most of the Sun's visible light is emitted into space. These layers make up a very thin shell, only a few hundred kilometers thick. The opacity of the photosphere increases rapidly with depth; thus the intensity of emitted radiation drops off rapidly with depth into the photosphere. A photon emitted outward from the lower photosphere has a large probability of being absorbed or scattered by the atoms and free electrons within the photosphere. Because of this, photons are likely to escape into space only from the photosphere's uppermost layers. For this reason, the edge (or limb) of the Sun is sharply defined, and the Sun appears to have a definite surface.

The photosphere produces the continuous radiation observed across the entire solar spectrum. The intensity peaks in the yellow-green part of the visible spectrum and falls off toward both longer and shorter wavelengths. That is why the IR contributes such a small percentage of the Sun's total electromagnetic radiation. This spectral distribution of intensity is similar to that of a blackbody (an ideal thermal radiator) with a temperature of about 5,800 kelvins. In reality, each layer making up the photosphere emits its own blackbody spectrum, which in turn depends on the temperature of that layer. The sum total of all the emissions from all the layers is similar to one imaginary layer emitting at approximately 5,800 kelvins; this is referred to as the Sun's "effective temperature." Those layers that are deeper in the photosphere emit at higher temperatures, and those that are higher in the photosphere emit at lower temperatures, as low as 4,400 kelvins.

The blackbody spectral distribution results from the high opacity of the layers. (Perfect blackbody radiators are perfectly opaque.) The source of this high opacity is the presence of negative hydrogen ions, first proposed by Rupert Wilt in the late 1940's. The negative hydrogen ion is a hydrogen atom with an additional (second) electron weakly attached. It easily absorbs radiation in the visible spectrum and especially in the infrared. The second electron is bound to the hydrogen atom very weakly—with a bond about 3.5 percent as strong as that of the first electron. Since the bond of the second electron is so weak and since the temperature drops off rapidly toward the top of the photosphere, the electron density also diminishes rapidly in the photosphere's upper reaches. In other words, the number of free electrons and negative hydrogen ions per unit volume—known as their "number densities"—are very sensitive to temperature and thus height in the photosphere.

The phenomenon of limb darkening can be observed on any good white-light solar image (an image of the Sun obtained over the entire range of visible wavelengths, the combined colors from violet to red giving a white image with a slight yellow tinge). The limb of the Sun is noticeably darker than the center of the solar disk because of the temperature gradient in the photosphere. As an observer's line of sight moves toward the limb, it passes through only the upper, cooler layers, whereas deeper, hotter layers are seen near the disk center. Thus the intensity of light decreases toward the limb due to the drop in temperature with height. (In contrast, the Sun's limb appears brighter than the disk center in observations of the chromosphere and lower corona, because of the reversal of the temperature gradient in those layers.)

Since intensity of radiation can be measured as a function of both distance from the disk center and wavelength, much can be learned about the way in which physical properties change with depth, once the processes of photon emission, absorption, and scattering within the atmosphere are understood. These processes control the flow of electromagnetic radiation through the solar atmosphere by radiative transfer. Using the processes of radiative transfer and observed details of the solar spectrum and limb darkening, astronomers can calculate mathematical models of the solar atmosphere. These models are tabulations of height versus variables of interest such as temperature, density, and pressure.

KNOWLEDGE GAINED

The entire solar photosphere emits radiation like a stacked system of glowing shells, each shining at its characteristic temperature. The net effect of all these glowing shells is similar to that of one thin shell emitting at a temperature of 5,800 kelvins, the "effective temperature" of the photosphere. The infrared and far-infrared spectra are emitted from the coolest layers (4,400 kelvins), located at the top of the photosphere. Most of the solar IR spectrum is in a continuous form, with few dark absorption lines. IR radiation originates mainly in the uppermost photosphere. At this level, the solar atmosphere is at its coolest temperature: about 4,400 kelvins.

CONTEXT

The study of the IR spectrum is important because these observations, along with limb-darkening measurements, are needed in modeling the solar atmosphere. These layers of gas are essentially in local thermodynamic equilibrium (LTE). Radiation is transferred in LTE by well-understood processes: photon emission, absorption, and scattering. The primary contributor to the opacity of all layers of the photosphere is the negative hydrogen ion. Sources of opacity and transport of energy by known nonthermal processes—such as mechanical mechanisms, sound waves, shock waves, and magnetic energy—are very small in the photosphere.

Furthermore, spectral studies of solar IR have illuminated details of the atomic and molecular composition of Earth's atmosphere. These studies, in turn, help scientists to understand the interaction of other bands of electromagnetic radiation, such as ultraviolet, with the terrestrial atmospheric constituents detected in the IR.

James C. LoPresto

FURTHER READING

Chaisson, Eric, and Steve McMillan. *Astronomy Today*. 6th ed. New York: Addison-Wesley, 2008. This well-written introductory astronomy textbook contains a good description of the Sun's atmosphere and energy output across the electromagnetic spectrum.

Foukal, Peter. *Solar Astrophysics*. 2d rev. ed. Weinheim, Germany: Wiley-VCH, 2004. A detailed look at the Sun and our understanding of it. Also covers the history of solar astrophysics. Suitable for undergraduates.

Frazier, Kendrick. *Our Turbulent Sun*. Englewood Cliffs, N.J.: Prentice-Hall, 1980. This book summarizes the research on and observational discoveries about the Sun from about 1950 to 1981. When possible, Frazier relates these developments to the impact on Earth of the various phenomena, such as whether the solar activity cycle affects Earth's climate. Suitable for general audiences.

Giovanelli, Ronald. *Secrets of the Sun*. Cambridge, England: Cambridge University Press, 1984. Spectacular black-and-white images from a number of observatories accompany an account of modern solar physics. Giovanelli highlights the physical nature of the various atmospheric components of the Sun: the photosphere, the chromosphere, and the corona.

Nicolson, Iain. *The Sun*. New York: Rand McNally, 1982. In an atlas format, Nicolson provides a historical perspective on solar studies; discusses the relationship of the Sun to the stars, the Galaxy, and the universe; and presents a detailed description of the solar interior and atmosphere. Solar activity and solar-terrestrial relationships are examined at length. For readers with a basic background in solar physics.

Schneider, Stephen E., and Thomas T. Arny. *Pathways to Astronomy*. 2d ed. New York: McGraw-Hill, 2008. Very thorough college textbook for introductory astronomy courses. Has several sections on the Sun's atmosphere and energy output.

See also: Coronal Holes and Coronal Mass Ejections; Earth-Sun Relations; Electromagnetic Radiation: Thermal Emissions; Infrared Astronomy; Red Giant Stars; Solar Chromosphere; Solar Corona; Solar Evolution; Solar Flares; Solar Geodesy; Solar Interior; Solar Magnetic Field; Solar Photosphere; Solar Radiation; Solar Radio Emissions; Solar Seismology; Solar Structure and Energy; Solar System: Origins; Solar Ultraviolet Emissions; Solar Variability; Solar Wind; Solar X-Ray Emissions; Sunspots; Thermonuclear Reactions in Stars; Ultraviolet Astronomy.

Solar Interior

Category: The Sun

The nuclear fusion reactions that power our Sun occur in its interior. The nature of this interior determines conditions at the surface of the Sun, affecting the rest of the solar system.

OVERVIEW

The Sun is composed primarily of hydrogen and helium. Though these substances are gaseous on Earth, the conditions in the Sun are such that they do not behave inside the Sun the way that they do on Earth. The Sun's gravity compresses and heats these gases in the Sun's interior. About one-fifth of the way below the surface of the Sun, the gases are so hot that they are ionized—that is, they have been stripped of their electrons. The closer to the center of the Sun these gases are, the more compressed and heated they are. Almost 94 percent of the Sun's mass is contained within the innermost half of the Sun's radius.

Near the center of the Sun, the temperature is more than 15 million kelvins. Gases are compressed to a density of about 160 grams per cubic centimeter (g/cm^3), more than fourteen times denser than lead. At such density and temperature, hydrogen nuclei begin to fuse into helium. Every second, close to 600 million tons of hydrogen is consumed in this fusion process. The result of this fusion is the production of about 596 million tons of helium. The difference in these two mass figures, 4 million tons, is converted into energy via the equivalence of mass and energy given by Albert Einstein's famous equation $E = mc^2$. This energy released in the deep interior of the Sun is what supports the Sun against gravity and keeps it from collapsing further under its own weight.

Density and temperature are greatest at the center of the Sun. This is where nuclear fusion occurs at the fastest rate. The density and temperature decrease with distance from the center of the Sun. Therefore, the rate of nuclear fusion decreases with increasing distance from the center of the Sun. Beyond about 25 percent of the distance from the Sun's center to its surface, the density and temperature are too low to support fusion at any appreciable rate. This innermost portion of the Sun, where nuclear fusion occurs, is called the core of the Sun. The temperature at the top of the core is about 7 million kelvins. However, there is no well defined edge to the core. Rather, the farther from the center of the Sun, the lower the rate of fusion.

Throughout the core, energy is produced through nuclear fusion. Much of this energy is initially in the form of high-energy gamma rays. The gamma rays produced in this manner travel only an extremely short distance before they collide with an electron. Gamma rays are a form of electromagnetic radiation. While often described as waves, electromagnetic radiation also acts like particles, called photons. These photons carry momentum. Therefore, when the gamma rays collide with electrons, they scatter off of the electrons in a process called Compton scattering. In the scattering process, the gamma rays lose momentum and energy and the electrons gain momentum and energy. These collisions between the gamma rays and the electrons are what support the interior of the Sun. Gamma rays continually rebound from electron to electron, scattering in random directions. Gamma rays continually lose energy, eventually becoming X rays. The radiation scatters in random directions with each collision. In this manner, called radiative diffusion, the energy from the nuclear fusion in the Sun's core gradually works its way outward from the middle of the Sun. Because of the large number of collisions in random directions, it takes a long time for the radiation to travel outward. On average, the radiation diffuses outward at a rate of about 50 centimeters per hour and takes nearly 170,000 years to make its way out of the Sun.

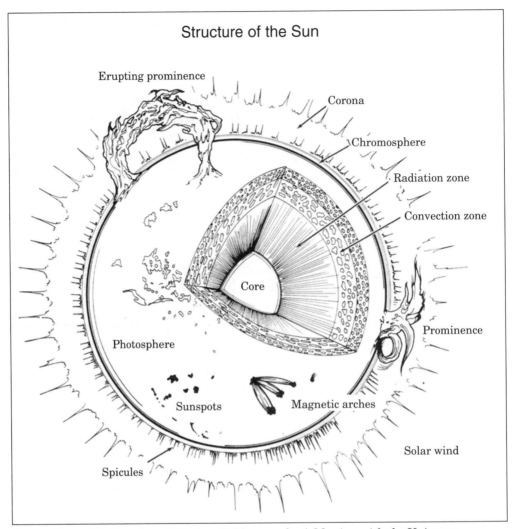

Structure of the Sun

Erupting prominence

Corona

Chromosphere

Radiation zone

Convection zone

Core

Photosphere

Prominence

Sunspots

Magnetic arches

Solar wind

Spicules

Source: Bevan M. French and Stephen P. Maran, eds. *A Meeting with the Universe:
Science Discoveries from the Space Program.* NASA EP-177. Washington D.C.:
National Aeronautics and Space Administration, 1981, pp. 68-69.

Radiative diffusion dominates until a distance from the center of about 71 percent of the Sun's radius. At that point, the temperature has dropped to about 2 million kelvins and the gas density has dropped to about 0.2 g/cm^3 (about 150 times denser than Earth's sea-level atmospheric density). Here some electrons are captured by the atoms. Instead of the light scattering off of electrons, it is absorbed by atoms, heating the gases. The hot gas then expands and rises. When the gas reaches the surface of the Sun, the photosphere, it cools by radiating light into space. The cooler gas then sinks until

it is again warmed by absorption of more energy. Heat transfer in this manner is called convection. Since convection dominates the energy transfer mechanism in the upper portion of the Sun, the top 29 percent of the Sun's radius is called the convective region. The lower 71 percent of the Sun's radius, including the core, is called the radiative region because radiative diffusion is the chief mechanism for energy transfer.

Between the radiative region and the convective region is a small region that acts as an interface between the radiative and convective regions. This zone is called the tachocline. Below the tachocline, atoms are almost all ionized, and radiative diffusion dominates. Above the tachocline, atoms have electrons, and convection dominates, but the transition is not sharp. There is no set distance from the center of the Sun where the transition between radiative diffusion and convection occurs. The lower portion of the tachocline has more radiative diffusion than convection, and the upper portion of the tachocline has more convection than radiative diffusion. The tachocline is important to solar astrophysicists because this is the region where the Sun's magnetic field is believed to be produced.

The top of the convective region is the photosphere. This is often regarded as the visible surface of the Sun. However, because the Sun is not solid, it really has no "surface." Rather, when the density and temperature of the gases that make up the Sun drop to low enough values, the gases become transparent to light, and the heat energy then shines out into space as thermal (blackbody) electromagnetic radiation. The photosphere itself is not a sharp boundary, but a rather thin zone. The distance from the center of the Sun to the top of the photosphere is generally regarded as the Sun's radius.

METHODS OF STUDY

The interior of the Sun is difficult to study. It cannot be imaged directly. However, it is important because the Sun's magnetic field is produced deep inside the Sun, and the Sun's magnetic field and magnetic behavior are responsible for solar activity, solar storms, and considerable interaction of the Sun with the rest of the solar system.

Though the Sun's interior cannot be imaged directly, it can be studied by the way that it influences the surface of the Sun. The Sun resonates in certain vibrational modes. This can be thought of as being analogous to the ringing of a bell, only at far lower tones because the Sun is so large. These vibrations can be recorded on the surface of the Sun through observations of the rise and fall of the Sun's surface using the Doppler shift of spectral lines in the Sun's spectrum. Study of the motions of the surface of the Sun in

this manner is called helioseismology. As shock waves pass through the interior of the Sun, they diffract (or bend) through the different layers of the Sun. Shock waves can also reflect from different layers in the Sun. Motion of material within the Sun can also distort the shock waves. Helioseismology can therefore yield a great deal of information about the Sun's interior structure.

Solar astrophysicists have learned much about the motion of material in the Sun's convective region beneath the photosphere. The interior of the Sun has been found to rotate at a slightly different rate than the surface layers of the Sun. This is believed to play a role in the Sun's magnetic behavior.

A further probe into the Sun's interior is through the study of neutrinos streaming from the Sun. Nuclear fusion occurring at the Sun's core produces not only gamma rays but also tiny, weakly interacting particles called neutrinos. These particles then flow outward from the Sun and can be detected on Earth. Studies of those neutrinos can yield information about the nuclear fusion processes going on in the Sun's core. Early studies of the neutrinos appeared to show far fewer neutrinos than had been predicted by nuclear theory. However, more recent studies suggest that the neutrinos oscillate, or change form, between three types, and the early experiments only detected one type of neutrino.

CONTEXT

The Sun is the nearest star to the Earth, and it is the center of the solar system. It dominates everything else in the solar system. The interior structure of the Sun is determined by its composition and mass, and that interior structure determines the observational characteristics of the Sun. Understanding the interior structure of the Sun, therefore, is important to understanding the Sun itself and ultimately its interactions with its planets, including Earth.

For many years, it was assumed that because the Sun is hot and bright, like fire, it must be shining through some sort of burning process. Burning, though, was unable to account for the Sun's energy. Astronomers theorized that perhaps the Sun was shining through release of its gravitational energy. That, too, failed to account for the Sun's energy output. Finally, by the early twentieth century, physics had advanced to the point where astrophysicists understood nuclear fusion to be the Sun's energy source.

Likewise, the interior structure of the Sun could not be understood until physicists understood the physics of materials under conditions such as exist inside the Sun. Theoretical astrophysicists now describe the Sun's structure through a set of equations called a solar model. Helioseismological studies match the current theoretical models to a very high degree of accuracy.

Raymond D. Benge, Jr.

FURTHER READING

Bhatnager, Arvind, and William Livingston. *Fundamentals of Solar Astronomy*. Hackensack, N.J.: World Scientific, 2005. A complete introduction to solar astronomy that does not require extensive mathematics to understand. The book emphasizes solar observations rather than solar structure, but there is a chapter on solar structure. An extensive glossary is included.

Caroll, Bradley W., and Dale A. Ostlie. *An Introduction to Modern Astrophysics*. 2d ed. San Francisco: Pearson Addison-Wesley, 2007. A textbook designed for undergraduate astronomy majors, providing an excellent and thorough description of stellar astronomy. A knowledge of calculus is assumed.

Freedman, Roger A., and William J. Kaufmann III. *Universe*. 8th ed. New York: W. H. Freeman, 2008. An excellent and thorough introductory college astronomy textbook. An entire chapter is devoted to the Sun, and other chapters cover other topics in stellar astronomy.

Golub, Leon, and Jay M. Pasachoff. *Nearest Star: The Surprising Science of Our Sun*. Cambridge, Mass.: Harvard University Press, 2001. An easy-to-understand discussion at the popular science level. Most of the book is about observations of the Sun and solar activity, but there is some discussion of the Sun's interior structure.

Hansen, Carl J., Steven D. Kawaler, and Virginia Trimble. *Stellar Interiors: Physical Principles, Structures, Evolution*. New York: Springer, 2004. A graduate-level text on the astrophysics of stars. Technical; assumes a familiarity with physics and calculus.

McDonald, Arthur B., Joshua R. Klein, and David L. Wark. "Solving the Solar Neutrino Problem." *Scientific American* 288, no. 4 (April, 2003). An extremely good review of the research that finally led to resolving the early confusing results of solar neutrino observations. A brief history of solar neutrino studies is included.

Zirker, Jack B. *Sunquakes: Probing the Interior of the Sun*. Baltimore: Johns Hopkins University Press, 2003. A very good survey of the science of helioseismology and how it is used to study the solar interior. The book is well documented and thorough without being too technical for laypersons to follow.

_____. *Total Eclipses of the Sun*. Expanded ed. Princeton, N.J.: Princeton University Press, 1995. A very good survey of solar structure written at a level easily accessible to the layperson. This is one of the best books on solar structure for the nonspecialist. Includes numerous photographs.

See also: Coronal Holes and Coronal Mass Ejections; Electromagnetic Radiation: Thermal Emissions; Infrared Astronomy; Nuclear Synthesis in Stars; Red Giant Stars; Solar Chromosphere; Solar Corona; Solar Evolution; Solar Flares; Solar Geodesy; Solar Infrared Emissions; Solar Magnetic Field; Solar Photosphere; Solar Radiation; Solar Radio Emissions; Solar Seismology; Solar Structure and Energy; Solar System: Origins; Solar Ultraviolet Emissions; Solar Variability; Solar Wind; Solar X-Ray Emissions; Sunspots; Thermonuclear Reactions in Stars; Ultraviolet Astronomy.

Solar Magnetic Field

Category: The Sun

The solar magnetic field constitutes a powerful force throughout the Sun, causing flares, prominences, sunspots, and a host of other phenomena. Charged particles from the Sun and the solar magnetic field interact with the terrestrial magnetosphere, affecting life on Earth.

OVERVIEW

A magnetic field is produced by and associated with moving charges (electrical currents). In a simple ring current, a dipolar field resembling the field of a bar magnet results. Complex interactions of charges occur everywhere in nature, and the large-scale motion of charges is extensive and intricate. Thus, depending on the current, magnetic fields can be found almost anywhere in various strengths. (For example, Earth's dipolar magnetic field is about 0.6 gauss, rather weak compared to solar active regions with field strengths of a few thousand gauss and pulsars with field strengths of 1 trillion gauss.) Interactions between magnetic fields and charged particles result in the emission of X rays, the energizing and modulation of cosmic rays, and the acceleration of charged particles. For example, the strong solar magnetic field expels charged particles from the Sun, while the weaker terrestrial magnetic field interacts with these particles, forming a protective shield around Earth by deflecting many of them, and letting some into the upper atmosphere producing the auroras (the northern and southern lights).

On a laboratory scale, the properties of magnetic fields are well known. The solar magnetic field, however, is much larger and necessarily more complex. The solar magnetic field appears to be generated and sustained by the Sun's rotation and the turbulent convective motion of charged particles below the photosphere (the Sun's visible surface). The development of improved observing techniques and instruments, including the use of space probes and space telescopes, has yielded much information on the Sun's magnetic field and its effects.

Sunspots have been seen from Earth since at least the fourth century B.C.E., as evidenced by early descriptions of them. They appear as darker spots of various sizes on the brighter solar photosphere. The temperature of the photosphere is about 5,800 kelvins, while sunspots are cooler by about 1,000 to 1,500 kelvins, thus making sunspots about one-third as bright as the rest of the photosphere. The spots have diameters ranging from a few thousand kilometers to more than 150,000 kilometers. Generally, smaller spots last only a few days, while larger ones linger for several weeks.

The solar sunspot cycle was discovered in 1843 by Samuel Heinrich Schwabe, who observed that the average number of sunspots varies systematically with an approximate periodicity of eleven years. At the beginning of a sunspot cycle, the spots appear symmetrically at solar latitudes of about 30° to 40° north and south, gradually working their way toward the Sun's equator by the end of the eleven-year period. Then the pattern repeats during the next sunspot cycle.

In 1908, George Ellery Hale—the American astronomer responsible for establishing the observatories at Yerkes, Mount Wilson, and Palomar—observed that lines in the spectra of sunspots were split into several components. Earlier, in 1896, Pieter Zeeman, a Dutch physicist, had shown that such splitting occurs in the presence of a strong magnetic field. Hale thus established that the sunspots had intense magnetic fields (up to 5,000 gauss). Since the Sun is a sphere of hot gases, such magnetic fields can exist only as a result of powerful convective currents. Hale and his colleagues at Mount Wilson Observatory tried to measure a general solar dipolar magnetic field, similar to that of Earth (whose magnetic field resembles that of a huge bar magnet), but it was not until 1953 that Horace Babcock, using a specially designed solar magnetograph, succeeded. He showed that a weak, periodically varying dipolar field of 2 to 7 gauss exists over the entire Sun.

The solar magnetic field changes over a twenty-two-year cycle, twice the eleven-year periodicity of the sunspot cycle. Sunspots appear either in pairs or in clusters, but they are always organized in coherent pairs of opposite magnetic polarity. During a particular eleven-year sunspot cycle, for example, the leading members of the sunspot pairs may display positive polarity in the Sun's northern hemisphere and negative polarity in the Sun's southern hemisphere. At the end of the eleven-year sunspot activity cycle, the polarities in the two hemispheres reverse. The magnetic cycle then repeats after two sunspot cycles.

The Sun's diffuse, dipolar magnetic field also undergoes a reversal of polarity with a periodicity of twenty-two years. A positive magnetic pole appears around the solar north rotational pole at the peak of every other sunspot cycle, while a negative magnetic pole is found around the solar south rotational pole at that time. The polar fields subsequently expand to latitudes of 50° to 60° while the sunspot numbers decline. After the sunspot minimum passes, the increase in number of sunspots with their strong local magnetic fields forces the global field back toward the rotational poles, there to coalesce and form opposite poles from those of the previous cycle.

The sunspot cycle and the twice-as-long magnetic cycle are more or less successfully explained by the solar dynamo theory, which involves interactions between the convective motions of plasma (ionized gas) beneath the photosphere and the Sun's differential rotation. Energy produced by nuclear fusion in the Sun's central core is transported slowly outward through the deep interior by the process of radiative diffusion (repeated absorption, emission, and scattering of photons by particles). Higher up, the energy is transported by radial convection currents and eddies (transport of energy by the actual motion of matter). It is generally believed that, within these convection zones, powerful magnetic fields are generated by a dynamo process. Basically, a dynamo converts the energy of motion (kinetic energy) of an electrical conductor into electromagnetic field energy. The ionized matter (plasma) within the solar convection zone is a good conductor of electric current, with the ability to retain and "freeze" the magnetic field. The Sun rotates differentially, with a rotation period at the equator of about 25 days, increasing to 30 days above 60° latitude. This differential rotation modifies the magnetic fields generated by dynamo action in the convection zone.

According to the dynamo theory advanced by Babcock, Eugene Parker, and others, the solar magnetic cycle starts with the Sun's global, dipolar magnetic field lines of force extending from pole to pole and frozen into the plasma a few hundred kilometers beneath the photosphere. The greater rotational speed at the lower latitudes means that the frozen-in magnetic field lines are carried faster and thus farther than those at higher latitudes. In due course, after many rotations, the magnetic field lines become tightly wrapped around the Sun, producing an intense east-west azimuthal magnetic field parallel to the equator. Adjacent magnetic field lines tend to repel one another, and when the field strength is a few thousand gauss, the repulsive force is strong enough that the field lines burst through the solar surface and loop back to reenter at a neighboring point, thus creating sunspots of opposite polarity and intense magnetic fields. The temperature around sunspots is lowered (dropping about 1,000 to 1,500 kelvins) as a result of lowered pressure in the region, caused by adjacent field lines repelling each other.

The dynamo theory successfully explains the reversed polarities of leading sunspots in terms of the opposite directions of the azimuthal field lines in the two hemispheres during a given sunspot cycle. The local azimuthal field gradually neutralizes the general dipole field, eventually reversing the polarity and starting the next sunspot activity cycle. The polarities, both of the dipole field and of the leading sunspots, are reversed from those of the previous activity cycle, so this starts the second half of the magnetic cycle. After two sunspot cycles, the magnetic cycle repeats in polarity.

Sunspot pairs occur as a result of random bursting of magnetic field lines upward from within the photosphere and their reentering at a neighboring point to resume their path around the Sun. Each sunspot has a central darker umbra, with a temperature of about 4,500 kelvins, surrounded by a lighter penumbra, with a temperature of about 5,500 kelvins. Most sunspot umbrae have diameters ranging between 4,000 and 22,000 kilometers, and magnetic field strengths ranging between 2,500 and 3,500 gauss. Smaller sunspots, with umbrae ranging from 1,400 to 3,600 kilometers in diameter and field strengths averaging 2,000 gauss, are known as "pores." Compact magnetic structures with diameters less than 1,000 kilometers, field strengths less than 1,500 gauss, and no discernible penumbral region are called "magnetic knots." The lifetime of the sunspot and its associated magnetic field typically is days to weeks, and in general it is found to be proportional to the total field strength of a sunspot or an active region.

Large clusters of magnetically active regions tend to produce prominences and flares because of excessive magnetic buoyancy. Prominences are relatively cool masses of gases arching above the photosphere into the corona, following magnetic lines of force. According to the Skylab data of 1973, eruptive (or active) prominences, triggered by large-scale magnetic fields associated with major sunspot activity, widely dissipate the field and disperse charged particles out into interplanetary space. Flares (even more violent bursts of energetic charged particles and electromagnetic radiation) occur near complex groups of sunspots, and in the process, intense magnetic field lines emerge and dissipate into space.

When high-energy charged particles from large flares reach Earth, they interfere with Earth's geomagnetic field and terrestrial communications systems. Coronal mass ejections (CMEs) are giant magnetic bubbles of ionized gas that carry enormous amounts of energy into space. If they encounter Earth, they can dump enough energy into its magnetosphere to cause disruptions of communication and electrical power distribution systems. Such phenomena gradually finetune the solar dynamo mechanism for the succeeding cycle. The solar wind (the continuous flow of charged particles from the corona into the far reaches of interplanetary space) carries a tenuous magnetic field with it as well. Fluctuations in the solar wind, as well as in X-ray and ultraviolet emissions from the Sun, are closely correlated with the magnetic cycle.

Evidence for long-term variability in solar activity and corresponding fluctuations in the solar magnetic field comes from the quantitative analysis of radioactive carbon-14 deposition in tree rings. Cosmic rays from various sources normally reach Earth's upper atmosphere and produce several radioactive isotopes, including carbon 14, which results from a collision between energetic particles and nitrogen nuclei. During a prolonged period of reduced sunspot activity, the lack of turbulence in the Sun's magnetic field will allow more intense cosmic rays to reach Earth, thus increasing the rate of production of carbon 14, which is absorbed by vegetation and eventually deposited in tree rings. Historical records (often indirect) show a good correlation between climate variations and various indicators of solar activity (such as sunspot numbers, auroras, and the size of the corona seen during solar eclipses). The Spörer minimum of the mid-fifteenth century (corroborated by the recorded paucity of auroras), the Maunder minimum (a seventy-year hiatus in sunspot activity, beginning about 1645), and the Little Maunder minimum between 1800 and 1830 are examples of periods of solar inactivity and colder climates on Earth.

Searches for magnetic cycles in nearby Sun-like stars indicate that in earlier times, the solar cycle may have been more irregular, erratic, and intense than it is now. Stronger magnetic fields, causing a significant loss of electromagnetic and particle energy from the Sun, may have contributed to a loss of angular momentum, thus decreasing the rotation rate. In any case, as stars age, their magnetic cycles tend to become more regular and well established. The Sun's magnetic field strength is expected to decline as it slowly enters the red giant stage. The expanding Sun's rotation rate will slow down and the mass density of its convection zone will decrease, thus reducing the dynamo effect. Chromospheric emissions triggered by the magnetic field dissipation process are expected to subside to a low level; however, studies of chromospheric emission activities of some subgiants show a curious revival of phenomena associated with their magnetic fields.

KNOWLEDGE GAINED

Clearly, there is still much to learn about solar magnetism. What is known has been gleaned by both earthbound and space-based observations from the mid-twentieth century onward. The solar magnetograph invented in 1952 by Horace Babcock and his father, Harold Babcock, marks a milestone in the observational study of the Sun's magnetic field. Observations from space by the instruments onboard Skylab and the Solar Maximum Mission of regions of the electromagnetic spectrum that are blocked by Earth's atmosphere from reaching the ground have enhanced the understanding of solar magnetic phenomena and their astrophysical implications.

The advent of new and improved magnetometers has yielded a wealth of data concerning the Sun's magnetic field. Observations and measurements of small and large magnetic features and related solar atmospheric activities have led astrophysicists to develop an elaborate dynamo theory to explain the origin and variations of the solar magnetic field. Astrophysicists have established that the twenty-two-year magnetic cycle is responsible for most, if not all, of the phenomena collectively referred to as solar activity (such as sunspots, prominences, and flares), as well as the processes that result in the solar wind and the formation of the interplanetary medium.

The Skylab data of 1973 amply demonstrate the link between the magnetic activity cycle and stress-relieving eruptive phenomena such as prominences, flares, and coronal emissions. Solar plasma, carrying with it extensive unstable magnetic fields, interacts with terrestrial magnetic fields to affect the size and shape of Earth's magnetosphere and processes occurring in it. Data from the Solar Maximum Mission showed modulations in the output of solar energy attributable to the time-varying, churning action of the Sun's magnetic activity cycle.

Questions have been raised concerning the magnetic stability of a convective shell dynamo such as the one envisioned for the Sun. Pointing out the difficulty of initiating in the solar atmosphere the large electric currents associated with the observed magnetic field, numerous theorists have suggested that there may be a potent, primordial magnetic field frozen within the solar core. It has been hypothesized that when interstellar nebulae with tenuous magnetic fields collapse to form stars, they can retain intense primordial fields over a relaxation time of some 5 billion years, which is the present age of the Sun.

CONTEXT

Almost half a century after Hale's discovery of the strong magnetic fields associated with sunspots, Babcock and Babcock, in 1952, devised the modern solar magnetograph, opening up a new era of solar observational physics. Knowledge of the Sun's observed differential rotation, inferences about the Sun's convective zone drawn from computer models of its interior, and the laws of electrodynamics from physics were combined to develop the theory of the solar dynamo to explain observed magnetic phenomena.

While the larger pieces of the puzzle appear to be in place, many missing elements blur the picture. The solar magnetic field is not uniform, but filamentary or tubular. The precise manner in which the large-scale field produces observed emission features and expels unstable field elements, seemingly perpetuating the solar dynamo and creating the solar wind, is not well understood. It is known that a large-scale solar magnetic field can lead to a variety of mechanical wave modes. These may contribute to a heating mechanism, but this notion requires further observational confirmation. Virtually all activity in the solar atmosphere (except for the granulation in the photosphere) owes its existence to and is orchestrated by the turbulent and somewhat unpredictable solar magnetic cycle, yet many astrophysically important details are not known. Numerous attempts at improving the dynamo model, including the invocation of a strong primordial magnetic field, have proved to be only marginally successful. Long-term, detailed observations of the finer magnetic elements and processes, possibly through remote sensors, will be required, along with concurrent theoretical refinements.

An area of immediate and compelling interest is the further study of the long-term effects of solar activity cycles on terrestrial weather patterns and climate trends. Cosmic rays and their modulation by magnetic fields play an important role in genetic mutation and the evolution of life in general. Quantitative observation in this area will aid in evaluating the extent to which the deflecting action of the solar magnetic field has affected life. Further probing in this area may bring to light relationships between the evolution of life on Earth and the solar magnetic cycle.

Finally, research on the Sun's magnetic field has added to our insight into stellar magnetic fields in general. It has been firmly established that the Sun's magnetic cycles, flare phenomena, and coronal properties are common at least among main sequence stars similar to the Sun. For astronomers, then, the Sun is a cosmic laboratory, a window to the realm beyond the solar system.

V. L. Madhyastha

FURTHER READING

Chaisson, Eric, and Steve McMillan. *Astronomy Today*. 6th ed. New York: Addison-Wesley, 2008. This well-written introductory textbook for college astronomy courses offers a good description of the Sun's magnetic field and its effects.

Fraknoi, Andrew, David Morrison, and Sidney Wolff. *Voyages to the Stars and Galaxies*. Belmont, Calif.: Brooks/Cole-Thomson Learning, 2006. A well-written, thorough college textbook for introductory astronomy courses. Provides descriptions of infrared emissions and how the Sun produces them.

Freedman, Roger A., and William J. Kaufmann III. *Universe*. 8th ed. New York: W. H. Freeman, 2008. This introductory astronomy textbook contains a concise summary of the Sun's magnetic field and its effects.

Gibson, Edward G. *The Quiet Sun*. NASA SP-303. Washington, D.C.: Government Printing Office, 1973. Written by Skylab astronaut Gibson, this volume summarizes solar physics as understood at the time of the Skylab mission, addressing the solar activity cycle and myriad details involving the magnetic cycle and its effects. Aimed at solar researchers, but useful as a supplement for college and even high school students as well.

Jordan, Stuart, ed. *The Sun as a Star*. NASA SP-450. Washington, D.C.: Government Printing Office, 1981. This collection of review articles by noted authorities, although directed toward specialists in the field, incorporates a variety of observational results and theoretical models, showing clearly the complexities involved in understanding the ever-changing pattern of solar magnetic fields. With its numerous references at the end of each article, the volume serves as a valuable reference work on the subject at an advanced college level.

Newkirk, Gordon, Jr., and Kendrick Frazier. "The Solar Cycle." *Physics Today* 35 (April, 1982): 25-34. Using the solar magnetic-dynamo model, the authors discuss eleven-year and twenty-two-year solar cycles. The Maunder minimum in the seventeenth century and other historic periods of low sunspot activity are considered. Discussions of a possible clock in the Sun, the variation of luminosity correlated with sunspot activity, and photospheric pulsation are included, along with a list of useful references.

Parker, E. N. "Magnetic Fields in the Cosmos." *Scientific American* 249 (August, 1983): 44-54. The author, an authority on the subject of cosmic magnetic fields, presents the theory of the solar dynamo, the mechanism believed to be the source of magnetic fields of the Sun, the planets, and the galactic plane. The same mechanism, combined with the differential rotation of the Sun with latitude, affords a natural explanation of the intense magnetic fields associated with sunspots, flares, and other aspects of the sunspot cycle.

Schneider, Stephen E., and Thomas T. Arny. *Pathways to Astronomy*. 2d ed. New York: McGraw-Hill, 2008. A college textbook for introductory astronomy courses. Contains several sections on the Sun's magnetic field and its effects.

Walker, Arthur B. C., Jr. "Golden Age for Solar Physics: New Instruments for Astronomy." *Physics Today* 35 (November, 1982): 60-67. An overview of solar physics and of projected lines of investigation of the multitudes of problems remaining to be solved as of the early 1980's. Features a clear account of the solar activity cycle, magnetic explosions, and phenomena triggered by the magnetic field: flares, coronal loops, and solar winds. Includes a list of references.

Wilson, Olin C., Arthur H. Vaugan, and Dimitri Mihalas. "The Activity Cycle of Stars." *Scientific American* 244 (February, 1981): 104-119. After discussing the Sun's eleven-year cycle, the authors report on the ongoing study of some ninety nearby stars and the quest for similar cycles. The goal is to understand why such cycles arise in the first place and why they sometimes suddenly vanish, as in the case of the Sun, only to reappear decades later. Astronomers may eventually be able to predict the course of the solar activity cycle from systematic observation of a large number of Sun-like stars.

See also: Coronal Holes and Coronal Mass Ejections; Earth's Magnetic Field: Origins; Earth's Magnetosphere; Earth-Sun Relations; Interplanetary Environment; Jupiter's Magnetic Field and Radiation Belts; Neptune's Magnetic Field; Planetary Magnetospheres; Solar Chromosphere; Solar Corona; Solar Evolution; Solar Flares; Solar Geodesy; Solar Infrared Emissions; Solar Interior; Solar Photosphere; Solar Radiation; Solar Radio Emissions; Solar Seismology; Solar Structure and Energy; Solar Ultraviolet Emissions; Solar Variability; Solar Wind; Solar X-Ray Emissions; Sunspots; Uranus's Magnetic Field.

Solar Photosphere

Category: The Sun

The solar photosphere is the visible surface of the Sun. Because it is opaque, it is not possible to see the Sun's interior layers directly. Convection currents transport energy from the interior through the photosphere, and this convection, along with magnetic activity, causes surface features such as granules and sunspots. Solar magnetic activity may affect Earth's climate.

OVERVIEW

The Sun is a ball of gas and, as such, has no solid surface. The Sun's photosphere approximates its surface, marking the region between the solar interior and the chromosphere (the region that approximates the solar "atmosphere"). Because the photosphere is relatively opaque, it blocks any view of the solar interior. As a result, most images of the solar disk show only the photosphere. The photosphere is also the coolest layer of the Sun, with a temperature of about 6,600 kelvins near its bottom and a temperature of about 4,400 kelvins near the chromosphere at the top; in the 400 kilometers between, the photosphere's temperature gradually decreases with altitude above the solar interior. Astronomers know that the photosphere is relatively cool, because it displays absorption lines that are darker than the spectrum of the hot solar interior. The temperature of the overlying chromosphere is likewise hotter, climbing to about 30,000 kelvins as the distance from the photosphere increases. Where the chromosphere ends is the corona, and in the approximately 100-kilometer-thick transition between the chromosphere and corona, the temperature rapidly increases from about 6,000 kelvins to several hundred thousand kelvins. In the corona itself, the temperature can be millions of kelvins, but the gas is very thin.

Looking at a picture of the solar disk, one can notice that the edge of the disk appears darker than the central portions. This illusion is called limb darkening. In the central portion of the solar disk one sees light from the deepest and hottest layer of the photosphere. Near the Sun's edge—its limb—only the upper, cooler layers of the photosphere are visible, because the very edge of the solar disk does not have the base of the photosphere in our line of sight behind the top layers. The cooler upper layers of the photosphere emit less energy than the hotter, deeper layers, so the limb of the solar disk appears darker. It is not really dark; it is just less bright. Limb darkening, therefore, indicates that the lower layers of the photosphere are hotter than the upper layers.

Because it is cooler than the hot, compressed gas of the solar interior, the Sun's photosphere produces an absorption line spectrum (a continuous spectrum with dark absorption lines superimposed on it at certain wavelengths). In 1814, Joseph von Fraunhofer observed the solar spectrum, and—because he was unable to identify the elements producing the various spectral absorption lines he observed—labeled the prominent lines using capital letters.

The most obvious features of the Sun's photosphere are the granules, which give the photospheric surface a mottled light and dark appearance. Granules are bright regions that vary in size but are typically about 1,000 kilometers across. They also are temporary features, typically lasting from about five to thirty minutes. Granulation results from convection currents below the photosphere.

Nuclear reactions powering the Sun take place in the core, and the energy must be transferred from the core to the surface. In the deep interior of the Sun, the energy is transferred by radiation. In the upper portions of the interior, just below the photosphere, the energy is transferred by convection currents. The solar convection currents are similar to the convection currents that heat a room, if there is a radiator in one side and no fan to blow the warm air to the other side. These solar convection currents form convection cells, regions where the hot gas from the interior flows up and then flows back down after it cools. The granules that we observe are the tops of these convection cells. They appear brighter because they are still hot from the Sun's interior energy.

Supergranules are larger versions of granules. They might typically be about 35,000 kilometers across. A typical supergranule might last one or two days, much longer than the typical lifetime of a granule. Rather than being observed directly, like granules, supergranules are typically observed from Doppler maps of the Sun's surface. The upward-moving material is moving toward Earth, and hence toward the observer, so the wavelengths of common spectral lines are blueshifted to slightly shorter (higher-energy) wavelengths and hence appear brighter.

Granules are part of what astronomers call the "quiet Sun," which comprises the solar features that are always present. Features that are present or more common only during the maximum of the solar magnetic activity cycle comprise the "active Sun." Features of the active Sun found in the photosphere are sunspots and faculae.

Sunspots are dark areas on the Sun's photo-sphere. Sunspots appear dark because they are about 2,000 kelvins cooler than the rest of the photosphere. Although this temperature is still quite hot, sunspots are nevertheless cool when compared to the background of the rest of the photosphere. Large sunspots can extend to tens of thousands of kilometers, larger than Earth but still very small compared to the Sun's size. Sunspots will typically cover less than about 1 percent of the Sun's photosphere.

Magnetograms of the solar surface measure the magnetic field strength at points across the Sun's surface. Magnetograms show that sunspots are regions of intense magnetic fields, with magnetic field strengths up to a few thousand times stronger than the rest of the Sun's surface. Sunspots are darker and cooler than the rest of the photosphere because these strong magnetic fields conspire to deflect the convection currents bringing heat energy from the interior to the photosphere. The reduced heat flowing from the interior causes lower temperatures.

Faculae are regions on the photosphere that are hotter and therefore brighter than the rest of the photosphere. They might be thought of as the opposite of sunspots; like sunspots, faculae are regions of strong magnetic fields, but for faculae the magnetic fields concentrate, rather than deflect, the energy from the interior. Faculae form in regions surrounding sunspots and in lower boundaries between the elevated granules. When faculae extend up into the chromosphere they are called plages.

The numbers of both sunspots and granules wax and wane in an eleven-year sunspot cycle. This eleven-year cycle in the number of spots is actually half of the Sun's twenty-two-year magnetic activity cycle. Sunspots form in groups that contain leading and following spots. The magnetic polarities (north or south magnetic poles) of the leading and following spots in a group interchange each eleven-year cycle to produce a twenty-two-year magnetic activity cycle. The last sunspot maximum occurred around 2001.

KNOWLEDGE GAINED

The Sun's total energy output, or luminosity, is largely the energy from the photosphere. Hence if the photosphere's brightness changes, the Sun's total energy output and the solar radiation reaching Earth also change. Variations in the Sun's luminosity could cause, and very likely have caused, climate changes on Earth. Changes in the Sun's luminosity are so small that, from using ground-based measurements, it is difficult (but not impossible) to measure the Sun's luminosity accurately enough to measure those variations. With the advent of space-based observatories—which eliminated the need to correct for the amount of light absorbed by Earth's atmosphere—it became possible to measure the Sun's luminosity more easily and accurately. Beginning in the late 1970's, satellite data show that when there is a maximum number of sunspots, the Sun is a very small amount more luminous than at sunspot minimum.

These data, when combined with the long-term sunspot record, suggest that changes in the Sun's energy output may have caused the Little Ice Age in the seventeenth century. The Little Ice Age was a period of more than two centuries that was colder than normal and that coincided with a prolonged period of virtually no sunspot activity, from about 1645 to about 1715 C.E. There was also a prolonged medieval grand maximum in sunspot activity from about 1100 to about 1250 C.E. This period was the warmest of the past millennium. It is likely, therefore, that changes in the amount of magnetic activity have affected Earth's climate at other times in the past. It is also possible that such changes may be a contributing factor to current global warming. The Sun has been fairly active since the mid-1700's, except for minor, brief decreases in sunspots in the early 1800's and in the late 1800's.

CONTEXT

The Sun's energy is crucial to life on Earth. Without it we would not survive. Hence we want to understand the Sun and all its components, including the photosphere. As the visible disk of the Sun that we see, the photosphere is the immediate source of this energy. Changes in the photospheric energy output could have drastic effects for life on Earth and may even affect climate change.

Visible light, ultraviolet light, infrared light, radio waves, X rays, and gamma rays are all forms of electromagnetic waves. The significance of visible light is that it is the region of the electromagnetic spectrum that is detectable by the human eye. This range is determined by the Sun's photosphere. With a temperature of 5,800 kelvins, the photosphere is brightest at the wavelength region from red to blue, peaking at yellow. This is also the wavelength of peak sensitivity of the human eye and most animal eyes. Our eyes evolved to detect the wavelength region of the electromagnetic spectrum that is most plentiful.

Paul A. Heckert

FURTHER READING

Chaisson, Eric, and Steve McMillan. *Astronomy Today*. 6th ed. New York: Addison-Wesley, 2008. One chapter of this very readable introductory astronomy textbook covers the Sun.

Frazier, Kendrick. *Our Turbulent Sun*. Englewood Cliffs, N.J.: Prentice-Hall, 1980. This book gives a readable account of our knowledge of the Sun through its publication date.

Freedman, Roger A., and William J. Kaufmann III. *Universe*. 8th ed. New York: W. H. Freeman, 2008. All aspects of the photosphere are well covered in this excellent introductory astronomy textbook.

Golub, Leon, and Jay M. Pasachoff. *Nearest Star: The Surprising Science of Our Sun*. Cambridge, Mass.: Harvard University Press, 2001. Provides a detailed summary of our knowledge of the Sun. The section on the photosphere goes into considerable detail on sunspot activity.

Heckert, Paul A. "Solar and Heliospheric Observatory." In *USA in Space*. 3d ed. Edited by Russell Tobias and David G. Fisher. Pasadena, Calif.: Salem Press, 2006. Describes the SOHO mission, which was used to study the Sun's outer layers, including the photosphere, and helped us understand the Sun's magnetic activity and why the temperature increases in the chromosphere and corona.

Hester, Jeff, et al. *Twenty-First Century Astronomy*. New York: W. W. Norton, 2007. Chapter 13 of this very readable astronomy textbook is about the Sun. The discussion of sunspot activity and its effect on Earth is detailed yet accessible.

Morrison, David, Sidney Wolf, and Andrew Fraknoi. *Abell's Exploration of the Universe.* 7th ed. Philadelphia: Saunders College Publishing, 1995. The Sun is covered in chapter 26 of this classic astronomy textbook.

Zeilik, Michael. *Astronomy: The Evolving Universe.* 9th ed. Cambridge, England: Cambridge University Press, 2002. An extremely well-written introductory astronomy textbook. Chapter 12 is an overview of the Sun.

Zeilik, Michael, and Stephen A. Gregory. *Introductory Astronomy and Astrophysics.* 4th ed. Fort Worth, Tex.: Saunders College Publishing, 1998. Pitched at a level for undergraduate physics or astronomy majors, this textbook goes into more mathematical depth than most introductory astronomy textbooks. Chapter 10 covers the Sun, including the photosphere.

See also: Auroras; Coronal Holes and Coronal Mass Ejections; Electromagnetic Radiation: Thermal Emissions; Infrared Astronomy; Red Giant Stars; Solar Chromosphere; Solar Corona; Solar Evolution; Solar Flares; Solar Geodesy; Solar Infrared Emissions; Solar Interior; Solar Magnetic Field; Solar Radiation; Solar Radio Emissions; Solar Seismology; Solar Structure and Energy; Solar System: Origins; Solar Ultraviolet Emissions; Solar Variability; Solar Wind; Solar X-Ray Emissions; Sunspots; Thermonuclear Reactions in Stars; Ultraviolet Astronomy.

Solar Radiation

Category: The Sun

The total solar radiation that falls on Earth is the primary factor in determining Earth's weather and climate. Even the smallest variation in the solar irradiance, if sustained, could alter the terrestrial environment drastically.

OVERVIEW

Solar radiation spans the entire electromagnetic spectrum, from very short-wavelength, high-energy gamma rays given off by some solar flares to extremely long-wavelength, low-energy radio waves given off by magnetic disturbances associated with sunspots and other kinds of solar activity. Between these two extremes, there are X rays, ultraviolet light, visible light, and infrared radiation, all of which provide clues to the processes occurring in and on the Sun. Astronomers study solar radiation and the solar spectrum with the use of ground-based telescopes, high-flying aircraft, balloons, and spacecraft. Study of solar radiation addresses one of the most important problems in solar physics: Does the Sun change its radiation output over time, or is it constant?

The Sun's luminosity is the total electromagnetic energy emitted by the Sun per unit time into space, or the solar radiative power. Solar luminosity is approximately 3.8×10^{26} watts, meaning it radiates 3.8×10^{26} joules of electromagnetic energy into space every second. The solar radiation per unit time per unit area impinging upon the top of Earth's atmosphere is known as the total solar irradiance, also referred to as the "solar constant." This value is $1,368 \pm 7$ watts per square meter. Whether the Sun's luminosity is really constant, however, is in question. A decrease of as little as one-half of 1 percent over a period of one century could send the entire Earth into an ice age. An increase of the same magnitude could produce first a global tropical rain forest and eventually scorched desert continents.

Most of what is known about the Sun's radiation is derived from an analysis of its electromagnetic spectrum. The Sun's visible spectrum, like that of most other stars, consists mainly of a smooth distribution of intensity of emitted light known as the continuum (or the continuous spectrum), with narrow dips in brightness, termed dark absorption lines. An oversimplified explanation is that the continuum is emitted from the Sun's photosphere, an extremely thin shell of gas giving off the visible light that can be seen with the human eye (although damage to the retina is severe if the Sun is observed directly). Also known as Fraunhofer lines after their discoverer, solar absorption lines are produced when electrons in atoms in the Sun's atmosphere absorb photons with specific energies or wavelengths of light, causing the electrons to jump to higher energy levels in the atoms.

The visible colors are only a narrow band of wavelengths within the entire electromagnetic spectrum. The wavelengths of the visible light range from just under 400 nanometers in the violet to more than 700 nanometers in the far red. Approximate wavelength boundaries for the other regions of the electromagnetic spectrum outside the visible range include the near ultraviolet, from about 120 nanometers up to just below 400 nanometers, where visible violet begins; extreme ultraviolet, wavelengths between about 10 and 120 nanometers; soft X-ray wavelengths, between about 0.1 and 10 nanometers; hard X rays, between about 0.001 and 0.1 nanometers; and gamma rays, shorter than about 0.001 nanometers. At the other end of the spectrum, near-infrared emissions range from more than 700 nanometers (where visible red ends) up to about 1,000 nanometers; far infrared refers to wavelengths between about 1,000 and 1 million nanometers (about 1 millimeter, or 1,000 microns); and radio waves are longer than 1 millimeter and can be many kilometers in length. All these regions of the spectrum provide information about different layers of the solar atmosphere as well as about aspects of solar activity associated with sunspots, prominences, and flares.

The spectral distribution of the Sun's electromagnetic radiation approximates that of a blackbody, which is a hypothetical object that is opaque to all the radiation that falls upon it. It is a highly useful concept for describing the way stars radiate their continuous spectra. Blackbodies of higher temperatures emit much more electromagnetic energy at all wavelengths than those of lower temperatures, and the peak of the radiation distribution (spectrum) is at progressively shorter wavelengths for increasing blackbody temperature. The Sun's continuum radiation distribution is similar to that of a blackbody whose temperature is about 5,800 kelvins.

Changes in the Sun's output of electromagnetic energy and hence the solar "constant" could profoundly affect the climate on Earth. There is some evidence that short-term climatic variations, measured in terms of decades to millennia, may be brought on by changes in the Sun's luminosity and the solar constant. Astronomers have wondered whether even shorter-term climatic changes, such as extensive droughts, might be caused by changes in the Sun's energy output. (On the other hand, long-term variations, such as the advance and retreat of continental glacial ice sheets over periods of tens of thousands to hundreds of thousands of years, are best explained by seasonal and latitudinal variations of the solar energy input caused by changes in the geometry of Earth's orbit and axial tilt. Even longer intervals of hundreds of millions of years between major continental glaciations are thought to result from changes in ocean and continent geometries caused by continental drift.)

Claude Pouillet, who introduced the concept of the solar constant in 1837, tried to measure the solar constant by monitoring the temperature of a blackened box filled with water. The temperature increase of the water per unit of time would reflect the energy gained via sunlight. Samuel Pierpont Langley and Charles Greeley Abbot were later pioneers in measuring and monitoring the solar constant. In 1878, Langley invented a device that he named the bolometer. The bolometer measures the energy of incoming radiation, regardless of its wavelength. (Langley discovered far-infrared light while using his bolometer to study solar radiation.) Langley used the bolometer at high altitudes to try to measure the solar constant, making mathematical corrections for Earth's atmospheric extinction. His values of the solar constant are considered fairly accurate even by today's standards.

However, modern solar astronomers question the reality of the small variations Langley and Abbot detected in the solar constant and solar luminosity. Such doubts have arisen because of the lack of precision of nineteenth century equipment and the inability to estimate accurately experimental errors and uncertainties. It can be said that, if Langley and Abbot's study suggests anything, it suggests that the solar constant is indeed constant. Any small variations simply escaped the capabilities of their measurements, even though they were convinced that they detected small changes of about 0.5 percent or less.

Certain types of variations, however, have since been demonstrated, principally variations associated with the solar cycle. John Eddy conducted an exhaustive historical study, looking back many years for evidence of past solar cycles. Much of this work concentrated on eras before the invention of the telescope and thus necessitated gathering descriptions of the appearance of the solar corona during solar eclipses, accounts of sunspot observations, and descriptions of the aurora borealis.

Regular magnetic changes occur in the solar atmosphere over a period of about 11.2 years, the period of the solar cycle. Sunspots are observed to be at a maximum each time the cycle reaches its highest magnetic strength. The years 1969 and 1980 were such peak years; during the latter, the Solar Maximum Mission (SMM) satellite was launched. The solar cycle is driven by the Sun's interior magnetism. Magnetic fields are wrapped by the rotation of the Sun. The Sun is a differential rotator; that is, its rotation rate depends on latitude; its equator rotates with a period of about twenty-five days, but at latitudes of 75° north and south, the rotation period can be thirty days or longer. Magnetic field lines between the Sun's magnetic poles are dragged by the differentially rotating gas and coiled more and more tightly around the Sun. As the magnetic field lines become twisted and snarled, they arch into the upper layers of the solar atmosphere and produce a variety of solar magnetic activity, including sunspots, prominences (arching spires of gas), and flares (caused by magnetic trapping of thermal energy, which, at its breaking point, suddenly triggers a release of immense energy).

These cycles vary in intensity and are irregular in occurrence. The intensity of a cycle is measured most directly by the number of sunspots observed during the cycle. These numbers vary noticeably from one cycle to another. Several cycles have been observed to be deficient in sunspot production, and in a few cases almost no spots are produced. These "weak" cycles might be associated with a smaller luminosity and solar constant. Some solar physicists believe that the rate of solar rotation may vary from one cycle to another and that this rate may affect the solar constant. The magnetic centers of activity

on the Sun are thought to interfere with the normal outflow of radiation; they can act to obstruct or reduce the flow outward and thus slightly modify the solar constant.

Peter Foukal and Jorge Vernazza found a possible correlation between solar rotation and the solar constant variation by examining the data from the experiments by Langley and Abbot. Their statistical analysis found a change of 0.07 percent every twenty-seven days in the observations. This tiny variation, occurring suspiciously in phase with the solar rotation, is accepted by some solar physicists and doubted by others. Some hold that such a small change is difficult to verify.

The period of solar rotation seems to have suffered "glitches" in the past. There is even modern evidence for small changes in solar rotation on a very short timescale, weeks or even days long. A known, but presumably very small, connection between solar rotation and the solar constant is the fact that as solar rotation carries sunspots around the Sun, their appearance on or disappearance from the earthward side of the Sun alters the solar constant very slightly, because less light is emitted by the sunspot than from a comparable area of the photosphere.

Variable solar rotation, if real on both long and short timescales, may also be an indication of other types of changes in the Sun's deep interior that could lead to changes in the solar constant. A. Keith Pierce and James C. LoPresto, using the McMath solar telescope at Kitt Peak, Arizona, reported in 1984 that the Sun's rotation does quickly change; it speeds up and slows down in a large cap around the polar regions within periods as short as a day or two.

Many astronomers and solar physicists have not been particularly interested in measurements of the solar constant over periods of time long enough to include one or more solar cycles. They contend that such experiments are difficult and expensive and that it is unlikely that significant changes in the solar constant could be documented. However, in response to the need for good solar irradiation measurements, it was decided to include an experiment for measuring the solar constant on board the Solar Maximum Mission (SMM) satellite, launched in 1980. Although the primary mission of the satellite was to study the Sun's behavior during the peak (or "maximum") activity cycle, it was also decided that monitoring any changes in the solar constant during the maximum was of utmost importance. Richard C. Willson thus devised an instrument dubbed the active cavity radiometer irradiance monitor (ACRIM) that was carried onboard SMM. Willson's measurements with ACRIM from SMM gave a value for the solar constant of 1,368 ± 7 watts per square meter. Furthermore, it has been established that large sunspots decrease the solar irradiance by a few watts per square meter for very short periods (a week or so).

KNOWLEDGE GAINED

Despite all the studies of the solar constant, attempts to verify its constancy or variability were inconclusive until the last few decades. Richard C. Willson measured the solar constant from high-altitude balloons in 1969 and again from an Aerobee rocket in 1976. To within 0.75 percent, the solar constant during that period remained unchanged. This period included the last half of a solar cycle. In 1967, 1968, and 1978 David A. Murcray of the University of Denver measured the solar constant from high-altitude balloons. He found no change from 1967 to 1968, but he detected a 0.4 percent change between 1968 and 1978.

The SMM spacecraft, launched into Earth orbit by the space shuttle in 1980, carried a diverse, sophisticated payload to study solar activity. Among the instruments on board was the ACRIM, which was designed by Willson to detect changes in the solar constant of as little as 0.1 percent. ACRIM measured the solar input during half of each 131.072-second measurement interval and a known radiation source onboard the spacecraft during the other half of each measurement interval, during which time a shutter blocked solar radiation. The solar irradiance received by Earth is proportional to the difference between the heating rates with the shutter open and with it closed. The ACRIM results indicate that the average value of the solar constant is 1,368 ± 7 watts per square meter. It decreased slightly but steadily between 1980 and 1986, leveling off in 1987 and 1988. There are indications of a possible slow period vari-

ation connected with the solar cycle of about 2 watts per square meter. It was also determined that sunspots cause very short term variations, for periods equal to the life of the spot.

Solar astronomers working at the National Solar Observatory, Kitt Peak, Arizona, will continue to monitor certain absorption lines in the solar spectrum that are known to be sensitive to stellar luminosity. They have found that the strengths of these lines closely correspond to the variability first detected by ACRIM, both in magnitude and in the temporal variation associated with the sunspot cycle.

CONTEXT

The goal of solar spectral irradiance measurements is to measure solar absolute brightness and how much and how rapidly it changes. The visible spectrum dominates the energy emission, and both it and the infrared seem to be relatively constant. In contrast, radio, ultraviolet, and X rays show large fluctuations associated with solar magnetic activity during the solar cycle.

Prior to the space age, measurements of solar irradiance had a very incomplete time record, even over one solar cycle. Many astronomers and solar physicists were not particularly interested in measuring the solar constant over long periods of time, contending that such experiments were difficult and expensive and it was unlikely that significant changes in the solar constant could be documented. Since the solar spectrum covers an enormous wavelength range, from radio waves to X rays, taking measurements of absolute brightness was, and remains, very challenging, inasmuch as different detectors and different techniques are required in each wavelength band. Over the electromagnetic range, at least five different experimental techniques are required for imagery and dispersion of the solar spectrum.

Another difficult problem in attempts to measure the solar constant was the attempt to determine how much energy Earth's atmosphere blocked before the radiation reached detectors on the ground. Earth's atmosphere filters out radiation according to wavelength; its blocking of many parts of the electromagnetic spectrum is very effective. Ozone in the upper atmosphere completely absorbs all radiation shorter than the very longest near-ultraviolet waves, including gamma rays, X rays, and most ultraviolet rays. At the lower end of the spectrum, water vapor and carbon dioxide strongly absorb much of the infrared, and water vapor and oxygen absorb radio waves shorter than about 1 centimeter. The ionosphere, a thin layer of charged particles at an altitude of about 100 kilometers, reflects radio waves longer than about 10 meters. Even those wavelengths that are not completely blocked, still may suffer some absorption and scattering as they pass through our atmosphere.

The solution to this problem appeared when it became possible to to make observations above Earth's atmosphere, and just about the entire solar spectrum has been observed using the Orbiting Solar Observatories (OSOs), Skylab, the Solar Maximum Mission (SMM), and the Solar and Heliospheric Observatory (SOHO).

James C. LoPresto

FURTHER READING

Chaisson, Eric, and Steve McMillan. *Astronomy Today*. 6th ed. New York: Addison-Wesley. A well-written college-level textbook for introductory astronomy courses. Good description of the Sun's atmosphere and energy output.

Eddy, John A. *A New Sun: The Solar Results from Skylab*. Washington, D.C.: Government Printing Office, 1979. Eddy highlights what was learned about the Sun as a result of the Skylab mission and the use of the Apollo Telescope Mount. Spectacular photographs make clear the enormous advances in knowledge of the Sun that were made possible by Skylab. Suitable for general audiences.

Foukal, Peter. *Solar Astrophysics*. 2d rev. ed. Weinheim, Germany: Wiley-VCH, 2004. A detailed look at the Sun and our understanding of it. Also covers the history of solar astrophysics. Suitable for undergraduates.

Fraknoi, Andrew, David Morrison, and Sidney Wolff. *Voyages to the Stars and Galaxies*. Belmont, Calif.: Brooks/Cole-Thomson Learning, 2006. A well-written, thorough college textbook for introductory astronomy courses. Good description of the Sun's atmosphere and energy output.

Frazier, Kendrick. *Our Turbulent Sun*. Englewood Cliffs, N.J.: Prentice-Hall, 1980. This book summarizes the research and observational discoveries made about the Sun between about 1950 and 1981. When possible, the author relates these developments to the impact the various solar phenomena have on Earth; for example, he discusses the question of whether the solar activity cycle affects the climate of Earth.

Freedman, Roger A., and William J. Kaufmann III. *Universe*. 8th ed. New York: W. H. Freeman, 2008. College-level introductory astronomy textbook. Thorough and well-written. Good description of the Sun's atmosphere and energy output.

Gibson, Edward G. *The Quiet Sun*. Washington, D.C.: Government Printing Office, 1973. This excellent survey of solar physics details what is known about the Sun, starting with the solar interior and working outward through the atmosphere, giving a chapter each to discussions of the photosphere, the chromosphere, and the corona. An informative overview is provided in the first chapter. Solar activity and solar terrestrial relations are covered in the latter chapters. A somewhat technical presentation.

Giovanelli, Ronald. *Secrets of the Sun*. Cambridge, England: Cambridge University Press, 1984. An overview of modern solar physics, with the aid of spectacular black-and-white images from a number of observatories throughout the world. The presentation of the material highlights the physical nature of the various atmospheric components of the Sun, the photosphere, the chromosphere, and the corona.

Nicolson, Iain. *The Sun*. New York: Rand McNally and Co., 1982. This atlas of the Sun includes a historical overview of solar research, explores the relationship of the Sun to the stars, the Galaxy, and the universe, and provides a detailed description of the solar interior, the solar atmosphere, solar activity, and solar-terrestrial relationships. Solar flares, the solar neutrino problem, coronal holes, solar oscillations, solar wind, and even the pragmatic problems of solar energy are discussed.

Noyes, Robert W. *The Sun: Our Star*. Cambridge, Mass.: Harvard University Press, 1982. This excellent book reviews the research done in solar physics from ground-based telescopes and space satellites. Shows how scientists' knowledge of the Sun relates to their knowledge of stars in general. Studies of the solar spectrum are shown to have provided valuable information about the complex solar atmosphere. The contents are arranged according to regions or zones within the interior and atmosphere of the Sun. Much attention is given to solar activity and its relationship to Earth.

Schneider, Stephen E., and Thomas T. Arny. *Pathways to Astronomy*. 2d ed. New York: McGraw-Hill, 2008. Very thorough college textbook for introductory astronomy courses. Divided into lots of short sections on specific topics. Has several sections on the Sun's atmosphere and energy output.

White, Oran R., ed. *The Solar Output and Its Variation*. Boulder: Colorado Associated University Press, 1977. This highly technical volume details what was known about the measurement of the entire solar spectrum, from gamma rays to radio waves, as of the time of its publication. In addition to the measurement of the solar constant, the book takes up solar variability and the sunspot cycle. Gives ample attention to the effects on Earth's atmosphere and Earth's climate. Primarily aimed at technical audiences.

See also: Auroras; Coronal Holes and Coronal Mass Ejections; Electromagnetic Radiation: Nonthermal Emissions; Electromagnetic Radiation: Thermal Emissions; Infrared Astronomy; Interplanetary Environment; Nuclear Synthesis in Stars; Red Giant Stars; Solar Chromosphere; Solar Corona; Solar Evolution; Solar Flares; Solar Geodesy; Solar Infrared Emissions; Solar Interior; Solar Magnetic Field; Solar Photosphere; Solar Radio Emissions; Solar Seismology; Solar Structure and Energy; Solar System: Origins; Solar Ultraviolet Emissions; Solar Variability; Solar Wind; Solar X-Ray Emissions; Sunspots; Thermonuclear Reactions in Stars; Ultraviolet Astronomy.

Solar Radio Emissions

Category: The Sun

Specific solar phenomena lead to the production of radio waves that affect Earth. These can be broken down into several types of radio emissions, many of which appear to be more closely tied to certain solar events than others. Detecting and understanding these emissions are important to both human activities and life on Earth.

OVERVIEW

In 1942, investigations began to determine what was "jamming" British radar and allowing German battle cruisers to pass freely through the English Channel. J. S. Hey attacked this problem. Investigations led to recognition that the Sun, and not any German technology, was the source of this kind of radio interference. Hey learned that "jamming" happened during daytime and was worse when British receiving instruments were pointed in the general direction of the Sun. Hey also discovered that the Sun at that time sported a large group of sunspots. Indeed, solar radio waves were creating the interference.

It is necessary to understand the basic structure and dynamics of the Sun before seeking to understand why radio regions are produced. Two main solar processes emit energy. Beginning closest to the core, radiative diffusion is at work. Photons are created in the center of the Sun and move outward as a result of absorption and reemission by atoms and electrons that make up the interior of the Sun. After they move out past the radiative zone, they undergo the process of convection. Convection is caused by temperature differences that make hot and cool fluids circulate. Hotter gas rises, and cooler gas falls. As cooler gas sinks closer to the Sun's core, it begins to be heated, causing it to rise again. This progression repeats itself, forming convection cells.

The innermost layer of the Sun's atmosphere is called the photosphere, and is also known as the Sun's visible "surface." Most of the light emitted by the Sun escapes through this "surface." Furthermore, sunspots are found in this region. The umbra, or dark region of a sunspot, emits approximately 30 percent less light than an equally sized area without sunspots. Above the photosphere is the chromosphere. The chromosphere is characterized by an emission line spectrum, whereas the photosphere displays an absorption line spectrum. Vertical spikes of rising gas, or spicules, as well as plages, originate in the chromosphere. The outermost layer of the atmosphere is the corona. Atoms located here are highly ionized due to extremely high kinetic temperatures. This is responsible for nonthermal radio emissions.

Variations in the Sun's magnetic field often cause explosive emissions of energy, in the form of both streams of particles and electromagnetic radiation. Visually, these can be seen as solar flares erupting from the lower corona to the photosphere, specifically where sunspots are located. However, there is more complexity to this phenomenon than this simple picture might suggest. Flares create shock waves that excite the local plasma into oscillation. These oscillations produce radio waves of a frequency equal to that of the plasma oscillations that give rise to them. Occasionally, electrons near the location of a flare will get excited and interact with strong magnetic fields. This interaction will also create intense radio emissions, called radio bursts.

Detailed studies of these solar flares were done by the radio heliograph in Culgoora, Australia. It was discovered that the explosions originated from the corona. However, there were limitations to the effectiveness of the radio heliograph. It could observe at only three radio frequencies: 327 megahertz (MHz), 180 MHz, and 80 MHz. To help overcome this problem, the Clark Lake Radio Observatory was designed to use frequencies ranging from 15 MHz to 125 MHz. This new capability unveiled radio microbursts that are believed to be caused by coronal heating. Radio emission frequencies have been discovered to be a function of electron density. Typical ranges for the lower corona are approximately 100 MHz to 1 gigahertz (GHz). Near the middle of the corona, the frequencies become 10 MHz to 100 MHz. The frequencies for the lower coronal areas are larger than higher up in the Sun's atmosphere, because electron density is greater closer to the core.

The Sun also emits radio waves through synchrotron radiation, the same process by which auroras are produced. Electrons become caught spinning around magnetic field lines and have their motion restricted in two directions: one rotating around the field line and the other in the direction along the magnetic field line. This type of radiation is more intense at radio wavelengths. When plasma expands, it can overcome the Sun's inward gravitational pull. When this occurs, it is possible for electrons to escape, creating the solar wind.

Solar radio emissions come in several types. Type I is mainly narrow-band bursts that occur frequently and can last on the order of hours to days. Type II radio bursts have a wavelength of approximately 1 meter. These bursts are usually caused by solar flares. Along with solar flares, Type II bursts can be created from highly energetic particles. They are also slow "drift bursts" that begin at high frequencies and drift to lower frequencies. Type III bursts are fast drift radio bursts because their emission frequency decreases over time. They can occur from energetic electrons escaping along magnetic field lines. Typically, these bursts have wavelengths in the meter range, but have also been discovered in the decimeter range as well. Energies of the particles that create these bursts range from 1 to 100 kilo-electron volts (keV). Bursts that have broadband qualities are Type IV and Type V. Type IV are also related to solar flares. Type V most commonly last for a few minutes but can last for longer periods of time if the frequency is decreased; this type has also been observed simultaneously with Type III bursts.

KNOWLEDGE GAINED

It was once believed that Type II radio bursts correlated with energetic solar flares, in that these bursts occur only from high-energy flares. Investigation of radio bursts, however, led to the determination that Type II bursts do in fact occur with less intense flares as well as high-intensity flares.

The robotic Cassini orbiter was outfitted with a radio and plasma wave science (RPWS) instrument, designed to measure electric fields, magnetic fields, and electron density while that spacecraft orbits the planet Saturn. Even though Cassini's main mission was to explore Saturn and its large satellite Titan, in 2003, with the spacecraft still en route to the Saturn system (Saturn orbit insertion was in mid-2004), Cassini unexpectedly picked up two Type III radio bursts coming from the Sun, one on October 28 and the other on November 4. These bursts were caused by intense solar flares. The orbiter was approximately 8 astronomical units (AU) away from the Sun at the time when it detected these flares. That these solar flares were so intense was evidenced by the detection of those radio bursts at such a great distance from the Sun. Solar flares have a huge effect on the "space weather" of the entire solar system.

Ulysses is a joint National Aeronautics and Space Administration (NASA) and European Space Agency (ESA) spacecraft devoted to studies of the Sun. Launched in October, 1990, from the space shuttle *Discovery* on mission STS-41, Ulysses was sent on a trajectory rising up above the ecliptic plane in order to fly around the poles of the Sun and thereby study the Sun at all latitudes. Its mission was to understand what is happening during solar maximum, the period in the solar cycle when a copious number of sunspots are present.

Ulysses recorded coronal mass ejections and discovered an irregular solar wind that is not very periodic. This mission also made it possible to look closely at Type III radio bursts from the Sun, because Ulysses was equipped with a unified radio and plasma wave instrument (URAP). One of the functions of URAP is to reveal the direction, angular size, and polarization of radio sources located in the heliosphere, the area of space that encompasses the Sun's magnetic field and the solar wind. Investigating radio bursts is important because it can lead to mapping out the pattern of the Sun's magnetic field by the detection of energetic particles making up these long-wavelength electromagnetic waves. Scientists have questioned how beams of particles of Type III emissions can stay together long enough to be observed. Data from Ulysses addressed this by making observations of the solar plasma's electric field. It discovered that disturbed plasma can stabilize the beams and "hold" the particles and beams together. These bunches have been termed envelope solitons. Ulysses also recorded that more shock waves occur during periods when sunspot activity increases.

CONTEXT

If properly funded, the proposed Solar Orbiter will fly near the Sun, taking measurements of radio emissions and plasma waves that can later be compared to visual measurements to make correlations. One of the main tasks of Solar Orbiter will be to further investigate interrelationships between solar flares, coronal mass ejections, and radio bursts. Solar Orbiter will have a radio spectrometer (RAS) to measure solar radio emissions of frequencies in the range of 100 kilohertz (kHz) to 1 GHz. Because of the large range of frequencies that it can observe, RAS has three component spectrometers: one ranging from 100 kHz to 16 MHz, a second from 16 to 200 MHz, and a third from 200 MHz to 1 GHz. It can observe plasma variations originating at the low coronal level of the Sun's atmosphere.

NASA's Solar Dynamics Observatory (SDO) is another orbiter designed to explore the Sun. In 2008, it was under construction at the Goddard Space Flight Center. SDO's purpose is to focus on coronal mass ejections, sunspots, and solar flares. This idea of investigating so-called space weather is important because violent weather can endanger astronauts, satellites, probes, and airplanes flying near the poles of the Earth.

Jessica Lynn Bugno

FURTHER READING

Balogh, André, Louis J. Lanzerotti, and S. T. Suess, eds. *The Heliosphere Through the Solar Activity Cycle*. New York: Springer, 2007. Investigates the Sun during the sunspot cycle to analyze correlations between the cycle and the heliosphere. Discusses theory on the subject as well as actual data from spacecraft.

Golub, Leon, and Jay M. Pasachoff. *Nearest Star: The Surprising Science of Our Sun*. Cambridge, Mass.: Harvard University Press, 2001. Gives a detailed description of the solar atmosphere, discussing properties of the photosphere, chromosphere, and corona. Provides information on how the Sun affects the inner solar system's terrestrial planets and describes future plans to study the Sun.

Hilbrecht, Heinz, Klaus Reinsch, Peter Volker, and Rainer Beck. *Solar Astronomy Handbook*. New York: Willmann-Bell, 1995. Explains solar astronomy for amateurs. Goes into detail on necessary equipment, how to build certain instruments, and what aspects of the Sun should be the focus of observations. Provides pertinent theory as well.

Kitchin, Chris. *Solar Observing Techniques*. New York: Springer, 2001. Part of the series Patrick Moore's Practical Astronomy, aimed at amateur astronomers. Includes color photographs to help the observer. Focuses mostly on visible astronomy, but some radio astronomy is mentioned as well. Also describes how to observe the Sun safely.

Lang, Kenneth R. *The Cambridge Encyclopedia of the Sun*. New York: Cambridge University Press, 2001. Begins by describing the formation of the Sun and goes into detail on properties of the Sun, as well as data on solar activity. Uses considerable amounts of physics and mathematics.

_____. *Sun, Earth, and Sky*. New York: Springer, 1997. Covers a wide range of topics on the Sun, such as composition, radiation, sunspots, and other active regions. Describes the solar wind and its effect on Earth.

Severny, A. *Solar Physics*. San Francisco: University Press of the Pacific, 2004. Describes physical processes at work in the Sun. Includes information on electromagnetic radiation generated by the Sun, from small wavelengths to the long wavelengths of radio emissions.

Stone, Robert G., Kurt W. Weler, Melvyn L. Goldstein, and Jean-Louis Bougeret, eds. *Radio Astronomy at Long Wavelengths*. New York: American Geophysical Union, 2000. Provides information on solar radio astronomy, focusing on topics that relate to low energies and long wavelengths.

Verschuur, G. L. *The Invisible Universe*. 2d ed. New York: Springer, 2007. Explains what radio astronomy is and provides a history on how it started. Historical treatment includes initial serendipitous discoveries made by pioneering radio astronomers. Covers topics such as the Sun, active galactic nuclei, nebulae, and other radio sources.

Wentzel, G. Donat. *The Restless Sun*. Washington, D.C.: Smithsonian Institution Press, 1989. Describes the structure and evolution of the Sun. Also details areas of the Sun that have much activity, such as sunspots and solar flares, as well as how some of these events have been observed. For general readers.

See also: Coronal Holes and Coronal Mass Ejections; Electromagnetic Radiation: Nonthermal Emissions; Electromagnetic Radiation: Thermal Emissions; Interplanetary Environment; Nuclear Synthesis in Stars; Radio Astronomy; Red Giant Stars; Solar Chromosphere; Solar Corona; Solar Evolution; Solar Flares; Solar Geodesy; Solar Infrared Emissions; Solar Interior; Solar Magnetic Field; Solar Photosphere; Solar Radiation; Solar Seismology; Solar Structure and Energy; Solar System: Origins; Solar Ultraviolet Emissions; Solar Variability; Solar Wind; Solar X-Ray Emissions; Sunspots; Thermonuclear Reactions in Stars; Ultraviolet Astronomy.

Solar Seismology

Category: The Sun

Solar seismology, or helioseismology, is the study of the oscillations that take place within the Sun. These periodic vibrations originate in the Sun's convective zone. By analyzing the motion of these oscillations, scientists can image the solar interior and develop more accurate models of the Sun.

OVERVIEW

In 1960, Robert Leighton of the California Institute of Technology discovered that small areas of the Sun's surface oscillated up and down with a period of about five minutes. He observed that the spectra of various points on the Sun's surface were Doppler-shifted alternately toward shorter and longer wavelengths. The Doppler effect shifts electromagnetic radiation to shorter wavelengths (termed "blueshifts") when the source moves toward the observer and to longer wavelengths (termed "redshifts") when the source moves away from the observer. However, he was not able to tell whether the oscillations occurred only in small regions or over the entire Sun.

A few years later, it was found that the entire Sun vibrates. Like many other scientific discoveries, it happened while trying to measure something else. Astronomers were attempting to measure the oblateness of the Sun to discriminate between Einstein's general theory of relativity and a competing theory of gravity—the Brans-Dicke scalar-tensor theory. Since the mid-1800's, it had been known that Mercury's orbit around the Sun precesses; that is, the elliptical orbit itself slowly rotates about the focus occupied by the Sun. Most of the observed precession could be explained by the other planets gravitationally perturbing Mercury according to Sir Isaac Newton's laws of gravity and motion, but a small residual was unaccounted for. The existence of an unknown planet, the legendary Vulcan, inside the orbit of Mercury was suggested to provide the extra gravitational perturbation. Searches were made, and some even reported spotting Vulcan, but none of the "discoveries" was confirmed.

In the early 1900's, Einstein developed his general theory of relativity. One of its consequences is that mass distorts space-time in its vicinity. Thus Mercury's orbit would precess because of the curvature of space-time around the Sun. The amount predicted by general relativity for Mercury's precession matched exactly the observed discrepancy, and this was hailed as major evidence supporting the theory.

A half century later, Carl Brans and Robert H. Dicke, developed an alternative to general relativity called the Brans-Dicke scalar-tensor theory. It, too, predicted the precession of Mercury's orbit if the Sun were slightly oblate—that is, if the Sun had a small equatorial bulge due to its rotation. In the early 1960's, Dicke's team at Princeton University claimed to have measured a small oblateness of the Sun using a telescope of their own design and construction. A few years later, Henry Hill of the University of Arizona built a telescope that was designed specifically to detect a distortion in the shape of the Sun. He found no evidence of any distortion, but after further study, Hill and his colleagues discovered periodic oscillations of the Sun's entire surface.

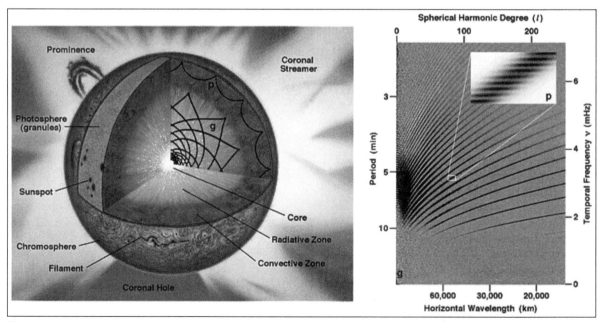

These diagrams show how sound waves propagate through the interior of the Sun. Understanding this helps scientists better understand the Sun's interior structure, composition, and rotation. (NASA/ESA/SOHO)

Since then, it has been determined that the entire surface of the Sun is in a state of constant oscillation, with periods varying between minutes and hours. It might be said that the Sun is ringing like a bell being struck continuously. The origin of these vibrations is the convective zone beneath the photosphere or "visible surface" of the Sun. Bubbles of hot gas rise in convection cells, carrying heat from the Sun's interior to the surface. The tops of these convection cells produce the granulation seen in images of the photosphere. The hot bubbles rise toward the surface, accompanied by a tremendous roar. These sound waves oscillate through the Sun and cause its surface to rise and fall periodically as they are reflected by the bottom of the photosphere. As the waves travel back downward into the Sun, they encounter higher temperatures and pressures. These changing physical conditions result in the waves' velocity being increased, which eventually causes the waves to refract or bend upward toward the surface. When they reach the surface, again the waves are reflected back downward. The depth that the wave reaches depends upon its wavelength. The wavelength also determines how far a wave will travel around the Sun between reflections from the surface.

The Sun's interior is conducting waves with virtually millions of different wavelengths and frequencies. Some waves have the exact wavelength necessary to make an even number of bounces before they return to where they began. Astronomers categorize these waves by the number of times that they reflect from the surface in one complete circuit of the Sun. For example, a wave with the designation I-4 strikes the surface in three places before it bounces back to its starting position. Once it returns to its origin, it has struck the surface of the Sun four times. Scientists have found that waves with low I numbers travel deep into the Sun and can reveal physical characteristics there, while waves with higher I numbers probe the shallower zones closer to the Sun's surface.

KNOWLEDGE GAINED

On Earth, geologists use seismic waves (produced naturally by earthquakes or artificially by setting off explosions or by "thumping" the ground) as probes of the Earth's interior. Astronomers use solar seismic waves in an analogous way to image the interior of the Sun, which cannot be observed directly. Prior to this new development in solar physics, knowledge of the processes that occur within the Sun and the locations of various boundaries within the Sun came only from computer-generated models of its interior.

Currently, the generally accepted model of the Sun's interior is called the Standard Solar Model (SSM). Observed frequencies and wavelengths of observed solar oscillations for the most part are consistent with the SSM to within 0.1 percent. For example, within the Sun's core, hydrogen nuclei are fused together to form helium nuclei. This process converts a small fraction of the input mass into energy that keeps the Sun shining. The SSM indicates the central temperature needed to maintain this reaction at the right rate to account for the Sun's luminosity is about 15 million kelvins. Vibration patterns observed on the Sun's surface but that travel through the deep interior are consistent with this temperature.

Direct observation shows that the Sun's surface rotates differentially—the equatorial region has a shorter rotation period than the polar regions. Helioseismology provides a way to study the internal rotation of the Sun, and it turns out to be quite complex. At shallow depths there is a "zonal" flow with alternating bands moving faster and slower than the average. Just below the surface are wide "rivers" moving more slowly near the equator and faster near the poles, just the opposite of what the surface itself does. The base of the convection zone oscillates in rotational speed with a period of 1.3 years, sometimes moving faster than the surface and sometimes more slowly than the surface by about 10 percent. The deeper radiative zone, including the core, rotates reasonably uniformly with a period of about 27 days.

CONTEXT

The discovery that the surface of the Sun is oscillating was made in the early 1960's. At the time, scientists were gathering data on the oblateness of the Sun. As it turned out, there was no measurable oblateness of the Sun, but subsequent observations revealed an alternating Doppler shift in the solar spectra taken from various points on the Sun's surface. This Doppler shift provided evidence of periodic oscillations. Further investigations revealed that the Sun is ringing as if it were a large bell that is continuously being struck. The millions of different wavelengths and frequency combinations of waves are believed to originate within the Sun's convective zone. In this region, the tremendous heat from the interior is carried outward toward the surface by rising bubbles of hot gas. The sound waves given off by this movement of hot gases cause the Sun to vibrate.

The Global Oscillations Network Group (GONG) has provided extensive observations of solar oscillations. GONG consists of an array of 50-millimeter-aperture refracting telescopes at various locations around the Earth. These locations were selected to ensure that at least two telescopes could gather data from the Sun at all times, enabling astronomers to differentiate real oscillations, together with their amplitudes and periods, from noise produced by instrumentation or Earth's rotation. Each of the telescopes contains a Fourier tachometer. This device is capable of measuring extremely small Doppler shifts at more than sixty-five thousand different points on the surface of the Sun. By observing these shifts, astronomers can determine oscillation periods of these various points and form a detailed picture of the solar disk. This large amount of data is necessary to determine the paths that the waves follow through the Sun and convert that information into a model of the solar interior. The Solar and Heliospheric Observatory (SOHO), launched by the European Space Agency (ESA) in 1995 and on station between the Earth and Sun, also provides continuous monitoring of solar oscillations.

Helioseismic waves act as probes of the solar interior that enable solar scientists to map the interior of the Sun and solve some of the perplexing problems in solar physics. By observing oscillations visible on the Sun's surface, astronomers are assembling a detailed picture of the solar interior, in much the same way that geologists gain knowledge of Earth's interior by the study of seismic waves traveling through our planet. For example, solar oscillations provide information about the temperature and density of the Sun from its surface down to its core. They reveal the rotational speeds of internal layers of the Sun. Helioseismic imaging helps delineate the boundaries between the convective zone, radiative zone, and the core itself. Solar seismology also provides clues as to how energy is transferred from the solar surface to the chromosphere and corona. (It is currently believed that intense magnetic fields, along with acoustic shock waves from the tops of convecting cells, are responsible for high temperatures in the chromosphere and the corona—up to 2 million kelvins in the corona.) In sum, the accumulation of helioseismic data over the years is helping astronomers form an accurate model of the conditions and processes occurring both on the solar surface and in the interior.

David W. Maguire

FURTHER READING

Chaisson, Eric, and Steve McMillan. *Astronomy Today*. 6th ed. New York: Addison-Wesley, 2008. Very well-written college-level textbook for introductory astronomy courses. Includes a good description of solar oscillations, including the GONG and SOHO projects.

Cox, A. N., W. C. Livingston, and M. S. Matthews, eds. *Solar Interior and Atmosphere*. Tucson: University of Arizona Press, 1991. A series of articles covering nuclear and atomic physics topics relating to the study of the Sun. Also discusses information gathered from the world's largest solar telescopes.

Fraknoi, Andrew, David Morrison, and Sidney Wolff. *Voyages to the Stars and Galaxies*. Belmont, Calif.: Brooks/Cole-Thomson Learning, 2006. A well-written, thorough college textbook for introductory astronomy courses. Includes some discussion of solar oscillations.

Freedman, Roger A., and William J. Kaufmann III. *Universe*. 8th ed. New York: W. H. Freeman, 2008. College-level introductory astronomy textbook that covers what we know of solar oscillations.

Gribbin, John R. *Blinded by the Light: The Secret Life of the Sun*. New York: Harmony Books, 1991. Discusses the Sun and our understanding of it. Suitable for general audiences.

Lopresto, James Charles. "Looking Inside the Sun." *Astronomy* 17 (March, 1989): 20-28. Discusses the origin of the study of helioseismology and its possibilities in solar research. Although very little mathematics is used, the article contains an abundance of technical terms, and readers need a background in basic physics and astronomy.

Mitton, Simon. *Daytime Star: The Story of Our Sun*. New York: Charles Scribner's Sons, 1981. A nontechnical volume accessible to the general reader. Mitton discusses the Sun, its structure, its processes, and its future.

Schneider, Stephen E., and Thomas T. Arny. *Pathways to Astronomy*. 2d ed. New York: McGraw-Hill, 2008. A thorough college textbook for introductory astronomy courses, including a short discussion of solar oscillations.

See also: Coronal Holes and Coronal Mass Ejections; Earth-Sun Relations; Solar Chromosphere; Solar Corona; Solar Evolution; Solar Flares; Solar Geodesy; Solar Infrared Emissions; Solar Interior; Solar Magnetic Field; Solar Photosphere; Solar Radiation; Solar Radio Emissions; Solar Structure and Energy; Solar Ultraviolet Emissions; Solar Variability; Solar Wind; Solar X-Ray Emissions; Sunspots.

Solar Structure and Energy

Category: The Sun

Advances in both theoretical physics and astronomical observing capabilities have provided an increasingly detailed description of the Sun. As the nearest star to us, the Sun can also function as a laboratory for investigations of stellar physics in general.

OVERVIEW

The Sun is composed mostly of hydrogen. When the Sun first formed, there was enough hydrogen in the estimated size of its core for this reaction to keep the Sun shining for approximately 10 billion years. Since the Sun is about 4.5 billion years old, it is only about halfway through with generating energy in its core by hydrogen fusion.

The Sun's mass is about 2×10^{30} kilograms. It consists primarily of the two simplest atoms: hydrogen (about 91 percent by number of atoms, about 71 percent by mass) and helium (about 9 percent by number of atoms, about 27 percent by mass). Other chemical elements account for a combined total of about 0.1 percent by number of atoms or 2 percent by mass. More than sixty chemical elements have been identified in the Sun's spectrum, and probably all the elements are present in minute amounts. Some of the more abundant of the other elements include oxygen, carbon, nitrogen, silicon, magnesium, neon, iron, and sulfur.

A major clue to the conditions inside the Sun is that it seems to be reasonably stable, with no large, rapid changes. Thus it must be in a state of hydrostatic equilibrium. This means that its self-gravity, which tries to make it contract, is balanced by its internal pressure, which tries to make it expand. If it were not in hydrostatic equilibrium (or at least very, very close to it), the Sun would either be contracting or expanding noticeably.

The Sun's luminosity—that is, the rate at which it emits electromagnetic energy—is approximately 3.8×10^{26} joules per second. The nuclear reaction that supplies the Sun's energy needs is the fusion of four hydrogen nuclei into one helium nucleus. In the production of this single helium nucleus, 4.8×10^{-29} kilograms of mass are converted into 4.3×10^{-12} joules of energy. Thus, to generate the Sun's luminosity (3.8×10^{26} joules per second), every second about 36×10^{37} hydrogen nuclei (with a total mass of 602,300,000 metric tons) are fused into 9×10^{37} helium nuclei (with a total mass of 598,100,000 metric tons), and the excess 4,200,000 metric tons of mass is converted into energy in the Sun's core.

From the center outward, the Sun has several layers or zones: the hydrogen fusion core, the radiative zone, the convective zone, the photosphere, the chromosphere, and the corona. However, other than the photosphere, these layers are either difficult or impossible to observe. Above the photosphere, the gas is so tenuous that it emits much less visible light. Below the photosphere, the gas is opaque—sort of like fog—so electromagnetic radiation cannot escape. Consequently, the layers below the photosphere must be studied indirectly via computer models of the Sun's interior. The layers above the photosphere can be seen visually on special occasions or observed anytime at various wavelengths outside the visible range.

The energy generated in the Sun's core initially makes its way outward by radiation—that is, as a flow of photons. However, because the Sun's interior is so opaque, a photon travels only about 1 centimeter before it is absorbed and reemitted by an atom. This occurs over and over again, so this method of energy transport is called radiative diffusion. The region of the Sun in which energy is transported entirely by radiative diffusion (or simply radiation) is its radiative (or radiation) zone, and this zone extends from the center out to about two-thirds of the Sun's radius.

Beyond this distance, the gas becomes even more opaque, making it more difficult for photons to travel. As a result, convection sets in, and energy is carried by rising bubbles of hot, lower-density gas. After transferring their heat, the cooler, denser bubbles sink. The region of the Sun in which convection occurs is its convective (or convection) zone, and it occupies the outer third of the Sun's radius. Even in this zone in which convection occurs, some energy is also carried by radiation, that is, by photon flow.

The next layer, the photosphere or visible surface of the Sun, is the lowest level of the solar atmosphere, with a radius of about 700,000 kilometers and a thickness of about 500 kilometers. Temperatures decrease from about 6,600 kelvins at its base to about 4,400 kelvins at its top; the overall effective temperature of the photosphere is 5,800 kelvins.

When examined with high-resolution imagers, the photosphere reveals a wealth of structure and detail. Most pronounced is the presence of granulation, an alternation of brighter spots with darker borders, resembling a mixture of salt and pepper. Each granule is a region of gas about 1,000 kilometers in diameter, larger than the state of Texas. The photosphere

is at the top of the Sun's convective zone, the part of the Sun's interior where energy is transported outward by convection. The brighter region at the center of the granule is a hot gas bubble rising, while the darker border regions of a granule are cooler gases sinking back down. The convection zone in the Sun produces waves of thermal energy that shoot up through the photosphere. These waves make the photosphere appear to oscillate, with periods ranging from minutes to hours.

Above the photosphere are two more layers of the solar atmosphere, called the chromosphere and the corona, that usually cannot be seen with the unaided eye. The photosphere emits so much light and is so much brighter than the rest of the solar atmosphere that usually it overwhelms the weaker visible light from the layers above it, making them difficult to see. During a total solar eclipse, however, the Moon blocks the bright photosphere, and the fainter chromosphere and corona are visible to the unaided eye, extending out around the silhouette of the Moon.

Chromosphere literally means "color sphere." When seen during total solar eclipses, it appears as a narrow red ring just beyond the Moon's silhouette. The red color is one of the wavelengths (656.3 nanometers) that can be strongly emitted and absorbed by hydrogen atoms. The thickness of the chromosphere is about 2,000 kilometers, with temperatures of 4,500 kelvins at its base and rising to 30,000 kelvins at its top. These temperatures result in strong ultraviolet emission, and the chromosphere can be observed any time (not just during eclipses) at ultraviolet wavelengths. However, our atmosphere is opaque to most of the ultraviolet, so ultraviolet observations must be conducted by spacecraft above our atmosphere.

The chromosphere contains hundreds of thousands of thin spikes, called spicules. These spicules are jets of hot gas, hundreds of kilometers across and thousands of kilometers tall. They rise dramatically and then fall over a lifetime of several minutes, thus creating a dynamic appearance, like the dance of many small candle flames. The chromosphere also has a granulated structure. This structure cannot be directly observed but has been deduced from spectroscopic studies of the motions of gas in the chromosphere using the Doppler effect (a change in the frequency and wavelength of electromagnetic radiation caused by the motion of its source toward or away from the observer). Such studies have revealed that the chromosphere contains large, organized cells of gas called supergranules that move in unison under the influence of convective forces. These supergranules are 30,000 kilometers in diameter and contain hundreds of normal granulation regions.

Above the chromosphere, the density rapidly decreases and the temperature abruptly increases to about 1 million kelvins in a thin layer called the transition region. Beyond is the corona, the outermost part of the solar atmosphere, The corona's density is so low that it emits relatively little visible light, but during a total solar eclipse, the corona can be seen as a broad, glowing white halo of light out to distances of several solar radii. The temperature of the corona is about 1 million to 2 million kelvins. Consequently, it strongly emits X rays. It can be observed at any time at X-ray wavelengths, but our atmosphere is opaque to X rays (just as it is opaque to most ultraviolet radiation), so X-ray observations of the corona must also be conducted from spacecraft above our atmosphere.

APPLICATIONS

Our Sun is a star, the only star in our solar system. It is often described as an "average" star. While it is average in the sense that it is a "main sequence star" and just about in the middle of the ranges of stellar luminosity, mass, diameter, surface temperature, and various other properties, it is not typical in the sense that the vast majority of stars are smaller, cooler, and less luminous than the Sun. Furthermore, the comparatively small number of stars that are larger and more luminous than the Sun are so bright that they account for most of the light emitted within a galaxy of billions of stars.

Nevertheless, the Sun is the closest star to us—at a distance of about 150 million kilometers (which defines one astronomical unit, or 1 AU)—and as such provides a "laboratory" where normal stellar processes can be observed and studied at relatively close range. The next closest star, Proxima Centauri (which probably is a member of the Alpha Centauri star system and thus sometimes is called Alpha Centauri C), is about 40 trillion kilometers away, or about 270,000 times the distance to the Sun. Light from the Sun takes only five hundred seconds (a little over eight minutes) to reach Earth, while light from Proxima Centauri takes more than 4 years to reach us.

Aside from the fact the most other stars are smaller and less luminous, they do share have much in common with the Sun. Most stars have compositions similar to the Sun's—mostly hydrogen, some helium, and very small amounts of other chemical elements. Most stars, like the Sun, are reasonably stable and thus are in hydrostatic equilibrium. The light we receive from other stars, like the light we receive from the Sun, comes from a layer that resembles the photosphere (and hence other stars are said to possess "photospheres"). Observational evidence suggests that at least some stars (and probably most if not all) possess chromospheres and coronas as well—like the Sun.

Also like the Sun, stars generate energy for most of their active "lives" by nuclear fusion reactions. The fusion of hydrogen into helium in its core is the first and by far the longest-lasting fusion reaction that a star can tap. Most stars thus are generating energy the same way the Sun is, by hydrogen-to-helium fusion in their cores. After a stellar core's hydrogen is exhausted, other nuclear fusion reactions are possible, such as the fusion of three helium nuclei into a single carbon nucleus, but these higher fusion reactions do not supply energy for long.

The energy generated in a star's core is transported outward in most stars by the same two processes operating in the Sun: radiation and convection. However, computer models of stellar interiors indicate that in some stars, the convective zone is in the central part of the star, and the radiative zone is farther out. Furthermore, in some parts of some stars in some stages of their lives, a third energy-transport mechanism, conduction, is more effective than either radiation or convection.

CONTEXT

Unraveling the mystery of how the Sun and most other stars shine is one of the great achievements of science in the twentieth century. Ever since Anaxagoras in the fifth century B.C.E. asserted that the Sun was not a god but simply a big ball of fire, the question of solar fuel was a mystery. An early suggestion for the energy source was some form of chemical combustion; in other words, the Sun was literally a ball of fire that was burning something, just as Anaxagoras said. The problem was the timescale. No matter what the Sun was made of, it would have to chemically consume (burn) its entire mass in a time on the order of thousands to tens of thousands of years to keep shining at its present rate.

Another suggestion was meteoritic infall; that is, the Sun was heated by matter falling into it. Again, there is a problem of scale. A rocky asteroid 11 kilometers in diameter would have to slam into the Sun every second to release enough energy to supply its present luminosity. Also, such a rate would have produced some observable immediate effects, but none had ever been detected.

By the middle of the nineteenth century, the answer seemed to be slow gravitational contraction. The Sun would only have to shrink by a few tens of meters each year (an insignificant amount compared to its radius of 700,000 kilometers) to provide for its present luminosity. Furthermore, slow gravitational contraction could keep the Sun shining for tens of millions of years. At the time, it seemed like a reasonable duration to astronomers and physicists, but geologists, paleontologists, and biologists were beginning to figure out that the Earth and its rock and fossil record were at least hundreds of millions if not billions of years old. Again, the timescale did not fit the theory.

The accepted energy source now is nuclear fusion, a process in which light atomic nuclei are fused into heavier atomic nuclei, releasing energy. The mass of the resulting nucleus is not quite as great as the sum of the masses of the nuclei that fused to form it. That small difference in mass gets converted into energy according to Einstein's famous equation, $E = mc^2$, which says that matter, m, and energy, E, are equivalent and are related by a physical constant (the speed of light, c, squared). To be able to overcome the electrostatic repulsion that pos-itively charged nuclei feel toward each other, nuclear fusion reactions can occur only at high temperatures (so the nuclei are moving rapidly) and high densities (so the nuclei are close together), exactly the conditions at the center of the Sun and most stars. According to computer models of the Sun's interior, the central temperature is about 15 million kelvins and the central density is about 150 times the density of water.

Our current understanding of the structure and physics of the Sun has been obtained in two ways: through observations of the Sun in all parts of the electromagnetic spectrum, and through computer models that calculate conditions inside the Sun. The processes that take place in the Sun are representative of processes that take place in stars generally. Thus our solar research has also contributed to our study of stars throughout our Milky Way galaxy and in galaxies beyond the Milky Way.

Karl Giberson

FURTHER READING

Chaisson, Eric, and Steve McMillan. *Astronomy Today*. 6th ed. New York: Addison-Wesley, 2008. Very well-written college-level textbook for introductory astronomy courses, with a full chapter about the Sun and its structure.

Emiliani, Cesare. *The Scientific Companion*. New York: John Wiley and Sons, 1988. A somewhat more technical book than others listed here, yet accessible to the general reader with a limited science background. Contains a full chapter dedicated to the physics of the Sun and related chapters on nuclear physics and stellar evolution.

Fraknoi, Andrew, David Morrison, and Sidney Wolff. *Voyages to the Stars and Galaxies*. Belmont, Calif.: Brooks/Cole-Thomson Learning, 2006. A well-written, thorough college textbook for introductory astronomy courses. Contains two chapters about the Sun and its structure.

Freedman, Roger A., and William J. Kaufmann III. *Universe*. 8th ed. New York: W. H. Freeman, 2008. Thorough and well-written college-level introductory astronomy textbook, with a chapter about the Sun and its structure.

Noyes, Robert W. *The Sun, Our Star*. Cambridge, Mass.: Harvard University Press, 1982. A thorough treatment of all aspects of the Sun. Contains hundreds of black-and-white photographs of solar phenomena. Nontechnical and accessible to the general reader. Comprehensive and detailed.

Schneider, Stephen E., and Thomas T. Arny. *Pathways to Astronomy*. 2d ed. New York: McGraw-Hill, 2008. A thorough college textbook for introductory astronomy courses, divided into many short sections on specific topics. Three units deal with the Sun and its structure.

See also: Coronal Holes and Coronal Mass Ejections; Earth-Sun Relations; Solar Chromosphere; Solar Corona; Solar Evolution; Solar Flares; Solar Geodesy; Solar Infrared Emissions; Solar Interior; Solar Magnetic Field; Solar Photosphere; Solar Radiation; Solar Radio Emissions; Solar Seismology; Solar Ultraviolet Emissions; Solar Variability; Solar Wind; Solar X-Ray Emissions; Sunspots.

Solar System: Element Distribution

Category: The Solar System as a Whole

The abundance and distribution of chemical elements in the solar system resulted from many events: the creation of hydrogen and helium in the big bang, the synthesis of heavier elements by nuclear fusion reactions in earlier generations of stars, differential condensation in the early solar nebula during the accretion of the planets, and physical and chemical processes within solar-system bodies after their formation. Elemental abundance and distribution within the solar system therefore point to the origin and history of the solar system and its constituent bodies.

OVERVIEW

The solar system consists of three rather distinct parts. At the center is the Sun, composed mostly of hydrogen and helium and containing most of the solar system's mass. Next comes an inner system of four small planets (the terrestrial planets, Mercury, Venus, Earth, and Mars), three moons, and many asteroids, all composed mainly of rock and metal. Last there is an outer system of four large gas/liquid/ice planets (the Jovian or "gas giant" planets, Jupiter, Saturn, Uranus, and Neptune), many small ice-rock moons and other bodies, and lots of icy cometary nuclei. The four inner planets are made predominantly of magnesium and iron silicates, with cores rich in iron, nickel, and sulfur. The four gas giants have thick envelopes of hydrogen, helium, and hydrogen compounds in the form of gases, liquids, ices, or a combination of these, with cores of silicates and metals; these large planets generally approximate the Sun in composition. The small ice-rock bodies may have rocky silicate cores with outer layers of various ices, or they may be homogeneous mixtures of ice and rock.

A little over 91 percent of the atoms in the solar system are hydrogen, and almost 9 percent are helium; together they account for almost 99.9 percent of the atoms. In terms of number of atoms, the next two most abundant are oxygen (about 0.08 percent) and carbon (about 0.04 percent). Then come nitrogen, silicon, magnesium, neon, iron, and sulfur. Each of the other elements accounts for less than 0.001 percent of the atoms, and their abundance generally falls off rapidly with increasing atomic number. There also is a general alternation in abundance, with atoms of even-numbered elements more abundant than odd-numbered elements.

The overall abundance of chemical elements in the solar system was established by events before the solar system formed. Hydrogen and most of the helium were created in the first few minutes after the big bang; it also created a trace of lithium, even less beryllium, and possibly still less boron. Most of the atoms in our solar system heavier than helium were formed by nuclear fusion reactions in earlier generations of stars that went through their "life-cycles" before the solar system began to form. The heavier atoms were dispersed throughout interstellar space when these stars ended their energy-producing "lives," either puffing off their outer

An artist's conception of the protoplanetary disk—the cloud of interstellar gas and dust with its proto-Sun—that would form into our solar system. (NASA/JPL-Caltech)

layers as planetary nebulae or exploding violently as supernovae. The heavier atoms enriched the interstellar clouds of gas from which new generations of stars and their accompanying systems of planets formed. Measurements of the solar system abundances for the heavier nuclei roughly match the expected abundances based on the types of nuclear reactions that are expected to occur in stars.

The solar system formed when part of a large cloud of interstellar gas and dust began to contract inward under its own gravity. As the cloud contracted, it began to rotate faster to conserve angular momentum; eventually it spun off an equatorial disk of matter. Most of the mass of the contracting cloud collected in the center to form the proto-Sun, while the rest of the solar system formed in the equatorial disk. Chemical reactions governed the events that occurred in

the "leftover" matter of the disk. Close to the developing Sun, only a few refractory materials condensed to form small grains of minerals and metals; far from the Sun, even gases such as ammonia and methane condensed to form grains of ice.

The first solids to condense in the early solar nebula, at temperatures around 1,600 kelvins, were oxides of calcium, aluminum, titanium, and some rare Earth elements. At about 1,300 kelvins, iron-nickel metal condensed, followed at about 1,200 kelvins by enstatite ($MgSiO_3$), the magnesium variety of the silicate mineral pyroxene. At about 1,000 kelvins, sodium, potassium, calcium, and aluminum reacted with silicate grains to create the feldspar minerals. At 680 kelvins, metallic iron reacted with hydrogen sulfide gas to create the iron sulfide troilite (FeS). Also, between 1,200 and 500 kelvins, metallic iron reacted with oxygen to create ferrous iron oxide (FeO), which reacted in turn with enstatite to create a variety of the silicate mineral olivine ($FeMgSiO_4$). Between about 600 and 400 kelvins, water reacted with calcium silicates and olivine to form various hydrous minerals. This sequence accounts for all the most common constituents of meteorites. A class of meteorites called chondrites matches the element abundance in the Sun, minus those elements which would have remained gases. Chondritic meteorites are thus considered to be the most primitive surviving materials in the solar system.

Ice sublimes quickly into gas in a vacuum. Only at and beyond the distance of Jupiter, about 5 astronomical units from the Sun, was it cold enough (about 175 kelvins) for ice to form in a vacuum and survive billions of years. At 150 kelvins, methane hydrate (a solid mixture of methane and water) condensed, followed by ammonia hydrate at 120 kelvins. Argon and pure methane solidified at about 65 kelvins.

The condensation of various solid grains at different distances from the early Sun explains much of the present distribution of materials in the solar system, and the accretion of the planets further modified their chemistry. Computer simulations of solar system accretion indicate that the small grains orbiting the Sun coalesced first into many small planetesimals (up to several hundred kilometers across) and then into protoplanets (2,000 to 3,000 kilometers across). As these bodies grew, their increasing gravity swept in matter from wider and wider swaths in the disk. The protoplanets became massive enough to attract one another, disturbing their orbits. Thus, the protoplanets swept up matter from rather wide bands in the disk, mixing materials from a variety of temperature zones. Late in the accretion process, the protoplanets collided to create larger planets. The collisions would have vaporized substantial parts of the impacting bodies, while perhaps mixing materials from two protoplanets of rather different composition.

The Earth-Moon system may be the result of such a collision. The Moon has a much lower overall density than does the Earth, suggesting that it has a lower proportion of metal to rock than does the Earth. Also, the Moon is poorer in volatile materials than is the Earth, and the more volatile a material is, the lower is its abundance on the Moon. Thus, it appears that the Moon formed in a hotter region of the solar disk than did the Earth. Many of the problems explaining the formation of the Moon can be resolved if the early Earth collided with a Mars-sized protoplanet. Much of the material of the impacting protoplanet, along with part of the Earth's crust and mantle, could have sprayed off into orbit around the Earth, later coalescing to form the Moon.

The large rocky/metallic objects possess a concentric-shell structure consisting of a dense metallic central core, a less dense silicate mantle, and a thin silicate crust. This layering reflects internal changes after the body formed. Gravitational separation of a dense metallic core occurred during or shortly after the accretion of a planet; this process may have been well under way during accretion. Collision of two large protoplanets with metallic cores might account for the unusually large core of the planet Mercury; the collision would have vaporized much of the outer rocky shells of the protoplanets, resulting in a large core but only a thin, rocky mantle and crust.

Deep within planets, high pressures crush minerals into new and denser crystal lattice structures. At depths below about 700 kilometers in the Earth, for example, magnesium silicates are crushed into densely packed cubic crystal structures. Water ice undergoes a remarkable series of changes in crystal structure as pressure increases; some of these high-pressure forms of ice are surely present in the interiors of large satellites in the outer solar system.

Internal processes may also concentrate materials on the surface of a planet or satellite to form a crust. For example, the Moon's crust formed fairly simply by melting of chondritic material. Early in the history of the Moon, its outermost 100 kilometers melted to produce a "magma ocean," probably from heat generated by the final impacts of the accretion process. As the molten material solidified, magnesium and iron atoms accumulated in dense minerals such as pyroxene and olivine, which settled to the bottom. Calcium and aluminum mostly went to form less dense feldspar, which was neutrally buoyant, neither rising nor sinking. Sinking of the olivine and pyroxene would have created a lower layer of peridotite, leaving behind an upper layer of anorthosite, matching the observed makeup of the ancient lunar crust.

When chondritic material is melted and then cooled, basalt or gabbro forms; basalt forms from rapid cooling and has small mineral crystals, while gabbro forms when the cooling is slower, allowing the mineral crystals to grow larger. Thus, it is reasonable to expect basalt and gabbro to be very common rocks in the solar system. They form the crust of the ocean basins on Earth, where they are derived from rocks of the underlying mantle. Basalt also forms the dark lava plains, or maria, on the Moon, and lava flows on Venus. The shield volcanoes on Earth are made of basalt, and similar shield volcanoes are present on Venus and Mars. The reflection spectra of some asteroids indicate that they probably have basalt on their surfaces.

In contrast to the simplicity of ocean crust, the continental crust of the Earth is chemically more complex, consisting of granite, a rock type relatively rare in the solar system; the only other possible identification of it in the solar system came from one of the Venera landers on Venus (Venera 8). Granite consists of quartz, potassium and plagioclase feldspar, micas, and amphibole. Compared with basalt and gabbro, it is greatly enriched in silica (SiO_2), potassium, and sodium, and depleted in iron and magnesium. The granitic crust of the Earth is also enriched in some less abundant elements, notably those whose atoms are unusually large (rubidium and some rare-Earth elements) and those with large atomic numbers (uranium, thorium, and lead). These elements do not fit easily into the dominant minerals of the mantle, where the principal metallic elements (iron, magnesium, and calcium) have moderate-sized atoms and atomic numbers. Granite has formed by chemical differentiation during repeated melting of the Earth's mantle and crust. Granite found on another solar-system object would be clear evidence that it had a high degree of internal activity.

Io, the innermost of Jupiter's large Galilean satellites, is a remarkable example of a body whose surface was created by internal processes. Io, though slightly smaller than the Earth's moon, is internally hot and volcanically active due to tidal flexing. Io has too little gravity to retain water vapor, the main propellant for terrestrial volcanic eruptions. In fact, Io is extremely dry, having lost most of its water to space. The only material that is cosmically abundant, volatile enough to power eruptions, and heavy enough to have been retained by such a small body is sulfur. Io has a spectacular white, yellow, and red surface, probably coated by sulfur dioxide frost and various crystalline forms of sulfur erupted onto its surface.

As protoplanets and planets grew larger, they eventually became massive enough to attract and hold gases. However, violent collisions such as those that appear to have occurred late in the accretion process would probably have driven off whatever atmospheres the protoplanets had, so smaller planets would have had great difficulty in accumulating atmospheres. In the inner solar system, the warm temperatures would have given gas molecules greater speeds, enabling them to escape more readily from small planets. Also, the early Sun probably underwent a time of intense activity (called the T-Tauri phase) during which it emitted intense streams of charged particles that swept the inner solar system free of its remaining gas. For these reasons, the inner planets have thin atmospheres or none. Earth and Venus are massive enough to have retained significant atmospheres, but Mars has only a thin atmosphere, and Mercury has only a bare trace.

In the outer solar system, the Jovian planets grew large enough to retain gases despite disruption by protoplanet collision and early solar activity. Jupiter and Saturn, which were massive enough to retain essentially all their gases, are quite close to the Sun in composition. Uranus and Neptune did not become massive enough to attract or retain hydrogen and helium quite as effectively as Jupiter or Saturn, and thus they contain somewhat less hydrogen and helium and a somewhat greater proportion of denser gases such as ammonia and methane. The heavy elements in all four Jovian planets probably accumulated into roughly Earth-sized solid cores at their centers.

Small bodies in the outer solar system accreted like those in the inner solar system, but with the addition of water ice as a major constituent. Except for Io, the small bodies of the outer solar system have silicate cores with icy mantles and crusts. Comets, whose orbits extend to great distances from the Sun, probably formed in the vicinity of Jupiter and Saturn and were expelled during close passages by the giant planets. They are mostly water ice, with other frozen gases in smaller amounts.

METHODS OF STUDY

The composition of many solar-system objects is known at least partially through spectroscopy. When materials are in the form of a diffuse gas, their atoms or molecules emit and absorb certain specific wavelengths of light and other forms of electromagnetic radiation. The specific wavelengths of electromagnetic radiation emitted or absorbed by a gas act like a chemical fingerprint of its composition. Solids have far more complex and less conclusive spectral patterns than do gases, but the spectra of light reflected from the solid surface of asteroids can be matched to laboratory measurements of the reflection spectra of various types of meteorites.

Since the generally accepted model is that the entire solar system formed during the same time from a homogeneous cloud of gas and dust, it would be expected that any solid grains that condensed from the solar nebula would have a composition similar to that of the Sun, minus those elements that would have remained gases. In the inner solar system, where the temperature of the solar nebula would have been above the freezing point of water, most hydrogen, nitrogen, and carbon would have remained gases. Most oxygen would combine with hydrogen to form water vapor, but some would combine with various metals to form oxides or would combine with silicon to form silicate minerals. Some sulfur would be available for iron-sulfide minerals such as troilite (FeS). The noble gases (helium, neon, argon, krypton, and xenon) do not chemically bond with other elements and would be nearly absent from any solid grains. This simple approximation predicts that the inner solar system bodies should consist of the most abundant elements in the Sun, minus the gases; thus the most abundant elements of the inner planets should include oxygen, magnesium, silicon, sulfur, and iron, and the actual compositions are consistent with the predictions just described.

Meteorites are particularly valuable because they provide samples of the actual materials that condensed in the early inner solar system. The type of meteorites called chondrites match the theoretical expected composition very closely and are believed to be samples of the primitive inner solar system.

Samples of rocks from the Earth's crust are easily obtained (even from ocean basins). Samples of the Earth's mantle are found in two geologic settings: as parts of ophiolite suites (fragments of displaced oceanic crust and underlying mantle) and as kimberlite pipes, or volcano-like vents that appear to have brought rocks (and occasionally diamonds) from the mantle to the surface with great speed and violence. These processes bring both shallow mantle rocks, made mostly of olivine and pyroxene, and deeper mantle rocks, or eclogites, to the surface.

Studies of the propagation of seismic waves through the Earth provide information about both the internal structure of the Earth and the physical properties of the Earth's interior. The observed physical properties match those of a magnesium and iron silicate mantle and a dense iron-nickel core.

One measure of solar system bodies can provide great insight into their chemistry, even for bodies not sampled directly. That measure is bulk density or average density, which is the mass of a body divided by its volume. The volume of a body can be readily computed once the diameter is known, and diameter can be obtained with high precision with spacecraft imagery. The same spacecraft that obtains imagery can also provide an accurate mass determination, through the gravitational effect of the body on the path of the spacecraft. Most common rocks from the Earth's crust have densities of about 2.7 to 3.0 grams per cubic centimeter. The rocks of the Earth's mantle are denser, about 3.3 grams per cubic centimeter for rocks from the upper mantle. Compression within the interior of a planet would result in higher densities, but bodies with bulk densities between 3 to 4 grams per cubic centimeter, such as the Moon, Mars, and Io, are probably made mostly of silicates of the sort found in the Earth's mantle. Denser bodies, such as Earth, Venus, and Mercury, with average densities between 5 and 6 grams per cubic centimeter, must have some denser material in addition to silicates. The most likely dense material is an iron/nickel core like that of the Earth.

The only solids that are abundant in the solar system and less dense than silicate rocks are various ices, with densities of about 1 gram per cubic centimeter. Small solid bodies with densities of about 1 to 3 grams per cubic centimeter are very likely made of varying proportions of silicate rock and ice. Most of the satellites in the outer solar system are of this composition.

The very large outer planets also have low bulk densities; Saturn has an average density of only 0.7 gram per cubic centimeter, and Jupiter, Uranus, and Neptune have average densities between 1 and 2 grams per cubic centimeter. These planets are known, from spectroscopic evidence as well as direct imaging by spacecraft, to have dense gaseous atmospheres, that, according to models of their interiors, liquefy with depth as a result of tremendous pressure. At their centers probably are solid cores of ice, rock, and/or metal.

CONTEXT

The principal value of knowing the chemical composition of the solar system is the insight it provides into its origin and evolution. For example, scientists know that the gases that formed in the early solar system did not include free oxygen. Free oxygen would have combined rapidly with other substances. Therefore, the present atmosphere of the Earth cannot be original but must have formed through chemical and biological processes on the Earth's surface.

Many elements essential to technological society, such as silver, lead, gold, and uranium, are rare in the solar system (and the universe generally). They can be extracted economically only because geologic processes on Earth have concentrated them into ore bodies. The Earth's diverse ore deposits evolved in many ways because the Earth is a dynamic planet both in its interior and on its surface. Chemical elements are concentrated in the Earth's crust by plate tectonic processes; they are further redistributed and concentrated by weathering and erosion, driven by Earth's oxygen-rich atmosphere and liquid water.

Understanding solar system chemistry also can offer a glimpse into a possible future. There has been speculation that extraterrestrial mining might someday contribute valuable resources to human civilization. Some asteroids are rich in iron and other metals; eventually it may be technologically feasible to mine these metals economically. On the other hand, the Moon lacks both the internal and surface activity of Earth, so many types of ore deposits are unlikely to form. Mining the Moon may be the most practical way to supply a future lunar colony, but it probably would involve extracting metals from common rocks, a very energy-intensive and expensive proposition. Thus it is unlikely we will mine the Moon to supply Earth.

Steven I. Dutch

FURTHER READING

Fairbridge, Rhodes W. *The Encyclopedia of Geochemistry and Environmental Sciences.* Stroudsburg, Pa.: Bowden, Hutchinson and Ross, 1972. A general reference on chemicals and chemical processes in nature, written at a moderately technical level. The article "Elements: Planetary Abundances and Distribution" deals with the chemistry of the solar system. Articles on individual chemical elements are also of interest.

Hartmann, William K. *Moons and Planets.* 5th ed. Belmont, Calif.: Thomson Brooks/Cole, 2005. A college textbook beyond the introductory level, its approach is based on comparative planetology. Provides an extensive description of all aspects of the origin of the solar system and its individual members, including their composition and evolution.

Kallenbach, Reinald, et al., eds. *Solar System History from Isotopic Signatures of Volatile Elements.* New York: Springer, 2003. A collection of articles written by specialists in a variety of scientific fields, presented at the 2002 workshop held by the International Space Science Institute. They analyzed the abundance of elements in the Sun, planets, comets, and dust particles, which can reveal much about the early stages and evolution of the solar system.

Manuel, Oliver. *Origin of Elements in the Solar System: Implications of Post-1967 Observations*. New York: Springer, 2001. A compilation of works from a symposium held by the American Chemical Society. Articles were written by astronomers, geologists, chemists, physicists, and other scientists. An excellent reference work focusing solely on the chemical composition of the solar system.

Schneider, Stephen E., and Thomas T. Arny. *Pathways to Astronomy*. 2d ed. New York: McGraw-Hill, 2008. This astronomy textbook contains units that provide a thorough discussion on the structure, composition, and origin of the solar system, as well as detailed descriptions of the planets and other objects in the solar system.

See also: Earth's Age; Earth's Atmosphere; Earth's Composition; Earth's Origin; Earth-Moon Relations; Earth-Sun Relations; Eclipses; Interplanetary Environment; Kuiper Belt; Oort Cloud; Planetary Interiors; Planetary Ring Systems; Solar System: Origins.

Solar System: Origins

Category: The Solar System as a Whole

The solar system formed about 4.6 billion years ago from a cloud of gas and dust that contracted due to its own gravity. Most of the matter went to form the Sun at the center. The planets, their satellites, and the asteroids and comets formed through condensation and accretion in an equatorial disk that developed around the contracting proto-Sun.

OVERVIEW

Although many models and theories of the solar system have been generated from ancient times, most scientists today accept a version of the nebular hypothesis, which states that the solar system formed from a cloud of gas and dust. This nebula probably was part of a much larger interstellar cloud of gas and dust that contained enough matter to form several hundred to several thousand stars like our Sun. Density irregularities probably caused the larger cloud to fragment into many smaller clouds that produced stars and accompanying planetary systems. One of these smaller clouds was the solar nebula, the protosolar system.

Although the solar nebula was only a small fragment of the much larger parent cloud, it probably was at least one light-year in diameter and contained at least as much matter (and maybe up to twice as much) as the solar system today. Its composition probably was similar to what we determine spectroscopically today for the Sun's atmosphere: about 71 percent hydrogen, 27 percent helium, and 2 percent all the other known chemical elements. The hydrogen and helium were formed in the aftermath of the big bang, which occurred about 13 to 14 billion years ago. The heavier chemical elements (those up to iron in atomic mass) were synthesized by nuclear fusion reactions in the cores of massive stars, and those heavier than iron were formed when massive stars exploded as supernovae. The explosive deaths of these massive stars dispersed the heavier elements through interstellar space, where they enriched the clouds of gas and dust from which new generations of stars and their systems of planets would form.

The trigger for the gravitational contraction of the solar nebula could have been its passage through the density wave associated with one of our Milky Way galaxy's spiral arms. As the cloud orbited the center of our galaxy, passing through a spiral arm density wave would have slowed the cloud a bit, compressing the gas and starting its gravitational contraction. Support for this idea comes from observing that the spiral arms of galaxies are highlighted by groups of very young stars that have just recently formed and glowing clouds of gas that have been excited by nearby hot young stars. Another possible trigger to compress the gas and start gravitational contraction could have been a shock wave from the explosion of a nearby supernova. In fact, there is evidence based on the abundance of certain isotopes that such a nearby supernova exploded not long before the solar system formed.

At first, the cloud of gas and dust was cold and rotated slowly. Gravity pulled the material toward the center, and as the cloud contracted, it rotated faster to conserve angular momentum. It continued to contract and consequently rotate even faster, and after about 100,000 years it spun off a disk of material in its equatorial plane. Much of the matter of the solar nebula collected at the center of the disk, becoming the proto-Sun. As the proto-Sun contracted, gravitational potential energy was converted into thermal energy, heating the proto-Sun to the point where it began to shine at infrared wavelengths.

The proto-Sun continued to contract, increasing in temperature and luminosity. This heated the inner part of the equatorial disk close to the proto-Sun, but not the part farther out. The radial difference in temperature across the disk led to differential condensation into small, solid grains. At the high temperatures of the inner part of the disk, only metals and silicate minerals could condense, but farther out, where it was cooler, more abundant materials such as water, ammonia, and methane could condense into ices. The solid grains that formed by condensation collided with each other as they orbited the proto-Sun; if the collisions were not too violent, they stuck together in a process called accretion, gradually building objects up to several kilometers in size called planetesimals. Exactly how the grains stuck together is not certain; perhaps they acquired electrical charges and were held together by static electricity, or if the grains were near their melting points, they might have been somewhat "sticky."

As the planetesimals grew in size, their increased mass gravitationally attracted additional material until they had swept clear the area around their orbits. The planetesimals continued to grow through collisions, becoming protoplanets. The planetesimals and protoplanets that grew in the inner part of the disk were mostly rocky and metallic in composition, while those that grew farther out in the disk were composed mostly of ices such as water, ammonia, and methane. Because water, ammonia,

and methane were so much more abundant than metals and silicate minerals, the outer protoplanets grew much larger than the inner ones; with their increased mass, they were able to capture hydrogen and helium gases, which were even more abundant in the disk, and hence grew larger still. The hydrogen and helium formed thick envelopes around the planet cores, with the material in the outer part of the envelope remaining gaseous, but liquefying with depth due to increasing pressure. Eventually, only a few planet-sized bodies remained: four small, rocky and metallic inner planets (the terrestrial planets) and four large gas/liquid/ice outer planets (the so-called gas giants).

The asteroid belt, a region of small rocky and metallic bodies between the orbits of Mars and Jupiter, probably is leftover material from an early stage of solar system development that failed to coalesce into a planet, perhaps because of the gravitational influence of Jupiter. The Kuiper Belt, a region of icy and rocky bodies beyond the orbit of Neptune, like the asteroid belt probably also is leftover material that never coalesced into a major planet; Pluto is one of the largest members of the Kuiper Belt. Ices in the vicinity of the Jovian planets accreted into small cometary nuclei that, through gravitational interactions with giant planets, were tossed randomly into the far outer reaches of the solar system to become the Oort comet cloud.

Most of the larger moons of the planets probably formed in a process similar to that of the planets themselves: by accretion in an equatorial disk around the protoplanet. Smaller moons may have formed separately from the protoplanet and were later captured gravitationally. The Earth's moon probably formed as the result of the collision of a large protoplanet with the early Earth, the impact blasting material from the Earth's crust and mantle as well as the impacting object itself into orbit around the Earth, there to accrete into our Moon.

As the proto-Sun continued to contract gravitationally, the density and temperature in its core eventually became high enough to initiate the fusion of hydrogen into helium. In this thermonuclear reaction, four hydrogen nuclei fuse to make one helium nucleus. The combined mass of the four hydrogen nuclei that go into the

Formation of the Solar System

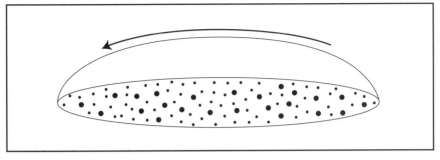

1. The solar system began as a cloud of rotating interstellar gas and dust.

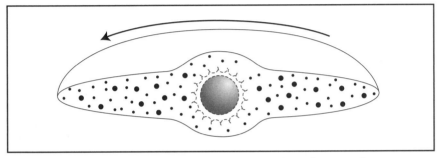

2. As the force of gravity contracted the nebula, rotational speed increased to conserve angular momentum and most of the mass formed a central proto-Sun.

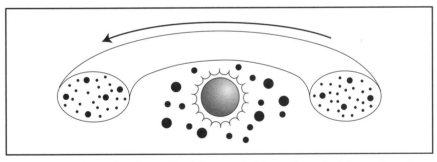

3. Rotation accelerated, and centrifugal force pushed icy, rocky material away from the proto-Sun. Small planetesimals formed around the Sun in interior orbits.

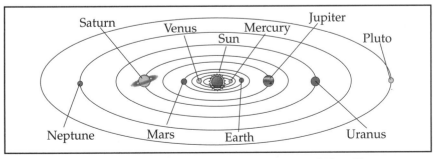

4. The interior, rocky material formed Mercury, Venus, Earth, and Mars. The outer gaseous material formed Jupiter, Saturn, Uranus, and Neptune. Pluto and other Kuiper Belt objects orbit beyond Neptune.

reaction slightly exceeds the mass of the single helium nucleus that results, and this small excess in mass gets converted into energy. Once hydrogen fusion was initiated in its core, the Sun became a full-fledged main sequence star.

The young Sun was much more active than it is today, shedding gases profusely from its surface into space. This became the solar wind, streams of electrically charged particles ejected from the atmosphere of the Sun. With a speed of at least several hundred kilometers per second, the early energetic solar wind blew any gas and dust remaining in the equatorial disk out into space.

The solar system took a few hundred million years to stabilize. Accretion ended with a period of heavy bombardment, when the remaining planetesimals and protoplanets, as well as icy cometary nuclei, smashed into the planets and moons; the scars left by thousands of impacts remain today on the surfaces of many of the planets and moons that are geologically inactive. It may be that much of the water on Earth came from icy cometary impacts early in Earth's history.

METHODS OF STUDY

To deduce how the Sun and planets formed, scientists are constrained by what the solar system is like today. In a sense, investigating the origins of the solar system is like a detective mystery in which ambiguous—and sometimes misleading—clues must be put together to assemble a plausible sequence of past events. Unfortunately, some clues are still missing, and still others are poorly understood.

Meteorites, asteroids, and cometary nuclei are among the smallest bodies of the solar system. These objects should have changed the least since their formation, and their composition should reflect the original material from which the solar system formed. The age of the solar system is determined by dating meteorites using the decay of radioactive isotopes they contain. Most meteorites have very nearly the same age, 4.6 billion years. The most abundant type of meteorite, the chondritic stony meteorites, also appear to be the most primitive and unprocessed. Their internal structure of small mineral grains along with somewhat larger glassy

chondrules is taken to be evidence of the early period of condensation and accretion. Asteroids seem to be the parent bodies of many meteorites, based on comparison of the spectra of the light both reflect. Asteroids are thought to be examples of the planetesimals that formed through accretion, and the compositions of different types of asteroids and meteorites probably represent the processing of solar nebular material that occurred as the planetesimals grew in size. Cometary nuclei probably are unprocessed samples of the ices that condensed in the outer part of the solar nebula.

Spectra of the light emitted by the Sun and reflected by other solar-system objects provides information about the composition of their atmospheres (if any) and their surface materials. Actual samples of surface material include Earth rocks and minerals, Moon rocks brought back by the Apollo lunar missions, and a few meteorites that are thought to have been blasted off the Moon and Mars by large impacts. In addition, landers on the Moon, Mars, and Venus have sent back information on surface composition. Clues about the overall composition of solar-system objects are provided by their average density, found by dividing their mass by their volume.

CONTEXT

Just about every culture has its creation myths—stories about the origin of the world. Naturalistic explanations for the origin of stars generally and the solar system in particular can be traced back to the late 1600's, after scientists came to accept the heliocentric model of the solar system and Sir Isaac Newton published his theory of gravitation. Newton himself suggested that the Sun and stars could have formed by gravitational contraction of initially diffuse matter evenly dispersed through an infinite space. About the same time, the French philosopher René Descartes introduced perhaps the first description of what has come to be called the nebular theory. He proposed that a large cloud of gas (a nebula) had contracted under its own gravity, with the Sun forming at the center and the planets forming in the cooler outer parts. This idea was developed by the German philosopher Immanuel Kant in the mid-1700's.

The French mathematician Pierre-Simon Laplace in 1796 added conservation of angular momentum to the model, concluding that, as the nebula contracted, it would spin faster.

Alternative models for the origin of the solar system were also proposed. Many involved a second star in addition to the Sun. One hypothesis was that another star passed near our planetless Sun and gravitationally pulled matter out of it to become the planets. Another hypothesis was that our Sun was a member of a binary star system; the companion star exploded, and its debris formed the planets. However, by the early to mid-twentieth century, these alternative models fell into disfavor because of difficulties in getting them to work the way they were supposed to when physics was applied to them to try to make them more rigorous. Meanwhile, the nebular model continued to be refined to the point where today it is the generally accepted explanation.

Any successful model for the formation of our solar system must explain several patterns and regularities observed today. First, the planets all orbit the Sun in nearly the Sun's equatorial plane and in the same direction that the Sun rotates. The orbits of the planets are nearly circular. The four inner planets (the terrestrial planets) all are small, dense, and composed mainly of rocky and metallic material. The four outer planets (the Jovian planets) are large, low in density, and composed mainly of gases, liquids, and ices. The Sun, the planets, and other solar-system objects all have about the same age—about 4.6 billion years. Current versions of the nebular model account for all these points.

Solar system oddities and irregularities also need to be explained. In the basic nebular model, one would expect that the planets would all rotate in the same direction as they revolve around the Sun, and their rotational axes should be perpendicular to their orbital planes. Indeed this is the case for a few planets, like Jupiter. However, most have their rotation axes tilted at moderate angles of 20° to 30° away from the perpendicular to the orbital planes; for example, Earth's rotational axis is tilted about 23.5°. The most glaring exceptions are Uranus, which is tilted over so much that its rotational axis lies almost in its orbital plane, and Venus, which rotates very slowly and opposite to its direction of revolution. These departures are explained by impacts during the final accretion of the planets. Off-center impacts by moderately sized protoplanets could account for the moderate axial tilts of planets like Earth, and off-center impacts by large protoplanets are invoked to explain the unusual rotations of Uranus and Venus.

A feature not fully understood is the spacing of the planetary orbits; the orbits of the four inner terrestrial planets are much more closely spaced than the orbits of the four outer Jovian planets. Of course the Jovian planets are larger, which means that they formed from more material spread over a greater range of distances. Did they just happen to form this way by chance, or was it because of gravitational interactions among the planets leading to long-term orbital stability?

A major problem with early nebular models concerned the distribution of angular momentum in the solar system. The Sun, with more than 99 percent of the mass of the solar system, accounts for only 2 percent of the total angular momentum; the planets, with less than 1 percent of the mass, together possess 98 percent of the angular momentum. There had to be some mechanism for the early Sun to transfer most of its angular momentum to the equatorial disk in which the planets formed. It now generally is assumed that magnetic braking between the magnetic field of the early Sun and the ionized gas in the inner part of the surrounding disk transferred the angular momentum. Also, much of the original angular momentum could have been carried out of the system entirely by the energetic early solar wind.

Until 1995, most scientists agreed that there was only one system of planets orbiting a star: our own. Then, in 1995, the first planets orbiting other stars were detected; now several hundred such systems are known. Many of these systems are quite different from our own in terms of their planets' masses and the distances of those planets from their parent stars. Comparing the properties of our own solar system with these others can help refine our models for the formation of planetary systems generally.

Divonna Ogier and Richard R. Erickson

FURTHER READING

Chaisson, Eric, and Steve McMillan. *Astronomy Today*. 6th ed. New York: Addison-Wesley, 2008. One chapter of this excellent introductory astronomy textbook provides a thorough description of the origin of the solar system.

Dermott, S. F., ed. *The Origin of the Solar System*. New York: John Wiley & Sons, 1978. A collection of papers aimed at scientists, using technical language and mathematics. Gives a broad overview of the many theories of how the solar system formed and the problems with each theory.

Fraknoi, Andrew, David Morrison, and Sidney Wolff. *Voyages to the Stars and Galaxies*. Belmont, Calif.: Brooks/Cole-Thomson Learning, 2006. A well-written, thorough college textbook for introductory astronomy courses. Part of one chapter contains a good description of the origin of the solar system.

Frazier, Kendrick. *Solar Systems*. Alexandria, Va.: Time-Life Books, 1985. A richly illustrated volume with an extremely well-written narrative designed for general readers. Concepts are explained in simple, conceptual ways and are supported by beautiful artwork, which graphically portrays the concepts.

Freedman, Roger A., and William J. Kaufmann III. *Universe*. 8th ed. New York: W. H. Freeman, 2008. College-level introductory astronomy textbook in which part of one chapter contains a good description of the origin of the solar system. Describes the solar system as a whole, its origin, and its future.

Hartmann, William K. *Moons and Planets*. 5th ed. Belmont, Calif.: Thomson Brooks/Cole, 2005. A college textbook beyond the introductory level, its approach is based on comparative planetology. Provides an extensive description of all major aspects of the origin of the solar system.

Henbest, Nigel. *Mysteries of the Universe*. New York: Van Nostrand Reinhold, 1981. A well-written and beautifully illustrated volume that discusses the edges of our contemporary understanding of the solar system and the universe. Although meant for the general reader, the text assumes a basic knowledge of astronomy. While dealing with very complex subjects, it keeps within the range of the interested nonscientist.

Sagan, Carl. *Cosmos*. New York: Random House, 1980. An easily understood and nicely illustrated volume that not only describes and explains many features of the universe and the place of humankind in it but also looks at humans' relationship to the universe culturally and historically. Places the study of the universe in perspective with modern times, stressing its importance to the future.

Schneider, Stephen E., and Thomas T. Arny. *Pathways to Astronomy*. 2d ed. New York: McGraw-Hill, 2008. Very thorough college textbook for introductory astronomy courses, divided into short sections on specific topics. Several units provide a thorough discussion on the structure and origin of the solar system.

See also: Earth's Age; Earth's Atmosphere; Earth's Composition; Earth's Origin; Earth-Moon Relations; Earth-Sun Relations; Eclipses; Interplanetary Environment; Kuiper Belt; Oort Cloud; Planetary Interiors; Planetary Ring Systems; Solar System: Element Distribution.

Solar Ultraviolet Emissions

Category: The Sun

The Sun's upper chromosphere, visible as a reddish ring around the Moon during a solar eclipse, has been found to be the principal source of solar ultraviolet radiation.

OVERVIEW

Ultraviolet (or UV) radiation is a form of electromagnetic radiation between visible light and X rays. Most UV radiation is absorbed by ozone in Earth's upper atmosphere; only the "near ultraviolet" (UV with wavelengths not too much shorter than visible violet light) can penetrate to the ground. (Solar near UV is what produces suntans and sunburns.) Since Earth's atmosphere acts as an efficient UV filter, astronomical observations in most of the UV range are best conducted by instruments on board a spacecraft, beyond the atmosphere of Earth.

Observations of both UV and extreme ultraviolet radiation (EUV radiation, with short wavelengths approaching "soft X rays") are important because of the effects these emissions have on Earth's upper atmosphere and also because they can be used to model empirically the layers of the solar atmosphere known as the chromosphere and lower corona. The UV spectrum is a good spectral diagnostic for inferring physical conditions in these layers, with the brightness and variations of the spectrum providing clues about the solar atmosphere.

Before the Orbiting Solar Observatories and Skylab were launched, observations of the Sun's ultraviolet region had been made by brief sounding-rocket observations, lasting for several minutes at best. The results of these observations made it clear that the study of the Sun's atmosphere and its UV radiation would be crucial to an understanding of the Sun. Observations of total solar eclipses had made solar astronomers somewhat aware that the chromosphere and corona, invisible to the human eye outside those eclipses, emit much of their light in UV and shorter wavelengths. It became apparent with the eight Orbiting Solar Observatory satellites that wavelength regions other than the visible had to be observed and mapped in detail if these layers were to be understood. It was also clear that solar EUV radiation over time underwent significant variations in brightness, variations that perhaps could be attributed to changing solar magnetic activity.

The solar atmosphere has a thin but significant layer known as the transition region, located between the upper chromosphere and corona. For a variety of reasons, the temperature, which steadily falls as the distance from the

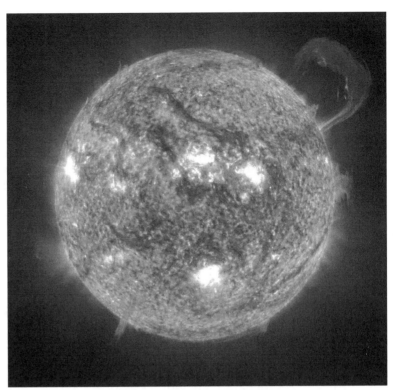

In September, 1999, the Extreme Ultraviolet Imaging Telescope captured this image of a large solar prominence extending like a handle from the Sun's atmosphere; the hottest areas appear as bright splotches on the surface. (SOHO/Extreme Ultraviolet Imaging Telescope Consortium)

Sun's center increases outward, reverses in the upper chromosphere and spikes upward in the transition region to very high temperatures in the corona, near 2 million kelvins. Solar astronomers had long wanted to use ultraviolet images to study the upper chromosphere and transition region. The complete study of the chemical composition, temperatures, densities, and physical processes within these regions would help astronomers to understand the complex mechanisms involved in the structure and dynamics of the solar atmosphere. How is heat transferred to the corona through the transition region? How are the transition region and chromosphere related to the corona above and photosphere below?

Skylab provided the first detailed pictures of the transition region, using the ultraviolet spectrograph associated with the Apollo Telescope Mount. At UV wavelengths, the edge, or limb, of the Sun appears brighter than the disk because of the increase of temperature with increasing height above the surface. This thin region has a temperature of about 150,000 kelvins in its lower layers and 300,000 kelvins at the halfway point between the surface and the coronal layers. (This behavior is opposite to that observed in ordinary visible light, where limb darkening occurs in the visible image because of a decrease in temperature with height in the photosphere.) The duration in space and the large size of the Apollo Telescope Mount observatory telescopes enabled the astronauts to obtain sensational images that would never be possible on the ground because of interference from Earth's atmosphere. Previously, these images had not been obtainable from space either, since the weight of instrumental packages on spacecraft was highly restricted.

Observations of the UV chromosphere give insight into the way the Sun's atmosphere changes with distance from the surface, as well as details of how changes come about as a result of solar magnetic activity. The chromosphere is covered by a mesh of bright spike-like jets of gas, shooting from the chromosphere into the low corona, called spicules. When viewed near the limb of the Sun, they resemble slanted spikes arranged in uniformly spaced circles, separated by distances slightly larger than the diameters of the circles, so that they do not intermesh. This arrangement is referred to as the "spicule forest." Many solar astronomers believe that spicules play a crucial role in transferring nonradiative energy (energy produced by mechanisms other than electromagnetic radiation) into the layers of the solar atmosphere at and above the transition region.

UV images show fluffy loops that join and link active regions together in a lacelike pattern. This pattern shows up in the images as a mottled network that is no doubt directly related to the spicule network. In higher layers of the chromosphere near the transition region, this mottled network takes on a more uniform distribution.

The Skylab full-disk pictures provided a better understanding of spicules and the chromospheric network and also revealed larger spicule-like features found at the Sun's poles. These observations provided new insights into the way energy is transmitted upward to heat the outer layers of the solar atmosphere. The network fades and becomes nearly indistinguishable at the poles, because the larger spicules protrude to roughen the appearance of the solar limb. The overall chromospheric network is also distorted and made into a nonuniform pattern near centers of activity in middle solar latitudes.

Fine brushes of lines sprout from bright UV regions of the chromosphere, revealing lower legs of magnetic loops that stand in magnetically active regions. The burning-bush appearance of these curving lines is formed by the roots of the magnetic loops that correspond to concentrated magnetic areas in the photosphere where sunspots and other aspects of solar activity are observed. Above the photosphere, in the chromosphere, corresponding to higher arch structures in these same loops, other forms of activity are observed, such as prominences (large flame-like tongues of gas that appear to be extensions of the chromosphere into lower regions of the corona) and flares (sudden, triggerlike releases of tremendous amounts of thermal energy trapped by these magnetic arched loops). They appear to occur in tubelike sections of these arched loops from the corona through the chromosphere and on occasion into the photosphere. The flares are believed to occur where two or more of the arches magnetically join one another and violently rearrange their magnetic fields. This entire chromospheric ultraviolet network fades in the lower corona and gives way to a looser, less uniform distribution of material.

Almost all the light emitted by hot coronal gas comes from loops that trace out and mark patterns of magnetic lines of force rooted in bright, active regions in the chromosphere. The entire corona appears to be made up of intertwined arrangements of many of these arched loops. Snarled and twisted magnetic fields show up as particular patterns in the UV chromospheric network where magnetic activity is located; these UV patterns also signal the beginning of energetic events associated with this activity.

In addition to revealing the overall network-like structure of the chromospheric layers and the location of solar activity, the UV images from Skylab provided great insight into the structure and physical properties of various forms of transient solar activity such as prominences, surges (sudden eruptions of gas from deeper layers), and flares. Skylab obtained many detailed pictures of prominences, showing their temperature and three-dimensional structure. The ghostly prominences are shaped and supported by magnetic forces and are made up of chromospheric material elevated and immersed in the hotter coronal gas. Detailed UV images of prominences show coronal tunnel-shaped refrigerators maintained by magnetic tubes, whose outside diameter is surrounded by the hot corona. The tube has a temperature that is lowest in a line running down its center. The temperatures of these structures are inferred from the brightness of the UV image. The brightness-temperature correlation is scientifically defined in terms of a quantity called "emission measure," but simply put, the basic principle is the hotter the gas, the brighter the glow.

The ultraviolet spectrograph on board the Solar Maximum Mission (SMM) satellite gave solar astronomers the ability to observe prominences at wavelengths, and hence temperatures, not seen from Earth. These observations established the existence of downflows at the footpoints (regions of strongest photospheric magnetic fields) of the arches of the prominences. After sufficient loss of mass to lower layers, the prominence can rise into the corona and bow up into an archlike shape. A prominence can be thought of as a long magnetic tube whose length twists and turns much like a rope. Tight wrapping of the ropelike structure can give rise to various magnetic activities associated with phenomena such as flares or eruptive prominences.

Observations of sunspots made with the ultraviolet spectrograph provided information on the mass flow in the umbra (darkened central area) of sunspots and measures of the emission of radiation from them. With the ultraviolet spectrograph, scientists detected upward-propagating acoustic waves above the umbra, created by either a subphotospheric cavity trapping plasma waves or the resonant transmission of magnetoacoustic waves by a chromospheric cavity.

KNOWLEDGE GAINED

In the near UV, radiation with a wavelength of about 300 nanometers originates in the photosphere, and at progressively shorter wavelengths the radiation originates at higher and higher layers throughout the chromosphere. Shorter than 140 nanometers, the spectrum changes from a bright continuum with dark Fraunhofer lines to a faint continuum with emission lines. At wavelengths near 140 nanometers, the solar material is very opaque to radiation from the lower levels. As a result, it becomes possible to see emission lines from the chromosphere, since the continuum becomes so faint.

Many new emission lines were discovered by the Skylab and SMM spectrographs. The strongest emission line, the Lyman alpha line of hydrogen at 121.6 nanometers, is a very good solar activity indicator much like the line of ionized calcium at 392 nanometers. Images at the wavelengths of both lines show a mottled structure that fills in and becomes patchier and more irregular near centers of magnetic activity. Study of the Sun in various emission lines allows scientists to study higher and higher layers of the chromosphere and transition region. The Lyman alpha emission line, the strongest line in the entire solar spectrum, emits more energy than the entire solar spectrum between 0 and 120 nanometers.

The total solar UV irradiance is very low, accounting for about 1 percent of total solar electromagnetic emission. There is some evidence that the entire UV irradiance varies with the solar cycle. This variation has been established for EUV but not for all of the UV range. EUV irradiance is very sensitive to solar activity.

Radiation wavelengths of 100 nanometers or less have important terrestrial effects. This radiation originates in the chromosphere, transition region, and lower corona. The elaborate network structure of the chromosphere was verified and understood in much more detail than had previously been known from ground-based observations.

The transition region is striking in UV images. Since the temperature rises with height quickly in the transition region, the images taken in the EUV band show a pronounced limb brightening. The edges of these images are similar to a lightened ring surrounding the disk.

UV and EUV images of prominences show them to be coronal refrigerators. They are essentially tunnel-shaped magnetic tubes twisted like a rope, often into an arch rising above the chromosphere into the corona. The motion of material in the tube can be very complex. Motion occurs predominantly downward near the footpoints of the arches, whereas material near the top of the arch appears to have a buoyant quality. The arches are often seen to rise and stretch simultaneously until dissipating into the corona.

The ultraviolet spectrograph on the SMM verified that material motion, which had been detected in ground-based observations, occurs in the umbrae of sunspots. Ground-based observations suggest a motion starting at the center of the spot and radiating outward into a circle. In some cases, the reverse of that phenomenon—radiating motion inward to the center of the circle—was observed. The motions observed by the ultraviolet spectrograph seemed to be essentially vertical, rising outward from the spot. The motions also seemed to have a wavelike, pulsing character, suggesting a form of solar oscillations.

CONTEXT

Solar UV striking the upper atmosphere dissociates diatomic oxygen molecules (O_2) into separate oxygen atoms. Between 15 and 50 kilometers above Earth's surface, ozone (O_3) is formed by a combination of single oxygen atoms and diatomic oxygen molecules. Ozone is destroyed by UV radiation, particularly between 210- and 310-nanometer wavelengths. This radiation is harmful to living tissue. Equilibrium is normally maintained between the creation and the destruction of ozone. Earth's ozone concentration peaks at an altitude of about 20 kilometers, the "ozone layer," where its concentration is about ten parts per million. Nevertheless, this thin concentration is essential for the preservation of human life.

Knowledge of the UV and the EUV has allowed astronomers and solar physicists to study the elusive transition region between the chromosphere and corona. The transition layers glow mostly in the UV and show a pronounced increase in temperature and decrease in density with height in the solar atmosphere. Within these layers, various waves (both mechanically and magnetically generated) dissipate energy, mostly in the form of mechanical heat (the vibration of atoms and particles) to the layers above. Thus, they account for the steep increase in temperature with height in the chromosphere and corona above the photosphere. The pronounced decrease in density within these layers is partly responsible for allowing only certain types of wave energy to move outward, retaining most of the thermal energy. Various heat mechanisms, such as sound waves, induce vibrations among the particles, which in turn introduce heat and other forms of energy to their surroundings.

Most solar physicists still consider the increase in temperature of the chromosphere and corona with height above the photosphere to be one of the most difficult problems in solar physics. The foregoing explanations account for some of the mechanisms responsible, but not all. UV and EUV observations show fine structural details that may also be observable in X-ray wavelengths. These details may reveal phenomena that might contribute to this heating and also provide clues as to the physical nature of solar flares. Most notably, these details reveal the interaction of many small magnetic loops giving rise to new local magnetic geometries having the ability to accelerate charged particles and waves into the chromosphere and corona. Solar oscillations of various modes and amplitudes, the "ringing" of the Sun three-dimensionally, no doubt send energy to the chromosphere and corona.

James C. LoPresto

FURTHER READING

Bhatnagar, Arvind, and William Livingston. *Fundamentals of Solar Astronomy*. Hackensack, N.J.: World Scientific, 2005. An intermediary work between basic and advanced textbooks. In addition to solar astronomy, the author also provides directions for building a solar telescope and details how to observe the Sun at various portions of the electromagnetic spectrum.

Buchler, J. Robert, and Henry Kandrop, eds. *Astrophysical Turbulence and Convection*. New York: New York Academy of Sciences, 2000. Approaches the subject of turbulence and convection from four different perspectives: those of theorists, experimentalists, astrophysicists, and computational physicists. Technical, for scientists and college students studying the Sun and its energy production and emission.

Eddy, John A. *A New Sun: The Solar Results from Skylab*. NASA SP-402. Washington, D.C.: Government Printing Office, 1979. This spectacular volume highlights what was learned about the Sun as a result of the Skylab mission, using the Apollo Telescope Mount. Using spectacular photographs, the book contrasts astronomers' knowledge of the Sun before and after Skylab and shows how space-acquired information corrected and greatly amplified previous knowledge.

Fraknoi, Andrew, David Morrison, and Sidney Wolff. *Voyages to the Stars and Galaxies*. Belmont, Calif.: Brooks/Cole-Thomson Learning, 2006. A well-written, thorough college textbook for introductory astronomy courses. Provides discussions of solar energy production, output, and the influence it has on Earth's magnetosphere and environment.

Gibson, Edward G. *The Quiet Sun*. Washington, D.C.: Government Printing Office, 1973. This excellent survey of solar physics uses a structural outline form to discuss what is known about the Sun. The author starts with the solar interior and works his way outward through the atmosphere, discussing the photosphere, chromosphere, and corona in separate chapters. The first chapter is a very helpful overview of the entire topic. Solar activity and solar-terrestrial relationships are covered in the latter chapters. Rather technical; recommended for the reader with some background knowledge of the subject.

Golub, Leon, and Jay M. Pasachoff. *Nearest Star: The Surprising Science of Our Sun*. Cambridge, Mass.: Harvard University Press, 2001. The authors explain the contemporary state of knowledge about the Sun, including historical and observational data. An excellent source for nonscientists.

Haigh, Joanna, et al. *The Sun, Solar Analogs, and the Climate*. New York: Springer, 2005. A set of lectures published by the Swiss Society for Astrophysics and Astronomy. Focuses on how the Sun changes and how Earth's climate responds. Changes in solar ultraviolet output can have dramatic effects on the biosphere. For undergraduate and graduate students.

Hille, Steele, and Michael Carlowicz. *The Sun*. New York: Harry N. Abrams, 2006. A vast collection of images taken by photographers, satellites, and observatories.

Nicolson, Iain. *The Sun*. New York: Rand McNally, 1982. In an atlas-style presentation, Nicolson provides an overview of what is known about the Sun. Includes a historical perspective; discusses the relationship of the Sun to the stars, the Galaxy, and the universe; supplies a detailed description of the solar interior, the solar atmosphere, solar activity, and solar-terrestrial relationships. Many problems are discussed in detail, including solar flares, the solar neutrino problem, coronal holes, solar oscillations, and the solar wind. Even the pragmatic problems of solar energy are discussed.

Noyes, Robert W. *The Sun: Our Star*. Cambridge, Mass.: Harvard University Press, 1982. This excellent book reviews the research done in solar physics from ground-based telescopes and space satellites. The material presented is related to how modern knowledge of the Sun increases knowledge of stars in general. There is an emphasis on how studies of the solar spectrum provide information about the complex solar atmosphere. The contents are arranged according to regions or zones within the interior and atmosphere of the Sun. Much attention is given to solar activity and its relationship to Earth. Suitable for general audiences.

Stix, Michael. *The Sun*. New York: Springer, 2004. Covers all aspects of solar physics. Assumes the reader has some mathematics and physics knowledge. Includes practice problems on most topics.

Zirker, Jack B. *Total Eclipses of the Sun*. Expanded ed. Princeton, N.J.: Princeton University Press, 1995. A delightful short study that discusses the history of observed eclipses as well as their significance to solar science. Current topics, such as using the Sun to test the general theory of relativity, the physics of the solar corona, and solar oscillations, are covered and explained well. There is also some discussion of the Sun's influence on Earth's atmosphere. An accessible presentation.

See also: Coronal Holes and Coronal Mass Ejections; Electromagnetic Radiation: Thermal Emissions; Interplanetary Environment; Nuclear Synthesis in Stars; Red Giant Stars; Solar Chromosphere; Solar Corona; Solar Evolution; Solar Flares; Solar Geodesy; Solar Infrared Emissions; Solar Interior; Solar Magnetic Field; Solar Photosphere; Solar Radiation; Solar Radio Emissions; Solar Seismology; Solar Structure and Energy; Solar System: Origins; Solar Variability; Solar Wind; Solar X-Ray Emissions; Sunspots; Thermonuclear Reactions in Stars; Ultraviolet Astronomy.

Solar Variability

Category: The Sun

There is evidence that the Sun's total energy output varies. The variations are much smaller than the brightness variations observed in other types of variable stars, but even small changes in the Sun's energy output could have significant effects on Earth's climate. Variations correlated with the eleven-year sunspot cycle have been reliably measured. The evidence for possible longer timescale variations is not as clear.

OVERVIEW

A significant fraction of the stars we observe are "variable stars," which means that they vary in brightness. Many of these variable stars vary by significant fractions of their total energy output. Fortunately, the Sun does not fall into this class. If the Sun's energy output were to vary significantly, Earth's climate would be very unstable. However, if the Sun were to vary a very small amount, there might be small but measurable effects on Earth's climate.

The amount of solar energy falling on a square meter at the top of Earth's atmosphere every second is called the solar constant. Accurate measurements of the solar constant can reveal whether the Sun's total energy output per second, its luminosity, varies or is constant. However correcting for Earth's atmospheric absorption is difficult, making it difficult to measure the solar constant from Earth's surface accurately.

Samuel Pierpont Langley invented an instrument called a bolometer in 1878. A bolometer is an extremely sensitive electrical thermometer that can be used to measure the amount of radiant energy from the Sun at all wavelengths. Using his bolometer, Langley, who was secretary of the Smithsonian Institution, pioneered efforts to measure the solar constant accurately. To minimize possible errors from atmospheric absorption, Langley traveled to the top of Mount Whitney in 1881 to make his measurements. In 1895 Langley offered Charles Greeley Abbot a job at the Smithsonian. Using an improved bolometer, Abbot again measured the solar constant. Abbot spent the rest of his career accurately measuring the solar constant and looking for solar variability. Under his direction, the Smithsonian repeatedly measured the solar constant from 1902 to 1957. This program remains the longest continuous search for solar variability. Abbot and his assistants made observations from various remote high-altitude sites in his quest for accuracy. Abbot maintained that he had observed variability in the solar constant, but few other scientists thought that his data were accurate enough to detect solar variability.

Sunspots are dark areas on the Sun's visible surface, where solar magnetic fields divert energy coming from the Sun's interior. There is an eleven-year cycle in the number of sunspots on the Sun. From the perspective of modern satellite observations, it seems that solar variability is too small for Abbot's instruments to have detected. However, Abbot did turn out to be correct in his conclusions that when a major sunspot group faces Earth directly, the Sun's luminosity is slightly reduced and that paradoxically the Sun has a slightly higher energy output during the maximum sunspot activity. The Sun can have a greater luminosity during periods of maximum sunspots because during these times the number of faculae, which are brighter areas on the Sun, increase. Most scientists still thought, however, that it was just too difficult to correct for the variable absorption from Earth's atmosphere, and the question of solar variability remained controversial.

In 1978, the National Aeronautics and Space Administration (NASA) launched the Nimbus 7 satellite with a new, accurate radiometer for measuring the Sun's energy output. In 1980, the Solar Maximum Mission satellite also contained an accurate radiometer. These and later similar satellites confirm two of Abbot's conclusions. When a large sunspot group faces Earth directly for a short time, the Sun's irradiance decreases a small amount because the sunspots block some of the Sun's radiant energy. Over decade-long timescales, however, the Sun's total luminosity is about 0.15 percent higher during sunspot maximum than during sunspot minimum. This result is just slightly less than the estimated accuracy limit of Abbot's observations. During maximum sunspot activity the effect of the brighter faculae is slightly greater than the effect of the fainter sunspots, and the Sun's energy output is slightly higher.

In 1894, E. Walter Maunder called attention to the fact that very few sunspots were observed from 1645 to 1715. Few astronomers paid much attention to Maunder's work until the mid-1970's. In 1976 John A. Eddy published an article confirming this seventy-year paucity of sunspots and naming it the Maunder minimum. During this time, the Sun had about one-thousandth the normal number of total sunspots. Galileo did not discover sunspots until the early 1600's, so modern scientists use indirect means to deduce the number of sunspots prior to the time of Galileo. These indirect methods show the Spörer minimum about 1500, another minimum around 1350, and a medieval grand maximum in sunspot activity around 1200.

If, as satellite data show, the Sun's luminosity varies with the eleven-year sunspot cycle, it is reasonable to ask if it also varies with these much longer cycles. Because these longer cycles have greater variations in sunspot activity than the eleven-year cycle, it is also reasonable to ask if the Sun's luminosity variations are larger than during the eleven-year sunspot cycle. If they exist, increases or decreases in the Sun's total energy output lasting for several decades or longer should affect Earth's climate. If the Sun is less luminous during extended sunspot minima and more luminous during sunspot grand maxima, then the Maunder and Spörer minima should be periods of lower solar luminosity and the grand maximum should be periods of higher solar luminosity. Are they?

It is not possible to travel back in time to measure the Sun's energy output during these periods. However, extended periods of variable solar luminosity should affect Earth's climate. We also lack accurate climate records this long ago. It is therefore necessary to rely on anecdotal and indirect evidence to determine the Earth's climate, and by extension the Sun's luminosity, during these minima and maxima.

Direct measurements are not available, but there is considerable indirect evidence that the Sun's luminosity has varied over the last millennium and that these variations have affected Earth's climate. The time of the Maunder minimum is often referred to as the Little Ice Age. Considerable anecdotal evidence suggests that Europe was much colder than normal during this period. The indirect evidence includes reports of the Thames River freezing in England, Dutch paintings of frozen landscapes in places that usually do not freeze, tree ring studies, and other biological indicators. The grand maximum corresponds to the time period when a Viking colony flourished on Greenland, which is normally much colder than it was during the grand maximum. It is plausible that the Sun had greater luminosity during the grand maximum, and Earth's climate was warmer.

KNOWLEDGE GAINED

Descendants of Langley's bolometer are still in use today, although they are much more sensitive than Langley's original instrument. Modern astronomers cool them with liquid helium to increase their sensitivity and mount them on the ends of large research telescopes to measure the infrared energy from distant stars and galaxies. They have added considerably to our knowledge of the universe.

Abbot and his coworkers at the Smithsonian spent fifty years systematically measuring the solar constant from a variety of locations on Earth as accurately as possible. (It helped that Abbot lived to be 101.) Because solar variability is so small, he was not able to convince many scientists that he was really measuring the solar variability that he claimed. However, even scientists who disagreed with Abbot's claim that the solar constant varied recognized the precision and long time span of Abbot's data. Many scientists have since analyzed Abbot's data to study changes in the absorption of Earth's atmosphere, such as might be induced by major volcanic eruptions and similar phenomena.

The understanding of solar variability made a great leap forward when astronomers were able to put radiometers on satellites to measure the solar constant from space. Because astronomers did not have to correct for the changing effects of light absorbed by Earth's atmosphere, the resulting measurements of the solar constant were more accurate. Scientists could finally agree that the variability in the solar constant is real rather than a measurement error. As a result, efforts to understand how solar variability may be associated with climate changes over past millennia will result in a better understanding of Earth's climate, including current climate changes.

Just as solar variability may affect Earth's climate, it can also affect the temperatures on other planets in the solar system. For example, the polar ice caps on Mars could grow or shrink in response to long-term increases or decreases in the Sun's energy output. Earth, however, is the only planet in the solar system known to have life, so such temperature changes are most serious on Earth. If one of the other planets or satellites in the solar system does indeed have life, then this life could be adversely affected by solar variability.

CONTEXT

Global warming is considered to be among the most serious environmental issues facing the world today. Effects include rising sea lev-els, melting glaciers, ecosystem damage or destruction, increasing frequency of violent storms, and, in some regions, more severe droughts. Many but not all scientists agree that the major cause of this global warming is the increase in carbon dioxide and other greenhouse gases in our atmosphere caused by human activity. However, the causes may be more complex. If the Sun's luminosity does indeed vary over time periods of decades or centuries, these variations can also affect Earth's climate. To be accurate, climate prediction models need to take into account any possible solar variability as well as changes in the amount of greenhouse gases. If the Sun's luminosity is decreasing slightly, that could help solve the global warming problem. If the luminosity is increasing, however, that could exacerbate the effects humans have on Earth's climate. Unfortunately, at this point, solar variability is still too poorly understood to answer these questions definitively.

Paul A. Heckert

FURTHER READING

Eddy, John A. "The Maunder Minimum." *Science* 192 (1976): 1189-1192. In this seminal article, Eddy demonstrates that the Maunder minimum is real and argues that the Sun's energy output varies with long-term cycles in the amount of sunspot activity.

Frazier, Kendrick. *Our Turbulent Sun*. Englewood Cliffs, N.J.: Prentice-Hall, 1980. A good, readable account of our knowledge of the Sun through the publication date. Early studies in solar variability are covered in considerable historic detail.

Freedman, Roger A., and William J. Kaufmann III. *Universe*. 8th ed. New York: W. H. Freeman, 2008. Chapter 16 of this introductory astronomy textbook is a complete, readable overview of our knowledge of the Sun.

Golub, Leon, and Jay M. Pasachoff. *Nearest Star: The Surprising Science of Our Sun*. Cambridge, Mass.: Harvard University Press, 2001. This well-written book gives a detailed summary of our knowledge of the Sun and includes discussion of solar variability.

Hester, Jeff, et al. *Twenty-First Century Astronomy*. New York: W. W. Norton, 2007. Chapter 13 of this astronomy textbook covers the Sun.

Hoyt, Douglas V., and Kenneth H. Schatten. *The Role of the Sun in Climate Change*. Oxford, England: Oxford University Press, 1997. Well written and extensively documented, this book completely covers the topic of solar variability and possible associated climate changes. The authors present both sides of controversial issues and try to appraise the role of solar variability in climate change realistically.

Maunder, E. Walter. "A Prolonged Sunspot Minimum." *Knowledge* 17 (1894): 173-176.

_____. "The Prolonged Sunspot Minimum, 1645-1715." *Journal of the British Astronomical Society* 32 (1922): 140ff. These two papers are Maunder's original publications arguing that there was a prolonged minimum in sunspot activity in the late seventeenth century. Maunder argues that the minimum was real and not an effect of few people observing the Sun.

Soon, Willie Wei-Hock, and Steven H Yaskell. *The Maunder Minimum and the Variable Sun-Earth Connection*. Hackensack, N.J.: World Scientific, 2003. This well-documented book explores the Maunder minimum and other long-term periods thought to result from solar variability, as well as the role this variability plays in Earth's climate changes.

Zeilik, Michael. *Astronomy: The Evolving Universe*. 9th ed. Cambridge, England: Cambridge University Press, 2002. An extremely well-written introductory astronomy textbook. Chapter 12 is an overview of our knowledge of the Sun.

See also: Auroras; Coronal Holes and Coronal Mass Ejections; Electromagnetic Radiation: Nonthermal Emissions; Electromagnetic Radiation: Thermal Emissions; Nuclear Synthesis in Stars; Red Giant Stars; Solar Chromosphere; Solar Corona; Solar Evolution; Solar Flares; Solar Geodesy; Solar Infrared Emissions; Solar Interior; Solar Magnetic Field; Solar Photosphere; Solar Radiation; Solar Radio Emissions; Solar Seismology; Solar Structure and Energy; Solar Ultraviolet Emissions; Solar Wind; Solar X-Ray Emissions; Sunspots; Thermonuclear Reactions in Stars; Ultraviolet Astronomy.

Solar Wind

Category: The Sun

The Sun emits streams of protons, electrons, and some heavier particles in all directions. Known as the solar wind, the outward flow of material in these streams comes from the outermost region of the Sun's atmosphere, the corona.

OVERVIEW

The Sun's corona (its outer atmosphere) does not end abruptly but gradually decreases in density, as it extends billions of kilometers into space. The outward movement and expansion of the corona are functions of distance from the Sun. Expansion close to the Sun is very slow, since the pull of gravity is dominant, but as the distance from the Sun increases, the outward flow increases. This flow of gas is the solar wind, the term originally devised by Eugene N. Parker in 1958 in his classic paper on the dynamics of the interplanetary gas. The solar wind is a stream of ionized gas constantly blown away from the Sun at high speed in all directions. It is composed primarily of electrons, protons, alpha particles (helium nuclei), and some heavier ions.

The rate of mass loss as a result of the outflow of material via the solar wind is only on the order of 10^{-14} or 10^{-15} of the solar mass per year (or about 10^{-9} of the Earth's mass per year). Since the Sun is losing mass, however, the total angular momentum of the Sun is decreasing. As a result, the angular speed of rotation of the Sun also decreases with time, at least on stellar evolutionary timescales. A study of surface rotation rates of young solar-type stars in the Pleiades and Hyades star clusters shows that they have rotation rates some ten times faster than that of the Sun.

At the distance of Earth's orbit from the Sun (150 million kilometers), the average number density of ions under "quiet Sun" conditions (periods when the Sun is not exhibiting high activity, as during solar maxima) is 5 particles per cubic centimeter. Varying solar activity can cause this number to vary widely from the average value; measurements from space probes yield a range of 0.4 to 80 particles per cubic centimeter. The temperature of the particles in the solar wind at the Sun is about 1 million kelvins. By the time they reach the Earth, their temperature has dropped to 200,000 kelvins on average, but because their density is so low, no appreciable heat transfer to Earth occurs. Again, there is considerable variation, from a minimum of 5,000 kelvins to a maximum of 1 million kelvins.

The use of the word "wind" is appropriate considering the speeds involved. At the Earth's orbit, the solar wind whips by at approximately 400 to 500 kilometers per second on average, though there are large fluctuations ranging between 200 kilometers per second minimum and 1,000 kilometers per second maximum. The solar wind is composed mostly of electrons and protons, with a helium abundance averaging 5 percent, but ranging from 0 percent up to a maximum of 25 percent.

Historically it was known that there is a cor-

relation between solar activity and geomagnetic storms. A large summary of historical data correlating geomagnetic activity and solar activity was published in 1940 by Chapman and J. Bartels. In 1931, 9 years prior to this comprehensive summary, the initial model to provide an explanation for the connection was proposed by Sydney Chapman and V. C. A. Ferraro. The model involved streams of ionized (electrically charged) gas ejected from the Sun at the time of solar flares. The interaction of these ionized gas streams, trapped in the magnetic polar regions of the Earth, with the Earth's atmosphere triggered the northern and southern lights (also known as the aurora borealis and australis). This initial model was updated in 1951 by Ludwig Biermann, who suggested that, rather than occasional streams, there was a continuous outward flux of charged particles from the Sun into interplanetary space. This revised model was designed to explain the antisolar spikes observed in some comets. The classic paper by

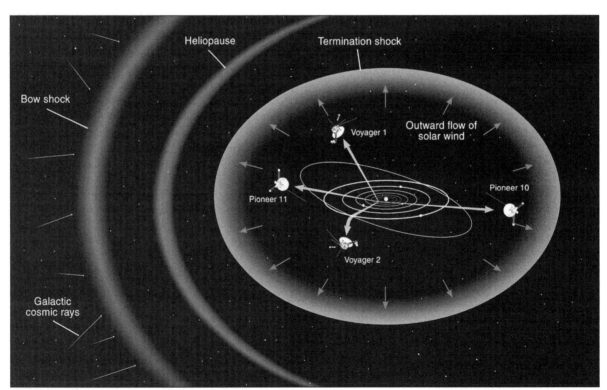

A schematic showing the influence of the solar wind, the extent of which largely defines the heliosphere, which begins to slow dramatically at the heliopause and ends at the bow shock, the outer boundary of the solar system where the pressure of interstellar matter checks the solar wind. (Lunar and Planetary Institute)

Parker in 1958 predicted that interplanetary space is filled with a solar wind. The content and speed of this predicted wind would be functions of the temperature of the corona. Parker predicted that the solar wind speed in the vicinity of the Earth would range from 400 to 800 kilometers per second.

Parker's paper came out just at the dawn of the space age. Since then, numerous space missions have gathered massive amounts of data on the solar wind that could not have been obtained any other way. Between 1959 and 1961, the Soviet interplanetary probes Luna 2 and 3, Venus 1, and Mars 1, using rudimentary (by today's standards) particle detectors, confirmed Parker's prediction. Also in 1961, the initial findings of the Soviet space probes were confirmed by the United States' Explorer 10 satellite using a Faraday-cup probe. The first long-term monitoring of the solar wind was conducted by the U.S. Mariner 2 spacecraft on its three-month mission to Venus in 1962. Mariner 2 again confirmed Parker's prediction by measuring a continuous solar wind with speeds ranging from 319 to 771 kilometers per second. However, the Mariner results disclosed something not predicted by Parker: gusts in the solar wind up to 1,000 kilometers per second that were phase-correlated with the rotational period of the Sun.

The reason for the correlation between the Sun's rotation, solar wind gusts, and geomagnetic perturbations with a recurrent twenty-seven-day period (which matched the Sun's synodic rotation period) was not determined until rocket and spacecraft X-ray data of the Sun revealed the existence of nonmagnetic, long-lived holes in the corona. The coronal holes rotate with the rest of the Sun, and, since they are nonmagnetic, they permit ions an easy exit from what normally are magnetically confined regions of the Sun. The streams of ions passing through the coronal holes have higher-than-normal speeds because they are able to escape without having to overcome the additional magnetic force effects usually present at the Sun's surface. Data from Skylab firmly established that coronal holes are the source of the high-speed streams (gusts) of the solar wind. This discovery is among the most well-established solar-terrestrial connections and is useful in making daily forecasts of geomagnetic disturbances.

Even though the solar wind is emitted in all directions from the Sun, the resultant plasma is not uniform. Observations indicate that the speed of the solar wind tends to be higher and more variable at high solar latitudes. The polar regions of the Sun are nearly always covered by coronal holes, and thus one would expect gusts from the higher latitudes at higher-than-normal speeds.

The Pioneer series of spacecraft carried instrumentation to detect and measure the solar wind. In August, 1972, Pioneer 9 (at a distance from the Sun close to Earth's) recorded solar wind speeds of 1,000 kilometers per second, while Pioneer 10 (which was 214 million kilometers from the Sun, nearing the orbit of Mars) recorded the solar wind at about half that speed. In 1983 Pioneer 10 detected the presence of the solar wind as far out as 4.5 billion kilometers, at the orbit of Neptune. All the Pioneer data show that the average speed of the solar wind changes comparatively little out to Jupiter's orbit, but the range of fluctuations in speed is remarkably diminished at Jupiter's orbit compared with the range at Earth's orbit. This can be seen clearly in the Pioneer data.

KNOWLEDGE GAINED

The bulk of the solar wind is composed of electrons, protons (hydrogen nuclei), and alpha particles (helium nuclei), but there are also traces of heavier elements, the most abundant of which are measurable at the distance of Earth's orbit. Solar wind abundances have been derived from Vela (3, 5, and 6) and Apollo (11, 12, 14, 15, 16, and 17) mission data for the elements that are most common in the entire solar system and in the Sun's corona. The agreement of relative abundance values is quite good considering the varying accuracies of the individual determinations (roughly a factor of two).

The Apollo 17 mission provided data to compute the abundance of iron in the solar wind. Other Apollo missions produced data that resulted in abundance ratios for the light noble gases neon and argon and their isotopes. A major finding of the Apollo program was the detection of the heavier noble gases krypton and xenon in the solar wind and, by implication, the presence of these two elements in the Sun. The anomalous presence of noble gases in meteoritic and lunar surface material is explained by the exposure of these materials to the solar wind.

The ancient solar wind is preserved in approximately the outermost micron of the surface of solid objects, since ions with kinetic energies greater than 1 kilo-electron volt (keV) are trapped on impact with solid surfaces. This record is an easy target for alteration or destruction by a whole host of events. Even so, analyses of lunar material and meteorites indicate that some sort of solar wind has been present at the distance of Earth's orbit for the past 3 to 5 billion years. A study of Apollo 15 deep-drill cores (as deep as 3 meters) by Donald D. Bogard and L. E. Nyquist concluded that the solar wind at the distance of Earth's orbit shows little variation over the last 400 million years.

Magnetic fields associated with the planets in the solar system are distorted by interaction with the solar wind. The magnetopause (outer boundary of a planet's magnetosphere) is where a balance exists between the solar wind and the planet's magnetic field. The magnetopause boundary appears to be stable for planets with large magnetic fields. An average value for the size of Earth's magnetosphere (a term in which the root "sphere" does not necessarily mean spherical in shape, but rather refers to the sphere of influence of Earth's magnetic field) is about 13 Earth radii. This is a sizable obstruction in the path of the solar wind. The result of this obstacle to the supersonic flow of the solar wind is a standing shock wave. In order to have a smooth flow about the periphery of the magnetosphere, the solar wind must make the transition from supersonic to subsonic speeds. In addition, when the plasma is so abruptly slowed, its effective temperature is increased (about tenfold). The point of closest approach by the solar wind to the center of a planet is called the stagnation point. The stagnation point for Earth is at a distance of about 10 Earth radii from the center of the planet (well above the atmosphere).

On December 27, 1984, the Active Magnetospheric Particle Tracer Explorers (AMPTE), a three-satellite cooperative venture by West Germany, Britain, and the United States, generated an artificial cloud of ionized barium. The project was designed to track ionized elements to determine how many of the solar wind's ionized particles actually enter the Earth's magnetosphere and to understand the formation and motions of the high-energy particles in the Van Allen trapped-radiation belts. Results from an earlier test in September, 1984 (with lithium as the tracer element), had shown that approximately 1 percent of solar-wind particles are transported into the magnetosphere surrounding the Earth.

The data from these tests were not definitive and provided only a measure for the specific conditions existing at the time of the test. They indicated that charged particles captured from the solar wind spill out of the outer Van Allen radiation belt (which is inside Earth's magnetosphere) and enter Earth's atmosphere around the north and south magnetic poles. The collisions of these charged particles with atoms of oxygen and nitrogen stimulate them to radiate pale greens and bright reds in the northern and southern skies. This colorful display is known as the aurora borealis or aurora australis (the northern or southern lights), most often seen in zones between 65° and 70° north and south magnetic latitude.

The heliopause is the boundary about 15 billion kilometers from the Sun (about three times the size of Neptune's orbit), where the solar wind merges with the interstellar medium and loses its identity. The interface is the outer limit of the heliosphere, the region influenced by the solar wind. Pioneer 10 data showed that the solar wind oscillates with the eleven-year solar cycle. Increases in solar activity and the solar wind result in decreasing numbers of entering galactic cosmic rays, since the more active heliosphere acts as a shield. Thus, the solar wind influences both the solar and galactic cosmic-ray fluxes at the Earth through a modulation process at the heliopause and through magnetic field-line reconnections at the magnetopause.

CONTEXT

The nucleus of a comet is a "dirty snowball" of various ices with embedded dust and grit. The outer layer of the nucleus sublimates when it nears the Sun, producing a gaseous cloud (the coma) around the nucleus and one or two tails—Type I and Type II. Type II tails are the dust tails. They are smooth, homogeneous, and point generally away from the Sun along a curve. Dust grains in Type II comet tails have a high area-to-mass ratio and thus the effect of solar radiation pressure is significant; Type II tails are blown away from the Sun by the transfer of momentum from solar photons to the dust grains. Type I tails are the ionic tails and are primarily composed of CO^+ along with some other ions, such as N_2^+, CO_2^+, CH^+, and OH^+. Type I tails are long, straight, patchy, and point radially away from the Sun.

Early theories postulated some sort of interplanetary medium to account for the orientation of Type I ionic tails. An extensive study of Type I comet tail orientations by C. Hoffmeister in 1943, long before Parker's predictions and the actual discovery of the solar wind, required the existence of an interacting resistive medium expanding outward from the Sun. A detailed analysis by John C. Brandt yielded an average expansion speed of 474 ± 21 kilometers per second, with a minimum near 150 kilometers per second. These values match very well the speeds measured for the solar wind by spacecraft. The interaction of the solar wind with the coma of a comet produces a bow shock, a hundred thousand to a million kilometers from the nucleus. This interaction in turn carries charged particles from the coma away from the Sun, forming the Type I tail.

In 1981, Brandt and Malcolm B. Niedner published an extensive photographic summary of Type I tail disassociation events (when the ionic tail disconnects from the comet). An angular sector structure of the interplanetary medium and solar wind was identified that rotates with the Sun, within which the direction of the dominant magnetic field is constant. The basic premise is that as a comet crosses a sector boundary, the sudden reversal of the dominant magnetic field direction causes the charged tail to break away from the rest of the uncharged comet, producing the patchy appearance of Type I tails. Thus, comets can be used as interplanetary magnetic probes to determine the spatial location of the sectors.

Theresa A. Nagy

FURTHER READING

Akasofu, Syun-Ichi, and Y. Kamide, eds. *The Solar Wind and the Earth*. Boston: D. Reidel, 1987. An advanced text that deals with a variety of aspects of the solar wind as it interacts with the Earth. The introductory portion of each of the fourteen chapters is quite easy to understand, and the degree of difficulty varies as a function of the topic. A very useful source book.

Brandt, John C. *Introduction to the Solar Wind*. San Francisco: W. H. Freeman, 1970. A good introduction to the subject of the solar wind with easy-to-read chapters on the historical summary, ground and space observations, and the interaction of the solar wind in the solar system. A few chapters are for the more advanced technical reader. Somewhat dated.

Chaisson, Eric, and Steve McMillan. *Astronomy Today*. 6th ed. New York: Addison-Wesley, 2008. Very well-written college-level textbook for introductory astronomy courses. Discusses the solar wind and its effects.

Foukal, Peter. *Solar Astrophysics*. 2d rev. ed. Weinheim, Germany: Wiley-VCH, 2004. New York: Wiley, 1990. A detailed look at the Sun and our understanding of it, including the solar wind. Also covers the history of solar astrophysics. Suitable for undergraduates.

Fraknoi, Andrew, David Morrison, and Sidney Wolff. *Voyages to the Stars and Galaxies*. Belmont, Calif.: Brooks/Cole-Thomson Learning, 2006. A well-written, thorough college textbook for introductory astronomy courses. Includes material on the solar wind and its impact on the planets and solar system.

Freedman, Roger A., and William J. Kaufmann III. *Universe*. 8th ed. New York: W. H. Freeman, 2008. College-level introductory astronomy textbook. Thorough and well-written. Covers the solar wind.

Gosling, J. T., and A. J. Hundhausen. "Waves in the Solar Wind." *Scientific American* 236 (March, 1977): 36-43. Variations of the solar wind are described. For a wide audience.

Meyer-Vernet, Nicole. *Basics of the Solar Wind*. Cambridge, England: Cambridge University Press, 2007. An introduction to solar wind. Covers the Sun's structure, interior, and atmosphere. Ideal for researchers and astronomy majors.

Moldwin, Mark. *An Introduction to Space Weather*. Cambridge, England: Cambridge University Press, 2008. A textbook designed for nonscience majors, giving a solid introduction to the Earth-Sun system, space physics, and space weather. Includes definitions of key terms, underlying physics concepts, and review questions.

Schneider, Stephen E., and Thomas T. Arny. *Pathways to Astronomy*. 2d ed. New York: McGraw-Hill, 2008. Thorough introduction to astronomy, divided into short sections on specific topics, including the solar wind and its effects.

See also: Auroras; Coronal Holes and Coronal Mass Ejections; Earth-Sun Relations; Interplanetary Environment; Planetary Magnetospheres; Solar Chromosphere; Solar Corona; Solar Evolution; Solar Flares; Solar Geodesy; Solar Infrared Emissions; Solar Interior; Solar Magnetic Field; Solar Photosphere; Solar Radiation; Solar Radio Emissions; Solar Seismology; Solar Structure and Energy; Solar System: Origins; Solar Ultraviolet Emissions; Solar Variability; Solar X-Ray Emissions; Sunspots.

Solar X-Ray Emissions

Category: The Sun

The Sun emits X rays and gamma rays, revealing the presence of extremely high temperatures (tens of millions of kelvins) and high-energy particles. Solar X rays are produced in the solar corona (the hot, tenuous outer atmosphere of the Sun), and both X rays and gamma rays are generated by the magnetic explosions or eruptions known as solar flares.

OVERVIEW

X rays and gamma rays are forms of electromagnetic radiation with very short wavelengths and very high energy photons. Electromagnetic radiation displays the properties of both waves and particles; the term "photon" is used to refer to electromagnetic radiation when it acts like a stream of particles. The wavelengths of X and gamma rays are less than about 10 nanometers, compared to visible light, which ranges from about 400 nanometers (violet) to 700 nanometers (red). Their photons have energies ranging from about 100 electron-volts (eV) upward; for comparison, the photons of visible light have energies from a little less than 2 to a little more than 3 eV. Because their photons are so energetic, the presence of X rays and gamma rays generally indicates high temperatures and interactions of high-energy particles.

The existence of temperatures of at least a million kelvins in the solar corona (the Sun's outer atmosphere) had been known since the 1940's because of the presence of emission lines of highly ionized elements in its optical spectrum. Why should the corona be so hot? The photosphere, or visible "surface" of the Sun, is much cooler (around 5,800 kelvins), and common sense had misled astronomers to expect a steady decrease of temperature outward, rather than the precipitous temperature increase that actually occurs.

The physical cause of the high temperatures of the solar corona now is thought to be the strong magnetic fields in the Sun. The Sun's magnetic field is complex, with loops or arches that extend beyond the photosphere out into the corona. Charged particles flow along these loops, gain energy from them, and transfer it to the corona. Observations show that the corona is not uniform at X-ray wavelengths. The strongest, "brightest" X-ray emission comes from regions where the magnetic field is the strongest and the gas is hottest. Coronal streamers of hot ionized gas follow the Sun's magnetic field lines outward from these areas. Coronal holes are large, dark (in X rays), cool regions with weak or absent magnetic fields.

Solar flares were originally discovered serendipitously in 1859, in ordinary visible light, by Richard Carrington, who had been making routine observations of sunspots. Subsequent observations at certain specific wavelengths (such as the red light at 656.3 nanometers emitted and absorbed by hydrogen atoms) disclosed flares more distinctly in the solar chromosphere (that part of the Sun's atmosphere in between the photosphere and the corona), so before the 1950's the general phenomenon was known as "chromospheric flares." They were not assigned much importance physically.

X-ray and gamma-ray observations changed this perception of the importance of solar flares. Whereas solar flares had been perceived as complicated but otherwise insignificant features of the solar atmosphere—some kind of solar cloud—they were discovered to be a fundamentally important phenomenon, the prototype object of high-energy astrophysics. This branch, which involves observations of X rays and gamma rays, brings together gravitational, atomic, nuclear, and plasma physics.

X-ray bursts accompany essentially every solar flare, and during times of great solar activity, this occurs many times per day. The "soft" (longer-wavelength, lower-energy) X-ray spectrum reveals the existence of dense plasmas, hot ionized gases composed of electrons and atomic nuclei with temperatures in the range of 10 to 20 million kelvins, some ten times the typical coronal temperatures. X-ray images obtained with grazing-incidence telescopes show the hot plasmas to be trapped in the magnetic tubes, or loops, that may rise about 100,000 kilometers above the photosphere of the Sun. The energy trapped in the hot plasma flows down the magnetic field lines over a period of minutes, feeding into the chromosphere. The plasma, shown by its soft X-ray emission, appears to be the central agent in producing many of the classical effects of solar flares.

What initially produces the hot flare plasma? Its creation is usually marked by a "hard" X-ray burst (shorter-wavelength, higher-energy X rays). These hard X rays come from the interactions of fast electrons, with energies far above those of the particles in the hot plasma responsible for the soft X rays. Furthermore, data from the Solar Maximum Mission (SMM) have shown that gamma-ray bursts also occur commonly in solar flares. The presence of gamma rays indicates that the high energy required to produce flares involves the acceleration of protons (and other ions) as well as electrons. X rays and gamma rays have revealed many of the mechanisms involved in flare events. Magne-

tism plays a crucial but ill-understood role in causing the plasma instabilities that put on such spectacular and dramatic displays in solar flares. However, the initial cause of these events—and hence a general theory of flares—remains elusive.

To summarize what we do know, a solar flare is now believed to originate as an instability occurring in the solar atmosphere, a heterogeneous magnetized plasma that extends upward from the photosphere (visible surface of the Sun) into interplanetary space. This instability results in the acceleration of high-energy particles, both electrons and ions, and the creation of plasmas with temperatures of tens of millions of kelvins. In essence, a flare is a magnetic explosion in the upper solar atmosphere, leading to rapid acceleration of high-energy particles and an outward eruption of denser solar material from near the photosphere. The flare features observable by ordinary techniques of optical spectroscopy from ground-based telescopes appear to be secondary products of this explosive release of magnetic energy.

KNOWLEDGE GAINED

Solar flares have connections to Earth that are both economically significant and scientifically important. Their high-energy radiation can perturb Earth's ionosphere (a layer of ionized atoms in the upper atmosphere), inducing surge currents to flow in electrical power grids, causing failures and sometimes extensive blackouts. Similar disturbance of the ionosphere can also interrupt radio communications, because the ionosphere reflects some types of radio communication. Unfortunately, solar physicists cannot predict the level of solar activity with much more precision than is afforded by the simple recognition of the well-known eleven-year sunspot cycle. They can predict large flares no better than seismologists can predict major earthquakes. The problem appears to lie in the complexity of the physics and the lack of adequate observations.

The high-energy radiation of a solar flare is produced in various ways. Soft X rays are produced by hot plasmas (with temperatures exceeding 10 million kelvins). In a major solar flare, the X-ray flux may reach a level as high as one-millionth of the total solar luminosity. Hard X rays are intense flashes of higher-energy X rays that occur near the onset of a solar flare, showing the acceleration of energetic electrons. Gamma rays are produced by nuclear interactions in solar flares, showing that particle acceleration in solar flares extends to protons and other heavier ionized particles. The spectra, time profiles, and spatial distributions of the high-energy radiation all serve as guides to the physics of the flare phenomenon. Fairly detailed observations of some of the properties of soft X rays have been obtained, but many observational gaps exist for hard X rays and gamma rays.

The introduction of grazing-incidence X-ray optics, first from sounding rockets and later (in 1973) from the Skylab crewed space station, showed that the hot plasmas responsible for X-ray emission from solar flares were trapped in magnetic loops. These structures have their "footpoints" anchored in the solar photosphere, but extend great distances up into the corona. Numerous X-ray observatories since the Skylab era have been used to study solar flares and the Sun's corona such as XMM-Newton and the Chandra X-Ray Observatory.

CONTEXT

The discovery of X rays and gamma rays from the Sun led directly to a new branch of astronomy, high-energy astrophysics, and to great changes in understanding solar behavior. These discoveries resulted from the use of instruments designed to detect these high-energy emissions, mounted on instrument platforms ranging from the V-2 rockets captured from Germany during World War II to high-altitude uncrewed balloons and eventually artificial Earth satellites and deep space probes. These vehicles were able to place these detectors above Earth's atmosphere, which blocks high-energy photons, and opened a whole new view of the Sun and other stars.

After World War II, when American researchers made observations above Earth's atmosphere, initially using V-2 rockets captured from the Germans. These observations showed the somewhat unexpected existence of "high-energy" radiation from the Sun, from ultraviolet (UV) to X rays. Herbert Friedman's group at the Naval Research Laboratory was instrumental in these observations, which paralleled those of James Van Allen, the discoverer of Earth's radiation belts. A most important discovery came in 1958, when Laurence E. Peterson and John Winckler flew a high-altitude balloon over Cuba for cosmic-ray studies and were able to detect gamma radiation from a solar flare that happened to occur during the balloon flight.

Over the decades since the 1950's, new launch vehicles, combined with remarkable progress in the technology of X-ray optics and X-ray and gamma-ray detectors, have led to a great expansion of research in this area. Solar observations in the X-ray and gamma-ray spectral regions have greatly broadened our knowledge of the physics of solar flares and hence the Sun's structure and physics.

In addition, X rays and gamma rays are now observed from a variety of other astronomical sources: white dwarfs, neutron stars, black holes, supernova remnants, galaxies, and clusters of galaxies, to name some. Observatories specifically designed to detect the X rays and gamma rays of such objects, including the Chandra X-Ray Observatory, the Compton Gamma Ray Observatory, XMM-Newton, the Swift mission, and Astro-E2, have been launched above Earth's atmosphere and have returned data that have rendered startlingly beautiful images of some of these sources. It can be speculated that some of the same physics of magnetized plasmas in the Sun may underlie the high-energy emissions from these objects.

Hugh S. Hudson

FURTHER READING

Chaisson, Eric, and Steve McMillan. *Astronomy Today*. 6th ed. New York: Addison-Wesley, 2008. Very well written college-level textbook for introductory astronomy courses, with material on X-ray observations, solar flares, and the Sun's corona.

Eddy, John A. *A New Sun: The Solar Results from Skylab*. NASA SP-402. Washington, D.C.: Government Printing Office, 1979. The focus here is on the abundant illustrations, mostly photographs obtained from the Skylab satellite, supplemented by discussion of the nature of the Sun. Includes some treatment of the phenomenon of solar flares and their radiation.

Fraknoi, Andrew, David Morrison, and Sidney Wolff. *Voyages to the Stars and Galaxies*. Belmont, Calif.: Brooks/Cole-Thomson Learning, 2006. This well-written college textbook for introductory astronomy courses covers X-ray observations of solar flares and the Sun's corona.

Freedman, Roger A., and William J. Kaufmann III. *Universe*. 8th ed. New York: W. H. Freeman, 2008. College-level introductory astronomy textbook, thorough and well written. Has sections on X-ray observations, solar flares, and the Sun's corona.

Giovanelli, Ronald. *Secrets of the Sun*. Cambridge, England: Cambridge University Press, 1984. A general introductory description of the Sun and of solar physics.

Noyes, Robert W. *The Sun: Our Star*. Cambridge, Mass.: Harvard University Press, 1982. Provides a general survey of knowledge of the Sun, with some material on processes leading to X-ray emission. This book is at a slightly higher technical level than the others in this bibliography, but is still suitable for a general readership.

Ripley, S. Dillon. *Fire of Life: The Smithsonian Book of the Sun*. Washington, D.C.: Smithsonian Exhibition Books, 1981. A beautiful, lavishly illustrated popular description of the Sun. Includes a discussion of societal impacts of solar phenomena.

Schneider, Stephen E., and Thomas T. Arny. *Pathways to Astronomy*. 2d ed. New York: McGraw-Hill, 2008. Very thorough college textbook for introductory astronomy courses. Divided into lots of short sections on specific topics, with several touching on X-ray observations, solar flares, and the Sun's corona.

See also: Coronal Holes and Coronal Mass Ejections; Electromagnetic Radiation: Nonthermal Emissions; Electromagnetic Radiation: Thermal Emissions; Red Giant Stars; Solar Chromosphere; Solar Corona; Solar Evolution; Solar Flares; Solar Geodesy; Solar Infrared Emissions; Solar Interior; Solar Magnetic Field; Solar Photosphere; Solar Radiation; Solar Radio Emissions; Solar Seismology; Solar Structure and Energy; Solar System: Origins; Solar Ultraviolet Emissions; Solar Variability; Solar Wind; Sunspots; Thermonuclear Reactions in Stars; X-Ray and Gamma-Ray Astronomy.

Space-Time: Distortion by Gravity

Category: The Cosmological Context

Space and time are linked together, and the fabric of space-time can be distorted or warped by the presence of matter. Albert Einstein's general theory of relativity reinterprets many diverse phenomena by replacing Newtonian gravity with distortions of space-time.

OVERVIEW

To the layperson, space is empty but time is full of activity. Time is perceived as a flow, carrying consciousness from one present moment to the next. To a physicist, however, the terms "space" and "time" denote quite different concepts: Space possesses physical properties and many levels of structure, and the flow of time is an illusion. Indeed, some theories hold that matter, rather than being located in space and time, is nothing more than disturbances in the fabric of space and time.

Although space and time can be perceived in radically different, separate ways, they are linked together by motion: Average speed is defined as distance divided by time. The motion of material objects and light signals is described in relativity theory in terms of a single, unified reality called space-time. Anything that happens in some particular place at some particular time is called an "event" in space-time. Its location in space-time is specified by its three spatial coordinates and one time coordinate. The series of events that trace the "where" and "when" of an object is called its world line through space-time. Objects at rest or moving with a constant velocity follow straight world lines, while the world lines of accelerated objects are curved.

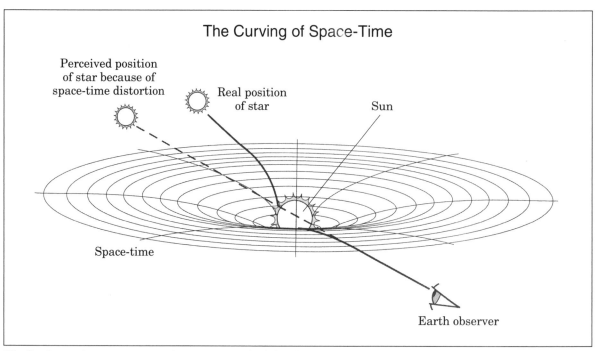

The Curving of Space-Time

Perceived position
of star because of
space-time distortion

Real position
of star

Sun

Space-time

Earth observer

The Sun's gravity causes space-time to curve, which bends the star's light and makes it appear to be located where it is not.

The structure of space can be described mathematically in terms of geometries. The geometry taught in most high schools is called Euclidean geometry. Developed (or at least described) by Euclid around 300 B.C.E., it deals with "flat" space. Some familiar features of flat space are that parallel lines never meet, even when extended infinitely far in both directions, and that the sum of the interior angles of a triangle equals 180°. During the mid-nineteenth century, the Russian mathematician Nikolai Ivanovich Lobachevsky (1793-1856) and the German mathematician Georg Friedrich Riemann (1826-1866) developed self-consistent non-Euclidean geometries. In these new geometries, the sum of the angles of a triangle could be more or less than 180°, and locally parallel lines could converge or diverge over large distances. A physical space described by a non-Euclidean geometry is called a curved space, as opposed to flat Euclidean space.

A familiar curved two-dimensional (2-D) space is the surface of a sphere, such as the Earth or a basketball. If one draws a triangle on the surface of a sphere, and that triangle is small (compared to the radius of the sphere), the angles of the triangle add up to 180° because the surface is nearly flat over small distances. However, if the triangle covers a large part of the surface, the angles add up to more than 180°. As an example, imagine doing this on the surface of the Earth. One side of the triangle is the 0° longitude line between the North Pole and the equator. The second side of the triangle is the 90° longitude line, also between the North Pole and the equator. The third side of the triangle is the equator from 0 to 90° longitude. Each of the three angles of this triangle is 90°, and thus the sum of the angles is 270°. Similarly, lines of longitude are parallel where they cross the equator, but because they are located on a sphere, they eventually meet at both poles.

These results—the angles of a triangle adding up to more than 180°, and locally parallel lines converging—are characteristic of, and can be used as tests for, the geometry developed by Riemann for positively curved spaces. The opposite properties—the angles of a triangle adding up to less than 180°, and locally parallel lines diverging—are characteristic of, and can be used as tests for, the geometry developed by Lobachevsky for negatively curved spaces.

The mathematics of curved space geometries took on new physical significance in 1915, when Albert Einstein (1879-1955) put forth his general theory of relativity. One consequence of it is that four-dimensional (4-D) space-time, first introduced by Einstein in his 1905 special theory of relativity, is curved by the presence of matter. Gravity appears to exist only because mass warps space-time in its vicinity. Physicist John Wheeler, who contributed enormously to relativity theory, provided this succinct description of the relation between matter and space-time: Matter tells space-time how to curve, and the curvature of space-time tells matter and light how to move. Where space-time is not flat, Euclidean geometry fails to provide an adequate description; non-Euclidean geometries must be used instead.

The fundamental postulate at the heart of general relativity is the principle of equivalence: The effects of a uniform gravitational field and the effects of a constant acceleration are indistinguishable. In a closed rocket accelerating through space at 9.8 meters per second per second (m/s^2), there is no experiment that an astronaut could perform that would distinguish this situation from one in which the rocket is at rest on the Earth's surface (where the acceleration of gravity is 9.8 m/s^2). Einstein realized that, if he could express this equivalence in a mathematical form, then he could relate both gravitation and acceleration to the curvature of space-time, thereby providing a geometric explanation of gravity. Non-Euclidean geometry provided the mathematical formalism to describe how the distribution of mass warps space-time. Einstein's field equations of general relativity enable one to compute, in principle, the curvature or warpage in space and time due to a given mass distribution.

Naturally, the bending of space-time profoundly affects the world lines of objects. As space-time warps, the world lines warp as well, since they are constrained to follow geodesics (locally straight lines) through a curved space. When an object moves on a world line that is curved, the object is being accelerated; that is,

its motion is not uniform. From the perspective of general relativity, gravity is not to be considered as a force, but rather as an effect of geometry—a distortion of space-time. The presence of mass will distort the fabric of space-time. What is experienced as gravitation is the warping of space-time because of mass.

APPLICATIONS

The idea that matter distorts space-time led Einstein to make several remarkable predictions. In Euclidean geometry, the shortest distance between two points is a straight line. In a curved space, the shortest path is curved because space itself is curved. In order to gain a clearer understanding of this concept, consider that New York and Tokyo have similar latitudes but that the shortest distance between them, along the Earth's surface, passes close to the North Pole. Although incomprehensible on a flat map, this fact is easily visualized on the curved surface of a globe.

Similarly, a light beam passing near a massive object will follow a curved path because the massive object distorts space-time in its vicinity. The bending of light in the presence of a massive object also is mandated by the equivalence principle. In a reference frame accelerating upward, a transverse light beam would appear to deflect toward the upwardly accelerating floor. Since the equivalence principle requires that there be no distinction between acceleration and gravitational fields, light has to be deflected toward a source of gravity—that is, a massive object that warps space-time around it.

In 1911, Einstein predicted that a ray of starlight passing near the Sun would be bent by the warpage of space-time around the Sun. (Stars can be seen near the Sun during total solar eclipses when the sky is darkened sufficiently.) Although the predicted deflection of starlight just grazing the Sun's limb (edge) is only 1.75 arc seconds, it was first confirmed at the 1919 total solar eclipse, and later at many subsequent total solar eclipses since then. The effect has also been verified at radio wavelengths when the Sun occults (passes in front of) the radio source 3C279 every October 9. A similar effect, called gravitational lensing, has been observed when light from very distant galaxies is bent when it passes near massive, closer galaxies.

A previously known but unsatisfactorily explained effect—the advance of Mercury's perihelion—also is explained by the warpage of space-time near the Sun. Mercury's orbit around the Sun is slowly precessing (or pivoting) so that Mercury reaches perihelion (the point when it is closest to the Sun) slightly later with each orbit. Observations of Mercury's motion were sufficiently accurate that its perihelion advance was know by the mid-nineteenth century. Almost the entire effect could be explained by the gravitational perturbations of the other known planets on Mercury, but a small residual was attributed to the gravitational perturbation of an as yet undiscovered planet (given the name Vulcan) inside the orbit of Mercury. Vulcan was searched for, and some claims of discovery even were made, though none was confirmed. In 1915, Einstein explained how the warpage of space-time around the Sun causes Mercury's orbit to pivot forward (or precess); the precession predicted by general relativity accounts almost precisely for the residual that once was attributed to Vulcan. This (along with the bending of starlight passing near the Sun) was one of the first predictions of general relativity to be confirmed observationally. Since then, residual perihelion advances have been measured for Venus, Earth, and the asteroid Icarus; all agree within their observational errors with the predicted values calculated from general relativity.

Since space and time are intrinsically bound together, the effect of mass on the geometry of space-time means that time will be warped as well as space. Einstein predicted that time would be slowed by the warped space-time in the vicinity of a mass. Thus, clocks at the Earth's surface should run slightly slower than clocks at higher altitudes, which are farther from Earth's mass, where the warpage is less. The effect is small (about one part in 10^{13} for each vertical kilometer of altitude), but this minuscule effect was first measured (indirectly) in 1960 by the physicists R. V. Pound and G. A. Rebka at Harvard, again verifying a prediction of general relativity.

CONTEXT

In 1827, the German mathematician Carl Friedrich Gauss published a paper in which he recorded his measurements of the interior angles of a large triangle formed by three mountain peaks. His measurement was an attempt to ascertain whether space was Euclidean. The experiment was inconclusive: He obtained 180° within the experimental accuracy of his measuring devices. Gauss's experiment may have been the first attempt to test the long-established assumption that the universe is best described by Euclidean geometry.

By the middle of the nineteenth century, two other self-consistent geometries had been devised: the geometry of negatively curved space, by the Russian mathematician Nikolai Ivanovich Lobachevsky, and the geometry of positively curved space, by the German mathematician Georg Friedrich Riemann. Until 1915, however, when Einstein published his general theory of relativity, most mathematicians and physicists assumed that non-Euclidean geometries were mathematical curiosities but had little to do with physical reality. In Einstein's theory, space-time is curved by the presence of mass, and gravity exists only because mass gives space a non-Euclidean character.

Several months after Einstein's theory was published, the German astrophysicist Karl Schwarzschild found rigorously exact solutions to Einstein's field equations for two different cases: an ideal point mass and a finite spherical mass. The first case predicted that, at a relatively small radius from the mass point, some of the mathematical terms become infinite. This condition represents such an intense warping of space-time that any signal (whether matter or electromagnetic radiation) within this boundary would be unable to escape. This radius, called the Schwarzschild radius, defines a surface (called the event horizon) such that any mass or energy within this surface is forever trapped. Most physicists of the time, Einstein included, believed that it would be impossible for any real object to contract to such a small size that its mass would be contained within this surface. (For example, an object with the mass of the Sun would have to shrink within a radius of about 3 kilometers, and an object with the mass of the Earth would have to shrink within a radius of about 9 millimeters.) In 1963, Roy Kerr developed a new exact solution to Einstein's equations for a rotating mass, and the Schwarzschild solutions were seen to be special cases of Kerr's solution. Since matter and energy could cross the event horizon in the inward direction but nothing could escape from inside the event horizon, the term "black hole" was coined.

By the 1970's, indirect evidence had begun to accumulate indicating the existence of just such black holes, and there now is compelling evidence for at least two classes of black holes. Stellar-mass black holes form when massive stars exhaust all the fuels they used for nuclear fusion reactions and finally explode as supernovae. Supermassive black holes have masses of millions of solar masses or more and are thought to exist at the centers of many galaxies, including our own Milky Way galaxy.

The application of the equations of general relativity to cosmology and the structure of the universe allows for the universe to be flat, negatively curved, or positively curved. Any curvature is related to how the average density of matter and energy in the universe compares to a value called the critical density, which is equivalent to about three hydrogen atoms per cubic meter.

In the simplest form of relativistic cosmology, any possible curvature of the universe is linked to its ultimate fate. If the average density of matter and energy equals the critical density, the universe is flat and it will continue to expand, but slowing down in such a way that it just barely expands forever. In this case the

The Schwarzschild Radius

At the start of World War I in 1914, Karl Schwarzschild, a young professor at the University of Göttingen, volunteered for military service. Craving action, he eventually managed to get transferred to Russia, where he heard of Albert Einstein's new general theory of relativity. Schwarzschild wrote two papers on that theory, both published that year. He provided a solution—the first to be found—to the complex partial differential equations fundamental to the theory's mathematical basis. Schwarzschild solved the Einstein equation for the exterior space-time of a spherical nonrotating body. This solution showed that there is an enormous, virtually infinite, redshift when a body of large mass contracts to that certain radius—a size now known as the Schwarzschild radius.

The value of that size is easily calculated by a simple astrophysical formula Schwarzschild derived, relating the radius to the universal gravitational constant, the star's mass, and the speed of light: $R = 2GM/c^2$. Surprisingly, he showed that the general theory of relativity gave basically the same results as Isaac Newton's more common theory of gravitation, but for different reasons. When the mass of the object is measured in units of the Sun's mass, the Schwarzschild radius is neatly given by three times the ratio of the mass to the Sun's mass, the answer expressed in kilometers: $R = 3 \times M/M(\text{Sun})$. If the Sun were contracted to a radius  of 3 kilometers, it would be of the right size to be labeled a "black hole." A body becomes a black hole when it shrinks to a radius of less than the critical radius; at that point, nothing, including light, will have enough energy ever to escape from the body—hence the name "black hole," since no light escapes and anything falling in remains. Earth would have to contract to a radius of approximately one centimeter to become a black hole.

While in Russia, Schwarzschild contracted pemphigus, an incurable metabolic disease of the skin. He was an invalid at home in 1916 when he died. He was forty-two years old. For his service in the war effort, he was awarded an Iron Cross. In 1960, he was honored by the Berlin Academy, which named him the greatest German astronomer of the preceding century.

density of matter and energy equals the critical density, the universe is flat and it will continue to expand, but slowing down in such a way that it just barely expands forever. In this case the universe is said to be critically open. If the average density of matter and energy is less than the critical density, the universe is negatively curved and it will easily expand forever, slowing just a little. In this case, it is said to be open. If the average density of matter and energy is greater than the critical density, the universe is positively curved. It will expand to some maximum size and then contract on itself in what some call the big crunch. In this case, it is said to be closed.

Various measures (mostly very indirect) indicate that the universe is extremely close to being flat, but the amount of matter and energy observed is substantially less than the critical density. This observation has led to the notion that most of the mass in the universe is in the form of "dark matter," which has not yet been detected. Moreover, it seems as if only a small part of the unobserved dark matter can be in the form of conventional matter; most of it must be something exotic and as yet unknown.

A completely unexpected discovery in the mid-1990's was that the expansion of the universe seems to be accelerating. This necessitates using a more general form of relativistic cosmology that includes an extra term involving a parameter called the cosmological constant (or some other modification similar to it). To account for the accelerating expansion of the universe, some unknown energy is required. Called "dark energy," it may represent 70 percent or more of the total matter and energy in the universe.

General relativity with its distortions of space-time has thus become an important tool for understanding the origin, structure, and future of the universe and its contents. Although its basic structure has remained unaltered since 1915, it continues to find applications in such diverse areas as the precession of orbits, the bending of light, gravitational redshifts, black holes, and cosmology. It is truly amazing that an abstract theory concerning the warping of space-time has turned out to be so powerful and useful.

George R. Plitnik and Richard R. Erickson

FURTHER READING

Davies, P. C. W. *Space and Time in the Modern Universe*. New York: Cambridge University Press, 1981. Written in an authoritative and lucid style, this book explores the changing ideas of space and time and their applications in astronomical and cosmological scenarios.

Gribbin, John. *Spacewarps*. New York: Delacorte Press, 1984. Describes the universal implications of Albert Einstein's general theory of relativity in physics and astronomy.

_____. *Timewarps*. New York: Delacorte Press, 1980. A clear and imaginative examination of some of the dramatic questions raised by the new concepts of time.

Misner, Charles W., Kip S. Thorne, and John A. Wheeler. *Gravitation*. San Francisco: W. H. Freeman, 1973. This comprehensive, massive textbook is still the fundamental reference on gravity, space-time, and general relativity. It begins with an exposition of the space-time concept of special relativity in a form that facilitates its natural extension to general relativity. Suitable for the diligent layperson with some background in physics, though many later parts are written at more advanced mathematical levels. Its mathematical rigor and sophistication is partly offset by its almost colloquial, folksy style of presentation.

Rucker, Rudolf. *Geometry, Relativity, and the Fourth Dimension*. New York: Dover, 1977. A highly readable and amusing exposition of four-dimensional space-time and the structure of the universe.

Wheeler, John Archibald. *A Journey into Gravity and Spacetime*. New York: Scientific American Library, 1999. Explores different phenomena to explain Einstein's theories of space-time and gravity. A good introductory book for the general reader.

Will, Clifford. *Was Einstein Right? Putting General Relativity to the Test*. 2d ed. New York: Basic Books, 1993. The renaissance of relativity is described with splendid clarity by one of the professional participants. Observations and theories that test the experimental basis for general relativity are presented without mathematics.

See also: Big Bang; Cosmic Rays; Cosmology; Electromagnetic Radiation: Nonthermal Emissions; Electromagnetic Radiation: Thermal Emissions; General Relativity; Interstellar Clouds and the Interstellar Medium; Milky Way; Space-Time: Mathematical Models; Universe: Evolution; Universe: Expansion; Universe: Structure.

Space-Time: Mathematical Models

Category: The Cosmological Context

In both the special and general theories of relativity, neither space by itself nor time by itself is independent of the state of the observer. Only a certain mathematical union of them, called space-time, has invariant properties. The geometry of space-time is the basis for relativistic physics, which are seen in our solar system. A full description of the advancement of the perihelion of Mercury, for example, requires the use of relativity. Also, the way the mass of the Sun can bend light coming from other stars and galaxies is described by relativity theory.

OVERVIEW

Space-time is a four-dimensional coordinate system or reference frame in which one mathematically describes the spatial location and temporal coordinate of an event. Such a frame of reference can be either inertial or non-inertial. An inertial frame of reference often is defined in Newtonian mechanics as a frame that is not accelerated, but then one must ask, "Accelerated relative to what?" A better definition consistent with relativity is that a local inertial frame (or LIF) is a frame in which a body subject to no external force moves at constant speed in a straight line, or alternatively, that it is a frame in which everything is weightless. The word "local" is added because in the space-time of general relativity, it is not possible to have a truly universal inertial frame, if any mass is present.

A wide variety of experiments have repeatedly confirmed that physical phenomena do not fundamentally differ from one inertial frame of reference to any other. The special theory of relativity asserts this as a basic postulate or principle: All the laws of physics are the same in every inertial frame of reference. This means that both the mathematical form of fundamental equations of physics and the values of the physical constants that they contain are the same in all inertial frames. When this principle is applied to the theory of electromagnetism, it requires that observers in all inertial frames of reference agree about the numerical value of the speed of electromagnetic waves in empty space. The universality of this speed—henceforth referred to as the "speed of light" and represented by the letter c—requires that space and time separately cannot be invariant but must change upon transformation from one inertial frame of reference to another.

An event is the name given to something that happens at a particular time and place. The collection of all events (the "whens" and "wheres") in the history of a particle is called its world line. Measurements of the time and place of an event will vary from one inertial frame to another, but there are equations called Lorentz transformations that relate the time and space coordinates of an event in one inertial frame to the time and space coordinates in another inertial frame, based on the relative velocity of the two frames. All the famous phenomena predicted by the special theory of relativity (relativity of simultaneity, length contraction, time dilatation) can be derived from these transformations.

Although the space and time coordinates of an event will differ from one inertial frame to another, there is a quantity called the space-time separation (or space-time interval) between the events that is invariant, meaning that it has the same value in all inertial frames. Let A and B be two events with time and space coordinates (t_A, x_A, y_A, z_A) and (t_B, x_B, y_B, z_B) in one inertial frame, and (t_A', x_A', y_A', z_A') and (t_B', x_B', y_B', z_B') in another inertial frame. The space-time separation s between the two events is defined as

$$s^2 = -c^2 (t_B - t_A)^2 + (x_B - x_A)^2 + (y_B - y_A)^2 + (z_B - z_A)^2$$

as calculated in one frame, and

$$s^2 = -c^2 (t_B' - t_A')^2 + (x_B' - x_A')^2 + (y_B' - y_A')^2 + (z_B' - z_A')^2$$

as calculated in the other frame. It can be shown using the Lorentz transformations that these two expressions for s are equivalent, demonstrating that the space-time separation is invariant between inertial frames.

The interval or separation between two events in space-time is somewhat analogous to the distance r between two points in space:

$$r^2 = (x_B - x_A)^2 + (y_B - y_A)^2 + (z_B - z_A)^2$$

However, there is a fundamental difference between the geometry of space-time and the geometry of space by itself. Notice that the square of the difference in the time coordinates appears in the space-time formula with the opposite sign from the squares of the differences in the spatial coordinates.

APPLICATIONS

Because the squares of the time and space coordinate differences have opposite signs in the formula for space-time interval or separation, the square of the interval or separation between two distinct space-time events can be positive, zero, or negative, depending on how the squared difference of the time coordinates compares to the sum of the squared differences of the space coordinates.

If the squared interval or separation is positive, the squared difference of the space coordinates dominates over the squared difference of the time coordinates, and the separation is called space-like. This means that there exists some inertial frame of reference in which the time coordinates of the two events are equal, so in this frame the two events are simultaneous and they differ only in spatial location. In particular, neither event can be the cause or effect of the other, since all physical influences require time to propagate. In other inertial frames, the events will have different time coordinates, and it is possible for either of the two events to have the larger time coordinate, meaning it occurred later. Consequently, "later" and "earlier" have no universal meaning for a pair of events with a space-like separation, since in some frames one event occurred later while in other frames the other event occurred later.

If the squared interval or separation is negative, the squared difference of the time coordinates dominates over the squared difference of the space coordinates, and the separation is

called time-like. This means that there is some inertial frame in which the spatial coordinates of the two events are equal, so in this frame the two events occurred at the same location and differ only in time. When a pair of events have a time-like separation, observers in all inertial frames agree on which event occurred first and which occurred second.

If the squared interval or separation is zero, the squared difference of the space coordinates equals the squared difference of the time coordinates (multiplied by the speed of light squared), and the separation is called null or light-like. This means that, in every inertial frame, the pair of events may be connected by the world line of a ray of light moving from one to the other. Such a ray of light could be the agent by which the earlier event causes the later event. Watching a pair of events that have light-like separation, observers in all inertial frames agree on which event occurred first and which occurred second.

Since a particle is always at its own location, intervals between events on the world line of a particle with mass must be time-like. On the other hand, intervals between events on the world line of a photon (a "particle" of light) must be light-like or null.

The sign of the squared intervals between events can be used to divide space-time into regions of different character. Suppose that event A is at the coordinate origin ("here and now") of space-time ($x_A = 0, y_A = 0, z_A = 0, t_A = 0$). As an aid to visualization, the z coordinate may be suppressed; then it is possible to draw a diagram illustrating these regions. The surface mapped out by all null intervals is a double cone. The upper cone (positive t) represents all events in the future that can be reached by a light ray emitted here and now, while the lower cone (negative t) represents all events in the past that could have sent a light ray to arrive here and now. All events that occur within the cone have squared space-time separations from event A that are negative or time-like.

Consequently the world line of any particle with mass that passes through ("coincides with") the event chosen as the "here" and "now" origin of the figure is confined to the interior of the light cone. All events outside the cone have

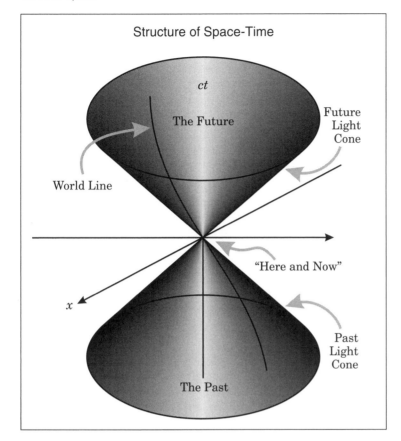

Structure of Space-Time

ct

The Future

Future Light Cone

World Line

"Here and Now"

x

Past Light Cone

The Past

squared space-time separations from event *A* that are positive or space-like. Consequently they can neither influence nor be influenced by event *A*, for to do so the influence would have to travel between event *A* and any event outside the cone at a speed greater than the speed of light.

The time axis of any space-time coordinate plot indicates the passage of time in the frame of reference with those coordinates. In an inertial frame, the time axis will be a straight line. The world line, curved or straight, of any particle can be considered a time axis for that particle. Intervals along its world line define its "proper time," which elapses on a clock carried by the particle.

The straight line between two points in Euclidean space has the shortest length of any curve joining them. The geometry of space-time is such that the straight line between two events with a time-like separation has the longest proper time of any world line joining them. This is the basis of a straightforward prediction

of relativity which is usually called the twin paradox. Effects of the special theory of relativity are analyzed from inertial frames of reference but may include accelerated objects, such as the twin who travels out and back, thus aging less than the twin remaining at rest in one inertial frame throughout the other's trip.

Analysis of motion from the point of view of observers in accelerated frames of reference is also possible, but it uses the mathematical concepts of differential geometry. As seen from accelerated frames of reference, the structure of space-time is not globally covariant but only locally covariant. This means that the light cones at various events may be tilted in relation to each other. Albert Einstein's general theory of relativity attributes such distortion of the geometry of space-time to gravity. This theory can be summarized in two intimately linked statements: (1) Matter warps space-time, and (2) warped space-time tells matter and light how to move.

CONTEXT

The root of the concept of space-time was the discovery by Hendrik Antoon Lorentz, published in 1898, of the rules of transformation of the coordinates of an event from one inertial frame of reference to any other inertial frame of reference. His derivation was carried out to find a transformation that does not change the form of the fundamental laws of electrodynamics, known as Maxwell's equations. However, Lorentz did not claim that the transformations he found which kept electromagnetism invariant had the broad applicability they are now understood to have. It remained for Einstein to formulate a comprehensive view, published in 1905 in his special theory of relativity, of space and time as measured in inertial frames moving relative to each other, and their dependence on the state of motion of an observer.

Even as he was establishing the foundations of what is now called the special theory of relativity, Einstein was aware of the incompatibility of these ideas with Newtonian gravitational theory. His early work on extending the principle of relativity beyond inertial frames of reference was hampered by mathematical complexities. Hermann Minkowski, a former math professor of Einstein at Zurich Polytechnic University, in an address presented in 1908, introduced the concept of unified space-time based on the new ideas expressed in Einstein's 1905 description of his special theory of relativity. Minkowski realized that Einstein's assertion of the constancy of the speed of light for all observers implied that space and time were fundamentally linked. He also introduced the powerful techniques of geometry, which provided both the mathematical formalism for dealing with noninertial frames as well as intuitive insights into the physical implications. By exploiting ideas first introduced to understand the differential geometry of curved surfaces, unified space-time became the natural way to understand all of physics. Thus, the mathematics of non-Euclidean geometry found application in Einstein's general theory of relativity, which was published in 1916. The general theory is a comprehensive synthesis of the relations among space, time, matter, and motion from the point of view of any frame of reference whatsoever.

In contemporary physics, space-time is accepted as the arena in which all things exist and move. The assertion that the laws of physics must be independent of arbitrary choice of a particular frame of reference in space-time is a powerful working tool of the theoretical physicist. This requirement puts limits on possible new hypotheses and the equations they imply in almost all areas of physics. The one branch of physics that has remained at odds with relativity theory is quantum mechanics. It remains to be seen whether quantum behavior can be unified with the space-time of relativity.

John J. Dykla

FURTHER READING

Ferington, Esther. *The Cosmos*. Alexandria, Va.: Time-Life Books, 1988. This profusely illustrated large-format book is a brief but surprisingly comprehensive introduction to modern cosmology. Includes illuminating presentations on the concept of space-time in both the special and the general theories of relativity.

Hawking, Stephen, and Roger Penrose. *The Nature of Space and Time*. Princeton, N.J.: Princeton University Press, 2000. A collection of lectures given by Hawking and Penrose, who pick up where Niels Bohr and Albert Einstein left off. Technical; for readers with strong physics and mathematics backgrounds.

Hawking, Stephen, Kip Thorne, Igor Novikov, Timothy Ferris, and Alan Lightman. *The Future of Spacetime*. New York: W. W. Norton, 2003. A collection of essays including a basic introduction to the concepts of relativity. Topics discussed include wormholes, gravity waves, and time travel. For the advanced reader.

Minkowski, Hermann. "Space and Time." In *The Principle of Relativity*. New York: Dover, 1952. This early presentation of the space-time concept to a technical audience uses for the most part elementary mathematics to explore some implications of the relativistic unity of space and time.

Misner, Charles W., Kip S. Thorne, and John A. Wheeler. *Gravitation*. San Francisco: W. H. Freeman, 1973. This comprehensive, massive textbook is still the fundamental reference on gravity and general relativity. It begins with an exposition of the space-time concept of special relativity in a form that facilitates its natural extension to general relativity. Suitable for the diligent layperson with some background in physics, though many later parts are written at more advanced mathematical levels. Its mathematical rigor and sophistication are partly offset by its almost colloquial, folksy style of presentation.

Rabinowitz, Avi. *Warped Spacetime, the Einstein Equations, and the Expanding Universe*. New York: Springer, 2009. This college text derives Einstein's equations using calculus-based physics. It also discusses space-time, black holes, worm holes, and cosmology. Designed for undergraduate and graduate science students.

Schwinger, Julian. *Einstein's Legacy: The Unity of Space and Time*. New York: Scientific American Books, 1986. A lively and well-illustrated account of the special and general theories of relativity, this volume includes a careful elementary presentation on the concept of space-time with helpful examples and applications.

Siegfried, Tom. "It's Likely That Times Are Changing." *Science News* 74, no. 6 (September 13, 2008): 26-28. Traces the changing views of time, from Einstein and Minkowski to the present.

Taylor, Edwin F., and John A. Wheeler. *Space-Time Physics*. San Francisco: W. H. Freeman, 1966. This brief text is a thorough treatment of the subject at an elementary mathematical level. Careful reading and working through its examples develops intuition for thinking relativistically. The few problems that require calculus for their solution are clearly identified.

See also: Big Bang; Cosmic Rays; Cosmology; Electromagnetic Radiation: Nonthermal Emissions; Electromagnetic Radiation: Thermal Emissions; General Relativity; Interstellar Clouds and the Interstellar Medium; Milky Way; Space-Time: Distortion by Gravity; Universe: Evolution; Universe: Expansion; Universe: Structure.

Stellar Evolution

Category: The Stellar Context

Stars go through a series of changes that are referred to as stellar evolution or stellar life cycles, in analogy to the life stages of living organisms. Stars are "born" in interstellar clouds of gas and dust called nebulae. They generate energy by various mechanisms for most of their "lives." They "die" when they finally run out of ways to produce any more energy.

OVERVIEW

For most of the twentieth century, one of the primary goals of astronomers was to determine how stars are born, how they live, and how they die. By the end of the century, a fairly complete picture finally had emerged. It is convenient to express most stellar parameters relative to the Sun: The Sun's mass is 1.99×10^{30} kilograms (330,000 times the mass of Earth), its radius is 696,000 kilometers (109 times larger than Earth's), its surface temperature averages about 5,800 kelvins, and its luminosity (rate of energy output) is 3.9×10^{26} watts, or 3.9×10^{26} joules per second.

Stars are born in nebulae, large clouds of gas and dust in interstellar space. The clouds typically are several tens of light-years in diameter and contain enough matter to form hundreds to thousands of stars. The gas is mostly hydrogen (about 71 percent by mass), with some helium (about 27 percent by mass) and small amounts of heavier elements. The dust particles are small, solid grains of carbon, silicate minerals, iron compounds, and ices, probably about 0.001 millimeter in size, on average. The nebulae in which stars form are cold, with temperatures on the order on 10 kelvins. At such low temperatures, atoms can bond together to form molecules, so these nebulae are referred to as molecular clouds. Also as a result of the low temperatures, the gas pressure is very weak and can barely keep the cloud from contracting from its self-gravity. In denser parts of the cloud, self-gravity may overcome the weak pressure and start that part of the cloud collapsing. The trigger for this increased density may be an encounter with a spiral arm density wave as the nebula orbits the center of its galaxy, strong stellar winds from a nearby star that already formed, or shock waves from a nearby supernova explosion.

A molecular cloud usually breaks up into separate clumps that collapse on their own. As each clump collapses, gravitational energy is rapidly converted into thermal energy, heating the gas and causing it to begin feebly emitting radio waves. When a sufficiently dense core has formed at the center of the clump, the contraction slows and the object is called a protostar. Usually many neighboring protostars form at about the same time. A protostar continues to contract gravitationally, growing hot enough to begin to shine in the infrared part of the electromagnetic spectrum. What little visible light the protostar emits is blocked by the shroud of dust that surrounds it. Eventually the dust shroud dissipates, some of it joining the growing equatorial disk around the protostar (in which planets may ultimately form) and some of it blown away as the protostar becomes hotter and brighter. As the protostar continues to contract and get hotter, collisions break molecules apart into individual atoms and electrons are stripped off the atoms, ionizing the gas.

When the temperature at the center of the contracting protostar reaches a few million kelvins, hydrogen nuclei start to fuse together to form helium nuclei. The energy released by this nuclear fusion reaction stops the contraction, and the protostar becomes a main sequence star. It takes higher-mass protostars less time to reach the main sequence stage than lower-

mass protostars, because their greater mass means stronger gravity and more rapid contraction. The protostar stage lasts on the order of a hundred thousand years for a star with 10 solar masses, a few tens of millions of years for a star like the Sun, with 1 solar mass, and from several hundred million to a billion years for a star with 0.1 solar mass.

In hydrogen fusion, four hydrogen nuclei fuse together to form one helium nucleus. The mass of the four hydrogen nuclei is slightly greater than the mass of the one helium nucleus produced, and the excess mass is converted into energy according to Albert Einstein's formula $E = mc^2$, which states energy E and mass m are equivalent and are related by a physical constant c (the speed of light squared). To overcome the electrical repulsion that the positively charged hydrogen nuclei (bare protons) have for each other requires high temperatures (so the nuclei are moving quickly) and high densities (so the nuclei are close together). These are the conditions in the cores of main sequence stars.

A main sequence star is in a state of hydrostatic equilibrium, which means the self-gravity trying to make the star contract is balanced by the pressure trying to make the star expand. The energy released by hydrogen fusion in the core provides the energy that the star radiates into space. The main sequence stage is the longest and most stable part of a star's energy-producing life. A star remains a main sequence star as long as it has hydrogen in its core to fuse to helium. Massive main sequence stars have more fuel, but they consume that fuel much more rapidly, so their main sequence lives are relatively short. For example, a 30-solar-mass star has a main sequence lifetime of about 5 million years. The Sun, with 1 solar mass, has a main sequence lifetime of about 10 billion years. (Since the Sun is currently about 4.5 billion

The nebula NGC 1333 in the Perseus constellation, seen here in this infrared image from the Spitzer Space Telescope, shows the birth of new stars "hatching" from the dust clouds in which they formed. (NASA/JPL-Caltech/R. A. Gutermuth, Harvard-Smithsonian CfA)

years old, it has progressed about halfway through its main sequence life.) Low-mass main sequence stars have less fuel, but they consume it much more slowly, so their main sequence lives are much longer, from about 30 billion years for a star with 0.5 solar mass to as much as a trillion years for a star with 0.1 solar mass. These main sequence lifetimes are longer than the age of the universe, so every low-mass main sequence star that ever formed is still a low-mass main sequence star.

When all the hydrogen in the core has been fused into helium, the core contracts and heats up. Hydrogen fusion is transferred to a still hydrogen-rich shell surrounding the contracting helium-rich core. This causes the outer layers of the star to expand and cool, and the expanding surface of the star becomes red in color. The star ceases to be a main sequence star and becomes a red giant or red supergiant. Stars similar to the Sun, with about 1 or 2 solar masses, become red giants, ten to one hundred times bigger than the present Sun. Massive stars become red supergiants, one hundred to one thousand times bigger than the present Sun.

When the temperature of the contracting helium-rich core reaches about 100 million kelvins, helium fusion is ignited. Three helium nuclei fuse to form one carbon nucleus. Add another helium nucleus, and an oxygen nucleus forms. The total mass of the three or four helium nuclei is a bit greater than the mass of the carbon or oxygen nucleus, and again this mass excess gets converted into energy. This fusion reaction supplies the star with energy for much less time than hydrogen fusion did, when it was a main sequence star.

When the helium in the core is exhausted, the core once again contracts and heats up. A star like the Sun is not massive enough for its core to shrink enough to get hot enough to start any more nuclear fusion reactions. Thermal pulsations and a strong stellar wind blow off its outer layers in one or more shells of gas. The expanding shell of gas surrounding what is left of the star is called a planetary nebula. (Planetary nebulae have nothing at all to do with planets. The name originated in the 1800's because, with the telescopes in use then, they looked round,

like planets, and fuzzy, like nebulae.) The central star of a planetary nebula is the exposed former core of the star. It contracts as much as it can and becomes a white dwarf star composed of carbon and maybe oxygen. White dwarf stars have about the mass of the Sun packed into a sphere about the size of the Earth. This makes them very dense, averaging about 1 metric ton per cubic centimeter. White dwarf stars can no longer generate energy; they cannot contract to release gravitational energy, and thus they cannot get any hotter to be able to tap other nuclear fusion reactions. They shine only because they are very hot. As they shine, they radiate their energy away and cool off, gradually becoming black dwarfs (cold, dark, dead stellar "embers").

The core of a massive star (more than about eight solar masses), because of the star's stronger self-gravity, can shrink enough to get hot enough to go through a series of nuclear fusion reactions, one after another, producing heavier atomic nuclei. This stops with the production of iron nuclei, since iron is the heaviest nucleus that can form through fusion reactions that release energy. To form heavier nuclei through fusion requires the input of energy. The iron core collapses, and the outer layers collapse on top of it and rebound, sending shock waves through the star. This tears the star apart in a Type II supernova explosion. In a few minutes, it releases more energy than it produced by nuclear fusion reactions during its entire preceding life. A Type II supernova becomes about a billion times more luminous than the Sun before it gradually fades away. Some of the prodigious energy released in the explosion goes into fusion reactions forming elements heavier than iron. Much of the star's matter is violently ejected into interstellar space at speeds ranging from thousands to a few tens of thousands of kilometers per second, thereby enriching the interstellar material in elements heavier than helium.

If the mass of the stellar remnant that remains after the Type II supernova explosion is less than approximately 2 to 3 solar masses, it becomes a neutron star. A neutron star has a radius of only 15 kilometers and a density of a billion metric tons per cubic centimeter. If the remaining mass of the supernova remnant is greater than about 3 solar masses, however, no

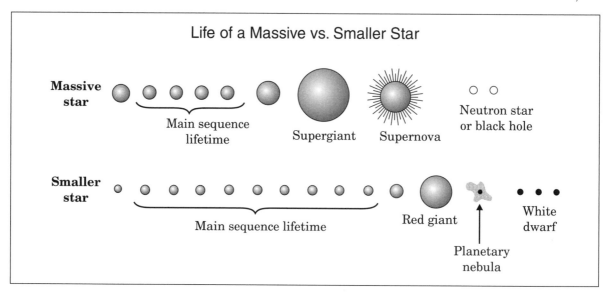

Life of a Massive vs. Smaller Star

known force can stop its collapse to a black hole. The gravitational field of a black hole is so great that nothing, not even light, can escape from it. Consequently, a black hole cannot be directly "seen" in any part of the electromagnetic spectrum. A black hole can be detected only by its effects, primarily through its gravity, on nearby objects.

KNOWLEDGE GAINED

The Earth and all life on it exist only because of the life and death of massive stars. Our current understanding of the big bang (the primordial "explosion" that created the universe about 13 to 14 billion years ago) is that it produced only hydrogen and helium with very small traces of lithium and beryllium. The first stars that formed in the earliest days of our galaxy consisted only of hydrogen and helium. Gas planets like Jupiter might have formed around those first stars, but not rocky/metallic planets like Earth, nor could carbon-based life have developed. The atoms of all the elements heavier than helium that make up our bodies and the Earth we live on were produced by nuclear fusion reactions in massive stars before and during their deaths as Type II supernovae. These heavy elements were spewed out into interstellar space by the supernova explosions, where they enriched the interstellar clouds of gas from which new stars formed. By the time the Sun

and solar system began to form in one of these nebulae about 4.5 billion years ago, about 2 percent of its mass consisted of elements heavier than helium. This provided the material for rocky/metallic planets to form in the inner solar system and for carbon-based life-forms to develop on one of them, our Earth. The production and dispersal of heavy elements by massive stars, with the resulting enrichment of nebulae, continues to the present time, so that today about 5 or 6 percent of interstellar matter consists of elements heavier than helium.

Our understanding of the evolution of stars like the Sun reveals the future of our solar system, including Earth. About 5 billion years from now, the Sun will run out of hydrogen in its core. It will expand until its radius is about one hundred times larger than it is now, swallowing the planet Mercury and possibly Venus. At that point it will be a red giant, about a thousand times more luminous than it is now. Temperatures on Earth will reach between 1,000 and 2,000 kelvins; our atmosphere will escape into space, our oceans will boil away, surface rocks will at least partly melt, and life will not survive here. Eventually, a strong solar wind and thermal pulses will blow the Sun's outer envelope back into space, producing a planetary nebula. What is left of the Sun will become a white dwarf, with perhaps one-half to two-thirds of its present mass. Initially, this white dwarf Sun will have a surface temperature of about 100,000 kelvins, but since it will be unable to generate more energy, it will cool, rapidly at first, then ever more slowly, until it becomes a cold, dark, dead black dwarf.

CONTEXT

The twentieth century witnessed the development of our modern understanding of stellar evolution. New and improved instrumentation, including detectors in space above Earth's atmosphere, made it possible to observe the solar system and deep space with sharper resolution and in all parts of the electromagnetic spectrum. Advances in theoretical physics and the widespread use of computers led to the construction of more and better models of the stellar interiors and the study of how they change with time.

In the late 1890's and early 1900's, George Ellery Hale oversaw the design and construction of several large telescopes—the 40-inch refractor at Yerkes Observatory in Wisconsin, and the 60-inch and 100-inch reflectors on Mount Wilson in California. For the next half century, they were the largest telescopes in the world, and their use led to many observational discoveries in all branches of astronomy.

The early 1900's also saw the development of the Hertzsprung-Russell (or H-R) diagram, a graph used for plotting stellar luminosity versus surface temperature. It turns out that the location of a star in this diagram indicates many things about the star; besides its luminosity and surface temperature, these include its radius, in some cases its approximate mass, and its evolutionary stage. Consequently, the H-R diagram has proved to be a powerful tool for tracing the life cycles of stars of various masses.

In 1926, Sir Arthur Stanley Eddington published *The Internal Constitution of the Stars*, a book that established much of the formalism still used today to construct models of stellar interiors. Its most serious deficiencies, of which Ediington was well aware, involved the sources of stellar opacities and energy generation. In the 1930's, Subrahmanyan Chandrasekhar used relativity theory to work out the structure of white dwarf stars, which, since their discovery in the middle to late nineteenth century,

had at best been a stellar enigma and which many astronomers considered to be a physical impossibility.

Continuing through the 1930's and into the 1940's, the problem of stellar energy generation began to be addressed by advances in nuclear physics, which led to the idea that stars generate energy through nuclear fusion reactions; a few specific fusion reactions were suggested for main sequence stars. Work in nuclear physics also suggested the existence of neutron stars, though many thought that even if they did exist, they could never be detected. (It was not until 1967 that the discovery of pulsars, identified as rapidly rotating, highly magnetic neutron stars, confirmed the existence of neutron stars.)

The late 1950's saw the publication of a number of seminal articles and books. A long article by Margaret Burbidge, Geoffrey Burbidge, William A. Fowler, and Fred Hoyle titled "Synthesis of the Elements in Stars" (1957) traced the various processes by which heavy elements could be produced by nucleosynthesis in stars. Martin Schwarzschild's book *Structure and Evolution of the Stars* (1958) described in an accessible way the methods for computing models of stellar interiors. About this time, electronic computers become more readily available, and Schwarzschild's book spurred increased use of them to rapidly calculate many models for the interiors of stars of different masses and in different evolutionary stages. His basic methods can still be applied today to personal computers and even programmable calculators.

The year 1957 also saw the dawn of the space age with the launch of Sputnik 1, the first artificial Earth satellite, by the Soviet Union. Visible light is distorted as it passes through Earth's turbulent atmosphere, resulting in images that are somewhat blurred. One way to get around this problem and produce sharper images is to observe from satellites above our atmosphere. The Hubble Space Telescope, placed in orbit at an altitude of 600 kilometers in 1990, has supplied incredible high-resolution images of various stages of stellar evolution. Since observing time on space telescopes is limited and sparingly assigned, techniques have been developed for ground-based telescopes to provide sharper images. In speckle interferometry, many short exposure images (with little blurring due to atmospheric turbulence, since the exposure is so short) are recorded electronically and then combined in a computer to produce a sharper final image. In adaptive optics, sensors monitor the effects of atmospheric turbulence and control small motors that slightly alter the telescope optics to compensate for it and reduce the blur. These techniques have given new life to large ground-based telescopes.

Observations from space are essential, however, for observations in certain nonvisible portions of the electromagnetic spectrum, because Earth's atmosphere is opaque to much of the electromagnetic spectrum, including gamma rays, X rays, and most ultraviolet and infrared radiation. Gamma rays, X rays, and ultraviolet radiation are emitted by hot objects; infrared radiation is emitted by cool objects. It is necessary to observe in all these wavelength regions to study both the cool births and the hot deaths of stars. Placing astronomical instruments on board satellites orbiting outside Earth's atmosphere has allowed astronomers to observe over all parts of the electromagnetic spectrum. Infrared observations are revealing regions of star formation and the embryonic stars developing there. Around many of them it is possible to detect disks in which planets might eventually form. Ultraviolet and X-ray observations have revealed that stars lose much more mass during the giant and supergiant stages of their lives than was previously thought, so computer models of stellar evolution are being revised to account for this discovery.

George E. McCluskey, Jr.

FURTHER READING

Asimov, Isaac. *The Exploding Suns: The Secrets of the Supernovas*. New York: E. P. Dutton, 1985. An excellent book, by the well-known physicist and author of science fiction and popular science books, which describes how massive stars die. Discusses the evolution of stars and explains how some of them reach their ultimate fate—the cataclysm of a supernova explosion. A highly accessible treatment.

Chaisson, Eric, and Steve McMillan. *Astronomy Today*. 6th ed. New York: Addison-Wesley, 2008. Very well-written college-level textbook for introductory astronomy courses. Contains a thorough description of the stages of stellar evolution, complete with transparent overlays.

Cohen, Martin. *In Darkness Born: The Study of Star Formation*. New York: Cambridge University Press, 1988. The author, a recognized authority in the field of star formation, has written the first general survey of the subject. The book discusses how stars form from interstellar material and how astronomers can use both ground-based and space-based astronomical instruments to observe this intriguing process. Appropriate for the person with little scientific background. Includes many black-and-white plates.

Cooke, Donald A. *The Life and Death of Stars*. New York: Crown, 1985. Offers general background on the subject and provides an overall picture of the Galaxy. There follow chapters on how stars form, how they shine and change as they burn their fuel supplies, and finally how they die as all energy sources are depleted. For the general reader.

Fraknoi, Andrew, David Morrison, and Sidney Wolff. *Voyages Through the Universe*. 3d ed. New York: Brooks/Cole, 2006. A well-written, thorough college textbook for introductory astronomy courses. Includes a good basic overview of stellar evolution.

Freedman, Roger A., and William J. Kaufmann III. *Universe*. 8th ed. New York: W. H. Freeman, 2008. College-level introductory astronomy textbook offering a good description of stellar evolution.

Genet, Russell M., Donald S. Hayes, Douglas S. Hall, and David R. Genet. *Supernova 1987a: Astronomy's Explosive Enigma*. Mesa, Ariz.: Fairborn Press, 1988. The full story of the brightest supernova in almost four hundred years: Supernova 1987a, discovered in March, 1987. This book discusses the evolution of this massive star from birth to its explosive demise in a supernova explosion. The book, which has more than one hundred illustrations, introduces the reader to the discoverers of Supernova 1987a and explains how astronomers used both ground- and space-based observatories to study this event.

Greenstein, George. *Frozen Star*. New York: Charles Scribner's Sons, 1983. Devoted to neutron stars, pulsars, and black holes, this book is clearly written and contains many diagrams. The structure of a neutron star is clearly illustrated, and the discussion of black holes is very helpful. To illustrate these bizarre objects, the author takes his readers on an imaginary trip to a pulsar and discusses what would happen if the Sun became a black hole.

Kippenhahn, Rudolf. *One Hundred Billion Suns: The Birth, Life, and Death of the Stars*. New York: Basic Books, 1985. Kippenhahn, who helped to pioneer the computer calculation of stellar evolution in the 1960's, has written an authoritative review of the subject of the life cycle of stars. The book is written in a delightful style and is one of the best treatments of the subject. Well illustrated.

Schneider, Stephen E., and Thomas T. Arny. *Pathways to Astronomy*. 2d ed. New York: McGraw-Hill, 2008. Very thorough college textbook for introductory astronomy courses. Divided into lots of short sections on specific topics. Contains a thorough discussion of the stages of stellar evolution.

See also: Brown Dwarfs; Gamma-Ray Bursters; Main Sequence Stars; Novae, Bursters, and X-ray Sources; Nuclear Synthesis in Stars; Protostars; Pulsars; Red Dwarf Stars; Red Giant Stars; Supernovae; Thermonuclear Reactions in Stars; White and Black Dwarfs.

Sunspots

Category: The Sun

Sunspots are small areas on the solar photosphere that appear darker than their surroundings because they are not as hot. They occur at sites with strong magnetic fields, which are their cause. Sunspots vary in number over an eleven-year cycle, and their associated magnetic fields switch polarity with each eleven-year cycle, so the full magnetic cycle is twenty-two years.

OVERVIEW

Sunspots appear as darker spots on the brighter photosphere of the Sun. Their temperature is about 1,000 to 1,500 kelvins cooler than the rest of the photosphere (which is about 5,800 kelvins); as a result, sunspots are only about one-third as bright, so they look dark in comparison. The spots have diameters ranging from a few hundred kilometers to occasionally more than 100,000 kilometers. Generally, smaller spots last only a few days, while larger ones linger for several weeks. At any given time, the Sun may display lots of sunspots or it may have none at all. Their numbers cyclically increase and decrease with a period of approximately eleven years. Especially large sunspots could occasionally be seen with the unaided eye, although the Sun should never be observed with the naked eye or directly through devices such as cameras or telescopes; such observation will severely damage the eyes.

In 1843, Samuel Heinrich Schwabe (1789-1875), a German pharmacist and amateur astronomer, announced after seventeen years of recording sunspots that the number of spots increased and decreased over an eleven-year cycle. At the beginning of each sunspot cycle, spots first appeared at latitudes around 40° north and south of the solar equator, progressing down to latitudes around 5° north and south at the end of that cycle; then the next cycle began.

A few years later, in 1859, the British astronomer Richard Carrington found that the motion of sunspots across the solar disk depends on their latitude. He concluded that the rotation period of the Sun varies with latitude, ranging from about 25 days near the equator to about 28 days at 40° north or south latitude. (This latitude range is the band in which sunspots are commonly seen. We now can measure the Sun's differential rotation in other ways, and have found that it increases to about 33 days at 75° north or south latitude.)

In 1908, George Ellery Hale, the American astronomer responsible for establishing Yerkes, Mount Wilson, and Palomar observatories, found that absorption lines in the spectra of sunspots were split into two or more components. He attributed this to the Zeeman effect, which describes the splitting of spectral lines in

811

the presence of strong magnetic fields. It had been noted that sunspots usually occur in close pairs, one ahead of the other in the direction of solar rotation. Hale discovered that the two spots in a pair have opposite magnetic polarities; that is, the magnetic field lines coming out of one spot return into its neighbor. Moreover, the leading spot in all the spot pairs in one solar hemisphere (north or south of the equator) have the same magnetic polarity, and the polarity of the leading spots is opposite in the other solar hemisphere. In 1924, Hale announced that the magnetic polarities of spot pairs reverse with each eleven-year sunspot cycle, so the complete magnetic cycle is twenty-two years, twice as long as the sunspot cycle.

Each sunspot has a central darker region called the umbra, which is surrounded by an outer not-as-dark region called the penumbra. The temperature is as low as 4,000 to 4,500 kelvins in the umbra and averages 5,000 to 5,500 kelvins in the penumbra, compared to around 5,800 kelvins in the adjoining photosphere; consequently the umbra is only about one-third as bright as the photosphere, so it looks dark in comparison. Within the central umbral region of the sunspot, the magnetic field is on the order of 3,000 gauss (0.30 tesla), strong enough to deflect away hot bubbles of ionized gas rising from the bottom of the Sun's convection zone up toward the photosphere. This inhibits the transfer of heat from below, making the spot cooler and thus darker. Furthermore, the cooler, denser gas in the umbra sinks, drawing in gas from inside and outside the surrounding penumbra.

The diameter of the umbra seldom exceeds 20,000 kilometers (although an exceptionally large one might on rare occasions exceed 100,000 kilometers), limiting its depth to about the same distance and large-scale mass movement to about half its size. Sunspots and their associated fields appear to be anchored at this depth. Vertical flow of matter downward in the umbra is found to be limited to 25 meters per second, while the horizontal flow from the penumbra into the umbra does not exceed 50 meters per second. Sunspots also show irregular patterns of bright points called "umbral dots" or "umbral granulation," in addition to solar filigrees caused by delicate, small-scale movement of magnetic field elements.

The average magnetic field strength in the penumbra is about one-half the field of the central umbra. The penumbral magnetic field has fine horizontal structures, giving the region a filamentary overlapping white and gray appearance. Penumbral matter flow is nearly horizontal, inward toward the umbra, with progressively decreasing velocities farther out.

"Pores," smaller regions with dark umbrae, also occur, having magnetic fields on the order of 2,000 gauss (0.20 tesla). "Mag-

Some very large sunspots can be seen in this 2001 image from the Solar and Heliospheric Observatory. (NASA)

netic knots" are compact magnetic structures, having fields on the order of 1,000 gauss (0.1 tesla).

The dynamo theory of solar magnetic fields was developed by Eugene N. Parker, Horace W. Babcock, and others in the 1950's to explain the Sun's magnetism and associated phenomena, such as sunspots, flares, and prominences. The outer zone of the Sun's interior, below its visible surface or photosphere, is in a state of convective motion; the convective cells that form are responsible for the observed boiling and bubbling (called granulation) on the solar surface. Along the granule boundaries, tubular magnetic fields emerge and disperse into the solar atmosphere. Basically, it is theorized that the rising and sinking motions of the ionized gas (plasma)

Sunspots on the solar surface are surrounded by glowing arcs of gas. (NASA)

in the convective zone constitute electrical currents that induce magnetic fields through a dynamo process. These magnetic fields get distorted and intensified by the Sun's differential rotation. As described earlier, the solar rotation period is about 25 days at the equator, progressively increasing to 28 days at 40° north and south latitudes, and 33 days at 75° north and south latitudes. The general background dipole magnetic field, produced by convective motion and "frozen" into the ionized gas of the convection zone, gets twisted and wound up around the Sun by this differential rotation, thus increasing the local magnetic field strength.

At the beginning of an activity cycle, the Sun's magnetic field is generally that of a dipole, with magnetic field lines running from near one rotational pole to the other. Differential rotation, after a few rotations, stretches these field lines that are frozen into the ionized gas near the surface so that they wrap around the Sun parallel to the equator. As the magnetic field lines are wound more closely together, the local field strength increases. The convective

motion of the plasma further twists the field lines like strands of rope. The increased intensity of the magnetic field lines at higher latitudes finally makes them break through the solar surface as sunspots roughly along the east-west direction. As the field weakens at higher latitudes by the dissipation of energy through sunspots, the field intensifies toward the equator, forcing the locations of new sunspots to migrate toward the equator. The sunspot cycle ends with the last spots fading near the equator and the magnetic field there weakening.

The next solar activity cycle begins with the dipole field reversed from the preceding activity cycle. Thus the solar magnetic cycle has a period double that of the activity cycle. The solar dynamo theory satisfactorily explains the incidence and location of sunspots during each activity cycle, and the reversal of spot polarity in the two solar hemispheres from one activity cycle to the next. However, the complexity of the proposed mechanisms warrants further investigation.

APPLICATIONS

The sunspot cycle seems to be related to long-term temperature fluctuations on Earth. There appear to have been long episodes of solar inactivity, as evidenced by historical records of reduced numbers of sunspots and other related manifestations, such as smaller coronas seen during total solar eclipses and fewer auroras. These generally occurred during times when the Earth's climate was cooler. An accumulating body of information indicates that the Sun may be a slightly variable star, possibly with complex periodicity. Such a prospect for solar luminosity changes would have far-reaching ramifications for life on Earth.

It is believed that the magnetic fields of the major planets, including the Earth, are products of a dynamo mechanism similar to that operating in the Sun. The study of sunspot cycles led to the solar dynamo theory, which in turn has contributed to progress in understanding planetary magnetic fields.

The Sun is the nearest star. Everything learned about the Sun can be extended to most other stars. Spectroscopic studies of magnetic fields in other stars have revealed that many stars possess fields similar to those found on the Sun, with intensities ranging from 2,000 to 20,000 gauss (0.20 to 2.00 tesla). The observed stellar dipolar fields appear to vary, with periods ranging from half a day to decades. Magnetic field activity, presumably similar to that of the Sun, appears to be stronger in younger stars, which rotate more rapidly in comparison with older and more slowly rotating stars. For example, the younger stars in Orion have magnetic field intensities three times that of similar-sized older stars.

Among many stars, "starspots" are suspected to exist. They tend to occur mainly in relatively younger stars. It is possible that sunspot activity and associated magnetic field intensity may have been stronger at an earlier time in our solar system, when the Sun may have been spinning more rapidly.

CONTEXT

Recorded descriptions of sunspots date back to at least the fourth century B.C.E. In 1612, Galileo (1564-1642) was the first to observe them with a telescope. He noticed their positions on the Sun's bright disk changed over a few days, and from this he concluded that the Sun rotates about once a month. These observations, along with many others Galileo made with his telescopes, contradicted the prevailing views held then by scholars throughout Europe that dated back to the ancient Greeks.

Today, much has been learned about the Sun and its physics through both Earth-based and, more important, space-based observatories and instruments. Solar research is concerned with many broad areas of the Sun's structure and observable activity, such as sunspots and the solar magnetic cycle. Solar activity has been determined to be periodic, marked by the occurrence of sunspots and related phenomena, and caused by complex variations in associated magnetic field intensities. It has become possible to study the Sun at virtually all wavelengths of the electromagnetic spectrum (from high-energy gamma rays to low-frequency radio waves) with the aid of telescopes on the ground and spacecraft outside the Earth's atmosphere. The advent of supercomputers has permitted very detailed modeling of the structure and development of sunspots and other solar activity. Advances in plasma physics and magnetohydrodynamics continue to refine our understanding of the solar dynamo mechanism and the related activity cycle, including sunspots.

V. L. Madhyastha

FURTHER READING

Balogh, André, Louis J. Lanzerotti, and S. T. Suess, eds. *The Heliosphere Through the Solar Activity Cycle*. New York: Springer, 2007. Investigates the Sun during the sunspot cycle to analyze correlations between the cycle and the heliosphere. Discusses theory on the subject as well as actual data from spacecraft.

Brody, Judit. *The Enigma of Sunspots: A Story of Discovery and Scientific Revolution*. Edinburgh, Scotland: FlorisSunspots, 2002. Provides an overview of the history of sunspot research, from earliest times to the present.

Chaisson, Eric, and Steve McMillan. *Astronomy Today*. 6th ed. New York: Addison-Wesley,

2008. Very well-written college-level textbook for introductory astronomy courses. Good description of sunspots and the Sun's magnetic field.

Fraknoi, Andrew, David Morrison, and Sidney Wolff. *Voyages to the Stars and Galaxies.* Belmont, Calif.: Brooks/Cole-Thomson Learning, 2006. A well-written, thorough college textbook for introductory astronomy courses. Good description of sunspots and the Sun's magnetic field.

Freedman, Roger A., and William J. Kaufmann III. *Universe.* 8th ed. New York: W. H. Freeman, 2008. College-level introductory astronomy textbook that includes a good description of sunspots and the Sun's magnetic field.

Gibson, Edward G. *The Quiet Sun.* NASA SP-303. Washington, D.C.: Government Printing Office, 1973. Although somewhat out of date, this NASA publication contains many useful photographs, illustrative graphs, and charts. Technical, but written so that a nonscientist will not experience any difficulty with the text.

Giovanelli, Ronald G. *Secrets of the Sun.* New York: Cambridge University Press, 1984. This book, which contains many photographs of sunspots and their progression in the eleven-year cycle, is specially written for scientists and lay readers alike. A rare example of a serious attempt to introduce the reader to a complex set of ideas.

Jordan, Stuart, ed. *The Sun as a Star.* NASA SP-450. Washington, D.C.: Government Printing Office, 1981. Although technical, this volume is a complete source of solar physics. Includes a large number of references at the end of each chapter. A nontechnical reader can skip over the occasional mathematical equations without experiencing a sense of loss in logic or continuity.

Maunder, E. Walter. "A Prolonged Sunspot Minimum." *Knowledge* 17 (1894): 173-176.

_____. "The Prolonged Sunspot Minimum, 1645-1715." *Journal of the British Astronomical Society* 32 (1922): 140ff. These two papers are Maunder's original publications arguing that there was a prolonged minimum in sunspot activity in the late seventeenth century. Together these papers form a seminal work on sunspots.

Newkirk, Gordon, Jr., and Kendrick Frazier. "The Solar Cycle." *Physics Today* 35 (April, 1982): 25-34. Using the solar magnetic-dynamo model, the authors discuss eleven-year and twenty-two-year solar cycles. Considers a possible clock in the Sun, the variation of luminosity correlated with sunspot activity, and photospheric pulsation.

Parker, E. N. "Magnetic Fields in the Cosmos." *Scientific American* 249 (August, 1983): 44-54. The author, an authority on the subject of cosmic magnetic fields, presents in precise, qualitative terms the theory of the solar dynamo, the mechanism that, combined with the differential rotation of the Sun with latitude, affords a natural explanation of the intense magnetic fields associated with sunspots, flares, and other aspects of the sunspot cycle.

Schneider, Stephen E., and Thomas T. Arny. *Pathways to Astronomy.* 2d ed. New York: McGraw-Hill, 2008. This introductory astronomy textbook offers several sections on sunspots and the Sun's magnetic field.

Wentzel, G. Donat. *The Restless Sun.* Washington, D.C.: Smithsonian Institution Press, 1989. This excellent volume contains photographs, illustrations, and information on all types of solar activities. Geared for beginners and nonspecialists.

See also: Auroras; Coronal Holes and Coronal Mass Ejections; Earth's Magnetic Field at Present; Earth-Sun Relations; Solar Chromosphere; Solar Corona; Solar Evolution; Solar Flares; Solar Geodesy; Solar Infrared Emissions; Solar Interior; Solar Magnetic Field; Solar Photosphere; Solar Radiation; Solar Radio Emissions; Solar Seismology; Solar Structure and Energy; Solar Ultraviolet Emissions; Solar Variability; Solar Wind; Solar X-Ray Emissions.

Supernovae

Category: The Stellar Context

Supernovae are stars that explode violently at the end of their lives. They are spectacular events, emitting light at rates up to ten billion times greater than the Sun. They are responsible for the synthesis of chemical elements heavier than iron and dispersing them through interstellar space. Shock waves from supernovae can act as triggers for new star formation.

OVERVIEW

Novae and supernovae have been observed and recorded since ancient times as "new" stars that appeared suddenly in the night sky (hence the name "nova" which is from the Latin *novus*, "new"). Originally there was no distinction between novae and supernovae, since superficially both display the same behavior: a nova or "new star" suddenly appears in the sky, brightening dramatically in no more than a few hours, and then slowly fades from view over a period of weeks to months. It was not until the 1920's that astronomers began to realize there was a difference; some of the outbursts were much brighter intrinsically than the others, up to a million or more times. The "super" outbursts came to be called supernovae. Today it is known that both novae and supernovae are not new stars but explosive events involving stars in the last stages of their lives. A nova is a relatively minor hydrogen fusion explosion on the surface of a white dwarf star, while a supernova is a much more violent explosion that tears a star apart.

A supernova is one of nature's most spectacular events. The total energy released is on the order of 10^{46} joules. Most of the energy is carried by a flood of neutrinos produced in the explosion, while a much smaller part goes to drive the explosive outburst that destroys the star. The electromagnetic radiation emitted by the blast is only on the order of 10^{43} joules, but this is almost as much as the Sun emits during its entire life. At its peak, a supernova shines as bright as 100 million to 10 billion Suns.

Supernovae are rare events. The last one to occur in our Milky Way galaxy was observed in the year 1604. Thus historical records of supernovae are invaluable resources for modern astronomy, even though those who recorded their observations long ago had no real understanding of what they witnessed. Descriptions of past supernovae can reveal when and where they occurred, how bright they got, and how long they remained visible in the sky. Their frequency (or rather infrequency) provides estimates of the death rates among stars.

Chinese records of celestial events date back to several hundred years B.C.E. They believed events in the sky were omens that forecast events on Earth, and woe to the astronomer who failed to correctly warn of an impending catastrophe. They recorded "broom stars" (comets with tails that "swept" the sky like brooms) and "guest stars" (novae and supernovae that suddenly appeared and eventually faded slowly away, like visiting guests). "Guest stars" were sighted and recorded in what would have been these years (all C.E.) in our current calendar system: 185, 386, 393, 1006, 1054, 1181, 1572, and 1604.

The supernova that appeared in May, 1006, was particularly bright. More than twenty reports of it have been identified from all around the world. Because of its position in the southern sky and thus below the horizon in northern Europe, the best records of it come from Chinese, Japanese, and Middle Eastern sources. It was a brilliant spot of light perhaps one hundred times brighter in appearance than Venus, visible during the day and bright enough to cast shadows at night. A gaseous remnant of the explosion, now very faint, was identified in the 1970's in the constellation of Lupus, the wolf. The rate of expansion of its filaments indicates it is about a thousand years old, consistent with its being a remnant of the supernova of 1006.

The supernova that appeared in July, 1054, was probably brighter than Venus (though not as bright as the supernova in 1006), visible in the daytime sky for nearly a month and in the night sky for two years. Interestingly, this bright object, although surely visible, went unmentioned in European historical records, possibly because its appearance in the sky somehow conflicted with church doctrine in medieval

Europe. The event definitely was recorded in China and the Middle East. A rock painting made at approximately that time by Native Americans in New Mexico shows a ten-pointed object next to a crescent and may represent this supernova when it first appeared next to a thin, waning crescent Moon. The descriptions of the position of this supernova match approximately the location of the Crab nebula in the constellation of Taurus the bull. The measured expansion rate of the Crab nebula is consistent with its origin in the supernova of 1054. This supernova left behind, in the center of the Crab nebula, a rapidly spinning neutron star called a pulsar.

On November 11, 1572, a supernova brighter than Venus appeared in the constellation of Cassiopeia, gradually fading from view over the next eighteen months. Most European scholars at that time still adhered to the teachings of Aristotle and Plato that the heavens were perfect and unchanging, and therefore this bright new thing could not be a star but had to be something close to Earth, perhaps in the air overhead. The young Danish astronomer Tycho Brahe made careful measurements of its position for as long as it was visible. Since it showed no motion relative to other stars, he concluded it was far from Earth, like the stars, but had to be a new star no one had seen before. (Brahe's work so impressed King Frederick II of Denmark that he financed the construction of two observatories for Brahe on the Danish island of Ven. It was from these observatories that Brahe measured—without a telescope, since the telescope had not yet been invented— the changing positions of the planets that Johannes Kepler used to derive his three laws of planetary motion.) An expanding shell of gas ejected by the 1572 supernova and plowing into the surrounding interstellar gas has been detected at visible, radio, and X-ray wavelengths. No central stellar object has been found, suggesting that the star was completely destroyed by the explosion.

Kepler, who had been Brahe's assistant for a year prior to his death in 1601, sighted a supernova on October 17, 1604. He described the star as similar to Jupiter in brightness, with the color of a diamond. It remained visible through October, 1605, although the star briefly disappeared behind the Sun from November, 1604, until January, 1605. Sightings of this supernova were also recorded in China and Korea. A remnant from this supernova was first photographed in 1947. This was the last supernova to be observed in the Milky Way.

Fortunately, because supernovae are so bright, hundreds have been observed in galaxies outside our own. Modern measurements of

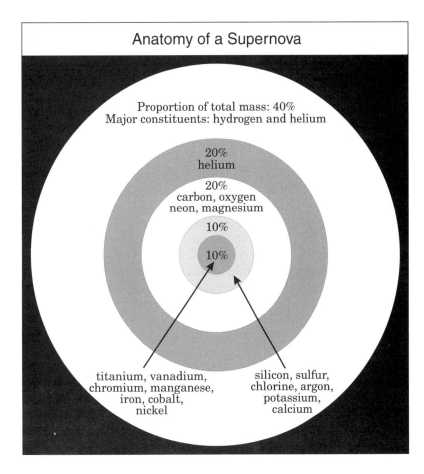

Anatomy of a Supernova

Proportion of total mass: 40%
Major constituents: hydrogen and helium

20% helium

20% carbon, oxygen neon, magnesium

10%

10%

titanium, vanadium, chromium, manganese, iron, cobalt, nickel

silicon, sulfur, chlorine, argon, potassium, calcium

their properties combined with historical records of past supernovae have yielded our current understanding of supernovae.

Several types of supernova spectra can be identified. Type II supernova spectra have emission lines of hydrogen, while Type I supernova spectra do not. Type I spectra are subdivided into Type Ia, with a strong absorption line of ionized silicon; Type Ib, with a strong absorption line of neutral helium; and Type Ic, with neither of these absorption lines. There are two distinct ways that supernovae explode. Types II, Ib, and Ic are core-collapse supernovae, in which the iron cores of massive supergiant stars collapse, triggering the supernova explosion. In contrast, Type Ia supernovae occur in white dwarf stars in close binary systems that acquire so much matter from their companions that their mass is raised above the Chandrasekhar limit (the maximum mass a white dwarf star can have); the whole white dwarf then collapses and explodes.

To understand how stars explode as supernovae, one must understand the earlier stages of their lives. Stars form from nebulae, clouds of gas and dust in interstellar space, by gravitational contraction. When the center of a contracting protostar reaches a temperature of a few million kelvins, hydrogen nuclei begin fusing into helium nuclei in the star's core, releasing energy. This stops the contraction, and the star becomes a stable main sequence star. It remains a stable main sequence star as long as it has hydrogen in its core to fuse into helium. This is the longest-lasting stage in a star's life. (The Sun is a main sequence star; it has been one for most of the 4.5 billion years of its existence, and it probably has enough hydrogen left in its core to last several billion years more.) When hydrogen is exhausted in its core, the star expands to become a red giant or supergiant and begins to fuse helium into carbon and maybe oxygen in its core.

Stars with up to about 8 solar masses cannot go further in nuclear fusion reactions. When their core helium is exhausted, they puff off their outer layers as expanding shells of gas called planetary nebulae (a purely descriptive term, having nothing to do with planets), and their cores become electron-degenerate white dwarfs (meaning their electrons are packed as tightly as quantum mechanics allows), shining only because they are hot. As they radiate their energy away, they grow cooler and fainter, eventually becoming cold, dark black dwarfs. This usually is the end of life for stars of this size.

However, a white dwarf in a close binary star system can end its life in a much more dramatic way if its binary companion transfers enough matter to the white dwarf. Since white dwarfs are electron-degenerate, there is a maximum mass they can have, the Chandrasekhar limit, which equates with about 1.4 solar masses. If enough matter is transferred from the companion star to the white dwarf so its mass is raised above the Chandrasekhar limit, its self-gravity exceeds the pressure of the degenerate electrons and it collapses in on itself. The rapid compression drives the temperature up to billions of kelvins, igniting a whole series of nuclear fusion reactions that release tremendous amounts of energy. This is a Type Ia supernova. At its peak, it shines as brightly as nearly 10 billion Suns. The star apparently is completely destroyed, its matter flung into interstellar space as a cloud of rapidly expanding gas containing the heavier chemical elements produced in the fusion reactions. The companion may become a runaway star, no longer held in orbit by the mutual gravitational attraction between the two stars. The last two supernovae seen in the Milky Way—in 1572 and 1604—apparently were Type Ia.

A different kind of supernova explosion, called a core-collapse supernova, is in store for stars more massive than about 10 solar masses. They are massive enough to continue to form heavier elements through various nuclear fusion reactions while they are supergiant stars. Each element formed in one fusion reaction becomes the fuel for the next reaction. Eventually the supergiant builds up a small central core of iron surrounded by shells of lighter nuclei, similar to the thin layers in an onion. In each of these shells, nuclear fusion reactions form heavier nuclei that become part of the next shell inward, except that iron is the heaviest nucleus that can form in fusion reactions that release energy. To form heavier nuclei than iron requires the input of energy. The iron core shrinks, trying to tap another nuclear fuel, but none is available. The temperature climbs as high as a few hundred billion kelvins, causing the iron nuclei to begin to break up. When the central density reaches about 400 billion times the density of water, free electrons are forced into the nuclei, where they combine with protons to form neutrons and neutrinos. The loss of free electrons reduces the pressure so much that the core collapses at speeds as fast as one-fourth to one-third the speed of light, shrinking from a diameter about as large as the Earth to less than 20 kilometers in less than one second. The outer layers collapse inward on top of the core, where some of the matter rebounds and collides with the still infalling gas above. This collision heats the gas to billions of kelvins, igniting fusion reactions that release energy. Even more energy is added by the outward flow of neutrinos. Although neutrinos do not ordinarily interact very much with other forms of matter, the density near the core is so great that the neutrinos transfer some of their energy to the collapsing outer layers, reversing their fall and blasting them outward at speeds greater than 10,000 kilometers per second. This core-collapse supernova has a spectrum of Type II, Ib, or Ic. (The differences in the spectra depend on how much of their outer layers the stars lost as a result of strong stellar winds while they were supergiants.) The huge amounts of energy now available allow fusion reactions to occur that form the chemical elements heavier than iron. The outer layers become a rapidly expanding supernova remnant that disperses into interstellar space the heavy elements formed both in the supergiant before it exploded and in the supernova explosion itself.

If the neutron-rich core at the center of this explosion is less than about 3 solar masses, it becomes a neutron star, a neutron-degenerate sphere about 20 kilometers in diameter with an average density of about a billion metric tons per cubic centimeter or about 10^{15} times the density of water. (Neutron degeneracy means that the neutrons are packed as tightly as quantum mechanics allows.) Neutron stars typically rotate very rapidly, up to several hundred times per second, a consequence of conserving their angular momentum as they collapsed. Their magnetic fields become concentrated, with two very strong magnetic poles that do not necessarily lie at the two rotational poles. Electromagnetic radiation beamed outward from the two magnetic poles can appear to blink on and off as seen from Earth in a "lighthouse beacon" effect as the star rotates, if the rotation and magnetic axes are aligned the right way. Such blinking neutron stars are called pulsars, and their blinking can be detected throughout the electromagnetic spectrum, from radio waves (where they were first discovered) to visible light to X rays. There is a pulsar spinning 30 times per second at the center of the Crab nebula, produced by the Type II supernova in 1054. The fastest-spinning pulsar known rotates more than 600 times per second.

However, if the core at the center of the explosion exceeds about 3 solar masses (called the Oppenheimer-Volkoff limit), neutron degeneracy will not stop its ultimate collapse to a black hole. "Black hole" is the name given by the physicist John Wheeler in 1969 to matter with such strong gravity that nothing, not even light, can escape from it. Although black holes cannot be observed directly in any part of the electromagnetic spectrum, precisely because no form of electromagnetic radiation can escape from them, they can be detected by the gravitational effects they have on matter nearby. Several likely stellar-mass black holes, the kind that can be formed from supernovae, have been identified, and it is strongly inferred that a black hole lies at the center of the Milky Way.

The final mass of the core, whether it is below or above the Oppenheimer-Volkoff limit, is what determines whether a neutron star or black hole is left by the supernova explosion. Computer models of the late stages of massive stars suggest that the final core mass is determined by the mass the star had when it was born. Stars with initial masses less than about 20 to 40 solar masses probably produce neutron stars, while those with initial masses greater than 20 to 40 solar masses probably produce black holes.

APPLICATIONS

Because supernovae emit so much light, they can be observed at very large distances, up to billions of light-years away in very distant galaxies. Core-collapse supernovae (Types II, Ib, and Ic) come from stars with a wide range of large masses and consequently somewhat different interior structures and histories. The explosions are not all identical and thus there is a range of peak luminosities; they do not all become equally bright. On the other hand, almost all Type Ia supernovae reach nearly the same peak luminosity, nearly 10 billion solar luminosities, because they have essentially identical progenitors: white dwarfs with masses that just exceed the Chandrasekhar limit of 1.4 solar masses. (Occasionally a Type Ia supernova may noticeably exceed the normal maximum brightness. This is thought to be due to a collision between the white dwarf and its binary companion.) The vast majority of Type Ia supernovae therefore make very reliable standard candles; their assumed "standard" real brightness, combined with their measured apparent brightness, yields their distance. If a Type Ia supernova is observed in a galaxy, then the distance to the galaxy can be determined. Type Ia supernovae were used to find the distances of galaxies in two independent projects that show the expansion of the universe is accelerating.

Supernova SN 1987A was first seen before dawn on February 24, 1987. It was bright enough to see with the unaided eye, since it was relatively close, only 160,000 light-years away in the Large Magellanic Cloud (LMC), a small galaxy that forms a companion to our own, much larger Milky Way galaxy. SN 1987A provided the first opportunity to study a relatively nearby supernova with a whole array of modern instruments. It was also possible to identify its progenitor in earlier images of the LMC, so astronomers know what kind of star produced it: a blue supergiant with about 20 solar masses. Hence, SN 1987A was a core-collapse, or Type II, supernova. Based on theory and computer models, astronomers formerly thought that core-collapse supernovae occurred in red supergiants; SN 1987 led astronomers to conclude that any massive supergiant could explode as a core-collapse supernova.

Neutrinos from the explosion were detected almost a day before the supernova was seen at visible wavelengths; the neutrinos were able to escape directly from the collapsing core, while the light was absorbed and reradiated by matter between the collapsing core and the star's surface, emerging almost a day later with the supernova shock wave. The time difference from the detection of the neutrinos to the first observed light from the supernova provided information on the size of the star just prior to the explosion, confirming that it was a blue supergiant rather than a red supergiant. Comparing the observed neutrino flux with the observed visible light, astronomers noted that the neutrinos carried more than one thousand times the energy of the visible light. Overall, less than 0.1 percent of the energy of the explosion was emitted in the form of visible light, about 1 percent of the energy was used to propel most of the star's matter outward, and the rest of the energy was carried away by neutrinos.

SN 1987A brightened rapidly at first, becoming a thousand times brighter in less than a day. Then it continued to brighten slowly, eventually reaching about 100 million solar luminosities after about a hundred days. In appearance it looked about as bright as some of the stars of the Little Dipper. Its peak luminosity was not as bright as would be expected for a Type II supernova produced by a 20-solar-mass supergiant. However, a blue supergiant is more compact and not as extended as a red supergiant of the same mass, so more energy would be needed to break up a blue supergiant, leaving less energy to appear as visible light.

For about the first forty days, the energy being radiated came from the explosion itself; afterward it came from the radioactive decay of unstable isotopes created in the explosion. The radioactive decays produced gamma rays that were absorbed by the expanding gas, heating it, and the energy was reemitted at visible wavelengths. Some of the gamma rays were detected directly by instruments in Earth orbit at the energies emitted by radioactive decays of nickel 56 and cobalt 56, isotopes expected in the aftermath of a supernova, thus confirming some theories about supernova nucleosynthesis.

Several months after the initial outburst of SN 1987A, spectroscopic observations with an infrared telescope flown at an altitude of 12,000 meters on a jet transport identified many more expected elements, including iron, nickel, cobalt, oxygen, neon, sodium, potassium, silicon, and magnesium. The intensity of the infrared spectral lines proved that the quantities of these elements were larger than what could have been present initially in the star at birth.

A few years later, a bright ring of gas about one light-year in diameter was observed around the star, and it remained bright for several years. This was not a ring of material ejected by the supernova. Rather it was matter blown off the star by stellar winds about 30,000 years earlier, when it was a red supergiant. The ring lit up when high-speed gas ejected by the supernova plowed into it, compressing and heating the matter in the ring. Meanwhile, an expanding "blob" was observed at the center of the ring, thought to be slower-speed gas ejected by the explosion.

CONTEXT

Supernovae are significant for a number of reasons. Perhaps the most important is that we owe our existence to massive stars that exploded as supernovae long ago. The big bang that created the universe about 13 to 14 billion years ago produced significant amounts only of hydrogen and helium. All the chemical elements heavier than helium—the carbon in our cells, the nitrogen and oxygen in the air we breathe, the silicon and iron in the Earth, the gold and silver in our jewelry, and everything else heavier than helium—were produced by nuclear fusion reactions in massive supergiants and the supernovae that destroyed them. These elements were dispersed into interstellar space by the supernova blasts, where they enriched the clouds of gas from which new generations of stars and accompanying planets formed. The first stars to form after the big bang would have contained only hydrogen and helium. Any planets that developed around them would have been gaseous planets composed of hydrogen and helium, maybe similar to Jupiter and Saturn, but not rocky/metallic planets like Earth. By the time our Sun and solar system formed about 4.5 billion years ago, enough massive stars had already run through their life cycles and exploded as supernovae that the nebula in which our Sun and solar system developed had become enriched in elements heavier than helium, so

that they accounted for about 2 percent of its mass. This provided the raw materials for rocky/metallic planets like Earth and for life as we know it.

Second, supernova shock waves propagating through large interstellar molecular clouds may provide the trigger for new bursts of star formation, compressing the gas and starting the gravitational contraction that leads to new stars and solar systems. (Supernova shock waves are only one of several possible ways to initiate the gravitational contraction that forms new stars.)

Supernovae may contribute to the evolution of life on Earth in two ways. Supernovae probably are a major source of high-energy cosmic-ray particles. Trapped by the magnetic field of our Milky Way galaxy, they continually bombard Earth. Although most never reach the Earth's surface, enough do to possibly cause genetic mutations in the life-forms on Earth, resulting in the diversity of life here. Although most genetic mutations are not advantageous and some are outright deadly, life could not evolve and adapt to changing environmental conditions without a steady supply of mutations. If it were not for mutations, then all organisms today would be mere replicas of the initial simple forms of life.

However, if a supernova were to explode too close to us, the large burst of electromagnetic radiation and high-energy particles would pose a serious danger to life. A supernova within 50 light-years of our solar system probably would wipe out all life on Earth, and one within 100 light-years could result in major extinctions. Although the mass-extinction event 65 million years ago that wiped out the dinosaurs and many other forms of life was probably the result of an asteroid impact, other extinctions in the history of Earth might be due to nearby supernovae. A relatively minor extinction of some marine species about 2 million years ago might have been related to a supernova explosion about 120 light-years away. Of course, an extinction of only some species opens up new ecological niches to those organisms that do survive, which in turn can lead to rapid evolution of new species as the survivors rush to fill the environmental holes.

Supernovae are very rare. The last one seen in our Milky Way galaxy was in 1604 C.E. According to estimates, up to five supernovae may occur in the Milky Way Galaxy each century, most escaping detection because they are obscured by interstellar dust in the galactic plane. In other galaxies, the rate varies from several each century in the largest spirals to one every few centuries in the smallest galaxies.

Since supernovae are so rare, it is hard to predict when or where the next one is likely to occur or exactly what the progenitor star will look like. The star will be either a massive supergiant in its last stages before collapse or a white dwarf in a close binary system. Fortunately, the suspected possible candidates are sufficiently far away that they should not pose a threat to life. One of the closest is Canopus (Alpha Carinae), the second brightest star in appearance in the night sky, located about 200 to 300 light-years away. Two other possible candidates are located in the constellation of Orion the hunter: the red supergiant Betelgeuse (Alpha Orionis), about 430 light-years away, and the blue supergiant Rigel (Beta Orionis), about 770 light-years away. Betelgeuse, in particular, seems unstable and pulsates. Another possible candidate is the supergiant star in the Eta Carinae nebula. With a mass of more than 100 solar masses, it is very luminous and is expelling mass at a prodigious rate. It has had a number of outbursts; the one observed in 1843 is the largest single loss of mass that any star is known to have survived.

Michael L. Broyles

FURTHER READING

Asimov, Isaac. *The Exploding Suns*. New York: Dutton, 1985. This famous author of major science books and science fiction covers many aspects of supernovae, from the observations of earlier civilizations to modern theories. An intriguing chapter outlines the formation of elements in a supernova and describes how life may have been affected by such random catastrophes.

Cooke, Donald A. *The Life and Death of Stars*. New York: Crown, 1985. Spectacular photographs and diagrams superbly illustrate all chapters. The Crab nebula is among several case studies highlighted. The bright star in the Eta Carinae nebula is mentioned as a possible future supernova, along with several other stars.

Gribbin, John, with Mary Gribbin. *Stardust: Supernovae and Life, the Cosmic Connection*. New Haven, Conn.: Yale University Press, 2000. The authors explain Carl Sagan's famous assertion, "We are made of star stuff," from dust particles in nebulae through supernova explosions. Also covers the nucleosynthesis theory and the role supernovae play in it. For the general reader.

Gussinov, Oktay, Efe Yazgan, and Askin Ankay, eds. *Neutron Stars, Supernovae, and Supernovae Remnants*. New York: Nova Science, 2007. Describes the core-collapse theory for supernovae, remnants of supernovae that can be found in nearby galaxies, and the connection between neutron stars, supernovae, and remnants.

Marschall, Laurence A. *The Supernova Story*. New York: Plenum Press, 1988. The search for supernovae in the night sky is presented from the early records to those discovered in modern times. A documentary chapter on Supernova 1987A discovery is included, with emphasis on the observational techniques used.

Mezzacappa, Anthony, and George M. Fuller, eds. *Open Issues in Core Collapse Supernova Theory*. Hackensack, N.J.: World Scientific, 2005. Proceedings from the Institute for Nuclear Theory conference on supernovae and the core-collapse theory. Includes papers on computer simulation, quantum chromodynamics, magnetic fields during core collapse, and radiation diffusion. For scientists, graduate students, and advanced undergraduates.

Mobberley, Martin. *Supernovae and How to Observe Them*. New York: Springer, 2007. A how-to book for those intersted in observing supernovae using modest telescopes and charge-coupled devices to record the images.

Murdin, Paul, and Lesley Murdin. *Supernova*. New York: Cambridge University Press, 1985. An excellent collection of photographs and diagrams is included, with discussions of the ancient sightings. Tables are compiled for the magnitudes of Brahe's and Kepler's supernovae. A unique energy-flow diagram for a supernova illustrates the nuclear processes of the evolving star.

North, Gerald. *Observing Variable Stars, Novae, and Supernovae*. New York: Cambridge University Press, 2004. Includes a CD-ROM by Nick James. A guide for amateur astronomers on how to observe supernovae and variable stars. Includes information on telescopes, detectors, and observational techniques. For the general reader.

Wheeler, J. Craig. *Cosmic Catastrophes: Supernovae, Gamma-Ray Bursts, and Adventures in Hyperspace*. New York: Cambridge University Press, 2000. Discusses the birth and death of stars, as well as gravity, wormholes, and supernovae. For the general reader.

See also: Brown Dwarfs; Cosmology; Gamma-Ray Bursters; General Relativity; Hertzsprung-Russell Diagram; Interstellar Clouds and the Interstellar Medium; Lunar History; Main Sequence Stars; Neutrino Astronomy; Novae, Bursters, and X-Ray Sources; Nuclear Synthesis in Stars; Planetary Formation; Protostars; Pulsars; Red Dwarf Stars; Red Giant Stars; Stellar Evolution; Thermonuclear Reactions in Stars; Ultraviolet Astronomy; Universe: Expansion; White and Black Dwarfs.

T

Telescopes: Ground-Based

Category: Scientific Methods

In the age of the Hubble Space Telescope and other space-based observatories, ground-based telescopes are often perceived as antiquated. However, ground-based telescopes are still playing key roles in the search for extrasolar planets and other fields of cosmological research, and larger telescopes are always under construction or development.

OVERVIEW

Ground-based telescopes have a long history of advancing various areas of astronomy research. Starting in the 1960's, large ground-based telescopes (which, for the purposes of this discussion, are those telescopes situated in advantageous locations on Earth's surface and having apertures larger than 250 centimeters) were used to explore and further study objects in our solar system. For example, in the 1950's, the Martian atmosphere was believed to be thin, and, based on the intensity and polarization of Martian reflected light, Martian surface pressure was determined to be only 5 to 10 percent of that on Earth. Designers of preliminary Martian landers used this value when deciding whether a descent to the Martian surface should be effected by balloon, glider, or downward-pointing rocket engine. Also, the presence of this much atmosphere suggested that Martian life might exist, as there would be enough air to breathe (if free oxygen were present) and enough protection from ultraviolet light and extreme temperature changes.

However, a University of California scientist, Hyron Spinrad, using ground-based telescopes, found evidence that the correct value might be very different. When he was able to obtain good high-dispersion spectra using the Lick Observatory's 3-meter telescope in the early 1960's,

Spinrad concluded that the true value for the Martian surface pressure must be only about 5 millibars. This is only 0.5 percent of Earth's atmospheric pressure—a very small value. New designs had to be considered for Martian spacecraft, and at that time it seemed less likely that anything remotely resembling terrestrial life could exist under such harsh conditions.

The oldest observatory in the United States is Mount Wilson, which overlooks Pasadena, California. George Hale founded the observatory in 1904 and used funding from the Carnegie Institute of Washington, D.C., to create a solar research laboratory. The Snow Solar Telescope was moved to Mount Wilson in 1904 from Yerkes Observatory in Wisconsin. Long-term studies of the Sun began in 1905. Hale ordered the mirror for his 100-inch telescope in 1906. Two years later, the 60-foot solar telescope became operational. Hale used it when he detected the Sun's magnetic field. In May, 1912, a solar telescope was mounted atop a 60-foot-tall tower, and work began on the foundation for Hale's 100-inch mirror telescope. The latter telescope achieved first light on November 2, 1917, to much fanfare. Several discoveries of great importance in the history of astronomy were made using the telescopes at Mount Wilson. These include detecting the first Cepheid variable star; measuring the distance to M31 (the Andromeda nebula, now called the Andromeda galaxy), thereby proving that the Milky Way is but one of many galaxies in the greater universe; determining extragalactic distances and providing evidence that the universe is expanding; detecting remnants of the 1604 supernova; and discovering four additional satellites of Jupiter.

Mount Wilson remains one of the top observatories conducting astronomical research. In 1995, it became home to Georgia State's CHARA stellar interferometer array. The 150-foot-tall tower telescope is operated by the Astronomy and Astrophysics Division of the University of California at Los Angeles. It is used to study solar magnetic activity and look for long-term changes that may provide an understanding of the Sun's variability and its implications for the solar radiation received by the Earth and its biosphere. The solar telescope is part of a worldwide network monitoring the Sun's surface (its photosphere). The 60-inch telescope, which became operational in 1908, is also the largest in the world available for approved use by the general public.

One of the oldest large optical telescopes in the United States is the Palomar Observatory's 5-meter Hale reflector. Completed in 1948, for thirty years it was the world's largest. Located in the mountains northeast of San Diego, California, Palomar was used primarily for deep space cosmology and stellar studies. With the advent of the space program and the accompanying renaissance of planetary astronomy, the Palomar telescope came to be used occasionally for solar-system work. It was used, for example, to analyze the Martian atmospheric gases and, in the 1950's, to examine the Martian surface for signs of chlorophyll to see if greenish surface regions might indicate plant life; they did not. With infrared detectors, the telescope was used to make some of the first temperature maps of the observed disk of Jupiter and to sample the temperatures of the surfaces of the other major planets.

The Palomar telescope has played an important role in space research with its exploration of objects that were first identified by the Infrared Astronomical Satellite (IRAS) as being anomalous infrared sources. Many of these objects are obscured stars and interstellar clouds within the Galaxy. These are not easily detected and studied at optical wavelengths, but Palomar observers have used infrared detectors, working at short enough wavelengths that are transmitted through Earth's atmosphere. There is still a considerable amount of research to be

done before the place of infrared-emitting galaxies in the general scheme of the extragalactic universe is understood thoroughly.

The largest telescope at the Lick Observatory—located at Mount Hamilton, east of San Jose, California—is the Shane 3-meter telescope, completed in 1959. It has been used by University of California astronomers to make numerous discoveries related to space research. An exotic example of such discoveries has to do with the planet Mercury. In 1961, there were reports from the Soviet Union that Mercury appeared to have an atmosphere; according to one Soviet astronomer, it was composed of hydrogen, but another report claimed that it was carbon dioxide. Space scientists were concerned about these reports. Because Mercury's surface was known to have high temperatures, astronomers had deduced that the planet should not be able to retain any appreciable atmosphere, because gases would be so hot that they would escape into space in a relatively short time. In 1962, to explore the question further, the Shane telescope was used to obtain high-dispersion spectra of Mercury. Because Mercury is always quite close to the Sun (never more than 28° from it in the sky), the telescope was used during the daytime for these observations. The results, af-

The 200-inch telescope at Mount Palomar, California. (NASA)

ter analysis, were clear. The better data provided no sign whatsoever of an atmosphere. Years later, in 1979, Mariner 10 confirmed this conclusion by returning data that showed that, because of its high temperature, Mercury does not have an appreciable atmosphere. It only retains temporarily a minute and tenuous envelope of hydrogen gas captured from the solar wind.

In 1979, the National Aeronautics and Space Administration (NASA) built a 3-meter infrared telescope to be used especially for space-related research. It was put on Mauna Kea in Hawaii to take advantage of the mountain's high altitude (4,200 meters) and lack of water vapor. It has been employed successfully in a large number of infrared projects, including studies of asteroids, comets, planetary satellites, the galactic nucleus, and infrared-bright and active galaxies.

The 2.7-meter telescope of the McDonald Observatory, near Austin, Texas, was built in 1968 with NASA sponsorship and with the intent that it be used largely for space-related research. A particularly interesting example of this has been its lunar-ranging measurements. A powerful laser at its focal point is used to beam visible light to the lunar surface, and 2.5 seconds later the telescope detects the reflected light. The light's travel time can be measured so accurately that the distance to the Moon could almost instantaneously be determined to an accuracy of about 2.5 centimeters.

In the mid-1980's some astronomers felt that ground-based astronomy had reached its limit. However, advances in computers and technology would prove them wrong. The first telescope using a new technique known as active optics was the New Technology Telescope (NTT) at the European Southern Observatory (ESO); it was installed in 1988. NTT has a 3.58-meter mirror that is only 24 centimeters thick and is therefore flexible. Traditional telescopes hold their shape by the thickness of the mirror. This thickness increases the mirror's weight and limits its size. The shape of NTT's mirror is maintained by a series of supports known as actuators, which are computer-controlled. Success with NTT design led to the construction of ESO's Very Large Telescope (VLT), which is an array of four 8-meter mirrors.

Two of the most famous telescopes using active optics are located at the Keck Observatory in Hawaii near the summit of the dormant Mauna Kea volcano. These telescopes (Keck I and Keck II, completed in 1993 and 1996) each have mirrors measuring 10 meters in diameter. The two can also be linked together to form an even larger interferometer. Keck I and II are used for investigations of brown dwarfs, globular clusters, black holes, Jupiter's atmosphere, and distant young galaxies, for example.

Active optics helps minimize the limiting effects of the telescope itself, increasing light collection and image quality. Adaptive optics, on the other hand, reduces or eliminates the effects of the Earth's atmosphere. Telescopes using adaptive optics (such as Keck) can achieve image resolutions of 30 to 60 milli-arc seconds at infrared wavelengths. Without adaptive optics, they would be able to produce resolutions of only one arc second. One method of adaptive optics uses a guide or reference star. The telescopes are equipped with a wavefront sensor that measures the atmospheric distortion of the light. Information is sent to a computer, which alters the telescope's deformable mirror to correct the image. When astronomers study very distant objects, the target is often too faint for this process. Instead, astronomers use a brighter reference star that is located near the target. Light from the guide star is then used to determine how to adjust the telescope to negate effects of atmospheric distortion of the target object. A second method for adaptive optics is using a laser beam in place of a reference star. The laser guide star (LGS) is directed toward the upper atmosphere and often is pulsed. Reflected light is then detected as it travels back down through the Earth's atmosphere, where it is used as an "artificial" reference star. LGS has advantages for scientists, because a reference star of sufficient brightness is not always found in all parts of the sky.

Technological advances of both active and adaptive optics have led to an ever-increasing number of larger ground-based telescopes. The largest optical telescope in the United States is the Large Binocular Telescope (LBT) located near Safford, Arizona. The telescopes are a joint project with the Italian and German astronomical communities, the University of Arizona, Arizona State University, Northern Arizona University, Ohio State University, Research Corporation of Tucson, and the University of Notre Dame. The LBT consists of two 8.4-meter mirrors with a common mount. Therefore the two telescopes together have the equivalent power of an 11.8-meter telescope. The first telescope was built in Italy and was shipped to Arizona in 2002. In March, 2008, LBT produced its first binocular images, making it the most powerful optical telescope in the world. The first images produced were false-color pictures of NGC 2770, a spiral galaxy 102 million light-years away. One image shows hot young stars concentrated in the spiral arms; the second shows older stars that are more evenly dispersed. The third image is a composite photograph that shows the full range of stars from cool to hot. An instrument is being developed to work with the telescopes at near-infrared wavelengths that would produce images ten times sharper than ones taken by the Hubble Space Telescope. A high-resolution spectrograph is also planned, which would help astronomers study the magnetic field of the Sun and other stars by looking at Zeeman splitting of spectral lines.

Several other extremely large telescopes (ELTs) are being designed or are currently under construction. The 25-meter Giant Magellan Telescope (GMT) is a joint endeavor of the Carnegie Institute of Washington, Harvard University, the Massachusetts Institue of Technology, the University of Arizona, the University of Michigan, the Smithsonian Institution, Australian National University, the University of

These two Keck telescopes at the top of Mauna Kea, Hawaii, can be linked together to form the Keck Interferometer, the world's most powerful ground-based optical telescope. (NASA/JPL)

Texas at Austin, and Texas A&M University. Located in Las Campanas, Chile, GMT will have six mirrors surrounding a central one. Each mirror will measure 8.4 meters, giving GMT a diameter of 25.2 meters. In August, 2008, grinding and polishing of the primary mirror began. When GMT is completed in 2017, scientists will use it to study the origins and evolutions of planetary systems, the formation of black holes, and dark matter.

The Thirty Meter Telescope (TMT) will have 492 hexagonal mirrors, each measuring about 1.44 meters in diameter. The 30-meter telescope will use six laser guide stars to study distant star systems and galaxies. Sites in Chile and Hawaii are being considered, and TMT could be constructed and operational by 2013.

The Large Aperture Mirror Array (LAMA) will combine a series of eighteen 10-meter liquid mirrors. The mirrors of the LAMA telescope are coated with a very thin layer of liquid mercury. The layer is initially 2-3 millimeters thick. The mirrors spin constantly, and after cleaning are started spinning manually. After the mirror is closed, a layer of oxide forms on top, which seals the surface. This allows the extra mercury to be skimmed off and can achieve a liquid mercury layer about 1 millimeter thick. Liquid mirrors are favored over glass ones, because on average they cost 95 percent less. Another benefit of liquid mirrors is the fact that the primary mirror does not tilt. This means that the support structure does not need to be as massive, thus also reducing cost. Together the mirrors that form the array will have the power of a 42-meter telescope. Sites are being considered in New Mexico and Chile. Scientists plan to use LAMA to study distant galaxies, stars, and extrasolar planets.

The most ambitious telescope in development is the European Southern Observatory's 100-meter Overwhelmingly Large Telescope (OWL). The primary mirror will be composed of 3,042 segments, each 1.6 meters in diameter. OWL's spherical primary mirror technology was originally used in the 1990's on the Hobby Eberly telescope. The ESO estimates OWL's cost to be 1.2 billion euros. Locations in Chile, Argentina, the Canary Islands, and near the South Pole are being considered. When fully operational, OWL will be able to observe stars with magnitudes as low as 38 (the faintest stars that can be seen with the unaided eye have a magnitude of 6). OWL will be used to study stellar and planetary system formation, extrasolar planets, and the mysterious dark matter.

Radio telescopes form another category of ground-based telescopes. These telescopes, used to receive and analyze extraterrestrial radiation at radio wavelengths, have been of great importance to the development of modern astronomy and space science. Most large radio telescopes have been one of three distinct types: large single dishes, interferometric arrays of dishes, and millimeter telescopes.

Centimeter-wavelength radio waves can best be detected by means of a very large parabolic reflecting surface (a "dish") that can move in position to follow the celestial object as it moves across the sky. The largest of these actually does not move, however, but instead has a moving receiver suspended above the dish. The 300-meter radio telescope of the Arecibo Ionospheric Radio Observatory was made by smoothing out a depression in the hills of Puerto Rico and lining it with a parabolic metal surface suspended a few feet above the ground. Three pylons rise from the rim of the valley. These hold cables that span the depression and support the small laboratory, dangling above the center of the dish, that contains the receiver. Motors pull the receiver building from west to east to compensate for the diurnal motion of the image of the celestial object being observed. Only a limited range in position in the sky is available to the Arecibo telescope, but its immense collecting area and high resolution have made it an important instrument for many years.

A different type of single-dish antenna is the 100-meter Effelsberg radio telescope near Bonn, West Germany. In this case, the parabolic metal dish is fully movable; it can turn to any place in the sky and can follow celestial sources at the diurnal rate. Although it is smaller than the Arecibo telescope, its flexibility has made it useful for many projects that would have been impossible with a large, fixed dish.

Similar, though smaller, single dishes that have been important are a 76-meter telescope at

Jodrell Bank, England, a 64-meter telescope at Parkes, Australia, and a 43-meter one at Green Bank, West Virginia. A 90-meter transit-type radio telescope at Green Bank collapsed suddenly in 1988; it was replaced by the Robert C. Byrd Green Bank Telescope (GBT), which is the world's largest fully steerable single-dish radio telescope. The 100-meter telescope became operational in 2000.

High resolution at radio wavelengths is hard to achieve. Resolution is proportional to wavelength; to achieve the same resolution afforded by an optical telescope, a radio telescope working at a long wavelength such as the 21-centimeter neutral hydrogen line must be many kilometers in diameter. This size would be impractical for a single-dish design, so radio astronomers have constructed giant arrays of radio dishes, connected electronically so that the signals received are blended and analyzed as if from an immense single telescope. These instruments are called "interferometric arrays" and range in size from 1 kilometer or so for the pioneer instrument in Cambridge, England, to intercontinental arrays.

The largest interferometers that are located in physical proximity are the Very Large Array (VLA) near Socorro, New Mexico, and the Westerbork Array in the Netherlands. The former consists of twenty-seven 25-meter dishes spread out in a Y-shaped pattern on the dry lake bed of the St. Augustine plain; the latter consists of a linear array of telescopes. All radio interferometers of this type have at least some of their antennae on wheels and tracks so that the spacing can be adjusted according to the needs of different observing projects.

The Australia Telescope array was designed with some of the properties of both the VLA and Westerbork; a continental array of telescopes planned to span North America would permit extremely high resolution. Ad hoc interferometers, made up of existing single dishes that are coordinated to simulate an array, have utilized even larger baselines (for example, from Australia to Canada), achieving radio images that are more detailed than even optical telescopes can produce, using normal detectors. Spaceborne radio telescopes can achieve even wider separations, as large as the solar system.

Millimeter-wavelength radio telescopes tend to have characteristics midway between those of standard radio telescopes and those of optical telescopes. Most are housed in domes of some sort and are fully steerable. For example, the 9-meter telescope on Kitt Peak in Arizona has a dome-shaped housing with a large slit that can be opened for observing. Other large millimeter telescopes are located in Hawaii (on Mauna Kea), in Japan, and in Massachusetts. They are especially powerful for detecting and analyzing emissions from cool molecular gas in star-forming regions.

The world's largest millimeter telescope is located on the Sierra Negra volcano, 350 kilometers southeast of Mexico City. The Large Millimeter Telescope (LMT) has a 50-meter dish. LMT is a joint effort between the National Institute of Astrophysics, Optics, and Electronics (of Mexico) and the University of Massachusetts. Construction was completed in 2006, after which followed a two-year testing period before initial research began. Scientists plan to use the large telescope to study extrasolar planetary atmospheres, compositions of comets, and the origins of the universe.

First conceived in 1995, the Atacama Large Millimeter/Submillimeter Array (ALMA) is a joint effort of organizations from North America, Europe, Japan, and Chile. ALMA is located in the Atacama Desert of the Andes Mountains in Chile. When completed in 2012, the array will include up to eighty radio antennas operating at 0.3 to 9.6 millimeters, at a cost of $1.3 billion. The dishes range in size from 7 to 12 meters in diameter. ALMA should produce resolutions about ten times better than the Hubble Space Telescope. Radio astronomers will use ALMA to study galactic nuclei, quasars, distant stellar compositions, and galactic, stellar and planetary formation.

The largest radio telescope presently in development is the Square Kilometer Array (SKA). When finished, this international facility will be fifty times more sensitive than any existing radio telescope. SKA will consist of thousands of radio dishes measuring 10 to 15 meters in diameter. Sites in South Africa and Australia are being considered. If built in South Africa, the Karoo Array Telescope (MeerKat) will play a central role, with other dishes spiraling outward across Africa.

Several other very large or extremely large telescopes are operational, under construction, or in development. Some notable ones are the Expanded Very Large Array (EVLA), the European Extremely Large Telescope (E-ELT), the Multi Mirror Telescope (MMT), South Africa's Large Telescope (SALT), Gran Telescopio Canaras (GTC), and the Allen Telescope Array (ATA, designed for the Search for Extraterrestrial Intelligence, SETI).

KNOWLEDGE GAINED

In the past, ground-based telescopes have helped scientists learn more about our solar system. Among the most important ways are the following: gathering of basic data on planetary celestial mechanics; discovery and orbit determination of comets and asteroids; mapping of galactic X-ray and infrared sources; discovery of quasars and radio galaxies; detection of cosmic background radiation; and the search for extrasolar planets.

In order to determine the continuously changing orbits of the planets well enough to make interplanetary spaceflight possible, it is important to map planetary positions very accurately. Laser ranging of the Moon, for example, allows astronomers with large ground-based telescopes to measure its distance to within a few centimeters. Radar-ranging measurements of the planets, especially Venus, Mercury, and Mars, have led to very precise determinations of their positions.

Spacecraft exploration of comets depends on ground-based telescopes, to discover comets in the first place and then to monitor them in their somewhat unpredictable paths near the Sun and Earth. Exploration of asteroids by spacecraft similarly depends on ground-based telescopes for tactical support.

Most of the cosmic objects found at X-ray and infrared wavelengths by orbiting detectors would be unexplained were it not possible to study them at other wavelengths from the ground. For example, two of the objects thought to be black holes, LMC X-1 and Cygnus X-1, would merely be mysterious, unidentified X-ray sources if it had not been possible, using large ground-based telescopes, to discover that each is a binary star, with a normal star in orbit around a dark, massive object.

The Keck telescopes in Hawaii have been used to study seasonal variations on Uranus. In 2007, astronomers were able to photograph the change of seasons on Uranus for the first time. Still in its early stages, this study found significant changes in some cloud features that had previously appeared to be unchanging. Scientists were also able to calculate wind speeds on the planet more extensively, up to 901 kilometers an hour. That year astronomers were also able to use the Keck telescopes to look at Uranus's ring system edge-on. They found that Uranus's dusty rings have changed since Voyager 2 visited the system in 1986. This ring crossing was also observed using the Very Large Telescope (VLT) in Chile, and the Palomar Observatory in Southern California. Such observations would not be possible, or economically feasible, without ground-based instruments.

As ground-based telescopes increase in size, they also increase in power. The newer generations of VLTs (very large telescopes) and ELTs (extremely large telescopes) allow scientists to study objects at greater distances and therefore look back in time, since their light takes years (light-years) to reach us.

CONTEXT

Astronomy seeks a better understanding of cosmology. Ground-based telescopes have provided the basic list of objects and phenomena (including quasars, radio galaxies, and cosmic background radiation) that allow astronomers to penetrate to the edge of the universe and the beginning of time. New generations of ground-based telescopes are designed to push farther into deep space and to determine the true story of how the universe came about. They study distant galaxies, stars, and planetary systems to learn about the formation and evolution of our own solar system.

Despite the advent of space-based observatories, ground-based observations will continue to play a major role in astronomy so long as the risks and costs associated with space-based telescopes continue to be high. Ground-based observations will therefore continue to provide data and images complementary to space-based observations, which together will aid astronomers in understanding the nature and evolution of the physical universe.

Jennifer L. Campbell

FURTHER READING

Anderson, Geoff. *The Telescope: Its History, Technology, and Future*. Princeton, N.J.: Princeton University Press, 2007. Anderson does a good job of summarizing the history of telescopes from their earliest days to those currently under construction. Intended to give the general reader the basics behind telescope design and a notion of current advancements in the field.

Brunier, Serge, and Anne-Marie Lagrange. *Great Observatories of the World*. Buffalo, N.Y.: Firefly Books, 2005. A photographic tour of fifty-six of the world's most impressive observatories. Decribes how telescopes work, as well as new advances in ground-based telescopes and the higher-resolution photographs they produce. For the general reader.

Kenyon, Ian. *The Light Fantastic: A Modern Introduction to Classical and Quantum Optics*. New York: Oxford University Press, 2008. An introductory work on optics aimed at advanced undergraduate and first-year graduate students. Also gives the reader practical examples and real-world applications. Includes a separate chapter on telescopes.

Kirby-Smith, Henry T. *U.S. Observatories: A Directory and Travel Guide*. New York: Van Nostrand Reinhold, 1976. A complete description of U.S. observatories, with some history and an account of the types of research and the equipment available. Notes on public availability for some are included. Suitable for general readers.

Kloeppel, James E. *Realm of the Long Eyes: A Brief History of Kitt Peak National Observatory*. San Diego: Univelt, 1983. A well-illustrated account of the development of the optical observatory on Kitt Peak, Arizona. Offers a complete history of the arguments for a national ground-based observatory and of the process of its establishment as one of the world's most productive scientific installations.

Krisciunas, K. *Astronomical Centers of the World*. Cambridge, England: Cambridge University Press, 1988. Provides thorough coverage of the world's major astronomical centers, including important historical installations. Includes material on advanced concepts, observatories of the future, and space observatories. Features useful tables and an index. Suitable for scientifically oriented general readers.

Kuiper, Gerard P., and Barbara M. Middlehurst, eds. *Telescopes: Stars and Stellar Systems*. Chicago: University of Chicago Press, 1978. A compendium of fairly technical chapters on various aspects of telescope design, with an emphasis on traditional large reflecting telescopes with equatorial mounts. Contains many tables, photographs, and graphs. For college-level readers.

Tucker, Wallace, and Karen Tucker. *The Cosmic Inquirers: Modern Telescopes and Their Makers*. Cambridge, Mass.: Harvard University Press, 1986. An excellent book about ground-based telescopes, including radio and space telescopes, and about the astronomers who brought them into existence. Includes narrative accounts and biographical data based on interviews with key people. For general readers.

Zirker, J. B. *An Acre of Glass: A History and Forecast of the Telescope*. Baltimore: Johns Hopkins University Press, 2005. Zirker's work provides a detailed study of telescope construction from the past to the future. It also explains their scientific significance. Answers any questions the general reader might have about telescopes.

See also: Archaeoastronomy; Coordinate Systems; Earth System Science; Gravity Measurement; Hertzsprung-Russell Diagram; Infrared Astronomy; Neutrino Astronomy; Optical Astronomy; Radio Astronomy; Telescopes: Space-Based; Ultraviolet Astronomy; X-Ray and Gamma-Ray Astronomy.

Telescopes: Space-Based

Category: Scientific Methods

Since the 1960's, robotic observatories placed in space have become a key part of the overall effort by astronomers to understand the solar system and universe. Able to view the universe without the interfering effects of Earth's atmosphere, these space telescopes have revealed sights and wonders that were previously unimagined.

OVERVIEW

With the exception of relatively few objects found within the solar system, objects of study outside our planet are too far away to visit and sample. We cannot simply travel to distant stars, nebulae, and galaxies to learn about them. Thus, if humans wish to study these distant objects, they must find another way to do so.

Electromagnetic Radiation. The primary method of learning about distant objects is by observing various kinds of electromagnetic radiation. One must bear in mind that since light travels at a finite speed (300,000 kilometers per second, or 186,000 miles per second), objects can be studied only as they were when they emitted the light we observe. Thus the telescope, whether in space or on Earth, can be thought of as something like a time machine. For example, if an object is determined to be 10 billion light-years away (a light-year being the distance traveled by light in vacuum in one year, that is, 9.467 trillion kilometers), the telescope provides a view of what that object looked like ten billion years ago. Depending on how distant objects are, they probably are radically different today from what they were in the past. For example, considering the presently accepted value for the age of the universe, a galaxy 12 billion light-years away is observed as it was only a little less than 2 billion years after the big bang.

Sir Isaac Newton demonstrated that visible white light is actually composed of an infinite sequence of wavelengths from red to violet in color. If one disperses white light with a prism or diffraction grating, then a rainbow band of colors called a spectrum is seen. Visible light is actually only a very small part of the total electromagnetic spectrum; the rest of it is made of other radiation, "invisible" light. If humans could see the other parts of the spectrum, they would find in turn beyond the violet end of the spectrum, at ever shorter wavelengths, ultraviolet, X-ray, and gamma-ray radiation. Beyond the red end, at ever longer wavelengths, are infrared, microwave, and radio radiation. Radio radiation includes frequencies used for broadcast, radar, and television signals. All of these different kinds of "light" are electromagnetic waves, or electromagnetic radiation, with specific frequencies and wavelengths. Electromagnetic radiation is such that all waves share the property that the product of their wavelength and frequency is the speed of light, approximately 300,000,000 meters per second (yielding the per-year results that define the light-year described previously). When taken together, these waves of spatially and temporally oscillating crossed electric and magnetic fields (or streams of photons) comprise the electromagnetic spectrum.

The instrument commonly used to collect light is the telescope. The telescope can be defined as a "light bucket," the function of which is to gather as much light as possible from a given region of the sky and direct it to a focal point. Today telescopes exist that study every part of the electromagnetic spectrum. The first telescopes, optical telescopes, collected light only in the visible region. These are still the most common type of telescopes today, especially on Earth's surface.

The opening in the telescope permitting entry of light is called the aperture, and its size serves two functions. The larger the aperture, the more light can enter, enabling one to see fainter objects. Larger apertures also result in greater resolution. "Resolution" pertains to the ability to "see" something in sharp focus. The smaller an object, the more resolution one needs to see it clearly. However, if greater resolution were the sole requirement for viewing faint objects, it would simply be necessary to make larger and larger telescopes in order to see them. The biggest problem facing astronomers, however, is Earth's atmosphere. Because of its chemical and physical composition, most electromagnetic frequencies are blocked out. Only certain frequencies have the ability to penetrate the atmosphere all the way to Earth's surface. One obvious frequency band is visible light. Another is radio waves. Other frequencies are absorbed or reflected by the atoms and molecules in the atmosphere, by water vapor, or by the natural or industrial dust and pollution in the air. If it were possible to see in any of these absorbed frequencies, such as ultraviolet light or X rays, the air would be opaque, like a wall. However, when instruments can be positioned beyond the atmosphere, rising above all the components of the atmosphere that block out invisible light, they are able to "see" in these invisible frequencies.

For these reasons, telescopes have historically been built in high places such as mountaintops. Telescopes have also sent aloft in airplanes, balloons, and rockets for short-term observations. However, it was not until the later half of the twentieth century, with the advent of the space age, that scientists were able to add telescopes and other detectors of electromagnetic radiation to orbiting spacecraft and interplanetary space probes in order to capture imaging data beyond Earth's atmosphere for long periods of time. As a result, new areas of astronomy—such as infrared, X-ray, and gamma-ray astronomy—blossomed. For example, infrared measurements from space are important to astronomers because infrared radiation is related to heat, and star-forming regions of the universe release large amounts of infrared radiation that is not detectable from Earth's surface. Similarly, high-energy astrophysics is studied by X-ray and gamma-ray observatories placed in Earth orbit.

Orbiting Solar Observatories. The idea of placing telescopes into orbit is an old one. The early space theorist Hermann Oberth wrote about this in the 1920's. In 1946 the astronomer Lyman Spitzer, Jr., composed a report titled *Astronomical Advantages of an Extra-Terrestrial Observatory*. Under the direction of its first chief astronomer, Nancy Roman, the National Aeronautics and Space Administration (NASA) prepared for the development of space-based observatories, an idea that was not initially welcomed by large portions of the astronomical community for fear that money would be di-

verted from the construction of ever larger ground-based observatories. NASA launched eight Orbiting Solar Observatories (OSOs) between 1962 and 1971, designed specifically to make observations of the Sun in visible, ultraviolet, and X-ray wavelengths. Seven of the eight OSOs were successful and provided data that added to our understanding of the eleven-year sunspot cycle, as well as characterizing the solar corona. An OSO produced the first full-duration image of the solar corona. Some of the OSO instruments were test beds for later space-based X-ray telescopes, such as those incorporated into the Skylab space station's Apollo Telescope Mount. OSO studies also investigated cosmic rays, neutron emissions, and extrasolar X-ray sources.

Orbiting Astronomical Observatories. The OSOs were followed by a sequence of Orbiting Astronomical Observatories (OAOs). The first OAO was launched on April 8, 1966. However, its pointing system failed when activated and caused the onboard batteries to explode. A successful OAO-2 was launched in December, 1968, making observations for four and a half years in the infrared, ultraviolet, X-ray, and gamma-ray portions of the electromagnetic spectrum. This advancement was followed by several other very successful orbiting observatories, including OAO-3 in 1972, the International Ultraviolet Explorer (IUE) in 1978, and the European Space Agency's (ESA's) Infrared Astronomical Satellite (IRAS) in 1983.

Granat. The Russians operated an X-ray/gamma-ray observatory named Granat from December, 1989, to November, 1998. It probed the center of the Milky Way, recorded spectral and temporal variability in black holes, and detected 511-kiloelectron volt (keV) X rays produced by electron-positron annihilation.

Extreme Ultraviolet Explorer. NASA's Extreme Ultraviolet Explorer (EUVE) was launched on June 7, 1992. This observatory was designed to detect and record radiation ranging in wavelength from 7 to 76 nanometers, a portion of the electromagnetic spectrum that had not been investigated by many space-based instruments previously. The telescope was outfitted with an imaging microchannel plate detector and three spectrometers available at its focal plane.

EUVE's mission continued through January 31, 2001.

The Great Observatories. In the 1980's NASA announced its Great Observatories Program, the purpose of which was to place in orbit a series of telescopes to observe across the electromagnetic spectrum. Observatories under this program include the Hubble Space Telescope (HST), which was deployed in orbit from the space shuttle *Discovery* (mission STS-31) on April 25, 1990; the Compton Gamma Ray Observatory (GRO), which was deployed in orbit from the space shuttle *Atlantis* (mission STS-37) on April 8, 1991; the Advanced X-Ray Astrophysics Facility (AXAF), which was deployed from the space shuttle *Columbia* (mission STS-93) on July 23, 1999; and the Space Infrared Telescope Facility (SIRTF), launched on August 25, 2003, by an expendable Delta II booster. After deployment, GRO was commissioned as the Compton Gamma Ray Observatory in honor of the American physicist Arthur Compton; AXAF was commissioned as the Chandra X-Ray Observatory (CXO) in honor of Indian astrophysicist Subrahmanyan Chandrasekhar; and SIRTF was commissioned as the Spitzer Space Telescope (SST) in honor of the American astronomer Lyman Spitzer, Jr.

Many of the observatories described thus far represented revolutionary steps in expanding humanity's investigations of the universe. Many opened up entirely new windows on the universe, since the types of radiation they detected could not penetrate Earth's atmosphere to reach ground-based telescopes. In order to gain a clear picture of the physical processes in stars, galaxies, and exotic objects such as quasars, active galactic nuclei, gamma-ray bursters, and black holes, it is necessary to examine those objects simultaneously across the different regions of electromagnetic spectrum. That indeed was the overriding aim of NASA's Great Observatories Program: to construct several cutting-edge telescopes that, once in space, would provide means to observe across the spectrum, from gamma rays to infrared radiation. No radio telescope in space was included in this project, since the majority of the radio spectrum is observable from ground-based facilities. Nevertheless, for reasons of having an electromagnetic quiet zone away from earthly sources, many astronomers have long dreamed of a radio telescope on the far side of the Moon.

Hubble Space Telescope. The crown jewel of all these Great Observatories is certainly the Hubble Space Telescope (HST); it has had the greatest appeal to the public. Hubble was launched aboard the space shuttle *Discovery* into the highest attainable space shuttle orbit, at 615 kilometers above Earth's surface, with an orbital inclination of 28.5°. After requiring a heroic on-orbit repair by space shuttle mission STS-61 astronauts, Hubble began producing a steady stream of impressive images that fascinated the public and rewrote astronomy textbooks. For example, by early 1997 Hubble had taken more than 100,000 images of various objects in the solar system and universe, amazing scientists and the general public alike with the incredible sights it had captured. Hubble is expected to continue providing new insights about the universe well into the second decade of the twenty-first century.

A telescope's ability to see is controlled by the size of its aperture, but it is also controlled by the medium through which the telescope must look (or what the radiation must travel through) to capture its images. Even though visible light can penetrate the air, the atmosphere has several detrimental effects on a telescope's ability to see. These effects include twinkling, which is caused by movement in the air; weather; the moisture content of the air; dust and air pollution; the natural glowing of the atmosphere, known as airglow; and light pollution caused by city lights. Because of all these factors, observatories are built on remote mountaintops. The largest optical telescopes in the world are the two 10-meter (393.7-inch) Keck telescopes on Mauna Kea in Hawaii and the 11-meter (433-inch) Hobby-Ebberly telescope at the McDonald Observatory on Mount Locke, Texas. Even these instruments, however, contend with atmospheric effects. By placing a telescope in orbit, all of these atmospheric factors are eliminated, making it possible to obtain much clearer images. The Hubble Space Telescope has a primary mirror of only 2.4 meters (94.5 inches), yet it can see many times better than either of these two earthbound telescopes.

Hubble was designed to provide the clearest view ever of the cosmos, providing astronomers with vision that was ten times clearer and fifty times more sensitive than that of the best ground-based telescopes. This provides astronomers with the largest boost in viewing capability since Galileo first used a telescope to view the sky in 1610. Hubble is 13.3 meters long, 4.3 meters in diameter, and 12 meters across with its solar array deployed. It weighs 11,200 kilograms (12.3 tons). The 2.4-meter primary mirror is a Cassegrain mirror. That is, the light reflected by the primary mirror is then reflected by a secondary mirror, after which the light travels through a hole in the primary mirror to the instruments positioned behind it.

The primary mirror was manufactured with greater precision than anything previously. Yet, shortly after the telescope was deployed

in 1990, it became apparent that there was a problem. Although the mirror was extremely smooth, it was not shaped correctly, exhibiting what is known as a spherical aberration—that is, the curvature of the mirror was slightly rounder than the parabolic shape required. As a result, the images were blurry. However, this problem was rectified by means of correcting lenses and mirrors known as the Corrective Optics Space Telescope Axial Replacement (COSTAR). COSTAR, along with several new gyroscopes and instruments, was installed by the STS-61 space shuttle servicing mission in December, 1993. A second servicing mission (STS-82) occurred in February, 1997. Additional servicing missions were conducted in late 1999 (STS-103) and in 2002 (STS-109). These servicing missions allowed new instruments to be emplaced within the telescope and accom-

The Hubble Space Telescope in orbit around Earth, 2002. (NASA)

The Compton Gamma-Ray Observatory at its deployment by the space shuttle Atlantis *in 1991.* (NASA)

plished maintenance and repair work to keep Hubble operational for its designed fifteen-year life span, until 2005. In May, 2009, a shuttle-based repair mission, STS-125, was flown to keep the Hubble Space Telescope operational at least through 2011-2013, when the next-generation space telescope, the James E. Webb Space Telescope (JWST), was expected to supplant Hubble.

Compton Gamma Ray Observatory. The Compton Gamma Ray Observatory (CGRO) was deployed in a high orbit (450 kilometers above the Earth's surface), yet still underneath the Van Allen radiation belts to preclude interference with the observatory's detectors. Compton was outfitted with four instruments: the Burst and Transient Source Experiment (BATSE), the Oriented Scintillation Spectrometer Experiment (OSSE), the Imaging Compton Telescope (COMPTEL), and the Energetic Gamma Ray Experiment Telescope (EGRET). Together these scintillation detectors spanned

six orders of magnitude of wavelengths in the gamma-ray portion of the electro-magnetic spectrum, specifically from 20 keV to 30 gigaelectron volts (GeV) in energy.

When Compton was on the verge of losing attitude control, NASA decided to deorbit the heavy observatory in a controlled fashion so that any debris, such as portions of its very massive detectors, would not impact populated areas. Compton was deorbited on June 4, 2000, and safely rained debris over open regions of the Pacific Ocean.

Infrared Space Observatory. ESA followed up its highly successful IRAS cataloging mission with the Infrared Space Observatory (ISO). After launch on November 17, 1995, ISO produced more than twenty-six thousand infrared observations before its liquid helium cryogenic coolant supply ran out after a bit more than twenty-eight months. ISO was outfitted with a high-resolution infrared camera (ISOCAM), an infrared photo-polarimeter (ISOPHOT), a shortwave spectrometer (SWS), and a long-wave spectrometer (LWS) in order to observe wavelengths between 2.4 and 240 microns. Among ISO's achievements were the finding of planet formation around old stars nearing their end of life, the detection of dust between galaxies, finding a number of new chemical compounds in interstellar gas clouds, locating protoplanetary disks around stars in early stages of formation, and determining chemical abundances in planets within this solar system.

Far Ultraviolet Spectroscopic Explorer. NASA launched the Far Ultraviolet Spectroscopic Explorer (FUSE), an Explorer-class satellite (specifically, Explorer 77), on June 24, 1999, atop a Delta II booster. FUSE, an observatory in the agency's Origins program, was placed in a low Earth orbit. The spacecraft was designed to last at least three years. Scientists at the Johns Hopkins University used the spacecraft's Wolter-type grazing incidence telescope and its spectrograph to observe far-ultraviolet emissions ranging from 90.5 to 119.5 nanometers. FUSE was designed specifically for determining the distribution of deuterium (the isotope of hydrogen having a nucleus consisting of a proton and a neutron) in the aftermath of the big bang. FUSE covered a region of the electromagnetic spectrum not detected by any of the members of the Great Observatory program and investigated astrophysical and cosmological questions that added to and complemented discoveries made by Hubble, Chandra, and other space-based telescopes.

FUSE suffered failures in its pointing system and for a time was on the verge of being shut down, despite the fact that its science hardware functioned perfectly. Engineers developed alternate means of pointing the telescope, and FUSE got a reprieve for more science. However, on July 12, 2007, the spacecraft lost its last reaction wheel. Efforts to restore FUSE to science operations failed, and two months later the telescope was abandoned.

Chandra X-Ray Observatory. Because X rays cannot be focused in the same manner as visible light, the design of the Chandra X-Ray Observatory required a quartet of nested pairs of mirrors in the shape of paraboloids and hyperboloids. At normal incidence (that is, perpendicular to the surface of the mirror), X rays are much more likely to be absorbed by than reflected off a mirror; hence, there the design for any X-ray telescope must allow for low grazing angles. Chandra's X-ray resolution was over a thousand times greater than any previous space-based X-ray detector. Placed in a highly elliptical orbit ranging from 10,000 to 140,000 kilometers, Chandra is capable of collecting data during as much as 85 percent of its sixty-five-hour orbital period. Chandra is outfitted with four instruments: an imaging spectrometer, a high-resolution camera, and high-energy and low-energy high-resolution transmission grating spectrometers.

Spitzer Space Telescope. The Spitzer Space Telescope was placed in an Earth-trailing helio-centric orbit in 2003. The telescope was designed to be cooled to 5.5 kelvins and last a minimum of 2.5 years. Outfitted with an infrared camera (the Infrared Array Camera), an infrared spectrometer (the Infrared Spectrograph), and far-infrared detection arrays (the Multiband Imaging Photometer for Spitzer), Spitzer was designed to cover infrared emissions from 3 to 180 microns. Spitzer's supply of liquid helium cryogenic coolant lasted longer than a minimal mission, and, in late 2008, it was calculated that Spitzer could last between another six months and a year, or perhaps longer.

Swift Gamma-Ray Burst Mission. The Swift Gamma-Ray Burst Mission, usually referred to simply as Swift, was launched on November 20, 2004, to study and identify the origin of gamma-ray bursts. Aptly named, Swift had the capability to reorient itself quickly in order to place an incoming gamma-ray flux centrally within the observatory's field of view, and thereby alert the astronomical community rapidly so that other telescopes and detectors could record the afterglow of a gamma-ray burster (GRB), even short-period ones.

Gamma-Ray Large Area Space Telescope. The Gamma-Ray Large Area Space Telescope (GLAST) was launched on June 11, 2008. It was a next step in gamma-ray astronomy after the loss of the Compton Observatory. GLAST was designed to investigate black holes, the high-speed jets of gas and energetic particles emitted from black holes, the physics of dark matter, solar flares, pulsars, cosmic-ray production, and gamma-ray bursts.

Kepler Observatory. The Kepler Observatory was designed to survey more than 100,000 stars in four years to detect transits of exoplanets across those stars' photospheres. Such a crossing would result in a slight diminishment of the star's light curve, and additions to the stellar spectrum of absorption lines from the planet's atmosphere, if it were to have one. Planned for launch in early 2009, Kepler's primary objective was to detect terrestrial planets within habitable zones around a wide variety of stars. The location of habitable zones is determined by a star's spectral class and luminosity. For a G2-class star such as the Sun, the habitability zone—where life as humanity knows it is permissible due to the possibility of water existing in liquid form—extends from 0.72 astronomical units (AU) out to 1.5 AU, essentially from Venus to Mars. However, for a red dwarf star, the habitability zone would be closer. The chance of detecting an Earth-class planet transiting a star is a complicated combination of star spectral class, the possibility of planets forming about that particular star, and the planet's mean distance from the star.

For example, for the star Tau Ceti (spectral class G8), if a terrestrial planet similar to Earth were located 1 AU away from the star, the possibility of a transit being observed is approximately 1 in 210. By the sheer number of survey stars Kepler would examine, the chances of finding evidence of terrestrial planets would be high. It was estimated that if Kepler were able to examine its planned 100,000 stars, then 480 or more terrestrial planets should be observed by this transit method.

James E. Webb Space Telescope. Originally called the Next Generation Space Telescope (NGST), the James E. Webb Space Telescope (JWST, named for James E. Webb, who served as NASA Administrator from 1961 to 1968) has often been described as the replacement for the Hubble Space Telescope. This is not strictly accurate, as JWST is designed as an exclusively infrared observatory, to be placed in a position where the Sun's gravitational attraction on the telescope is balanced precisely by that of the Earth. This location is called the L2 Lagrangian point. It is located 1.5 million kilometers from Earth.

JWST is designed to be constructed of eighteen hexagonal mirror segments that must deploy from a folded configuration at launch. If successfully deployed, JWST will have a light-collecting area six times greater than Hubble's. Instruments incorporated into JWST's design included a near-infrared camera, a near-infrared spectrograph, a mid-infrared instrument, and fine guidance sensors. These will provide coverage of the infrared between wavelengths of 0.6 to 28 microns. Launch of JWST is planned for sometime in the 2010's. Should Hubble still be available at the time, early studies conducted by JWST could be coordinated with those supplied by Hubble. Among JWST's primary objec-

tives are an investigation of the earliest epoch of star and galaxy formation, observation of evolutionary processes in galaxies, examination of star-forming regions and developing planetary systems, and searches for planetary systems that may have materials and conditions necessary for the evolution of life.

KNOWLEDGE GAINED

Findings from OAOs. Among the achievements of the Orbiting Astronomical Observatories were the finding that halos of hydrogen surround comets, the discovery that novae may increase in ultraviolet brightness while diminishing in visible luminosity, the discovery of long-period pulsars, and the collection of several hundred high-resolution spectra of stars in X-ray and ultraviolet wavelengths.

Findings from IRAS. The Infrared Astronomical Satellite was the first fully dedicated infrared observatory in space. Its greatest achievement was the location of an enormous number of infrared sources to be studied in detail by later generations of infrared detection systems placed in space, such as ISO and Spitzer.

Findings from IUE. Among the scientific discoveries of the International Ultraviolet Explorer were auroras in Jupiter's atmosphere, sulfur in comets, stellar spots, the progenitor star for Supernova 1987A, the extent of active regions in Seyfert galaxy nuclei, and galactic halos. IUE produced the first light curves collected undisturbed for more than twenty-four hours, and it detected the first emissions of wavelengths less than 50 nanometers. More than 104,000 observations were made by IUE before its decommissioning.

Findings from Hubble. One of the primary goals for Hubble was to measure the brightness of certain kinds of stars, called Cepheid variables, in several distant galaxies, making it possible to determine how far away these galaxies are. The importance of this is related to a discovery by the American astronomer Edwin Hubble, the man after whom the Space Telescope is named. Hubble discovered that the universe is expanding and that the distance to an object is proportional to just how far away the object is. This proportionality is known as Hubble's constant. If one finds a reliable value for Hubble's constant, it is possible to determine just how far away a galaxy is. It is also possible to determine how old the universe is. The problem is finding a reliable value. Using measurements made with the Hubble after the installation of COSTAR, astronomers were able to determine a definitive value for this constant. In 2008 the accepted value for the age of the universe was determined to be 13.7 billion years, and for the Hubble constant, 77 kilometers per second per megaparsec.

Hubble was not necessarily expected to make significantly new discoveries, because it observes in the same visible light in which observations from the Earth's surface are made. Rather, it was expected to see with unequaled clarity and resolution the objects previously observed. Yet, it has still succeeded in making previously unsuspected discoveries, mainly because it is able to see much fainter objects than could ever be seen using ground-based telescopes. One such discovery occurred during a ten-day exposure of a section of the sky that was previously thought to be empty and was found by Hubble to be filled with distant galaxies. Two such Hubble deep field surveys were eventually made. The other Great Observatories in orbit made radically new discoveries from the very beginning, because they observe in light frequencies in which humans were previously unable to observe due to atmospheric influences.

While it is the mirror that actually gathers the light, it is the instruments on the Hubble Space Telescope that provide its tremendously exciting scientific advances. One of these instruments is the Wide Field/Planetary Camera-2 (WF/PC-2). WF/PC-2 serves as the primary imaging camera. This instrument has provided us with the startling images of towers of interstellar matter and other phenomena that have become so familiar.

Another instrument is the Space Telescope Imaging Spectrograph (STIS), which gives the Hubble unique and powerful spectroscopic capabilities. A spectrograph separates light gathered by the telescope into its spectral components so that the composition, temperature, motion, and other chemical and physical properties of astronomical objects can be analyzed.

STIS's two-dimensional detectors allow the instrument to gather thirty times more spectral data and five hundred times more spatial data than existing spectrographs on the Hubble that observe one location at a time. The STIS is a particularly powerful tool for studying supermassive black holes. STIS searches for massive black holes by studying the star and gas dynamics around galactic centers. It also measures the distribution of matter in the universe by studying quasar absorption lines, using its high sensitivity and spatial resolution to examine star formation in distant galaxies and performing spectroscopic mapping of solar-system objects.

Another instrument is the Near Infrared Camera and Multi-Object Spectrometer (NICMOS), which promises to gather valuable new information on the dusty centers of galaxies and the formation of stars and planets. Consisting of three cameras, NICMOS is able to perform infrared imaging and spectroscopic observations of astronomical targets. Because these detectors perform more efficiently than previous infrared detectors, NICMOS has given astronomers their first clear view of the universe at near-infrared wavelengths between 0.8 and 2.5 micrometers. These views in the infrared are important because expansion of the universe has shifted the light from very distant objects toward longer red and infrared wavelengths. Hence, NICMOS's near-infrared capabilities provide views of objects too distant for research by Hubble's optical and ultraviolet instruments.

Astronauts installed the Advanced Camera for Surveys (ACS) inside Hubble in 2002, replacing the Faint Object Camera. ACS was designed to observe from far-ultraviolet to visible wavelengths, and therefore to collect images that would shed light on some of the earliest galaxies. ACS possesses a field of view twice that of the Wide Field Planetary Camera 2, giving it the capability to accomplish very broad surveys of the early universe. ACS lost performance in its charge-coupled devices in early 2007 and was slated for repair during the STS-125 final Hubble servicing mission.

Included as a part of a final planned shuttle service mission to the Hubble (the shuttle flight designated STS-125) was installation of the Cosmic Origins Spectrograph (COS), designed to perform spectroscopy between 115 and 320 nanometers in ultraviolet wavelength. According to the Space Telescope Science Institute, this instrument's principal research would center on investigating the large-scale structure of the universe, the early formation of galaxies and their subsequent evolution, formation of stellar and planetary systems, and the interstellar medium. Several spacewalks would be required on STS-125 to achieve a life extension for Hubble. The mission was set up to install six new gyrodynes, six new solar batteries, the COS, and the Wide Field Camera 3, a new cooling system, and a replacement Fine Guidance Sensor. The astronauts would also repair both the thermal insulation and STIS, in order to leave Hubble in an almost new configuration that would last at least five years.

Among the numerous major discoveries from ACS research was the collection of gravitational lensing data sufficient to determine the mass of seventy distant galaxies. Gravitational lensing is the process whereby massive objects bend and focus light passing near them so that a telescope will see a ring of identical images, something referred to as an Einstein ring. The amount of lensing present in an image is determined by the mass of the lensing objects, such as these seventy galaxies, which divert the light and form the Einstein ring.

Findings from Compton. Among the Compton Gamma Ray Observatory's impressive achievements over its nine years in orbit were the amassing of an all-sky survey of emissions over 100 megaelectron volts (MeV). Compton identified 271 individual sources, compiling an all-sky map of gamma-ray emissions from decays of the radioactive isotope Al^{26} of aluminum. Compton also detected gamma-ray burster 990123, the brightest object recorded to that time, along with gamma-ray emissions from the tops of terrestrial thunderstorms.

Findings from EUVE. The Extreme Ultraviolet Explorer's mission consisted of two phases. During the first, which lasted only half a year, the telescope was used in imaging mode to generate a full-sky atlas of sources of extreme ultraviolet radiation. The telescope's second phase of investigation then involved pointed studies of individual sources using the spectroscopic capabilities of EUVE. Among EUVE's discoveries and observations with regard to the solar system were extreme ultraviolet/X-ray emissions in Comet P/Encke, changes in interplanetary helium wind, day glow at Venus, mechanisms whereby solar flares promote coronal heating, and images of the full Moon in extreme ultraviolet wavelengths.

Findings from Chandra. The Chandra X-Ray Observatory's contributions to X-ray astronomy include the first detection of X-ray emissions from the Milky Way's supermassive black hole located at the galactic center; detecting a mid-mass-range black hole in galaxy M82; providing data which might be evidence of a star composed of matter collapsed to quarks (a quark star); demonstrating that virtually all main sequence stars also emit somewhat in the X-ray region of the electromagnetic spectrum; helping to determine the Hubble constant (and hence the age of the universe); and collecting evidence of dark matter involved in supercluster collisions.

In 2008 Chandra and the Very Large Array of Earth-based radio telescopes located near Socorro, New Mexico, collaborated to observe the most recent supernova in the Milky Way. This supernova had occurred only 140 years earlier. Prior to this discovery, the most recently noted supernova in the galaxy was Cassiopeia A. The data on this young supernova allowed astronomers to investigate the stellar explosion process and creation of a central neutron star or black hole. This discovery indicated how useful coordination of observations in both X-ray and radio emissions as well as coordination of space-based and Earth-based telescope facilities can be.

Another unique way Chandra was used involved developing a new method for determining the mass of supermassive black holes, such as those found at the centers of many galaxies. Looking for X-ray emissions from hot gas in the central region of the elliptical galaxy NGC 4649, Chandra was able to determine the peak temperature of that hot gas. The temperature of the hot gas is determined by the gravitational compression of the gas from the black hole. That, in turn, is dependent on the black hole's mass. This method was checked against earlier methods of "weighing" black holes. The results from Chandra for NGC 4649 agreed nicely.

Findings from Spitzer. The Spitzer Space Telescope greatly expanded on previous infrared research produced by IRAS and ISO, and collaborated with NASA's other Great Observatories for coordinated studies of important objects and also deep-sky field surveys. Among Spitzer's major accomplishments are the first direct detection of light from the hot-Jupiter extrasolar planets HD 209458b and TrES-1; detection of the youngest star ever found; determination that the Milky Way galaxy's core has a bar structure; capturing the glow from stars formed only 100 million years or so after the big bang; and the first determination of an extrasolar planet's atmospheric temperature (that of HD 189733b).

Spitzer was used to study Messier 101, otherwise known as the Pinwheel galaxy, detailing the infrared signature of the galaxy's spiral arms and central region. Data revealed that polycyclic aromatic hydrocarbons were present throughout much of the galaxy but disappeared in the outer region. This suggested that hydrocarbons, and hence organic materials, had a threshold as one looked far away from the central region of this galaxy.

In 2008 Spitzer observed a pair of young stars, DR Tau and AS 205A, located 457 and 391 light-years from Earth, respectively, particularly concentrating on their protoplanetary disks using the telescope's spectrographs. Both stars' disks displayed emission lines of water vapor within their innermost portions. This discovery, although the amount of water found was still considerably less than the amount of water found in Earth's oceans, indicated that water can be available and abundant in the inner regions of forming solar systems.

Findings from Swift. The Swift Gamma-Ray Burst Mission located a gamma-ray burster (GRB) with a burst duration of only 0.05 second. Swift also found the brightest object ever seen, GRB 080319B, located 7.5 billion light-years from Earth with a luminosity 2.5 million times that of the most brilliant supernova seen previously.

Findings from European and Russian Observatories. Although this essay has primarily concentrated on NASA observatories in space, the Europeans and Russians have placed a number of important and productive space-based telescopes in orbit, including the previously discussed IRAS and ISO. Also important are XMM-Newton, an X-ray observatory; the Salyut 6 space station's KRT-10 radio telescope; and COROT. These are only a few of the international astrophysical facilities in space.

One of the many discoveries made by the European Space Agency's XMM-Newton X-ray observatory was the finding of an exploding star in the Milky Way that had been previously missed. This object was once so bright it could have been seen by the naked eye during the period of its initial explosion and subsequent nova phase. XMM-Newton accidentally found this nova on October 9, 2007, as it slewed from one planned target of opportunity to the next planned observation. The discovery reminded astronomers that serendipity in space-based astrophysics is just as important as it has been during the history of earthbound astronomical observations.

COROT was launched in December, 2006, to survey stars and search for extrasolar planets using a transit method. In this type of investigation, light from a star is seen to diminish by a very small but measurable amount when a planet passes in front of that star. Absorption of light by the planet's atmosphere reveals something about the nature of that atmosphere. By mid-2008, COROT had surveyed more than fifty thousand stars. A discovery in 2008 was particularly intriguing. COROT detected the presence of an extrasolar planet the size of Jupiter orbiting a star similar in mass to the Sun. Located close to this star, the Jupiter-class exoplanet completes an orbit in just 9.2 days. The star rotates at exactly the same rate, thereby synchronizing the planet and star.

CONTEXT

Earthbound observatories have not been made obsolete by space-based telescopes. The former continue to be needed in order to perform some of the most important investigations in contemporary astrophysics. Space-based telescopes, however, have opened the eyes of scientists to a surprising, wondrous, energetic, and violent universe that could not have been envisioned based solely on earthbound observations. Our view of the solar system has been dramatically changed by widening the portion of the electromagnetic spectrum to which astronomers gained access once observatories were operational in space. One is reminded of the poetic verse of T. S. Eliot: "We shall never cease from exploration, and the end of all our exploring will be to arrive where we started and know the place for the first time."

Among the best examples of space-based telescopes that have altered the general public's view of the universe, and the utility of studying it, is the Hubble Space Telescope. Hubble's dramatic images of towering columns of interstellar matter and other phenomena have replaced the common notion of a sterile and relatively empty universe with a new understanding of the endless variety and dynamism of the cosmos. Besides adding to our rapidly advancing scientific understanding of the universe, Hubble has directly contributed to the health, safety, and quality of people's lives through a variety of technological spin-offs. For example, a nonsurgical breast biopsy technique using a device originally developed for Hubble's Imaging Spectrograph is now saving women pain, scarring, radiation exposure, time, and money. Called stereotactic automated large-core needle biopsy, this technique enables a doctor to locate a suspicious lump precisely and use a needle instead of incisional surgery to remove tissue for pathology. This precise procedure has been rendered possible because of a key improvement in digital imaging technology known as the charge-coupled device, or CCD. In addition to such practical applications, perhaps Hubble's greatest contribution has been in the realm of education, where it has been extremely successful in generating enthusiasm for astronomy among both students and the public alike.

Christopher Keating and David G. Fisher

FURTHER READING

Bely, Pierre, ed. *The Design and Construction of Large Optical Telescopes*. New York: Springer, 2003. An engineering history of the design and construction of large telescopes, including the Hubble Space Telescope. Explains the astronomical capabilities of Hubble.

Chaisson, Eric. *The Hubble Wars*. New York: HarperCollins, 1994. A well-written account of the inside story on the development and deployment of the Hubble Space Telescope, written by someone who was there. Careful attention is paid to explaining the scientific issues in terms the layperson can understand. Suitable for all levels.

Christensen, Lars Lindberg, Robert A. Fosbury, and M. Kornmesser. *Hubble: Fifteen Years of Discovery*. New York: Springer, 2006. Details the accomplishments of Hubble over its first fifteen years. Explains the ability of Hubble to examine the universe from ultraviolet through the near-infrared portion of the electromagnetic spectrum.

Devorkin, David, and Robert W. Smith. *Hubble: Imaging Space and Time*. Washington, D.C.: National Geographic Society, 2008. An oversized picture book including a large number of high-quality color images taken by the Hubble over its first seventeen years of research. Published just in advance of the final servicing mission to the telescope. Thoroughly readable.

Kerrod, Robin, Carole Stott, and David S. Leckrone. *Hubble: The Mirror on the Universe*. New York: Firefly Books, 2007. An overview of the Hubble Space Telescope's achievements. Fully illustrated with color prints of spectacular Hubble images. Describes the astronomical significance of the greatest achievements of Hubble.

Tobias, Russell R., and David G. Fisher, eds. *USA in Space*. 3d ed. Pasadena, Calif.: Salem Press, 2006. This three-volume set consists of a series of well-written, in-depth articles on all American space missions, both crewed and uncrewed. A number of articles discuss individual space telescopes in detail. A major library reference designed for the nonspecialist audience, with tables, lists, and indexes, and more than 400 photos. Suitable for all levels.

Trumper, Joachin, and Gunther Hasinger. *The Universe in X Rays*. New York: Springer, 2008. Reviews the development of X-ray astronomy and its impact on advancements in astrophysics and cosmology. Covers results from ROSAT, RXTE, BeppoSax, Chandra, and XMM-Newton.

Voit, Mark. *Hubble Space Telescope: New Views of the Universe*. New York: Harry N. Abrams, 2000. Details the saga of Hubble from concept to the flawed telescope needing orbital repair to its status as a frontier astrophysics research facility. Includes large-format high-resolution images taken by Hubble which demonstrate the tremendous capability of this observatory to examine objects at all distances, from those within the solar system to those more than 12 billion light-years from Earth—nearly back to the beginning of the universe.

Weeks, T. C. *Very High Energy Gamma Ray Astronomy*. New York: Taylor & Francis, 2003. Covers gamma-ray astronomy through results from the Compton Gamma Ray Observatory. Designed for students of either theoretical or experimental high-energy astrophysics.

Wilkie, Tom, and Mark Rosselli. *Visions of Heaven: The Mystery of the Universe Revealed by the Hubble Space Telescope*. London: Hodder & Stoughton, 1998. Presents numerous color images of Hubble's greatest discoveries with narrative to explain their astrophysical significance.

Zimmerman, Robert. *The Universe in a Mirror: The Saga and the Visionaries Who Built It*. Princeton, N.J.: Princeton University Press, 2008. Focuses not only on the stunning images produced by Hubble but also on the history of astronomical investigations and the development, construction, launch, and on-orbit repair of Hubble. Not only a technological story, Zimmerman's book is a human saga of scientists and the political process of obtaining funding for such a big science project.

See also: Archaeoastronomy; Coordinate Systems; Earth System Science; Gravity Measurement; Hertzsprung-Russell Diagram; Infrared Astronomy; Neutrino Astronomy; Optical Astronomy; Radio Astronomy; Telescopes: Ground-Based; Ultraviolet Astronomy; X-Ray and Gamma-Ray Astronomy.

Terrestrial Planets

Category: Planets and Planetology

The terrestrial planets are the four inner planets of the solar system. Mercury, Venus, Earth, and Mars are respectively the closest to the farthest from the Sun. These planets are composed of mostly silicate minerals and a dense inner core composed mostly of iron.

OVERVIEW

In broad terms, the planets of the solar system can be divided into two basic types: terrestrial and gas giants. The terrestrial planets (Mercury, Venus, Earth, and Mars) are by and large rocky with varying degrees of atmospheric envelopes and different amounts of water, ranging from very little to a great deal. The gas giants (Jupiter, Saturn, Uranus, and Neptune), as the name implies, are larger than the terrestrial planets and have tremendous gas envelopes surrounding their cores, which remain hidden from direct observation.

Mercury is the smallest of the terrestrial planets, 4,880 kilometers in diameter. The density of Mercury is relatively high, 5.4 grams per milliliter, implying that it has a bigger iron core than that of the Earth. Mercury also has a weak magnetic field. The magnetic field has been difficult to explain, since models for heat flow predict that the iron core ought to be solid. The surface of Mercury contains many impact craters, so it looks much like the surface of the Earth's moon. These craters probably formed by many meteorites (planetesimals) bombarding Mercury early in the formation of the solar system. The sparse data on the composition of the surface of Mercury suggest that the iron content is fairly low, so it cannot be composed of volcanic rocks, common on the other terrestrial planets, called basalts. Instead, the low iron content suggests that the surface is composed mostly of material formed from the planetesimals hitting the surface. There is no evidence that volcanic activity occurred on Mercury. Mercury has almost no atmosphere. The trace of hydrogen or helium in the atmosphere is most likely derived from the Sun. Temperatures vary from about 623 kel-

The terrestrial planets, in this composite image, are shown together to feature their relative sizes. From left: Mercury, Venus, Earth, and Mars. (Lunar and Planetary Institute)

vins in sunlight down to 103 kelvins in darkness.

Venus has a diameter very similar to that of the Earth (12,100 kilometers). The density of Venus is high enough (5.2 grams per cubic centimeter) to suggest that it also has a dense iron-rich core. Predictions of the planet's heat flow suggest that this iron-rich core ought to be molten like that of the Earth, which on the Earth produces the strong magnetic field. Venus, however, has no magnetic field, and the reason for this lack has not been determined. The surface of Venus is sparsely cratered, with about 1,000 randomly distributed impact craters. The reason for the distribution of these few craters is that large lava flows of basalt were extruded over most of the surface about 500 million years ago, covering craters formed earlier, so only the most recent impact craters are exposed at the surface. Many other surface features have been observed on Venus. For example, highlands and lowlands form 20 percent of the surface, and midlands form 80 percent of the surface. Each of these areas includes features such as folds, fractures, lava flows, volcanoes, and features resulting from weathering. The composition of the rocks is similar to those of basalts found on the Earth. The atmosphere of Venus has about ninety times the pressure at the surface as that on Earth. The main constituent in the atmosphere is carbon dioxide (96 percent), with lesser amounts of nitrogen and sulfur dioxide. The surface temperature of Venus is 733 kelvins, a result of extreme greenhouse warming due to the carbon-dioxide-rich atmosphere, which absorbs infrared radiation. If the atmosphere were similar to that of Earth, the temperature would likely be about 348 kelvins.

The diameter of Earth is 12,756 kilometers. Earth has a density of 5.52 grams per cubic centimeter, and it has a dense, iron-rich core. The planet's inner core is believed to be solid, and the outer core is believed to be liquid. The moving liquid iron is believed to produce Earth's strong magnetic field. This magnetic field helps to protect life by deflecting many of the charged particles (the solar wind) coming from the Sun. Few impact craters can be found on Earth because its crust is continuously re-formed as a result of plate tectonics and weathering. Earth's crust consists of about twenty plates of varying sizes, averaging around 100 kilometers thick. These plates move laterally at less than about 10 centimeters per year from areas where material is being formed in volcanic mountain ranges to regions where one plate is being subducted under another plate. Other volcanism occurs in the subduction zones. In contrast to the other terrestrial planets, Earth contains abundant liquid water, which covers about 70 percent of the surface, mostly in the oceans. Certain organisms in the water have helped to remove much of the carbon dioxide from the atmosphere by forming carbonate rocks containing carbonate minerals. Removal of the carbon dioxide from Earth's atmosphere has helped to avoid the absorption of infrared energy from the Sun, so the planet's temperature has remained cool and stable relative to that of Venus. The Earth's atmospheric pressure is, by definition, 1 atmosphere at sea level, and it contains 78 percent nitrogen and 21 percent oxygen. There are small amounts of other constituents, including varying amounts of water vapor.

The diameter of Mars is 6,788 kilometers, which is considerably less than that of Earth or Venus. The density of Mars (3.92 grams per cubic centimeter) is considerably less than that of the other terrestrial planets. The core of Mars still consists mostly of iron, but it appears to have fewer dense elements such as sulfur. Mars has no magnetic field, so the core is believed to be solid. Older surface rocks in the southern hemisphere of Mars are magnetized, suggesting that the core was liquid early in the planetary history. Some of the surface of Mars, especially in the southern hemisphere, contains impact craters. Huge volcanoes have formed in places, mostly from the buildup of lava. There are large plains that are composed of lava flows, sedimentary rocks, and material ejected from volcanoes through the air. The surface also displays large fractures in the rocks in places. There is no evidence, however, that plate tectonics occur on Mars as on Earth. In some locations, sinuous valleys suggest that running water may have existed on the surface in the distant past. If water did exist on Mars, some form of life might also have existed. However, no evidence has yet been found that life currently exists on Mars. The main obstacles to the current existence of life on Mars arise from its low atmospheric pressure (only 1 percent that of Earth) and an atmsophere composed mostly of carbon dioxide. Also, air temperatures are cold, ranging from only 190 to 240 kelvins from night to day. The low atmospheric pressure of Mars means that no liquid water can currently be stable on the Martian surface; any water in solid form would quickly sublimate (change from solid directly to

gas). Thus, the sinuous valleys could have been produced in the ancient past only when the atmospheric pressure was much higher.

KNOWLEDGE GAINED

Much of the information about Mercury, Venus, and Mars comes from observations from spacecraft that have flown by or landed on these planets. Also, some observations from Earth have been important.

Much of the information about Mercury came from the Mariner 10 spacecraft, which made three flybys in 1974 and 1975. Mariner 10 obtained information about the lack of an atmosphere, absence of a magnetic field, images of the abundant impact craters, and great variations in temperature.

Venus, with its continuous and thick cloud cover, has been difficult to study from Earth-based observatories. Mariner 2 flew by Venus in 1962, and the surface temperature and the lack of a magnetic field were detected. The Soviet Union's Venera missions determined that the atmosphere was mostly carbon dioxide. Venera 7 and 8 landed on the surface of Venus in 1970 and 1972, respectively. They measured the high temperature and pressure of the atmosphere and analyzed some of the surface rocks. Because of these conditions, however, the Venera spacecraft ceased to function within a few hours of landing. Later Venera spacecraft sent back color images of the Venusian surface, obtained better chemical analyses of the rocks, and made some high-resolution radar images of the surface. The Magellan spacecraft from the United States obtained almost complete high-resolution radar images of the surface, and they collected data on the gravity field.

Earth and its atmosphere have been studied in great detail by geologists, geophysicists, and atmospheric scientists for many centuries, so vast amounts of information have been collected compared to the other terrestrial planets. Geologists have mapped Earth's surface geology in great detail, and they have determined the mineralogy and chemical composition of the rocks at and near the surface of the Earth. Geophysicists have used seismic waves from earthquakes and data on variations in gravity and magnetism to estimate the composition of the interior of the Earth. Atmospheric scientists have determined the composition and variations of the present and ancient atmosphere of the Earth.

Initially Mars was viewed by telescope from the Earth, and some astronomers believed that they saw "canals" (or, from the Italian, "channels") on the surface. As a result, some nineteenth and early twentieth century astronomers speculated that Mars could have life. When Mariner 4 flew near Mars in 1965, it revealed that the southern hemisphere of Mars had impact craters and no canals. It also found that Mars had no magnetic field. Mariners 6 and 7 also flew by Mars in 1969, and they obtained more surface images. Mariner 9 orbited Mars in 1971. Detailed images of the surface revealed such features as large volcanoes and sinuous canyons. Vikings 1 and 2 landed on Mars in 1976, and they sent back information about the soil and air and searched for life until 1982, when they ceased to function. Numerous spacecraft since 1999 have extended the examination of Mars's surface and atmosphere. Chief among these have been the Mars Climate Orbiter, Mars Polar Lander, Mars Odyssey, Mars Pathfinder, Mars Exploration Rovers, Mars Express, Mars Reconnaissance Orbiter, and Mars Phoenix.

CONTEXT

There are still many questions about the terrestrial planets that need to be answered which should at least partially be answered by future space missions. For instance, the MESSENGER mission, launched on August 3, 2004, reached Mercury in 2008. It is designed to obtain further information about Mercury after it eventually enters orbit about the planet. The spacecraft needs to slow down considerably to be able to orbit Mercury, so the National Aeronautics and Space Administration (NASA) had it fly by Venus, Mercury, and Earth several times to slow it sufficiently to facilitate orbit of Mercury in March, 2011. As a result, MESSENGER flew by Venus in June, 2007, and took pictures to calibrate its cameras. It flew by Mercury in January 14, 2008, and again on October 6, taking a number of pictures of the surface. When MESSENGER eventually orbits Mer-

cury, it will take more pictures of the surface to help determine the planet's geologic history, study the weak magnetic field to help determine its origin and the nature of the core, and conduct experiments to determine the composition of some of the rocks and atmosphere, which in turn will lead to a better understanding of their formation.

Another mission to Mecury, a joint effort between the European Space Agency and the Japanese Aerospace Exploration Agency, is the Bepi Colombo mission, to be launched in 2013 and expected to arrive in 2019. Bepi Colombo is expected to include not only an orbiter but also a lander, which would allow more detailed observations of the surface of Mercury.

Mars is undergoing continual intense investigation through both orbiting spacecraft and surface experiments conducted by rovers. In early 2009, Earth-based observations led to the discovery of a methane signature in the Martian atmosphere. The planned Mars Science Laboratory mission, expected to launch in 2011, may be repurposed to focus on this intriguing discovery.

Robert L. Cullers

FURTHER READING

Faure, Gunter, and Teresa M. Mensing. *Introduction to Planetary Science: The Geological Perspective*. New York: Springer, 2007. The authors have summarized the information about all bodies in the solar system in this textbook. Includes many photographs, graphs, and tables, as well as a glossary and an index. Best for those with some background in geology.

Hansen, V. L., and D. A. Young. "Venus's Evolution: A Synthesis." In *Convergent Margin Terranes and Associated Regions*, edited by W. Carlson et al. Boulder, Colo.: Geological Society of America, 2007. This paper summarizes much of what is known about Venus, including the formation of the impact craters and theories about the planet's formation. Contains figures to illustrate the discussion.

Marvin, Ursula. "Geology: From an Earth to a Planetary Science in the Twentieth Century." In *The Earth Inside and Out: Some Major Contributions to Geology in the Twentieth Century*, edited by D. R. Oldroyd. London: Geological Society, 2002. This article describes the geology of the solar system, starting with that of Earth, and how the understanding of other planetary bodies has elucidated our understanding of how Earth formed. Includes fourteen figures and several tables.

Selley, Richard, Robin Cocks, and Ian Plimer, eds. *Encyclopedia of Geology*. 5 vols. Oxford, England: Elsevier Academic Press, 2005. Contains a vast amount of information about geology, including articles about the planetary bodies of the solar system.

Solomon, Sean. "Mercury: The Enigmatic Innermost Planet." *Earth and Planetary Science Letters* 216 (2003): 441-455. Summarizes what is known about Mercury and what needs to be studied further. Outlines planned space missions. Provides many figures and other illustrations, including photos of the surface.

See also: Europa; Jovian Planets; Neptune's Interior; Planetology: Comparative; Planetology: Venus, Earth, and Mars; Solar System: Origins.

Thermonuclear Reactions in Stars

Category: The Stellar Context

Thermonuclear reactions are the way stars generate energy for most of their lives. The various reactions are related to the changes stars undergo. They explain how the Sun has continued to shine for the past 4.5 billion years, and they predict how the Sun will change in the future.

OVERVIEW

Our Sun is a main sequence star, about halfway through its lifetime (an estimated 10 billion years); as such, it is the closest stellar "laboratory" for research into the astrophysics and astrochemistry believed to take place in similar stars throughout the universe. The nature of the Sun's energy source baffled scientists for

centuries. In 1905, however, while Albert Einstein was developing his special theory of relativity, an equation unexpectedly emerged: $E = mc^2$, where E is energy, m is mass, and c is the speed of light. This equation indicated that energy and matter are equivalent and can be transformed into each other. Another implication is that a small amount of mass is equivalent to a large amount of energy; for example, a 1-kilogram mass, if converted totally into energy, would yield 9×10^{16} joules. A joule is the energy needed to lift that 1-kilogram mass about 10 centimeters. The idea that mass could be converted into energy became the basis for thermonuclear weapons and for uranium-fueled fission reactors. It also is the mechanism that produces the energy of the Sun and other stars for most of their lives.

The Sun's power output, the rate at which it produces energy, is 3.83×10^{26} joules per second. One second of the Sun's energy output is equivalent to 10 million times the annual electricity consumed within the United States. To radiate at this rate, the Sun converts 4.25 billion kilograms of matter into energy every second.

The solar system, with its Sun, planets, satellites, and associated objects, developed from a cloud of gas and dust several light-years in diameter. That slowly spinning cloud started to contract under its gravitational attraction. As it shrank, the cloud rotated faster and spun off an equatorial disk. At the center, a ball of gas called the proto-Sun continued to contract and grow hotter. It was not yet a full-fledged star, since it was not converting mass into energy. Its rising temperature was caused by gravitational contraction of the gas and conversion of its gravitational potential energy into thermal energy. As the contraction continued, temperature, density, and pressure in the interior of the proto-Sun rose. When these became high enough, nuclear fusion ignited, and the proto-Sun became a star.

Since one cannot physically measure conditions within the interior of a star, astronomers mathematically model the Sun's internal structure. Using known conditions at the surface and the Sun's total mass, diameter, and energy output, computers can be used to calculate the changing conditions from the surface into the

Proton-Proton Chain Reaction

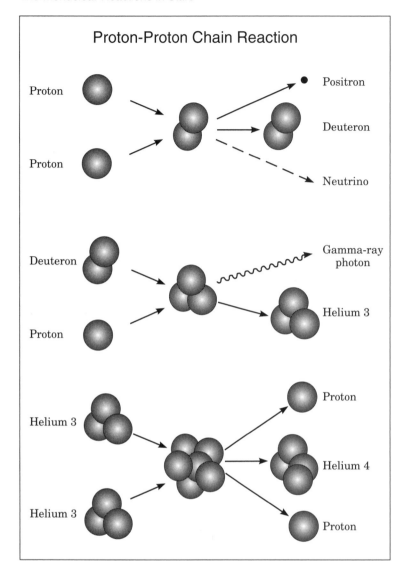

Reactions start with the collision of two high-speed protons. Normally, as two protons approach each other, their positive electrical charge produces a mutual repulsion, and they slow down, stop, and then move away in a nearly elastic collision. If they are moving sufficiently fast, however, they can come close enough for the short-range strong nuclear force to allow the protons to attract each other. At this point, one of the protons is converted into a neutron, and a deuterium nucleus (hydrogen 2, an isotope of hydrogen with one proton and one neutron) results. Since electrical charge must be conserved, the change of a proton into a neutron results in the emission of a positron, a positively charged electron, which is the antimatter form of the electron. When a positron and an ordinary electron collide, they mutually annihilate each other, and their mass is completely converted into 1.6×10^{-13} joules, as described by Einstein's mass-energy equation. The energy is released as gamma rays, a high-frequency, high-energy form of electromagnetic radiation. A neutrino, a neutral particle with negligible mass, is also emitted by the nuclear fusion reaction to conserve energy and momentum simultaneously.

Next, a fast-moving proton collides with the deuterium nucleus, yielding a helium-3 nucleus and a gamma ray. The helium-3 nucleus has two protons and one neutron. Finally, two helium-3 nuclei collide to produce one helium-4 nucleus (with two protons and two neutrons) and two free protons. The net result of the interactions is that four protons fuse to yield one helium-4 nucleus, two positrons (which are annihilated when they encounter two free electrons), two neutrinos, and gamma rays.

interior. The model's values for mass and energy output must match those of the Sun for the model to hold validity. Calculations suggest that the core temperature is 15 million kelvins, pressure is as high as 200 billion atmospheres, and the density reaches 150 grams per cubic centimeter. It is under these conditions of high temperature, high pressure, and high particle density that the fusion of four protons (each a nucleus of ordinary hydrogen) into helium 4 occurs. It is a series of thermonuclear reactions, known as the proton-proton chain, in the 350,000-kilometer-diameter core that powers the Sun.

The mass of four protons is 6.6942×10^{-27} kilograms, while the mass of a helium-4 nucleus is 6.6466×10^{-27} kilograms. That mass difference of 0.0476×10^{-27} kilograms is converted into 4.28×10^{-12} joules of energy according to Einstein's mass-energy equation. The amount of matter converted per second by the Sun into energy is equivalent to 4 billion kilograms, or the mass of about 2 million automobiles. Each second, 8.9×10^{37} nuclear reactions transform 610 billion kilograms of hydrogen into 606 billion kilograms of helium and 3.8×10^{26} joules of energy.

The gamma-ray photons resulting from the reactions are absorbed and reradiated many times during their journey to the Sun's photosphere. This energy requires from 100,000 to 1 million years to make the 700,000-kilometer trip from the core of the Sun to its "surface," where most of the energy departs as photons of ultraviolet, visible, and infrared electromagnetic radiation.

The released radiative energy also prevents the Sun from collapsing. The gravitational attraction of the Sun's mass produces an inward pull, but the radiative energy of the Sun heats the interior gases. This produces an outward gas pressure to counter the inward pull of gravity. Fortunately for life on Earth, this has resulted in a star that has been stable for the past 4.5 billion years. With the remaining hydrogen in the core, our Sun should last for about another 5 billion years. At the end of that time, the core's hydrogen supply will be depleted, and the Sun's helium "ash" core will start to contract.

This core contraction stage will raise the internal temperature to the point at which the hydrogen in a shell around the contracting helium core becomes hot enough to fuse into helium 4. The Sun's outer layers will expand because of this increased energy production, and the Sun will enter a red giant stage in which it will engulf the inner planets of the solar system at least out to Venus and perhaps as far out as Mars. Later, as the core temperature increases to more than 100 million kelvins, two of the core's helium-4 nuclei will fuse into a beryllium-8 nucleus. Another helium-4 nucleus will fuse with the beryllium to produce carbon 12. In each case, a gamma-ray photon is emitted. Core reactions will cease as the helium 4 is depleted, but

hydrogen and helium fusion in shells around the carbon core will continue. This ultimately will lead to expulsion of the Sun's outer layers to form a planetary nebula.

The Sun's carbon core will shrink to become a white dwarf about the size of the Earth. The Sun as a white dwarf will radiate only by its residual heat, since no thermonuclear reactions will occur within it. It will slowly cool to its final black dwarf phase, a burnt-out star with less than the original mass of the Sun.

Stars that form with masses greater than about 1.4 solar masses also fuse hydrogen into helium in their cores, but with a series of thermonuclear reactions known as the carbon-nitrogen-oxygen (CNO) cycle, rather than the proton-proton chain, due to core temperature, pressure, and particle density higher than those of a 1-solar-mass star. These conditions are caused by the larger mass of the star producing a greater gravitational pull inward. Because of higher temperatures and densities, the reactions will proceed at a faster rate. The star has a larger mass but consumes itself more rapidly. This results in a shorter lifetime for such a star. As the core depletes its hydrogen, it contracts and heats, and helium fusion commences. Hydrogen-to-helium fusion is initiated in the hydrogen shell around the core.

When the core's helium is expended, the core contracts again, with a resulting increase in temperature and density, and a new fusion reaction is ignited. This cycle of fusion, depletion, contraction, and new fusion continues until the nuclei in the core are ultimately converted into iron nuclei and the core is surrounded by several concentric shells of different fusing nuclei. The outer shell consists of hydrogen fusing to helium. Iron is the core's end product, since further fusion requires input of additional energy rather than its exothermic release. With no source of thermal energy to prevent the further gravitational collapse of the star, it implodes. Material bounces off the core, and the star violently explodes into a core-collapse supernova.

Energy is so abundant during the supernova explosion that nuclei are fused into elements with atomic numbers higher than iron, elements even as heavy as uranium and thorium. The remnant of the star collapses to as little as

10 kilometers in diameter and becomes a neutron star, because at such high pressures the electrons and protons of the star combine into neutrons. If a supernova leaves a stellar remnant with more than 3 solar masses, an even more dramatic end state occurs. The collapse continues beyond the neutron-star stage into a black hole. Nothing can stop the gravitational attraction that collapses the star to a size from which not even light can escape.

APPLICATIONS

The study of thermonuclear reactions is applicable to humanity's most urgent need: energy. The standard of living is governed by the availability of easily obtained energy, and humanity's present control of energy permits people to perform feats that previously were thought impossible. Human and animal muscle power, wind and water, coal, natural gas, petroleum, nuclear fission: Each has proved to be inadequate in one way or another, either because it is insufficient in amount or intensity, or because it produces harmful waste. Humanity needs a reliable source of energy, and thermonuclear reactions, either directly or indirectly, may well be that source.

Solar energy is the product of a natural thermonuclear reactor 150 million kilometers from Earth, and humans use solar energy in many forms without knowing it. In fact, except for tidal and geothermal energy, all energy sources are implicitly solar related. Coal, oil, and natural gas, for example, are forms of "fossilized" energy based originally on solar illumination of the planet. The direct application of solar energy is difficult, since solar energy is a diffuse energy resource. Large arrays of solar panels are needed to supply the required energy, and they operate only when the Sun shines.

One method of avoiding the earthbound problems of solar power is to construct in geostationary orbit around the Earth large solar power satellites. One such station, 5 kilometers by 15 kilometers in area, could supply the needs of New York City with plenty of energy to spare. Its solar cells would convert light into electricity, which would produce microwave radiation. That would be beamed to Earth, where a receiving antenna would convert the microwaves back into electrical power. One of these satellites would provide six to ten times the energy that an array of the same size on Earth would provide. Several hundred of these satellites would supply a large portion of humanity's energy needs.

Another possibility is to design machines that would permit the control of fusion reactions on Earth. For decades, engineers and scientists have struggled with the problem of obtaining sufficiently high particle densities, temperatures, and pressures. The main problem is the design of a vessel that will constrain deuterium long enough for the fusion to occur. Magnetic fields and lasers have been employed to initiate the fusion process, but the goal of a net energy output on an economic scale remains elusive. When practical fusion reactors are developed, hydrogen from the water of the Earth's oceans will provide an abundant source of fuel.

Scientists who construct models of the thermonuclear reactions and other processes that occur in the cores of stars use the Sun as their test case. Certain predictions about the Sun can be made from the model. One of these involves the number of neutrinos emitted by the reactions in the Sun's core. Neutrinos are particles with little mass and no electrical charge, traveling at speeds close to that of light. They do not interact readily with matter; a neutrino can easily pass through a light-year-thick layer of lead shielding. The model predicts the number of solar neutrinos that should be counted by experiments designed to detect them. The first measurements in the 1970's detected only one-third of the expected number of neutrinos. Scientists questioned if the models of the Sun's interior processes were incorrect, or if there was something unknown happening in the Sun's interior that the models had not taken into account. After decades of uncertainty over the Sun's missing neutrino flux, it has been determined that solar neutrinos change into other forms or flavors during their trip between the Sun and Earth, and the original apparent discrepancy seems to have been resolved.

Since neutrinos make the trip from the core to the detectors at nearly the speed of light, they provide information about what is occurring now in the Sun's core. The electromagnetic radiation from the Sun's photosphere discloses what occurred in the core 100,000 or more years ago. Further theoretical and experimental work should produce an answer on determining if the Sun's energy output varies substantially over time. Indeed, many scientists think of the Sun as a variable star, but fortunately that level of variability is quite low compared to other "traditional" classes of variable stars.

CONTEXT

In the eighteenth century, Immanuel Kant estimated that if the Sun were composed of coal, it would burn only for several thousand years. In the nineteenth century, Hermann von Helmholtz and Lord Kelvin independently reasoned that gravitational contraction of the Sun could provide its energy by converting gravitational potential energy into thermal energy. This theory allowed the Sun's age to be increased to 20 to 50 million years, but it did not satisfy geologists and the biologists, who argued that hundreds of millions to billions of years were necessary for the evolution of life and the geophysical development of the Earth.

In the nineteenth century, physicists conducted experiments to determine how the speed of light changed as the speed of the medium through which it traveled was varied. The conclusion was that there is no medium through which light moves. It was also observed that the speed of light is constant, no matter how fast its source moves. Classical physics was unable to accept this seemingly nonsensical result for the speed of light. Einstein, however, stated that the speed of light is a constant in any inertial frame of reference, and went on to investigate the consequences of this postulate. While making those derivations, his famous mass-energy equivalence emerged. In 1939, Hans Albrecht Bethe and Carl von Weizsäcker hypothesized that nuclear reactions could generate the Sun's energy. They suggested that four protons could fuse into helium 4, and that the mass difference was converted into energy.

In the early 1950's, the thermonuclear or hydrogen bomb was developed, in which ignition of a fission bomb trigger produces high temperatures that lead to fusion of deuterium and the release of even more energy. This produces the hydrogen bomb's greater destructive power and in a sense replicates in an uncontrolled fashion the tremendous power of the Sun's fusion process.

Stephen J. Shulik

FURTHER READING

Chaisson, Eric, and Steven McMillan. *Astronomy Today*. 6th ed. New York: Addision-Wesley, 2008. This astronomy textbook contains a thorough description of the stages of stellar evolution, complete with transparent overlays, and an exceptionally lucid treatment of thermonuclear energy generation.

Dinwiddie, Robert, et al. *Universe*. New York: DK Adult, 2005. A remarkable collection of articles written by science writers and professional astronomers on a wide range of topics that span the discipline of astronomy. Heavily illustrated and filled with high-quality photographs. For the general reader.

Fraknoi, Andrew, David Morrison, and Sidney Wolff. *Voyages to the Stars and Galaxies*. Belmont, Calif.: Brooks/Cole-Thomson Learning, 2006. A well-written, thorough college textbook for introductory astronomy courses. Provides a good description of stellar evolution and thermonuclear energy generation.

Freedman, Roger A., and William J. Kaufmann III. *Universe*. 8th ed. New York: W. H. Freeman, 2008. College-level introductory astronomy textbook. Offers a good description of stellar evolution and thermonuclear energy generation.

Hansen, Carl, Steven Kawaler, and Virginia Trimble. *Stellar Interiors: Physical Principles, Structure, and Evolution*. 2d ed. New York: Springer, 2004. Covers the fundamentals of stellar physics, structure, and life cycle. Newer research topics such as astroseismology and the effect of magnetic fields are also covered. For advanced undergraduates.

Prialnik, Dina. *An Introduction to the Theory of Stellar Structure and Evolution*. Cambridge, England: Cambridge University Press, 2000. An undergraduate textbook covering all aspects of stellar evolution and the structure of stars. Full solutions to exercises as well as basic physics and mathematics are included.

Schneider, Stephen E., and Thomas T. Arny.

Pathways to Astronomy. 2d ed. New York: McGraw-Hill, 2008. Very thorough college textbook for introductory astronomy courses. Divided into lots of short sections on specific topics, it presents a thorough discussion of stellar evolution and thermonuclear energy generation.

Thornton, Stephen T., and Andrew Rex. *Modern Physics for Students and Engineers.* 3d ed. New York: Brooks/Cole, 2005. A comprehensive presentation of the development of relativity, quantum mechanics, and nuclear and particle theory and experimentation. For undergraduates and serious scientific researchers.

Young, Hugh D., and Roger A. Freedman. *University Physics with Modern Physics.* 11th ed. New York: Addison-Wesley, 2003. An undergraduate text that spans classical mechanics, thermodynamics, Maxwell's electrodynamics, optics, and modern physics. Offers a good introduction to aspects of elementary particle physics as well.

See also: Brown Dwarfs; Gamma-Ray Bursters; Main Sequence Stars; Novae, Bursters, and X-ray Sources; Nuclear Synthesis in Stars; Protostars; Pulsars; Red Dwarf Stars; Red Giant Stars; Stellar Evolution; Supernovae; White and Black Dwarfs.

Titan

Categories: Natural Planetary Satellites; The Saturnian System

Saturn's largest satellite, Titan, is the only satellite in the solar system with a thick atmosphere. Astronomical observations made from Earth established some time ago that Titan has a density slightly greater than compressed ice, indicating a composition primarily of ice but with a relatively small rocky core. Observations made by the Cassini-Huygens spacecraft show a surface with multiple hydrocarbon lakes that could be breeding grounds for primitive living organisms.

OVERVIEW

Titan is Saturn's largest satellite, with a diameter of about 5,150 kilometers. Its atmosphere (whose density is several times that of Earth's atmosphere) was discovered in 1944 from spectral analyses of sunlight reflected from the cloud cover. The spectral data indicated the presence of methane gas (CH_4). Additional Earth-based observations in 1973 showed a reddish, hazy atmosphere, which was assumed to be photochemical smog created by ultraviolet light from the Sun acting on the methane and other hydrocarbon compounds.

Because Titan is such an unusual satellite, the Pioneer 11 spacecraft flew by Titan in 1979, followed by Voyager 1 and 2 in 1980 and 1981, respectively. Unfortunately, their instruments were not sensitive enough to penetrate Titan's thick atmosphere, although it was learned that its major constituent is nitrogen, with methane and smog making up less than 10 percent of the atmosphere. Hazy smog is formed as the methane is catalyzed by ultraviolet light from the Sun to form more complex organic molecules, similar to the manner in which photochemical smog is produced from unburned fuel in the exhaust emitted by vehicles in Earth's large cities. Because Titan is quite far from the Sun and relatively little of the available light would penetrate the thick clouds, the surface temperature was predicted to be about 94 kelvins.

Calculations indicated that at these temperatures methane should condense from the clouds and fall as rain. The denser organic molecules created in the atmosphere, such as ethane (C_2H_6), would also eventually settle on the surface as a layer of malodorous slime. It was thus assumed that the icy surface was covered by either an ocean of liquid methane or a hydrocarbon swamp, with frequent rainstorms of methane.

In 1994, scientists used the Hubble Space Telescope at near-infrared wavelengths (where the haze is more transparent) to map some of Titan's surface features according to their reflectivity. Although details were not resolvable, light and dark surface features were recorded over Titan's sixteen-day rotation period, and one bright area the size of Australia was documented. Definitive conclusions about the nature of the dark and bright areas could not be ascertained, but images proved that the surface was not a global ocean of methane and ethane, as had been assumed; at least part of the surface is solid. Although definite conclusions could not be made, it was thought that the bright areas were major impact craters in the frozen surface. Information gleaned from this research provided important background information for the Cassini mission, the program of National Aeronautics and Space Administration (NASA) that sent a robotic spacecraft to study Saturn and its satellites. In addition to data to be gathered from flybys, Cassini would release the European Space Agency's Huygens probe, which would parachute to the surface. Images of Titan from the Hubble telescope were used to locate an optimum landing site and to predict how Titan's winds would affect the parachute as it descended through the atmosphere.

A composite image of Titan from the Cassini spacecraft. (NASA/JPL/University of Arizona)

The Cassini spacecraft was launched in October, 1997, for its seven-year voyage to rendezvous with Saturn. Beginning to orbit Titan in 2004, it flew 1,192 kilometers above the surface, using infrared cameras and radar to produce detailed maps. It detected irregular highlands and smoother dark areas, including one large region, about the size of Lake Ontario (232 by 72 kilometers), so reminiscent of a lake that its perimeter even exhibited sinuous drainage channels leading to an apparent shorelike boundary. Because the surface temperature was so cold (94 kelvins), the lakes were presumed to be liquid methane and ethane fed by streams of dark organic gunk washed by precipitation from the highlands. Methane evaporating from the lakes would replenish the methane in the atmosphere, from which it would eventually precipitate and return to the surface as rain, mimicking the hydrologic cycle on Earth. The fact that this feature appears in Titan's cloudiest region, where presumably storms are

intense enough that methane rain reaches the surface, gave credence to the lake hypothesis. Furthermore, Titan's cold temperature would require a long time for liquid methane on the surface to evaporate; thus, a methane-filled lake would remain stable for a long time.

The Huygens probe was released on January 14, 2005. As it descended, it recorded the temperature, pressure, wind speed, and atmospheric composition at regular time intervals. It also radioed back more detailed images of the surface, showing dark drainage networks leading into the smooth areas but relatively few craters, as expected. The bright spots appeared to be "islands" around which dark material had flowed in the past. Other images showed areas evocative of water ice extruded onto the surface and short, stubby, dark channels which could be springs of liquid methane. Although these data suggested running liquids, no clear evidence of liquid methane was detected on the surface. After landing, Huygens probed the surface, which had the consistency of wet sand covered with a thin crust, possibly consisting of ice mixed with small amounts of solid methane. First pictures of the surface showed a plethora of small erosion-rounded pebbles, assumed initially to be rocks or granite-hard ice blocks, on an orange-colored surface. They later were determined to be mixtures of water and hydrocarbon ice. One

The Huygens lander took this image on January 14, 2005, from Titan's surface. The rocks at the front are about 4 to 6 inches in diameter and may be the remains of a lakebed. (ESA/NASA/JPL/University of Arizona)

image pictured tendrils of surface fog, presumed to be ethane or methane.

The concentration of methane in Titan's atmosphere is puzzling, because ultraviolet light from the Sun would dissociate the methane into carbon and hydrogen, which would either react with the nitrogen to form ammonia (NH_3) or be dissipated into space. More complex organic molecules, such as ethane, would also be created; being heavier, they would settle to the surface. It has been calculated that atmospheric methane should remain in the atmosphere for fewer than one million years. Consequently, methane must be injected from some surface source. Perhaps it is outgassed from the methane in the icy crust. Another possibility is a methane volcano. Detailed observations have identified one area where ice and methane may be rising to the crust from a subterranean heat source, to form a methane volcanic caldera emitting methane gas. Since Titan is too small to have a molten interior, the heat source driving the release of methane gas is suspected to be tidal heating, the frictional force generated as this massive satellite revolves in its elliptical orbit about Saturn. Several dark surface markings having straight boundaries with preferred orientations suggest the presence of internal tectonic processes.

Analyses of close flybys of Titan made by the Cassini orbiter in 2008 as well as conclusions based on data gathered by flybys made between 2004 and 2006 provided suggestive evidence for the possibility that Titan's surface may have active cryovolcanoes, perhaps spewing water, ammonia, and methane. Cassini recorded variations in brightness and reflectance (the ratio of reflected light to the incident light upon a surface) in two separate regions of Saturn's largest satellite using its Visible and Infrared Mapping Spectrometer. In one region, the reflectance increased significantly and remained at the elevated level; in the other, it rapidly increased and then tailed off again. Both would indicate vapor or liquid being ejected out of an active vent. Such a mechanism would explain why Titan continues to maintain a thick methane atmosphere when, without replenishment, it should have been greatly diminished over the passage of geologic time.

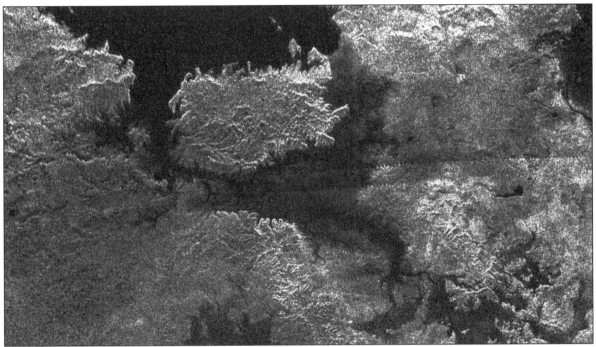

Cassini captured this image of Titan's surface on February 22, 2007, with a resolution of about 700 meters. The view looks directly down on an island (possibly a peninsula), about the size of the Big Island of Hawaii, in the middle of one of the moon's hydrocarbon lakes. (NASA/JPL)

KNOWLEDGE GAINED

Using radar, Cassini was still mapping Titan's surface in 2009 and was expected to continue doing so for some time to come. The hundreds of observed dark areas are believed to be lakes of liquid methane or ethane more than 12.2 meters deep, while shadowy dunes running along the equator are assumed to consist of complex solid organics. Titan's surface seems to contain many gigantic organic chemical factories producing complex hydrocarbons in an abundance surpassing all of Earth's oil reserves. The amount of liquid on Titan's surface is important to ascertain, because methane is a strong greenhouse gas; without atmospheric methane, Titan's surface would be even colder. Liquid methane on the surface could remain, at most, a million years before dissociating and reacting to form heavier hydrocarbon compounds. It is believed that the atmospheric methane is constantly being supplied by volcanic eruptions from the mantle.

In late December, 2008, after much analysis, a group of researchers were ready to publish a scientific article in the research journal *Icarus* reporting the first image taken of a liquid on a planetary surface other than the Earth's. The Mars Phoenix lander had provided clear evidence about six weeks earlier of subsurface water ice at its far north landing site, but when that ice was exposed, it fairly quickly sublimated into the gas phase and became part of the Martian atmosphere. However, an image taken by the Huygens probe after reaching the surface of Saturn's satellite Titan appeared to have clearly recorded a droplet of methane near the edge of the robotic spacecraft itself. The small droplet might have been created by heat emanating from the probe when it condensed humid air to temporarily form liquid methane. In several other images, splotches that appeared and then were not seen in subsequent images of the same area were believed by the authors of the *Icarus* paper also to be droplets of methane.

Titan has an extensive atmosphere, including a methane layer extending 696 kilometers above the surface. There the methane molecules are dissociated by ultraviolet light to form ethane (C_2H_6) and acetylene (C_2H_2). Cassini images showed two thin haze layers. The outer haze layer, floating about 400 kilometers above the surface, is where additional molecules (such as hydrogen cyanide) are formed from carbon, hydrogen, and nitrogen. About 200 kilometers above the surface, there is a thick global smog of complex organic molecules, produced by chemical reactions among the hydrocarbons dissociated by ultraviolet light. This haze layer absorbs about 90 percent of the incident sunlight, leaving only an orangish haze to reach the surface. It is not currently understood why two separate haze layers are present. Although Titan has a dense atmosphere, it is relatively inefficient at reradiating infrared radiation, thus producing negligible greenhouse warming.

CONTEXT

The Cassini-Huygens mission was a joint venture of NASA, the European Space Agency, and the Italian Space Agency. Enough data were gleaned to keep researchers occupied for years to come.

Titan's surface temperature (94 kelvins) appears to make the satellite a place inhospitable for life to evolve. This environment, although colder, is remarkably similar to that found on Earth billions of years ago, before life began adding oxygen to the atmosphere. Four billion years ago, Earth was covered with warm, shallow seas containing hydrogen, ammonia, and methane gases. From this primordial soup, driven by ultraviolet light and lightning discharges, complex hydrocarbons, including amino acids, formed. Over time, the amino acids linked together to form proteins, eventually creating one that was able to replicate itself. At that point, life was created and molecular evolution became biological evolution.

If life could evolve in Earth's primordial soup, it seems reasonable to suppose that the pools of organic gunk on Titan's surface could form amino acids, if not self-replicating proteins. Studying Titan's prebiotic chemistry can therefore facilitate the understanding of how life may have originated in the universe.

George R. Plitnik

An artist created this image of a lake and smoggy atmosphere of Titan based on data that led scientists to conclude the moon has lakes of liquid hydrocarbons. (NASA/JPL)

FURTHER READING

Chaisson, Eric, and Steve McMillan. *Astronomy Today*. 6th ed. New York: Addison-Wesley, 2008. This easily accessible work, written for laymen with inquisitive minds, has an excellent summary of the latest knowledge about Titan as well as pictures from the Huygens landing and an instructive graph of the variation of pressure and temperature as a function of altitude.

Coustenis, Athena, and Fredric W. Taylor. *Titan: Exploring an Earthlike World*. Hackensack, N.J.: World Scientific, 2007. A revised and expanded edition of the 1999 title *Titan: The Earthlike Moon*, this volume summarizes all that is known about Titan through the Cassini-Huygens mission, by two of the project's investigators. Aimed at a general audience, but scientifically rigourous nonetheless.

Hartmann, William K. *Moons and Planets*. 5th ed. Belmont, Calif.: Thomson Brooks/Cole, 2005. This authoritative and regularly updated text considers all the major planetary objects in our solar system. The material is presented by grouping objects under unifying principles, thus elucidating their similarities and their differences as well as the physical processes behind their evolution. Although most of the material is descriptive, some algebra and elementary calculus are included.

Lorenz, Ralph, and Jacqueline Mitton. *Titan Unveiled: Saturn's Mysterious Moon Explored*. Princeton, N.J.: Princeton University Press, 2008. This illustrated tome was the definitive work covering everything known about the surface and atmosphere of Titan at the time of publication. Because Cassini-Huygens was still years away from its encounter with Titan, the authors had to predict—but with some accuracy it turned out—features and conditions on the surface based on limited data.

Sagan, Carl. *Cosmos*. New York: Random House, 1980. Based on the television series of the same name, this lavishly illustrated book includes not only information about Titan's atmosphere but also speculation about the possibility of a methane-based life-form evolving in this environment.

Seeds, Michael A. *Foundations of Astronomy*. 9th ed. Belmont, Calif.: Thomson Brooks/Cole, 2007. This well-illustrated text commingles experimental evidence and theory to provide deep, but well-explained, elucidations of many fascinating facets of the universe. Although only two pages are devoted to Titan, there are four pictures, one of which was taken from the surface by the Huygens probe.

See also: Enceladus; Iapetus; Planetary Ring Systems; Planetary Satellites; Saturn's Magnetic Field; Saturn's Ring System; Saturn's Satellites; Solar System: Element Distribution.

Triton

Categories: Natural Planetary Satellites; The Neptunian System

Triton, Neptune's largest satellite, is the solar system's only major satellite that is in a retrograde orbit. It has smoke plumes that astronomers and planetologists cannot explain. It appears to be younger than most of the satellites and planets in the solar system.

OVERVIEW

The Neptunian satellite Triton is 2,706 kilometers in diameter. The satellite's density is 2.07 grams/centimeter^3. With this density, models of the satellite can be constructed. It is thought that there is a metallic core, a silicate mantle, a layer of ice, a possible ocean, and a top layer of ice. The core is expected to have a radius of about 600 kilometers and the mantle a thickness of about 350 kilometers, with a 150-kilometer layer of ice below the ocean and a 250-kilometer layer above the ocean.

William Lassell, a brewer by trade, found Triton on October 10, 1846, using a telescope he built himself. An amateur astronomer, Lassell had been asked by Sir John Herschel to look for satellites of the newly discovered planet, Neptune. Triton is 355,000 kilometers from Neptune. The eccentricity of Triton's orbit is zero, meaning that the orbit is circular. Triton is a most unusual satellite. Its orbit is retrograde, that is, it rotates around Neptune in the opposite direction to the rotation of Neptune. It is synchronous with Neptune, meaning it presents the same face to Neptune at all times. Being synchronous also means that the rotation time of Triton is the same as the time for one orbit around Neptune, 5 days and 21 hours. The angle of inclination is 23°. The angle between the orbit of Triton and the equator of Neptune is 23°.

This large angle of inclination, coupled with the retrograde direction of Triton's orbit, suggests that Triton was captured by Neptune's gravitational field. The method by which this capture occurred is unknown. Triton may have collided with another satellite, causing Triton to slow down enough to be captured and destroying the other satellite at the same time. Another idea is that the other satellite was knocked out of orbit, and Triton was then captured. A third idea is that Triton was part of a bi-

nary system. Triton was captured; however, the partner escaped. When Triton was captured, the orbit was probably highly elliptical. The gravitational force of Neptune gradually changed Triton's orbit into the current circular orbit. During the period of this change, the strong pull when Triton was close and the weaker pull as Triton was far away would have caused tidal flexing, or movement within Triton's structure. Such motion would have caused internal friction, generating heat. The heat would have caused differentiation, that is, separation of the components of Triton. Heavier materials, such as metals, would have sunk to the core; medium-mass materials, such as silicates, would have formed a mantle; and lighter materials would have been forced to the surface.

Many believe that Triton is a volcanic satellite because of plumes of dark material that appear to be blown from its surface into the air. This material is concentrated enough to be easily seen. The plumes are about 2 kilometers across, rise as high as 8 kilometers, and consist of particles that probably are less than two millimeters in diameter. The size of the particles can be inferred because they do not settle to the surface. These plumes may be putting out 10 kilograms of material per second and may last for years.

Triton has a thin atmosphere, which is composed predominantly of nitrogen at a pressure of 14 millibars (Earth's atmospheric pressure is about 1 bar). There are clouds composed of condensed nitrogen. A diffuse haze can also be seen. The haze probably consists of hydrocarbons and nitriles, produced by the action of sunlight on methane. Wind-driven streaks are oriented in an east-west direction. The wind causes some of the streaks by material blown from the plumes. When all of the streaks, clouds, and plumes are considered, the winds blow northeast close to the surface, eastward at intermediate levels, and westward at the top of the troposphere.

Triton (foreground) and Neptune appear together in this montage of images from Voyager 2. (NASA/JPL)

Infrared technology gave scientists the first look at Triton's surface. The spectra could be modeled only by a combination of solid methane (CH_4), called methane ice; liquid nitrogen; and water ice. Later spectra showed nitrogen (N_2) in a solid or liquid form. One idea is that there is a sea of nitrogen with small amounts of dissolved methane. More likely, there is a layer of solid nitrogen with contaminants of methane, carbon dioxide (CO_2), and carbon monoxide (CO). Even at the measured temperature of 38 kelvins, the nitrogen ice will sublime, forming the thin atmosphere first found by Voyager 2. The nitrogen refreezes at the winter pole of Triton, causing a polar ice cap. Solid nitrogen is very transparent; therefore, the sunlight that does reach Triton can heat the interior of the solid nitrogen in a greenhouse effect. Nitrogen, in either gaseous or liquid phase as a result of heating nitrogen originally in solid form, will flow to the surface, where it then freezes. Some of the nitrogen will escape into the atmosphere. The layer of nitrogen thus moves, or at least thins, with the seasons. There is plenty of time for the seasonal shift, because Triton has a 688-year climate cycle due to its unique rotational and orbital motion. Thus, there is a higher albedo (0.7) on the winter end of the satellite, where the layer of nitrogen ice is thick, and a lower albedo (0.55) on the summer end of Triton. The summer end will show methane ice, which has a reddish color. Radiation will eventually turn the methane dark. The fact that the methane is not dark means that the methane ice is refreshed on a short timescale.

Triton displays at least three different types of surfaces: a bright polar area; areas of dark patches surrounded by lighter material; and high, walled plains. The bright polar area seems to be nitrogen ice on top of a cantaloupe-like, dimpled surface. The dimples, called cavi, are caused not by volcanoes but by diapirism. Diapirism is generated by a gravitational instability in which less dense material flows up through denser material. The density gradient

Voyager 2 was 530,000 kilometers from Triton when it took this image in 1989 through green, violet, and ultraviolet filters. (NASA)

may be caused by temperature or by a difference in composition. The implication on Triton is that the crust has distinct layers, and that the top layer is no more than 20 kilometers thick. The area also has linear ridges across the cantaloupe terrain.

Dark patches within the lighter material are called maculae and probably are composed of carbonaceous material, such as methane ice. Bright material is probably nitrogen ice. Maculae may mark spots of heat that have lost the nitrogen ice layer, allowing the methane ice, with its lower albedo, to show through.

The high plains are caused by a flow of volcanic ice. Some of the plains are smooth plains with a flat-to-undulating structure. Other plains are surrounded by a terraced wall or steppes, called scarps. These plains are very flat, implying that they were filled with liquid at one time. Scarps may be the remainder, as the material on the plains sublimed.

One unique feature of Triton is its small numbere of craters. There is one crater that is 27 kilometers across, named Mazomba, but the small number of craters suggests either that the surface of Triton is very young or that the surface must have been refreshed fairly recently—or both. Part of Triton's surface is considered cryovolcanic instead of silicate-magma volcanic. Cryovolcanic activity is the eruption from the subsurface of icy-cold liquids, which then refreeze on the surface in a more or less smooth structure.

KNOWLEDGE GAINED

Much of what has been discovered about Triton has come from Earth-based instruments and the Hubble Space Telescope (HST). Hale Observatory used narrow-band spectrophotometry to determine that Triton had a constant spectral reflectance. Astronomers have compared data from HST, the Infrared Telescope Facility at the University of Hawaii, and Voyager 2 over a period of years to see if there is a seasonal change in Triton's surface. It appears that there is a change, but since the climate cycle is so long, the data remain inconclusive. Both types of surface composition, methane ice and solid nitrogen, were detected by infrared spectra from an Earth-based instrument.

Voyager 2 provided more information in a short time than the land-based instruments had been able to gather in the years since Triton's discovery. The spacecraft's small changes in flight path caused by Triton's mass allowed that mass to be calculated. Pictures allowed the size to be determined. Density could then be calculated. Models of the structure of the satellite could then be developed. The density, 2.07 grams/centimeter^3, indicates that there is a large component of silicate materials, even though they do not show in infrared spectra because they are under ice. Voyager also detected the nitrogen atmosphere.

Pictures showed the effects of wind on Triton, a phenomenon that was unexpected. Varied terrain and plumes were also noticed in the pictures. The temperature measured was the coldest of any surface measured. Even with the cold temperature, the different terrains indicated that the surface had been refreshed more recently than any other moon or planet except those planets or moons that are geologically active.

CONTEXT

The density of Triton is close to that of the Pluto-Charon system. Is that a coincidental fact, or are Triton and Pluto related? Could Pluto at one time have been a satellite of Neptune that was knocked off by Triton? The orbital inclination, rotational speed, and retrograde motion all point to some cataclysmic occasion that produced Triton as a satellite.

Triton's surface features raise interesting questions about its energy source. Triton's is the coldest surface in the solar system, yet it also appears to be active, given the plumes and smoothness observed. How can these conditions coexist? The idea that enough sunlight penetrates a deep sheet of solid nitrogen to produce a greenhouse effect under the ice is startling but

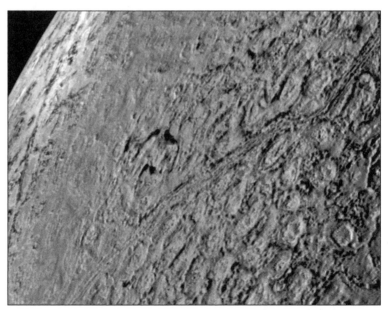

Voyager 2's flyby of Triton produced this close-up image (from 40,000 kilometers), showing the moon's unique northern hemispheric terrain of more or less regularly spaced circular depressions ringed by ridges. These are not impact craters but more likely areas of collapsed ice. (NASA/JPL)

may be true; it does appear that the ice sublimes and then refreezes in another place. Yet where does Triton get the energy to produce plumes rising 8 kilometers into the satellite's tenuous atmosphere? Even given the satellite's rather low gravitational acceleration, this phenomenon completes the satellite's overall mystery. Is Triton's interior heated radiogenically? Is there another heat source?

It is possible that scientists do not understand as much about the effect of very low atmospheric pressure, low temperature, and heavy mass as was previously thought, because the planet's heat is generated somewhere. The theory that Triton is heated radiogenically will have to wait until the subsurface can be monitored for radioactive isotopes, or for their daughter isotopes, before it can be confirmed or disproved. Certainly the answer to the question of Triton's energy source will add both to our understanding of the origins of the solar system and to our knowledge of energy physics.

C. Alton Hassell

FURTHER READING

Bond, Peter. *Distant Worlds: Milestones in Planetary Exploration*. New York: Copernicus Books, 2007. The author discusses each of the planetary systems, including planets, moons, and rings. Exploratory space missions and how they have developed our knowledge of each system are also addressed. Illustrations, bibliography, appendix, index.

Corfield, Richard. *Lives of the Planets*. New York: Basic Books, 2007. The author takes the reader through the different planets and the information gathered by space missions that investigated them. Index.

Croswell, Ken. *Ten Worlds: Everything That Orbits the Sun*. Honesdale, Pa.: Boyds Mills Press, 2007. Basic information on each system is presented separately. Illustrations, bibliography, index. For younger readers.

Cruikshank, Dale P., ed. *Neptune and Triton*. Tucson: University of Arizona Press, 1995. Voyager 2's 1989 encounter with Neptune revealed Triton to be a frozen, icy world with clouds, haze layers, and vertical plumes of particles rising high into the thin atmosphere. Originally presented as papers at a 1992 conference, the chapters in this volume are all by experts on Neptune, its many satellites, and its near-space environment. Until engineers can design propulsion systems for the next mission to the outer solar system, this 1,249-page tome will remain the most authoritative one-volume resource on Neptune and its satellites.

Dasch, Pat. *Icy Worlds of the Solar System*. Cambridge, England: Cambridge University Press, 2004. This book discusses ice, first on Earth, then on other solar-system bodies, including Triton. Illustrations, bibliography, index.

Hartmann, William K., and Ron Miller. *The Grand Tour: A Traveler's Guide to the Solar System*. 3d ed. New York: Workman, 2005. Focusing on the Voyager missions, this volume addresses each major planet and the major moons, including Triton. Includes outstanding illustrations. Illustrations, bibliography, index.

Irwin, Patrick G. J. *Giant Planets of Our Solar System: An Introduction*. 2d ed. New York: Springer, 2006. Suitable as a textbook for upper-level college courses in planetary science. Focuses on Jupiter, Saturn, Uranus, and Neptune and their satellites, rings, and magnetic fields. Filled with figures and photographs. Accessible to the serious general audience.

Lopes, Rosaly M. C., and Michael W. Carroll. *Alien Volcanoes*. Baltimore: Johns Hopkins University Press, 2008. The focus is on volcanism throughout the solar system, including the possibilities on Triton. Illustrations, bibliography, index.

McFadden, Lucy-Ann Adams, Paul Robest Weissman, and T. V. Johnson, eds. *Encyclopedia of the Solar System*. San Diego: Academic Press, 2007. The editors have collected articles written by many experts. It is one of the best surveys of material about the solar system. Illustrations, appendix, index.

See also: Neptune's Atmosphere; Neptune's Great Dark Spots; Neptune's Interior; Neptune's Magnetic Field; Neptune's Ring System; Neptune's Satellites; Planetary Satellites; Uranus's Satellites.

U

Ultraviolet Astronomy

Category: Scientific Methods

Ultraviolet astronomy explores the universe by focusing on wavelengths of the electromagnetic spectrum that are shorter than those of visible light. This portion of the spectrum is particularly important to astronomy, as practically all stars and many of the most abundant elements in the universe emit energy in the ultraviolet range.

OVERVIEW

Any material with a temperature above absolute zero emits electromagnetic radiation, and that radiation carries with it information about the nature of the event that produced it. An object will emit radiation over a range of wavelengths, with a concentration at a single wavelength. Very hot objects produce shorter wavelengths, while cooler objects emit longer wavelengths. For example, as metal is heated, it first glows red (the longer wavelengths of visible light), then, as its temperature increases, it begins to glow in the shorter-wavelength yellow light. In space, objects that are very cold, perhaps only a few degrees above absolute zero, will emit radiation in the very long infrared and radio wavelengths. At the other extreme, very hot stars give off ultraviolet radiation, X rays, and gamma rays.

Ultraviolet astronomy focuses on the area of the spectrum that is beyond violet light—the shortest wavelengths the eye can see. The ultraviolet portion of the spectrum begins at a wavelength of 390 nanometers, ranges to the extreme ultraviolet at 90 nanometers, and merges into the X-ray portion of the spectrum at 10 nanometers.

While the visible portion of the spectrum can be observed from the surface of the Earth, observations at ultraviolet wavelengths must be done outside the Earth's atmosphere. Ultraviolet radiation is readily absorbed by gases, both in space and in the Earth's atmosphere. Only the longest wavelengths of ultraviolet light penetrate the atmosphere. It is this radiation that is responsible for the destructive tanning effects of the Sun on the skin. Screening effects of the atmosphere protect life on Earth from the more harmful, shorter-wavelength ultraviolet radiation. At the same time, the atmosphere prevents astronomers from easily collecting information about the universe from this important range of the spectrum.

Practically every object in the universe emits some radiation at ultraviolet wavelengths. Any material that has a temperature between 10,000 and 1 million kelvins thermally emits most of its energy in the ultraviolet. This range of the spectrum is important to the knowledge of celestial objects, since the atmospheres of most stars, the surfaces of very massive stars, white dwarf stars, and regions of hot interstellar gas all fall within this temperature range. Every element emits and absorbs energy according to a characteristic pattern. By analyzing the pattern, or spectrogram, astronomers can determine the chemical composition of very distant objects. A spectrum is governed by the temperature, density, and chemical composition of the object emitting the energy, as well as how the energy has been altered by intervening processes en route to the instruments. In addition, the elements that are most abundant in the universe, such as hydrogen, helium, carbon, nitrogen, oxygen, and silicon, all have spectral features that are prominent in the ultraviolet. For this reason, ultraviolet astronomy can provide astronomers with important information about the universe.

Although instruments used in ultraviolet astronomy are designed to operate remotely, they can be very similar to optical instruments. Ordinary telescope mirrors will focus ultraviolet light. Electronic detectors record the image, or, in some cases, the image can be recorded on regular photographic film. A spectrograph can record such information by passing the radiation through a narrow slit and then through a prism, which separates the radiation into its component wavelengths. The result, a spectrogram, is then recorded on film.

The first ultraviolet telescopes were flown on high-altitude weather balloons. Later, they were launched on rockets, which raised them above the atmosphere for a few minutes at a time. The best way to gain access to ultraviolet information is to place an ultraviolet observatory in orbit. The first satellites to carry ultraviolet instruments were the Orbiting Astronomical Observatories (OAOs), a series of four identical satellites that carried different instrument packages to measure ultraviolet radiation from stars and interstellar gas. Only two of the four spacecraft proved functional. The first satellite failed after two days in orbit, and another failed to reach orbit. The second OAO achieved a general ultraviolet survey of the sky, discovering ultraviolet sources within the galaxy and measuring ultraviolet light from bright nearby galaxies. The final OAO, Copernicus, was the first to target specific ultraviolet sources. Carrying a 0.8-meter telescope, it was launched on August 21, 1972, and was functional for nine years. Copernicus took the first detailed look at objects in a wide range of the ultraviolet spectrum.

Detailed ultraviolet pictures of the Sun were produced with the OAO mission and later using the Solar Maximum Mission (SMM) satellite. Spectacular ultraviolet solar studies resulted from the American crewed space station Skylab, which was launched in 1973. Three crews of three astronauts inhabited the space station for a total of five and one-half months. Throughout this time, they kept a continuous surveillance of the Sun with the Apollo Telescope Mount (ATM). The ATM carried eight telescopes, which observed

the Sun in wavelengths ranging from visible light through the ultraviolet and into the X-ray range.

In 1978, two years before the OAO Copernicus ceased operation, the International Ultraviolet Explorer (IUE) was launched. The IUE was a joint venture by the National Aeronautics and Space Administration (NASA), the European Space Agency (ESA), and the British Science Research Council (SRC). Although the IUE was equipped with a telescope smaller than that of Copernicus (only 45 centimeters), it carried more modern ultraviolet detectors and was able to observe much fainter stars over a broader range of wavelengths. The IUE was run much like a traditional observatory and, as such, was designed to be used by visiting scientists rather than by a select group of researchers. The IUE remained stationed in an orbit that is geosynchronous (remaining constantly over one area of Earth), about 36,000 kilometers above the Atlantic Ocean, where it remained in continuous contact with at least one of two ground stations

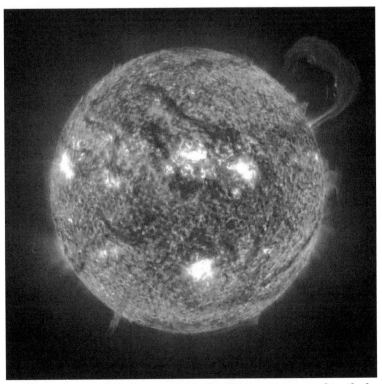

This 1999 image of the Sun was taken in 304-angstrom wavelengths by the Extreme Ultraviolet Imaging Telescope. (ESA/NASA/SOHO)

in the United States and Europe. This was an improvement over earlier satellites, which could not remain in continuous contact with ground stations. Their observations were also limited by low orbits, with a large percentage of the field of view blocked by the Earth. The IUE was decommissioned on September 30, 1996. Sadly it was merely shut off due to budget concerns, even though this workhorse observatory was functioning nearly at full capability.

Astro 1 was a Spacelab mission launched aboard the space shuttle in early 1990; it carried three ultraviolet telescopes. For the duration of the mission, project scientists conducted observations in the ultraviolet and X-ray regions. Other important ultraviolet satellites have been the European TD-1 and the Soviet Astron satellite. The TD-1, launched in 1972, measured the magnitudes of more than thirty thousand stars in four different spectral regions and gathered ultraviolet spectra from more than one thousand stars. The Soviet Astron satellite, launched in 1983, was similar in size and scope to the OAO Copernicus.

Ultraviolet astronomy advanced greatly with observatories such as the Hubble Space Telescope (HST), launched from the space shuttle Discovery on mission STS-31 on April 25, 1990; the Extreme Ultraviolet Explorer (EUVE), launched on June 7, 1992; and the Far Ultraviolet Spectroscopic Explorer (FUSE), launched on June 24, 1999. The HST has proved to be a powerful ultraviolet instrument, making exciting discoveries in the far ultraviolet, despite preliminary optical difficulties, which were repaired in 1993 during space shuttle mission STS-61. Thereafter the HST returned amazing images of distant galaxies, interstellar dust, and other objects from deeper in space than had ever before been observed. The EUVE was designed to survey the cosmos for objects emitting very energetic short-wavelength radiation in order to discover many new objects, including perhaps ten times as many white dwarf stars as previously known. FUSE was designed to carry a 2-meter ultraviolet telescope that would investigate objects in the far ultraviolet and extreme ultraviolet. NASA's Wind mission was conceived to use multiple instruments, including ultraviolet instruments, to study the solar

wind and sample plasma waves, energetic particles, and electric and magnetic fields. The European Space Agency's Lyman Observatory was designed to examine the dynamics of the Milky Way's halo as well as comets.

FUSE was designed for only a three-year primary mission but was still in reasonable operational condition in 2002, despite having difficulties with its reaction wheels (used for pointing the observatory at selected celestial targets). Budget issues almost resulted in cessation of FUSE observations, but the use of the orbital telescope was extended until July 12, 2007, when controllers lost the final reaction wheel. Despite efforts to restore the observatory, on September 7, 2007, FUSE investigations ceased. This relatively inexpensive Explorer-class satellite (also known as Explorer 77) led to more than four hundred published scientific papers and advanced the careers of many young astronomers specializing in ultraviolet astrophysics.

APPLICATIONS

Each energy region in the electromagnetic spectrum allows astronomers to "see" objects in a unique way. The more information that can be discovered about the nature of a celestial object in each of these energy areas, the more completely the object can be understood.

All objects known to exist in the universe—from comets and planets to stars, galaxies, and quasars—can be effectively studied in the ultraviolet range. Ultraviolet telescopes see the hottest stars of all; as a result, they tend to pick out the youngest star groups in the sky. Ultraviolet astronomy can thus focus on the youthful clusters of stars that lie close to regions of star birth. With this particular window, ultraviolet astronomy has been very useful for mapping regions of star formation, both in the Milky Way and in distant galaxies. Other galactic studies have shown that the Milky Way, as well as other galaxies, surrounded by a hot halo of gas.

There are excellent images of the Sun in the ultraviolet. Views of the ultraviolet Sun reveal different layers of its chromosphere, transition region, and lower corona. Bright, scintillating points of ultraviolet light in the Sun's atmosphere provide a measure of magnetic activity within the Sun, with perhaps even more accuracy than the sunspots that are seen on the visible photosphere. By observing other stars in the ultraviolet, astronomers have gained valuable knowledge about the nature of stars, which correlates with what is known about the Sun. It has been shown that many stars have hot outer atmospheres similar to the Sun. A new class of stars was discovered that had distinguishing characteristics visible only in ultraviolet light.

Ultraviolet astronomy has been very useful in studying binary star systems. A binary system is a pair of stars orbiting around their common center of mass. In a binary system, one star can be much brighter in optical wavelengths, and only a single spectrum can be observed. If the companion star is much hotter, however, it will dominate the spectrum of the system at ultraviolet wavelengths. A binary system gives astronomers a tool with which they can study the nature of these dimmer hot stars. Previously unobserved hot companions have been discovered in stars not suspected earlier of being binaries.

The supernova is another area in which ultraviolet astronomy can contribute significantly. A supernova is a stellar explosion in which all or most of the star's mass is expelled. Astronomers have studied the remains of supernova explosions as well as observing them in the beginning stages of development. Ultraviolet observations can determine the chemical composition of the layers of the star expelled by the explosion. Astronomers discovered that a nova explosion in Cygnus in 1978 produced much nitrogen, while the supernova that created the well-known Crab nebula threw out relatively small amounts of carbon. These facts are important clues to learning how new elements are formed, as well as in understanding the mechanisms behind supernovae.

The IUE observed Comet Kohoutek in 1976, finding a very bright image in ultraviolet wavelengths. Comet IRAS-Iraki-Alcock was observed in 1983 and Comet Halley was observed in 1986. In combination with observations at other wavelengths, it was found that the comets are similar in composition, suggesting that they have a similar origin.

Among IUE's legacy of ultraviolet investigations were 104,000 individual observations, the discovery of aurorae at Jupiter, the first determination of the water-loss rate in a comet, the production of the first orbital radial velocity curve for Wolf-Rayet stars, determination of the progenitor of Supernova 1987A, the first imaging of galactic halos, and the production of 44,000 stellar spectra per year, to name but a few. In a very real sense, IUE, more than anything before it, expanded ultraviolet astronomy from an interesting concept to a highly active and fruitful area of astrophysical observations.

Ultraviolet astronomy has confirmed some long-standing theories that previously lacked sufficient evidence. The theory of a "gravitational lens" had been predicted by the relativity theories of Albert Einstein but had never been supported by solid evidence. The gravitational lens is a process in which the gravitational field of a very massive object acts as a lens, bending the radiation from a more distant object behind it, distorting its image and often creating a double or multiple image. Observations by the IUE helped to indicate that such gravitational lenses exist. The first such lens system was discovered in 1979. Hubble routinely discovered new gravitational lensing objects.

A full list of science achievements of EUVE would be too long to include in this article. Among some of those discoveries were the production of an all-sky catalog containing a total of 801 ultraviolet-emitting objects (in wavelengths from 7 to 76 nanometers), participation in coordinated observations of numerous objects at different wavelengths across the electromagnetic spectrum, some of the first ultraviolet detections of extragalactic objects, measurements of quasi-periodic oscillations in dwarf novae, analyses of extreme ultraviolet spectral white dwarf star companions to main sequence stars, the detection of helium in a hot white dwarf, and the first extreme ultraviolet observations of the Coma Cluster.

Ultraviolet observations with the Hubble Space Telescope involved uses of the observatory's Goddard High Resolution Spectrograph (GHRS), Faint Object Camera (FOC), and Faint Object Spectrograph (FOS). GHRS and FOS were replaced during the second Hubble servicing mission (STS-82 in February, 1997). In their place were inserted the Space Telescope Imaging Spectrograph (STIS) and Near Infrared Camera and Multi-Object Spectrometer (NICMOS). During shuttle servicing mission 3B (STS-109 in March, 2002) the FOC was replaced by the Advanced Camera for Surveys (ACS). STIS suffered a malfunction in August, 2004, and portions of the electronics for ACS failed in 2006 and 2007. STIS and ACS were scheduled to be repaired during one final shuttle servicing mission, during which repairs would restore Hubble's capability to conduct ultraviolet astronomy. That shuttle mission was also intended to insert a new Wide Field Camera 3 and the Cosmic Origins Spectrograph to expand the observatory's capability to make ultraviolet measurements of objects at great distances (and hence tremendously early times in the cosmic past).

The July, 2008, edition of *Physics Today* presented research using ultraviolet data from both FUSE and HST that indicated that as much as 40 percent of the anticipated baryonic matter is missing from the portion of the universe along the line of sight to energetic objects such as quasars. The study involved collecting ultraviolet spectral signatures of quasars. What was anticipated was a typical quasar spectrum, which also incorporated into it absorption lines for the absorbing gas between the quasar source and the telescope in Earth orbit detecting that radiation. By analyzing the depth of absorption lines cutting into the quasar's spectral emissions, astronomers were able to calculate the density of the baryonic matter (originally formed within just a few minutes of the big bang) along the line of sight. It turned out to be too low. Obviously, more investigations of the intergalactic medium were necessary, or an adjustment of cosmological models would be in order.

FUSE provided insight into the abundance of deuterium in stars and studied a wide range of astrophysical objects, such as the intergalactic medium, cool stars, and galactic structures. By recording absorption and emission lines in the far-ultraviolet portion of the electromagnetic spectrum, FUSE increased our understanding of galactic, intergalactic, and extragalactic chemical processes.

CONTEXT

For thousands of years, the human eye was the only astronomical instrument. The eye evolved to be most sensitive to the range of visible light, the most abundant source of radiation at the surface of the Earth. Any celestial objects that were dim or emitted radiation at mostly nonoptical wavelengths remained invisible to the eye, which limited the range of information about the universe scientists could study.

The invention of the telescope radically altered astronomy, not only because of the fainter objects it allowed astronomers to see but also because it opened up the possibility that there was more to the universe than what the human eye was able to image. Astronomy improved dramatically over the next few centuries but remained optical. The first sign that there was another way to look at the universe with anything other than optical wavelengths came in 1800, when infrared radiation was discovered by Sir William Herschel, who placed a thermometer just outside the red range of the visible light separated by a prism.

The opening up of the wavelengths of electromagnetic energy got under way with the rapid growth of radio astronomy in the 1950's and 1960's and with the birth of the space program during the same period. The space program allowed ultraviolet astronomy to become an important new area of study. The potential value of ultraviolet observation from space was proposed to the U.S. Air Force in 1946 by the American astrophysicist Lyman Spitzer, Jr. With the establishment of NASA in 1958, the concept of placing orbiting observatories in Earth's orbit became a reality, and a series of orbiting observatories were launched over the next twenty years.

The result of more than forty years of observing in all ranges of the spectrum is that astronomers now have a more complete understanding of the processes occurring in the universe. Today's astronomers have taken images of the stars and galaxies that were unimaginable to the ancients, or even to the astronomers of a few decades past. Ultraviolet astronomy is now at a point where future missions will lose the "frontier" feel of the early missions, with increasingly complex and specialized missions. Even so, exciting new discoveries will continue to be made. In addition to new observations, research using information from the years of observations made by the satellites since the 1960's will allow astronomers to gain new insights into virtually every area of the universe.

Divonna Ogier

FURTHER READING

Arny, Thomas T. *Explorations: An Introduction to Astronomy*. 3d ed. New York: McGraw-Hill, 2003. A general astronomy text for the nonscientist. Includes an interactive CD-ROM and is updated with a Web site.

Barstow, Martin A., and Jay B. Holberg. *Extreme Ultraviolet Astrophysics*. Cambridge, England: Cambridge University Press, 2007. The universe in extreme ultraviolet (EUV) was revealed only when rockets in the late 1960's sent relatively primitive instruments capable of detecting EUV radiation briefly into space. Those initial investigations demonstrated the universe was rich in EUV-emitting sources. This work catalogs EUV sources, explains the cosmological importance of those sources, and describes the instrumentation that detected them.

Henbest, Nigel. *Mysteries of the Universe*. New York: Van Nostrand Reinhold, 1981. Explores the limits of what is known about the universe. Discusses theories about the origin of the solar system and the universe, exotic astronomy, and astronomy at invisible wavelengths.

Karttunen, H. P., et al., eds. *Fundamental Astronomy*. 5th ed. New York: Springer, 2007. A well-used university textbook in introductory astronomy. Contains some calculus-based treatments for those who need a more advanced textbook than the standard introductory work. Suitable for an audience with varied science and mathematical backgrounds. Covers all topics from solar-system objects to cosmology.

Marten, Michael, and John Chesterman. *The Radiant Universe*. New York: Macmillan, 1980. An overview of imaging that is not accessible in visible wavelengths. Discusses electronic processing as well as infrared and ultraviolet wavelength imaging. Beautiful pictures, along with easy-to-read and informative text, somewhat compensate for the age of this text.

Time-Life Books. *The New Astronomy*. Alexandria, Va.: Author, 1989. One volume of a series examining different aspects of the universe. Comprehensively covers all invisible astronomies, including high-energy astronomy and imaging techniques. Heavily illustrated and suitable for the general reader with an interest in astronomy.

See also: Archaeoastronomy; Coordinate Systems; Earth System Science; Gravity Measurement; Hertzsprung-Russell Diagram; Infrared Astronomy; Neutrino Astronomy; Optical Astronomy; Radio Astronomy; Solar Corona; Telescopes: Ground-Based; Telescopes: Space-Based; X-Ray and Gamma-Ray Astronomy.

Universe: Evolution

Category: The Cosmological Context

About 13 to 14 billion years ago, the big bang occurred, creating the space, time, matter, and energy of our universe. Since then, space has expanded, galaxies of stars have formed in that space, and the stars in the galaxies have evolved. The evidence indicates that the universe will continue to expand forever. Indeed, the expansion appears to be accelerating, due to some unknown cause referred to as "dark energy."

OVERVIEW

Evidence we have from observing radiation in all ranges of the electromagnetic spectrum—particularly the infrared—has led astrophysicists and cosmologists to conclude that, approximately 13 to 14 billion years ago, an explosive event dubbed the big bang created space and time, matter and energy from an unimaginably hot, dense state whose origin is unknown. Ever since the big bang, the universe has been expanding and evolving. Events following the big bang have been reconstructed in great detail by using known physical laws to predict the behavior of matter and energy as the universe expands.

As the universe expanded, its temperature and the density of both matter and electromagnetic radiation all decreased. In the early, very hot, very dense universe, cosmologists think that all four fundamental forces of nature—gravity, the strong nuclear force, the weak nuclear force, and the electromagnetic force—were unified as one force, indistinguishable because as yet unseparated from each other. Models that would successfully combine these four forces are called theories of everything (or TOEs), but currently there is no workable TOE.

All TOEs, however, assume that the four forces were unified only at the extremely high energies, corresponding to extremely high temperatures, that characterized the first few seconds of the universe. Sometime between 10 and 43 seconds after the big bang (called the Planck time), the temperature dropped to about 10^{32} kelvins, and gravity separated ("froze out") from the other forces. After about 10^{-35} second, the temperature dropped to about 10^{28} kelvins, and the strong nuclear force separated from the remaining two forces. The "freeze-out" of the strong nuclear force may have been what initiated a period of rapid inflation, in which the universe expanded exponentially, increasing in size by a factor of about 10^{50} in the next 10^{-32} second. This rapid inflation explains why distant regions of the present universe appear so simi-

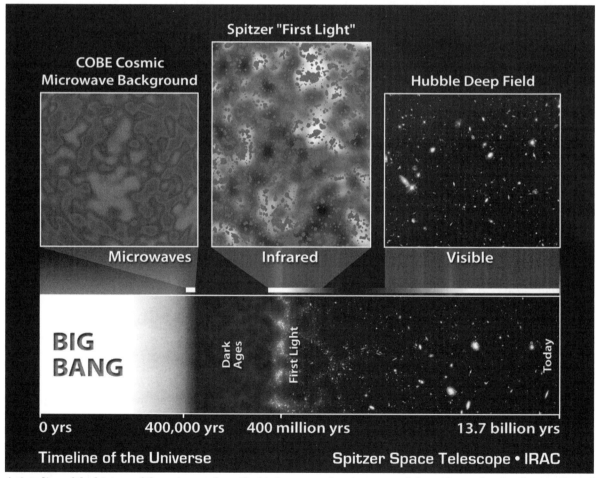

A time line of the history of the universe, from the big bang on, showing some of the evidence that has been collected by space telescopes capturing images in the microwave, infrared, and optical ranges. (NASA/JPL-Caltech/ A. Kashlinsky, GSFC)

lar, and why space on the large scale is so nearly flat. Then the universe resumed the slower expansion it had been undergoing before inflation occurred. After about 10^{-10} second, the temperature had dropped to 10^{15} kelvins, and finally the weak nuclear and electromagnetic forces assumed their separate characteristics.

The early universe was very hot and was filled with high-energy gamma-ray photons. When two gamma rays with sufficient energy collided, their energy could be converted to mass, as described by Einstein's famous equation, $E = mc^2$, which says that energy, E, and mass, m, are equivalent and related by the speed of light, c, squared. The mass appeared as a particle-antiparticle pair in a process called pair production. When a particle and its antiparticle encountered each other, they would mutually destroy each other in a process called annihilation, in which their mass would be converted back into energy as a pair of gamma-ray photons. The rates of pair production and annihilation were equal, and matter and radiation were in a state of thermal equilibrium. However, as the universe continued to expand, the temperature dropped to the point that the photons no longer had enough energy to produce specific particle-antiparticle pairs. Once the temperature had dropped below the threshold temperature for a particular type of pair production, no more of those particle-antiparticle pairs were formed, and those that had formed previously quickly annihilated each other.

The threshold temperature for proton-antiproton pair production is about 10^{13} kelvins; this temperature was reached when the universe was about 10^{-4} second old, after which no more proton-antiproton pairs were produced, and those protons and antiprotons that had been produced previously annihilated each other. The threshold temperature for electron-positron (another name for an antielectron) pair production is about 6×10^9 kelvins; that temperature was reached after a few seconds, after which no more electron-positron pairs were produced, and those electrons and positrons that had been produced previously annihilated each other. If exactly equal numbers of particles and antiparticles had been created, they all would have annihilated each other, and there would be

no matter or antimatter in the universe today. However, we live today in a universe composed predominantly of matter. Consequently, a slight excess of particles over antiparticles must have been created, by about one part in a billion; they survived and constitute the matter in the present universe.

Electrons could combine with protons to form neutrons, and protons and neutrons could combine to form nuclei of deuterium (also called heavy hydrogen), each deuterium nucleus consisting of one proton and one neutron held together by the strong nuclear force. High-energy gamma rays could break deuterium nuclei back into protons and neutrons as fast as they had formed, but by about three minutes after the big bang, the temperature had dropped to about one billion kelvins, and photons no longer had enough energy to break up deuterium nuclei. This began the time of nucleogenesis, when deuterium nuclei could fuse into helium nuclei and even form some lithium and beryllium nuclei. However, after about 15 minutes, the temperature had dropped to a few hundred million kelvins, and the nuclei no longer were moving fast enough to overcome their electrical repulsion. The nucleosynthesis of heavier elements by fusion would have to wait till much later, when it would occur in stars. This established the overall composition of the universe today—about one atom of helium for every ten atoms of hydrogen, with only very small amounts of all the other chemical elements.

In the early universe, the density of electromagnetic radiation was greater than the density of matter, but as the universe expanded the radiation density decreased more quickly than the matter density. Several thousand years after the big bang, at a time called the crossover time, the density of radiation and matter were equal. From that time on, the universe has been dominated by matter. At the high temperatures of the early universe, the matter was ionized, meaning it consisted of free electrons and bare atomic nuclei. Free electrons are very effective at scattering photons of electromagnetic radiation, so the early universe was opaque; photons could not travel far before encountering free electrons and being scattered in new directions. About 300,000 to 400,000 years after the big

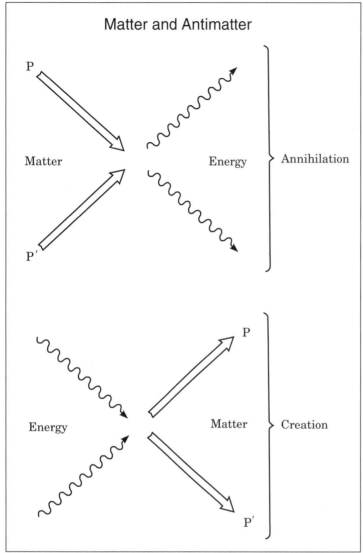

Matter and Antimatter

An elementary particle (P) and an antiparticle (P') may collide, annihilating each other and releasing energy (wavy arrows). The reverse may also occur, creating matter. Such collisions are relatively rare at the present stage of the universe's evolution.

After matter and radiation decoupled, small fluctuations in the distribution of matter started to grow; slightly denser regions gravitationally attracted matter from surrounding areas and became denser still. Within the first billion years, they developed into small protogalaxies or pregalactic fragments in which the first stars formed by gravitational contraction of clouds of gas. Through mergers, the protogalaxies formed larger systems of stars, the galaxies that make up the universe today. Within the galaxies, stars continue to form by the gravitational contraction of gas clouds, and at least some stars develop families of planets as a by-product of their own formation. It is stars that synthesize the heavier chemical elements. Some of the elements are formed by nuclear fusion processes during the active, energy-producing lives of the stars. Others are formed when stars explode as supernovae. Whether stars end their energy-producing lives violently as supernovae or more quietly, they expel some or most of their mass, and this disperses the heavier elements into the interstellar material, enriching the clouds of gas and dust from which new stars form.

The Milky Way galaxy formed more than 12 billion years ago. The Sun and its solar system formed from a cloud of gas and dust about 4.5 billion years ago. Without the nucleosynthesis of heavier elements by earlier generations of stars, there would be no carbon, oxygen, silicon, iron, or any of the many other elements needed to form a rocky planet like Earth and the life it supports.

Observations of distant galaxies indicate that the expansion of the universe is accelerating. The cause is unknown, but it is has been given the name "dark energy." If the expansion

bang, the temperature dropped to about 3,000 kelvins, and the electrons could combine with nuclei to form neutral atoms. Neutral atoms are able to absorb only certain specific photon energies, so the universe became transparent to most photons, and matter and radiation decoupled. The photons could travel freely through the universe, and today we observe them as the greatly redshifted cosmic microwave background radiation.

continues to accelerate, the distances between galaxies will grow ever greater. Eventually, all the matter in the galaxies will be processed into stars, the stars will all run out of energy and die, and the universe will grow dark and cold.

METHODS OF STUDY

An expanding universe is predicted by Albert Einstein's general theory of relativity. This conclusion was arrived at independently by Alexander Friedmann in 1922 and Georges Lemaître in 1927 from solutions they found to the field equations of general relativity applied to the structure of the universe. In 1929, Edwin Powell Hubble showed that the universe actually is (or at least appears to be) expanding when he discovered that the distances of thirty-one galaxies were correlated with the redshifts of their spectra. The cosmological explanation for this relationship is that, from the perspective of observers on Earth, the expansion of space stretches the wavelengths of electromagnetic radiation as it travels through space from the source to the observer. The greater the distance between source and observer, the longer it takes electromagnetic radiation to travel the distance, the longer the universe has been expanding, and the more the wavelengths are stretched. The long-wavelength end of the visible light spectrum is the red end, so the expansion of space shifts visible light to longer, redder wavelengths. The term "redshift" refers to a shift toward longer wavelengths of a photon of light in any portion of the electromagnetic spectrum, whether visible light or not, hence a shift toward the lower energy of that photon.

The cosmic scale factor $R(t)$ is defined to be a measure of how the universe changes in size as a function of time. Changes in the cosmic scale factor can be determined directly from the spectral redshifts. The amount the wavelengths are lengthened tells cosmologists how much the universe has expanded since the light was emitted; for example, if the features of some spectrum all have double their expected wavelengths, then the size of the universe and the cosmic scale factor have both doubled since the light was emitted; to put it another way, the universe and cosmic scale factor then both were half as big as they are now.

Lemaître, in 1927, was the first to propose that the expansion of the universe began from a compact, dense initial state—the "primeval atom" as he called it—which "fissioned" into all atoms in the universe today. Although wrong in the details, Lemaître's basic idea of expansion and evolution from a compact initial state has been developed into the big bang model of modern cosmology. Events following the big bang have been reconstructed in great detail by using known physical laws to predict the behavior of matter and energy as the universe expands.

Space is filled with electromagnetic radiation that has been traveling freely since the universe became transparent a few hundred thousand years after the big bang. Its serendipitous discovery by Arno Penzias and Robert Wilson in 1965 was a significant confirmation of a hot, dense big bang origin for the universe. This is thermal blackbody radiation, so the wavelength at which it is "brightest" is inversely proportional to the temperature. As the universe has expanded, the wavelength at which it is brightest has increased linearly with the cosmic scale factor R. Thus the temperature T decreases as the inverse of the cosmic scale factor R. Since the temperature of the cosmic background radiation now is about 3 kelvins and the temperature at which the universe became transparent was about 3,000 kelvins, the universe has expanded by a factor of about 1,000 since then.

The expansion of the universe causes the density of both matter and radiation to decrease with time. Since volume increases as R^3, the density of matter decreases as the inverse of R^3. However, the density of the energy of electromagnetic radiation decreases as the inverse of R^4. This is because the number of photons of electromagnetic radiation per volume of space decreases as the inverse of R^3, and the energy of each photon decreases as the inverse of R; the wavelength associated with each photon increases linearly with R, and wavelength and photon energy are inversely related. Thus the density of electromagnetic energy decreases as the inverse of R^3 times the inverse of R, which equals the inverse of R^4. This means that the density of radiation energy decreases more rapidly than matter density as the universe expands. Currently the density of matter is several thousand times the density of electromagnetic energy, but at an earlier time, when the universe and the cosmic scale factor were several thousand times smaller than they are now, the density of matter and electromagnetic energy were equal. The time when the densities of matter and electromagnetic radiation were equal is called the crossover time; before that, the density of electromagnetic radiation was greater than the density of matter. Current estimates of the crossover time place it several thousand years after the big bang.

The discovery that the expansion of the universe is accelerating was completely unexpected; in fact, it was discovered during an attempt to measure how much the expansion was decelerating. The models of the universe derived from the simplest form of general relativity all predict the expansion should be slowing. If it were slowing only a little, then the universe would be open and the expansion would continue forever. If it were slowing enough, then eventually the expansion would stop and the universe would begin to contract back on itself. It was expected that measuring the redshifts of very distant galaxies would show how much faster the expansion was in the long-ago past, when the light we now receive left those galaxies. However, the observational evidence indicates the expansion was slower in the past, meaning the expansion has been accelerating.

CONTEXT

Throughout history, humankind has wondered about the origin and fate of the Earth, its life, and the universe. This desire to understand our origins has led nearly every culture to form some kind of "creation myth." In Western culture, many religious and philosophical beliefs about the origin of the universe can be traced back thousands of years to the creation myths of the Middle East. Although science cannot explain the origin of what Lemaître called the "primeval atom," evidence from observations of phenomena billions of light-years old has now provided more definitive answers to many questions that humans have pondered for thousands of years.

Perhaps the most crucial of these questions is how the universe formed. Physics, coupled with

astronomical observations, has helped us work out the events and processes that likely occurred in the aftermath of the big bang. Our models of the evolution of the universe have profound implications for understanding life on Earth. The universe seems "fine-tuned" for the existence of life as we know it. If the physical laws and constants were changed slightly, the universe would be a very different place, and life as we know it could not exist. Some scientists explain this by invoking what is called the anthropic principle: The universe has to be the way it is because we are here; if conditions did not permit the development of life, we would not be here to speculate about it. Others argue that the odds are too great against it just being chance that the universe is the way it is, and suggest that it may have been deliberately designed that way. Still others speculate that our universe is just one of many universes, each with its own physical laws and constants; we live in the one in which life as we know it is possible.

Another aspect of the evolution of the universe with profound implications is its future and ultimate fate. If the expansion continues to accelerate, eventually, some billions of years from now, all matter will be processed into stars, all stars will run out of energy, and the universe will grow cold and dark.

Michael L. McKinney and Richard R. Erickson

FURTHER READING

Belusevic, Radoje. *Relativity, Astrophysics, and Cosmology*. Weinheim, Germany: Wiley-VCH, 2008. Addresses the interrelationship of the now entwined disciplines of astrophysics, particle physics, cosmology, and relativity theory, which have combined to form advanced theories of the origin and evolution of the universe.

Bennett, Jeffrey, et al. *The Cosmic Perspective*. 5th ed. San Francisco: Pearson/Addison-Wesley, 2008. Structured around a set of two-page figures dubbed "cosmic contexts," this interactive resource uses zoom-in illustrations to orient students to various images of the universe in relation to one another, while covering all the basics of cosmology. For general and introductory audiences. Chapters include "Light and Matter: Reading Messages from the Cosmos," "A Universe of Galaxies," "Dark Matter," "Dark Energy," and "The Fate of the Universe."

Drexler, Jerome. *Discovering Postmodern Cosmology: Discoveries in Dark Matter, Cosmic Web, Big Bang, Inflation, Cosmic Rays, Dark Energy, Accelerating Cosmos*. Boca Raton, Fla.: Universal, 2008. Cosmologist Drexler proposes a plausible and unique correlation of the seven mysterious areas of cosmological research listed in the book's subtitle. For all open-minded audiences.

Duncan, Todd, and Craig Tyler. *Your Cosmic Context: An Introduction to Modern Cosmology*. San Francisco: Pearson/Addison-Wesley, 2009. An introductory textbook for studies in modern cosmology that engages students by relating cosmological concepts to their own lives.

Ferreira, Pedro G. *The State of the Universe: A Primer in Modern Cosmology*. London: Phoenix, 2007. Oxford lecturer Ferreira presents a history of cosmology, examining the complexities that concepts such as dark matter and dark energy have imposed on a once "simple" Einsteinian universe ruled by relativity.

Gasperini, Maurizio. *The Universe Before the Big Bang: Cosmology and String Theory*. Berlin: Springer, 2008. A fascinating exploration of what might have happened prior to the explosion that formed the universe, which looks to string theory and other modern mathematical models to postulate that the universe was not born with the big bang but rather was well advanced in its overall evolution. Presented with nontechnical language for nonspecialists.

Lemoine, M., J. Martin, and P. Peter, eds. *Inflationary Cosmology*. New York: Springer, 2008. This collection of papers, by both venerable and younger astrophysicists and cosmologists, focuses on that period during which the early universe expanded at a greatly accelerated rate before settling down to a slower rate of expansion. Presents several different scenarios.

Liddle, Andrew, and Jon Loveday. *The Oxford Companion to Cosmology*. New York: Oxford University Press, 2008. An indispensable A-Z reference, consisting of more than 350 entries from antimatter to WIMPs, as well as individual physicists and other scientists who have advanced the field. Heavily illustrated with almost two hundred halftones and diagrams; includes cross-references and Web links.

North, John. *Cosmos: An Illustrated History of Astronomy and Cosmology*. Chicago: University of Chicago Press, 2008. Emphasizes the astrophysics of cosmology, with a focus on physical and mathematical concepts that are key to understanding classical field theory. Also describes experimental techniques and results. Technical.

Sagan, Carl. *Cosmos*. New York: Random House, 1980. One of the most popular science books of all time. It is superbly written and illustrated by a famous astronomer and is based on the popular television series of the same name. Fun and easy reading.

Schneider, Peter. *Extragalactic Astronomy and Cosmology: An Introduction*. New York: Springer, 2006. A textbook focusing on galaxies, clusters, and superclusters, beginning with the Milky Way. Covers their evolution, formation, and distribution, supported by beautiful color illustrations.

Weinberg, Steven. *Cosmology*. New York: Oxford University Press, 2008. The Nobel physics laureate presents the subject in two parts: one covering the isotropic and homogeneous "average" universe; the second, departures from the average universe. Provides detailed coverage of recombination, microwave background polarization, leptogenesis, gravitational lensing, structure formation, and multifield inflation. Includes mathematical calculations. Appendixes review general relativity, the Boltzmann equation, and sample problems. For advanced students and professionals.

Wudka, Jose. *Space-Time, Relativity, and Cosmology*. New York: Cambridge University Press, 2006. A history of relativistic cosmology from ancient times to Einstein. The nonmathematical approach emphasizes concepts over calculations yet explain the ideas clearly, applying them to research topics in cosmology. Designed for students from high school through college.

See also: Big Bang; Cosmic Rays; Cosmology; Electromagnetic Radiation: Nonthermal Emissions; Electromagnetic Radiation: Thermal Emissions; General Relativity; Interstellar Clouds and the Interstellar Medium; Milky Way; Novae, Bursters, and X-Ray Sources; Space-Time: Distortion by Gravity; Space-Time: Mathematical Models; Universe: Expansion; Universe: Structure.

Universe: Expansion

Category: The Cosmological Context

The universe is expanding, with the most distant galaxies receding at the greatest speeds. The rate of expansion and any change in the rate of expansion are critical data that provide constraints on the origin and ultimate future of the universe.

OVERVIEW

Galaxies are vast collections of millions to trillions of stars, all gravitationally bound together; our Sun and solar system are part of the Milky Way galaxy, which contains several hundred billion stars. There are billions of galaxies, and they are gravitationally bound together in galaxy clusters that contain between a few tens to a few thousand galaxies; our Milky Way galaxy is part of the cluster named the Local Group which contains over 40 known members, almost all of which are much smaller than our Milky Way galaxy.

Beginning in 1912, Vesto Melvin Slipher used the 24-inch refracting telescope at Lowell Observatory in Arizona to obtain spectra of objects called spiral nebulae. (At that time, it was not known that they actually were galaxies, although some astronomers thought they might be.) He found that, although a few of them had spectra that were shifted toward shorter, bluer wavelengths, most of their spectra were shifted toward longer, redder wavelengths, and some of the redshifts were surprisingly large. If the

spectral shifts were Doppler in origin (produced by motion of the spiral nebulae along the line of sight), then most of the spiral nebulae were moving away from us at high speed.

In the 1920's, using the Mount Wilson Observatory's 100-inch reflecting telescope (then the largest in the world), Edwin Powell Hubble succeeded in resolving individual stars in some of the spiral nebulae, including M31—the spiral nebula in Andromeda—and NGC 6822. He found that some of the stars were Cepheid variables. Earlier, Henrietta Swan Leavitt at Harvard had discovered that the average luminosity of a Cepheid variable is related to its period of light variation. The bigger the Cepheid variable, the longer it takes to pulsate and thus the longer its period of light variation. Also, the bigger the star, the brighter its average luminosity. Using the periods of the Cepheid variables, Hubble could determine their luminosities, or real brightnesses. By comparing their real brightnesses with their apparent brightnesses, Hubble was able to determine their distances, which placed them, and the spiral nebulae that contained them, far outside the Milky Way galaxy. This showed that spiral nebulae really were spiral galaxies, similar to the Milky Way galaxy, as had been suggested by the philosopher Immanuel Kant and others as early as 1755.

In 1929, Hubble, assisted by Milton Humason, announced a correlation between the distances and spectral shifts of about thirty galaxies. Only a few of the nearer galaxies had blueshifted spectra; most had redshifted spectra with the size of the redshift proportional to distance (a relation now known as Hubble's law). If the redshift is due to motion away from us, then the farther away a galaxy is from us, the faster it is moving away. Hubble and Humason discovered that the universe appears to be expanding, with the distances between galaxies continuing to grow larger.

Actually, an expanding universe had been predicted earlier. In 1922, the Russian mathematical physicist Alexander Alexandrovich Friedmann and, indepedently in 1927, the Belgian priest and cosmologist Georges Lemaître had found two classes of solutions to the field equations of Albert Einstein's general theory of relativity. In one type of solution, called open, the universe continues to expand forever. In the other type of solution, called closed, the universe will expand to some maximum size and then contract. In 1932, Einstein and the Dutch astronomer Willem de Sitter found a third solution, called flat or critical, in which the universe expands forever, but just barely; it is the boundary case between the other two.

Lemaître was the first to propose that the expanding universe had its origin in a small, super-dense state—the "primeval atom." Later, this idea was expanded into a sort of primordial "explosion" that created space and time, matter and energy. In the 1950's, it came to be called the big bang. In all such big bang models, the universe evolves; as the distance between galaxies increases, the density of matter and the density of radiation decrease.

An alternative, the steady state theory, was presented by the British cosmologists Hermann Bondi, Tom Gold, and Fred Hoyle in 1948. In this model, as the universe expands, new matter is spontaneously created in the space between the galaxies at just the right rate to keep the average density of matter constant. Such a universe has neither beginning nor end. Both the steady state and big bang models had supporters throughout the 1950's, but by the mid- to late 1960's, observational evidence mounted that the universe was expanding from an early high-density state, and the steady state theory gradually fell out of favor.

A simple estimate of the time since the big bang can be found from Hubble's law. Since the speed of recession is proportional to distance, then at some time in the past all space was infinitesimally small. The relation of the slope of the speed v versus distance r is called the Hub-

Fred Hoyle and the Steady State Theory

Troubled by the problems presented by George Gamow's early theory of the big bang, Fred Hoyle and the University of Cambridge developed an alternative and well-respected proposal: the "steady state" theory.

Ironically, the term "big bang" had been coined by its main adversary—Hoyle himself—during one of his series of BBC radio talks. Hoyle used the term to belittle Gamow's theory. Hoyle favored a different view: that the universe, although currently expanding, was infinitely old and in the long term existed in a steady state. Galaxies were not receding from each other as the aftermath of a primordial explosion (which defenders of the big bang held). Rather, space was being created between galaxies at a constant rate, and hydrogen was being created to fill that space, coalescing into nebular clouds that then formed young stars and galaxies among the old.

The problem with this theory was that it contradicted the law of the conservation of matter: namely, that matter could neither be created nor be destroyed without being converted into energy. In the 1950's, the discovery of radio galaxies by Sir Martin Ryle revealed that galaxies had evolved billions of years ago, supporting the big bang theory.

Once Arno Penzias and Robert Wilson discovered the cosmic microwave background radiation, Hoyle's steady state theory was largely abandoned in favor of the theory he himself had named: the big bang. Although Hoyle revised his theory to account for the background radiation, his once dominant view of the universe was out of favor. Hoyle, however, remained philosophical to the end: "The Universe eventually has its way over the prejudices of men, and I optimistically think it will do so again."

ble constant H, and $v = Hr$. The reciprocal of the Hubble constant, $1/H$, is called the Hubble time or Hubble age, tH: $tH = 1/H$. If nothing has sped up or slowed down the expansion, the Hubble age is the time since the expansion began. Most recent determinations of H put it in the range of about 70 to 75 kilometers per second per million parsecs, which gives a Hubble age of 13 to 14 billion years.

However, it was expected that the expansion should slow down or decelerate with time, due to the gravity of all the mass in the universe. The only question was whether the expansion would slow down just a little and the universe would keep expanding forever, or whether the expansion would slow down enough to stop, after which the universe would begin to contract. In any case, the actual age of the universe would be less than the Hubble age; if the universe were slowing down so much that eventually it would stop and then begin to contract, the actual age would be less than two-thirds the Hubble age. The difference in the cases depends, in part, on the density of matter in the universe. If the density were great enough, gravity would be strong enough to halt the expansion and cause the universe to contract; otherwise, the universe would continue to expand forever.

There are several observational tests for the ultimate fate of the universe. Attempts at measuring the density of matter that could be observed found much too low a density (by about a factor of 30) to stop the expansion, implying the universe would easily expand forever. On the other hand, the size of small fluctuations in the cosmic background radiation indicated the universe has a geometry that is very nearly flat, which is the geometry of the borderline case in which the universe just barely expands forever. Beginning in the 1990's, two independent research groups tried to detect a change in the rate of expansion, the presumed deceleration, by measuring the redshifts of very distant galaxies, and using Type Ia supernovae in them as "standard candles" to determine their distances. Contrary to all expectations, both groups found that the expansion has not been slowing down, but instead has been speeding up. This acceleration of the expansion has been attributed to some unknown mechanism called "dark energy," which seems to account for about 70 to 75 percent of all the matter and energy in the universe. If the acceleration continues, then the distance between clusters of galaxies will continue to grow at an increasing rate, and the universe will appear more and more empty.

METHODS OF STUDY

The expansion of the universe was initially predicted in the early 1900's by solutions to the field equations of Einstein's general theory of relativity. The observational demonstration of it comes from the redshift of galaxy spectra, which is interpreted to mean that galaxies are receding from each other. However, the expansion of the universe is not occurring because the galaxies are moving through space; the galaxies are moving with space as space itself expands.

Thus the redshift is not due to the Doppler effect, which is produced by the motion of objects through space; instead, the redshift is termed cosmological, meaning it is due to the expansion of space itself. This might seem to be just a semantic difference, but there is a fundamental physical and mathematical difference. As space expands, carrying the galaxies with it, the wavelengths of electromagnetic radiation are stretched by the expansion of space. (Think of the wavelengths being drawn on the surface of a balloon that is being inflated. As the balloon expands, the wavelengths get longer.) The equation converting redshift into recessional speed is not the same for Doppler and cosmological redshifts. At small redshifts and speeds slow compared to the speed of light, both equations reduce to the same approximate form. However, for large redshifts and large speeds, the equations are significantly different. Conceptually, the two types of cause for the redshift are completely different, even though both involve objects receding from each other.

Since all forms of electromagnetic radiation (including visible light) have the same speed in vacuum, about 300,000 kilometers per second, astronomers can look back in time simply by observing objects at greater distances. The farther away something is, the longer it has taken electromagnetic radiation to reach us. Therefore, light from distant galaxies is light from the distant past. This is why cosmology generally requires the use of the largest telescopes, including both ground-based and space telescopes, so as to be able to observe objects that appear very faint because they are very far away. Observing galaxies at greater distances provides views of what the universe was like in the past and what it was doing then. This is how astronomers have discovered that the expansion was not occurring as rapidly in the past as it is now, meaning that the expansion is accelerating.

The greatest uncertainty about observing objects at large distances is determining just how large the distances are. Very large distances generally are measured by techniques called standard candle methods. A "standard candle" is an object whose luminosity or real brightness is known more or less accurately by some means. The known (or estimated) real brightness is compared to the measured apparent brightness to yield the distance.

Hubble and countless others since then have employed Cepheid variables as standard candles to determine galaxy distances. Cepheid variables are supergiant stars, thousands of times more luminous than our Sun. The period-luminosity relation for Cepheids discovered by Henrietta Leavitt more than a century ago is used to determine the average real brightness by measuring the period of light variation. However, even though Cepheids are intrinsically bright, they can be detected out to distances on the order of only 100 million light-years.

To detect changes in the expansion rate of the universe requires comparing the expansion long ago with the expansion now. Measuring the expansion long ago means observing galaxies at very great distances, out to billions of light-years. Type Ia supernovae are the standard candles used for such large distances. A Type Ia supernova is a white dwarf star that explodes violently when it acquires enough matter from a nearby companion star to push its mass over the Chandrasekhar limit of 1.4 solar masses. When this happens, it becomes about 3 billion times brighter than the Sun and can be seen at distances of billions of light-years.

Galaxies at such large distances, as determined by Type Ia supernovae in them, have had the redshifts of their spectra measured. It was expected that these data would show the expansion of the universe in the past, when the light we now receive left those galaxies was greater than it is now; just how much greater would show how much the expansion has slowed down. Contrary to expectations, however, the data show that the expansion in the past was slower than it is now, indicating that the expansion of the universe is accelerating.

CONTEXT

The discovery of the expansion of the universe forced a revolutionary change in ideas about the universe. Before then, the origin of the universe was a subject for metaphysics or theology. The expansion discovered by Hubble and Humason and the moment of creation first proposed by Lemaître and later called the big bang moved such questions into the realm of science.

The redshifted spectra of galaxies showed that the universe was expanding. The relation between redshift and distance, Hubble's law, allowed astronomers to estimate the age of the universe. The finite speed of light means that looking out to greater distances is equivalent to looking farther back in time. A major factor spurring the building of larger telescopes has been the desire to observe fainter objects at larger distances and hence at earlier times closer to the origin of the universe. Furthermore, the development of new technology and observing techniques has permitted astronomers to study the universe and its origin not only with visible light but also over the radio, infrared, ultraviolet, X-ray, and gamma-ray parts of the electromagnetic spectrum. All this has revealed more information not only about the origin of the universe but also about its future.

During most of the twentieth century, a major question about the expansion of the universe was whether the universe would continue to expand forever or would someday stop and eventually collapse. Astronomers needed observations of more distant galaxies to determine how much the expansion of the universe is slowing. However, such measurements made in the 1990's showed the expansion is not decelerating but accelerating. This has led to new theories and models about what has come to be called dark energy, the name given to whatever might be making the expansion accelerate.

Questions about the origin and expansion of the universe bring together general relativity and quantum mechanics, but as they stand now, these two theories are not fully compatible. Also, it is hypothesized that at the very high temperatures and energies in the first moments after the big bang, all four fundamental forces—gravity, strong nuclear, weak nuclear, and electromagnetic—were unified as one; as the universe expanded and cooled, the forces separated. Attempts to unify the strong nuclear, weak nuclear, and electromagnetic forces are called grand unified theories (GUTs). Theories that try to unify gravity with the other three forces are called theories of everything (TOEs). The details of such unified theories have yet to be worked out. Furthermore, new observations made with new observing techniques may result in quite unexpected discoveries, leading again to new ideas about the universe.

Pamela R. Justice and Richard R. Erickson

FURTHER READING

Belusevic, Radoje. *Relativity, Astrophysics, and Cosmology*. Weinheim, Germany: Wiley-VCH, 2008. Addresses the interrelationship of the now entwined disciplines of astrophysics, particle physics, cosmology, and relativity theory, which have combined to form advanced theories of the origin and evolution of the universe.

Bennett, Jeffrey, et al. *The Cosmic Perspective*. 5th ed. San Francisco: Pearson/Addison-Wesley, 2008. Structured around a set of two-page figures dubbed "cosmic contexts," this interactive resource uses zoom-in illustrations to orient students to various images of the universe in relation to one another, while covering all the basics of cosmology. For general and introductory audiences. Chapters include "Light and Matter: Reading Messages from the Cosmos," "A Universe of Galaxies," "Dark Matter," "Dark Energy," and "The Fate of the Universe."

Drexler, Jerome. *Discovering Postmodern Cosmology: Discoveries in Dark Matter, Cosmic Web, Big Bang, Inflation, Cosmic Rays, Dark Energy, Accelerating Cosmos*. Boca Raton, Fla.: Universal, 2008. Cosmologist Drexler proposes a plausible and unique correlation of the seven mysterious areas of cosmological research listed in the book's subtitle. For all open-minded audiences.

Duncan, Todd, and Craig Tyler. *Your Cosmic Context: An Introduction to Modern Cosmology*. San Francisco: Pearson/Addison-Wesley, 2009. An introductory textbook for studies in modern cosmology that engages students by relating cosmological concepts to their own lives.

Eddington, Arthur. *The Expanding Universe*. Reprint. Cambridge, England: Cambridge University Press, 1987. Reissue of a classic early work supporting the theory of an expanding universe by a foremost scientist of the time. Contains a foreword by William McCrea, placing the work in its historical perspective.

Ferreira, Pedro G. *The State of the Universe: A Primer in Modern Cosmology*. London: Phoenix, 2007. Oxford lecturer Ferreira presents a history of cosmology, examining the complexities that concepts such as dark matter and dark energy have imposed on a once "simple" Einsteinian universe ruled by relativity.

Gasperini, Maurizio. *The Universe Before the Big Bang: Cosmology and String Theory*. Berlin: Springer, 2008. A fascinating exploration of what might have happened prior to the explosion that formed the universe, which looks to string theory and other modern mathematical models to postulate that the universe was not born with the big bang but rather was well advanced in its overall evolution. Presented with nontechnical language for nonspecialists.

Gribbin, John. *In Search of the Big Bang*. New York: Bantam Books, 1986. Explains the search for an understanding of the nature of the universe. "Part One: Einstein's Universe" gives a historical understanding of events leading to the discovery of the expanding universe. For the general reader.

Hawking, Stephen W. *A Brief History of Time: From the Big Bang to Black Holes*. New York: Bantam Books, 1988. Geared for the general reader by one of the leading theoretical physicists of the twentieth century. Provides a nonmathematical step-by-step explanation of the expanding universe and why the universe must expand.

Lemoine, M., J. Martin, and P. Peter, eds. *Inflationary Cosmology*. New York: Springer, 2008. This collection of papers, by both venerable and younger astrophysicists and cosmologists, focuses on that period during which the early universe expanded at a greatly accelerated rate before settling down to a slower rate of expansion. Presents several different scenarios.

Liddle, Andrew, and Jon Loveday. *The Oxford Companion to Cosmology*. New York: Oxford University Press, 2008. An indispensable A-Z reference, consisting of more than 350 entries from antimatter to WIMPs, as well as individual physicists and other scientists who have advanced the field. Heavily illustrated with almost two hundred halftones and diagrams; includes cross-references and Web links.

North, John. *Cosmos: An Illustrated History of Astronomy and Cosmology*. Chicago: University of Chicago Press, 2008. Emphasizes the astrophysics of cosmology, with a focus on physical and mathematical concepts that are key to understanding classical field theory. Also describes experimental techniques and results. Technical.

Schneider, Peter. *Extragalactic Astronomy and Cosmology: An Introduction*. New York: Springer, 2006. A textbook focusing on galaxies, clusters, and superclusters, beginning with the Milky Way. Covers their evolution, formation, and distribution, supported by beautiful color illustrations.

Weinberg, Steven. *Cosmology*. New York: Oxford University Press, 2008. The Nobel physics laureate presents the subject in two parts: one covering the isotropic and homogeneous "average" universe; the second, departures from the average universe. Provides detailed coverage of recombination, microwave background polarization, leptogenesis, gravitational lensing, structure formation, and multifield inflation. Includes mathematical calculations. Appendixes review general relativity, the Boltzmann equation, and sample problems. For advanced students and professionals.

Wudka, Jose. *Space-Time, Relativity, and Cosmology*. New York: Cambridge University Press, 2006. A history of relativistic cosmology from ancient times to Einstein. The nonmathematical approach emphasizes concepts over calculations yet explain the ideas clearly, applying them to research topics in cosmology. Designed for students from high school through college.

See also: Big Bang; Cosmic Rays; Cosmology; Electromagnetic Radiation: Nonthermal Emissions; Electromagnetic Radiation: Thermal Emissions; General Relativity; Interstellar Clouds and the Interstellar Medium; Milky Way; Novae, Bursters, and X-Ray Sources; Space-Time: Distortion by Gravity; Space-Time: Mathematical Models; Universe: Evolution; Universe: Structure.

Universe: Structure

Category: The Cosmological Context

Galaxies are gravitationally bound assemblages of millions to tens of trillions of stars. Galaxies are not scattered randomly across the universe but instead are grouped together in galaxy clusters, consisting of several tens to many thousands of individual galaxies. There is growing evidence of connective patterns between galaxy clusters stretching across regions at least as large as many hundreds of millions of light-years. Explaining the observed features of this large-scale structure is a challenge to cosmologists that puts limits on acceptable models of the origin and evolution of the universe.

OVERVIEW

The contents of the universe are arranged in a hierarchy of structures. Stars, of which our Sun is a familiar example, are hot balls of gas that generate energy through nuclear fusion reactions. There is growing evidence that many stars are orbited by families of planets, analogous to our solar system. Stars (and their families of planets, if they have them) are gravitationally bound together into galaxies, vast collections of millions to tens of trillions of stars, spanning thousands to hundreds of thousands of light-years. Our Sun and solar system are part of the Milky Way galaxy, a moderately large spiral galaxy consisting of several hundred billion stars along with gas and dust.

Galaxies in turn are gravitationally bound together into galaxy clusters, containing from several tens to many thousands of galaxies. The Milky Way galaxy is part of the Local Group, a small galaxy cluster containing about forty members. Galaxy clusters group together to form superclusters. The Local Group is part of the Virgo supercluster, containing several tens of thousands of individual galaxies and spanning more than 100 million light-years. On a still larger scale, galaxies, clusters, and superclusters appear to be arranged in a network of filaments and walls surrounding large, nearly empty voids or bubbles hundreds of millions of light-years across. This overall large-scale structure has been described as "frothy," or analogous to Swiss cheese with all its holes.

Understanding the structure and evolution of the universe requires evidence about the distribution of matter on the largest scales. Most models of the universe satisfy the cosmological principle. This is the assumption that, at any moment, the large-scale structure of the universe looks the same from all places (homogeneity) and in all directions (isotropy).

When Albert Einstein published his general theory of relativity in 1916, in which gravity is treated as a warping of space-time, almost immediately it was recognized that it would have a profound impact on cosmological models. Einstein himself tried applying it to a static universe and found that he needed to add an arbitrary term, which was called the cosmological term or cosmological constant, to obtain a static solution.

In 1922, the Russian mathematical physicist Alexander Alexandrovich Friedmann and in 1927 the Belgian priest and cosmologist Georges Lemaître independently derived two classes of homogeneous and isotropic solutions (without the cosmological constant) in which the universe expands. In one of these classes, space has a uniform positive curvature and finite extent though no boundary; the universe is said to be closed because it expands to some maximum size and then contracts. The two-dimensional surface (not including the interior volume) of a three-dimensional sphere is such a space; motion on the surface is never blocked by a barrier (perimeter line), but the area of the surface is finite. In the other class, space is negatively curved and of infinite extent; the universe is said to be open because it expands forever. In two dimensions, such a surface is termed "hyperboloid"; a saddle is an example of a finite part of such a surface which mathematically extends to infinity if it lacks a boundary curve. In 1932, Einstein and the Dutch astronomer Williem de Sitter proposed a third type of expanding model without the cosmological constant. In it, space is flat and infinite, geometry is Euclidean, and the universe just barely expands forever; it is the boundary case between the two Friedmann-Lemaître solutions. In two

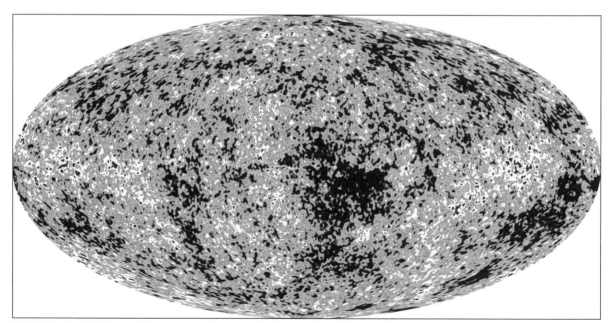

In 2003, the Wilkinson Microwave Anisotropy Probe produced this high-resolution map of the universe's microwave radiation only 380,000 years after the big bang. These data helped to prove that the age of the universe is 13.7 billion years; that 73 percent of the universe is composed of dark energy, 23 percent of cold dark matter, and 4 percent of atoms; and that the universe will continue to expand. (NASA/WMAP Science Team)

dimensions, the surface of a plane is flat and infinite. These are the only three possible spatial geometries and types of models that can be solutions of Einstein's gravitational field equations under the assumptions of homogeneity and isotropy.

The common feature of these models is that their scale factors (the distance between representative points, such as clusters of galaxies) change with time; the universe must be either expanding or contracting. In 1929, Edwin Powell Hubble published the first observational evidence that the universe is indeed expanding. Using the 100-inch reflecting telescope on Mount Wilson (then the largest telescope in the world), he found that the more distant a galaxy is, the greater its spectrum is redshifted. This means all features in the spectrum are shifted to longer than normal wavelengths, and this can be due to motion away from us.

If the universe is expanding, the scale factor was smaller in the past, and the local densities of both matter and radiation were greater. However, the mass density of matter varies inversely as the scale factor cubed, while the energy density of radiation varies inversely as the scale factor to the fourth power. At sufficiently early times when the scale factor was smaller, the density of radiation was greater than the density of matter, and the universe was dominated by radiation. With time, as the scale factor increased and the universe expanded, the density of radiation decreased faster than the density of matter did, and eventually the universe became dominated by matter, as it is today.

The universe today is "lumpy" on a variety of scales: stars, galaxies, galaxy clusters, superclusters, and "walls" and "voids." In recent years, there has been growing interest in trying to understand how an initially homogeneous big bang could produce the lumpy inhomogeneities observed today. This is called the homogeneity problem.

APPLICATIONS

Galaxy surveys carried out to great distances show that the density of galaxies in voids is typically a factor of ten less than average, and the density in the narrow but long walls and filaments is typically a factor of ten thousand greater than average. Explaining the origin of such variations in density from an early universe that was amazingly homogeneous is a major challenge in modern cosmology.

The cosmic microwave background radiation (called the CMB or CBR for short), coming from a time a few hundred thousand years after the big bang when the matter in the expanding universe became transparent, is remarkably uniform and isotropic. However, detailed observations of it made by the Cosmic Background Explorer (COBE) satellite, the Wilkinson Microwave Anisotropy Probe (WMAP) satellite, and the Balloon Observations of Millimetric Extragalactic Radiation and Geomagnetics (BOOMERANG) project in Antarctica do reveal small variations. The data perfectly fit a blackbody Planck curve for a temperature of 2.735 kelvins. Removing a slight asymmetry in temperature (about 0.007 kelvins) in opposite directions, presumably due to the motion of our solar system through the background radiation, leaves very small temperature fluctuations on the order of 10 micro-kelvins, about one degree in angular size. The hotter regions had slightly higher densities, and they are about the right size to develop into clusters and superclusters of galaxies.

The initial formation of the hotter, denser regions and their evolution into galaxy clusters and superclusters may involve dark matter. This is the name now used to refer to what formerly was called "missing mass." In many situations in astronomy, the amount of matter that can be detected through the electromagnetic radiation that it emits (whether radio waves, visible light, or other wavelengths) is much less—typically by a factor of about 5 to 50—than what is needed gravitationally to hold galaxies, clusters of galaxies, and superclusters together and to account for gravitational lensing of distant objects. Most astrophysicists consider the gravitational estimates reasonably well established, and thus believe the mass is not "missing" but simply is not emitting electromagnetic radiation. The challenge is to explain the nature of the dark matter, which is much more abundant than ordinary luminous matter.

Various observational tests suggest that nonluminous ordinary matter (perhaps in the form of boulders, planet-sized objects, black dwarfs, black holes, and other known nonemitting entities) can account for no more than about 10 to 12 percent of dark matter. Thus most dark matter must be in some more exotic form. One suggestion involves weakly interacting massive particles (WIMPs), subatomic particles that, like neutrinos, interact very rarely with ordinary matter (other than gravitationally) but are much more massive than neutrinos.

Whatever dark matter is, it seems to be necessary to give us the universe we observe today. Temperature (and hence density) fluctuations revealed in the background radiation are about the right size to develop into galaxy clusters and superclusters, but they do not contain enough ordinary matter to contract gravitationally into protogalaxies as quickly as galaxies seem to have formed after the big bang. Adding a lot of dark matter to these density fluctuations stimulates the rapid growth of galaxies because of the increased gravitational attraction.

Temperature and density fluctuations recorded in the background radiation probably are due to very small quantum fluctuations that occurred spontaneously during the first 10^{-35} second after the big bang, and then were magnified by many orders of magnitude during the era of cosmic inflation, when the universe expanded exponentially in the next 10^{-30} second. Since ordinary matter was opaque to electromagnetic radiation until a few hundred thousand years after the big bang, it would have been buffeted and kept smoothed out by the strong radiation field until finally the ordinary matter became transparent and decoupled from the radiation. However, since dark matter does not seem to interact with electromagnetic radiation, it would have been able to collect in the magnified quantum fluctuations as soon as inflation ended and thus built up density concentrations to attract ordinary matter later.

Another idea that may be relevant to the formation of large-scale structure in the universe is cosmic string theory. According to string theory, space-time has ten or eleven dimensions, but most are "rolled up" or compacted so that only the familiar four space-time dimensions (length, height, width, and time) are noticeable. Cosmic strings are hypothetical long, thin, linelike concentrations of unbroken symmetry left over from the spontaneous symmetry breaking that occurred when the electromagnetic, weak nuclear, and strong nuclear forces separated to bring the grand unification era to a close. Specific multidimensional modes of vibration of cosmic strings are thought to be manifested as all the particles and forces in the universe. In the early universe, cosmic strings may have served as "seeds" for the formation of long concentrations of matter, like droplets condensing on a wire, which later evolved into filamentary chains of galaxies.

CONTEXT

The history of the quest to understand the structure of the universe has been a progression toward recognition of ever more subtle organization at ever-larger scales. Just about all cultures and societies have divided naked-eye stars of the night sky into patterns we call constellations. In early Greco-Roman cosmology (and in many others), stars were attached to the inside of a hollow sphere—the celestial sphere—that enclosed the Earth fixed at the sphere's center. In the early 1600's, Galileo, in the first recorded use of telescopes to systematically study the sky, discovered there were many stars too faint to be seen with the unaided eye. Gradually during the 1600's, the idea developed that stars were similar to our Sun, and therefore their different apparent brightnesses meant that they were at different distances from us. Consequently the stars in a given pattern or constellation might not be a real grouping in space but could be at very different distances from us.

However, this idea could not be confirmed until the distances of stars could be measured directly. The first successful measurements of stellar distances were made in 1838 and 1839, independently by Friedrich Wilhelm Bessel in Germany, Thomas Henderson in South Africa, and Friedrich Georg Wilhelm von Struve in Russia. They all used the method of trigonometric parallax. That involved measuring small changes in the apparent positions of nearby stars relative to more distant stars as seen from

Leavitt, Shapley, and the Period-Luminosity Scale

In 1902, Henrietta Swan Leavitt became a permanent staff member at Harvard College Observatory. She studied variable stars, stars that change their luminosity (brightness) in a fairly predictable pattern over time. During her tenure at Harvard, Leavitt observed and photographed nearly 2,500 variable stars, measuring their luminosities over time. She was equipped with photographs of the Large and Small Magellanic Clouds collected from Harvard's Peruvian observatory. The Magellanic Clouds are very small galaxies visible in the Southern Hemisphere and close to the Milky Way. The Small Magellanic Cloud contained seventeen Cepheid variables having very predictable periods ranging from 1.25 days to 127 days. Leavitt carefully measured the brightening and dimming of the seventeen Cepheids during their respective periods. She collected photographs of other Cepheids in the Magellanic Clouds and made additional period-luminosity studies. In a circular dated March 3, 1912, she stated:

> The measurement and discussion of these objects present problems of unusual difficulty, on account of the large area covered by the two regions, the extremely crowded distribution of the stars contained in them, the faintness of the variables, and the shortness of their periods. As many of them never become brighter than the fifteenth magnitude, while very few exceed the thirteenth magnitude at maximum, long exposures are necessary, and the number of available photographs is small. The determination of absolute magnitudes for widely separated sequences of comparison stars of this degree of faintness may not be satisfactorily completed for some time to come. With the adoption of an absolute scale of magnitudes for stars in the North Polar Sequence, however, the way is open for such a determination.

Ejnar Hertzsprung of the Leiden University in the Netherlands and Henry Norris Russell of the Mount Wilson Observatory in Pasadena, California, had independently discovered a relationship between a star's luminosity and its spectral class (that is, color and temperature). Together, their experimental results produced the Hertzsprung-Russell diagram of stellar luminosities, the astronomical equivalent of chemistry's periodic table. According to their classification scheme, most stars lie along the "main sequence," which ranges from extremely bright blue stars ten thousand times brighter than the Sun to very dim red stars one hundred times dimmer than the Sun. Cepheid variables fell toward the cooler, red end of the main sequence.

Leavitt carefully measured the luminosities and cyclic periods of changing luminosity for each of many Cepheid variables from the Magellanic Clouds. From her careful measurements, she graphically plotted Cepheid luminosity against Cepheid period. She noticed "a remarkable relation between the brightness of these variables and the length of their periods. . . . the brighter variables have the longer periods." She had discovered that a Cepheid's apparent luminosity is directly proportional to the length of its period, or the time it takes to complete one cycle of brightening and dimming.

Harlow Shapley, an astronomer at the Mount Wilson Observatory, measured the distances of moving star clusters containing Cepheids, then related the Cepheid distances to Cepheid period-luminosity data. From these experiments, Shapley constructed a Cepheid period-absolute luminosity curve, which made it possible to plot a Cepheid variable having a specific measured period and obtain its absolute luminosity. Knowing the Cepheid's apparent and absolute luminosities, one can instantly calculate its distance and, therefore, the distances of all the stars in the star cluster containing that particular Cepheid variable.

The distances to Cepheid variables in the Milky Way and other galaxies were soon determined. Shapley used Cepheid distances to demonstrate that the center of the Milky Way is directed toward the constellation Sagittarius and that the Sun is located approximately thirty thousand light-years from the galactic center. Edwin Powell Hubble applied the technique to obtain estimates of the distances between our galaxy and others, which led to his monumental astronomical discovery that the universe is expanding.

Earth at various points in its orbit around the Sun. Trigonometric parallax is still the most assumption-free method of measuring distances, but it is limited by the ability to measure small angular shifts accurately. Hipparcos, the High-Precision Parallax Collecting Satellite operated by the European Space Agency, has been able to measure parallaxes with reasonable accuracy out to distances of about 1,600 light-years. Beyond that distance, other methods must be used, but their calibration ultimately is tied in to the distances obtained by trigonometric parallaxes. Thus constellations came to be seen as merely convenient direction indicators from our vantage point on Earth, not physical associations of stars.

In the early 1600's, Galileo discovered telescopically that the hazy white band of light known as the Milky Way actually consisted of lots of faint stars. Because the Milky Way forms a great circle band of light around the sky, by the 1700's its overall shape was described using terms like sheet, disk, millwheel, and grindstone. In 1784, William Herschel tried to determine its size and shape by counting the stars seen in various directions. These early models all placed the Sun near the Milky Way's center. It was not until 1918 that Harlow Shapley found that the Milky Way galaxy was much larger than previous estimates and the Sun was located far from the center.

At this time, there still was uncertainty about whether the Milky Way galaxy comprised the entire universe or not. As far back as the 1700's, there had been speculation, by Immanuel Kant and others, that the spiral nebulae might be other large collections of stars like the Milky Way. During the 1920's, Edwin Hubble, using the Mount Wilson 100-inch reflecting telescope, was able to resolve individual stars in some of the spiral nebulae; moreover, some of the stars were Cepheid variable stars, which could be used to determine distances. The spiral nebulae were shown to be located far beyond the Milky Way galaxy and hence were galaxies comparable to the Milky Way. A few years later, Hubble found that most galaxy spectra were redshifted, thus confirming the cosmological models of an expanding universe derived from Einstein's general relativity.

Ever since, central goals of observational and theoretical cosmology have been to produce accurate maps of the distribution of these galaxies and explanations of the origin of the features in this distribution. Pioneering work by George Abell and others, beginning in the 1950's, showed that galaxies were grouped into clusters, some rich with thousands of member galaxies and others poor with fewer than one hundred members. Some clusters are regular, with a spherical shape and many galaxies concentrated near the center; others are irregular, with galaxies scattered across an extended region of space. In the 1980's, Margaret Geller and John Huchra at Harvard-Smithsonian Center for Astrophysics began mapping the large-scale distribution of galaxies out to great distances. They, and teams at other institutions, have shown that out to distances of several billion light-years, superclusters of galaxy clusters are arranged in filaments and walls surrounding large, nearly empty voids.

Understanding the origin of structure in the universe on such large scales challenges the creativity of cosmologists. Although the exact details are not yet clear, it seems that all the patterns of organization seen in the universe today were determined by events and processes in its earliest moments after the big bang.

John J. Dykla and Richard R. Erickson

FURTHER READING

Belusevic, Radoje. *Relativity, Astrophysics, and Cosmology*. Weinheim, Germany: Wiley-VCH, 2008. Addresses the interrelationship of the now entwined disciplines of astrophysics, particle physics, cosmology, and relativity theory, which have combined to form advanced theories of the origin and evolution of the universe.

Bennett, Jeffrey, et al. *The Cosmic Perspective*. 5th ed. San Francisco: Pearson/Addison-Wesley, 2008. Structured around a set of two-page figures dubbed "cosmic contexts," this interactive resource uses zoom-in illustrations to orient students to various images of the universe in relation to one another, while covering all the basics of cosmology. For general and introductory audiences. Chapters include "Light and Matter: Reading Messages from the Cosmos," "A Universe of Galaxies," "Dark Matter," "Dark Energy," and "The Fate of the Universe."

Chaisson, Eric, and Steve McMillan. *Astronomy Today*. 6th ed. New York: Addison-Wesley, 2008. Very well-written college-level textbook for introductory astronomy courses. Several chapters address the development of structure in the universe.

Cohen, Nathan. *Gravity's Lens: Views of the New Cosmology*. New York: John Wiley & Sons, 1988. A book for the general reader by a researcher in general relativity and cosmology, this clear and well-illustrated volume features extensive discussion of the evidence for large-scale structure, and prospects for refined and more extensive observations in the future.

Drexler, Jerome. *Discovering Postmodern Cosmology: Discoveries in Dark Matter, Cosmic Web, Big Bang, Inflation, Cosmic Rays, Dark Energy, Accelerating Cosmos*. Boca Raton, Fla.: Universal, 2008. Cosmologist Drexler proposes a plausible and unique correlation of the seven mysterious areas of cosmological research listed in the book's subtitle. For all open-minded audiences.

Duncan, Todd, and Craig Tyler. *Your Cosmic Context: An Introduction to Modern Cosmology*. San Francisco: Pearson Addison Wesley, 2009. An introductory texbook for studies in modern cosmology that engages students by relating cosmological concepts to their own lives.

Ferreira, Pedro G. *The State of the Universe: A Primer in Modern Cosmology*. London: Phoenix, 2007. Oxford lecturer Ferreira presents a history of cosmology, examining the complexities that concepts such as dark matter and dark energy have imposed on a once "simple" Einsteinian universe ruled by relativity.

Gasperini, Maurizio. *The Universe Before the Big Bang: Cosmology and String Theory*. Berlin: Springer, 2008. A fascinating exploration of what might have happened prior to the explosion that formed the universe, which looks to string theory and other modern mathematical models to postulate that the universe was not born with the big bang but rather was well advanced in its overall evolution. Presented with nontechnical language for nonspecialists.

Lemoine, M., J. Martin, and P. Peter, eds. *Inflationary Cosmology*. New York: Springer, 2008. This collection of papers, by both venerable and younger astrophysicists and cosmologists, focuses on that period during which the early universe expanded at a greatly accelerated rate before settling down to a slower rate of expansion. Presents several different scenarios.

Liddle, Andrew, and Jon Loveday. *The Oxford Companion to Cosmology*. New York: Oxford University Press, 2008. An indispensable A-Z reference, consisting of more than 350 entries from antimatter to WIMPs, as well as individual physicists and other scientists who have advanced the field. Heavily illustrated with almost two hundred halftones and diagrams; includes cross-references and Web links.

North, John. *Cosmos: An Illustrated History of Astronomy and Cosmology*. Chicago: University of Chicago Press, 2008. Emphasizes the astrophysics of cosmology, with a focus on physical and mathematical concepts that are key to understanding classical field theory. Also describes experimental techniques and results. Technical.

Schneider, Peter. *Extragalactic Astronomy and Cosmology: An Introduction*. New York: Springer, 2006. A textbook focusing on galaxies, clusters, and superclusters, beginning with the Milky Way. Covers their evolution, formation, and distribution, supported by beautiful color illustrations.

Schneider, Stephen E., and Thomas T. Arny. *Pathways to Astronomy*. 2d ed. New York: McGraw-Hill, 2008. An introductory astronomy textbook. Divided into short units on specific topics, it offers extensive coverage throughout on the development of structure in the universe.

Weinberg, Steven. *Cosmology*. New York: Oxford University Press, 2008. The Nobel physics laureate presents the subject in two parts: one covering the isotropic and homogeneous "average" universe; the second, departures from the average universe. Provides detailed coverage of recombination, microwave background polarization, leptogenesis, gravitational lensing, structure formation, and multifield inflation. Includes mathematical calculations. Appendixes review general relativity, the Boltzmann equation, and sample problems. For advanced students and professionals.

Wudka, Jose. *Space-Time, Relativity, and Cosmology*. New York: Cambridge University Press, 2006. A history of relativistic cosmology from ancient times to Einstein. The nonmathematical approach emphasizes concepts over calculations yet explain the ideas clearly, applying them to research topics in cosmology. Designed for students from high school through college.

See also: Big Bang; Cosmic Rays; Cosmology; Electromagnetic Radiation: Nonthermal Emissions; Electromagnetic Radiation: Thermal Emissions; General Relativity; Interstellar Clouds and the Interstellar Medium; Milky Way; Novae, Bursters, and X-Ray Sources; Space-Time: Distortion by Gravity; Space-Time: Mathematical Models; Universe: Evolution; Universe: Expansion.

Uranus's Atmosphere

Categories: Planets and Planetology; The Uranian System

Uranus is the seventh planet from the Sun. It shares much in common with Jupiter and Saturn, but it is also significantly different from the larger Jovian, or "gas giant," planets. Its atmosphere is composed mainly of hydrogen and helium, but its color is governed by selective absorption of light by methane, which is abundant in greater measure in Uranus's atmosphere than in either Jupiter's or Saturn's.

OVERVIEW

The planet Uranus was the first to be discovered with a telescope. Its existence was declared by Sir William Herschel on March 13, 1781. Af-

ter several proposed names, the most curious of which was a proposed reference to the King of England, George III, the planet was named Uranus. From mythology Uranus is the father of Saturn and grandfather of Jupiter.

Uranus is the third largest planet in the solar system. With an orbit that varies from 18.4 to 20 astronomical units (AU, or the mean distance from the Earth to the Sun, namely 150 million kilometers), it takes Uranus eighty-four years to complete one revolution about the Sun. Naturally, at this greater distance from the Sun, Uranus receives far less solar radiation than Jupiter and Saturn. Nevertheless, its location suggested to early researchers that the composition and nature of Uranus were similar to those of Jupiter and Saturn.

Superficially, that is true. Uranus is composed largely of hydrogen and helium. However, the atmosphere of Uranus has been determined to be colder than that of Jupiter and Saturn and has a less dynamic structure than the turbulent atmosphere of Jupiter or the pastel banding of Saturn. Uranus's atmospheric temperature can drop to 49 kelvins. In addition to the preponderance of hydrogen and helium, the atmosphere has a larger amount of ices and hydrocarbon than do the atmospheres of Jupiter and Saturn. Ices include water, ammonia, ammonium hydrosulfide, and methane. Selective absorption of radiation, in good measure by methane, results in the planet's pale bluish-green appearance.

By compositional abundance, Uranus is 83 percent hydrogen, 15 percent helium, 2.3 percent methane. The total also includes other, low-concentration gases and hydrocarbons and does not add up precisely to 100 percent given significant uncertainties about the abundance of hydrogen and helium. Hydrocarbons that appear only in trace amounts include ethane, acetylene, methyl acetylene, and diacetylene. These and other hydrocarbons are thought to be produced in the upper atmosphere by photolysis of methane under incident solar ultraviolet light. Carbon monoxide and carbon dioxide have also been detected. Like the planet's water vapor, carbon dioxide and carbon monoxide must have been acquired by impacting comets and infalling dust. All totaled, Uranus has a carbon content, primarily found in the atmosphere,

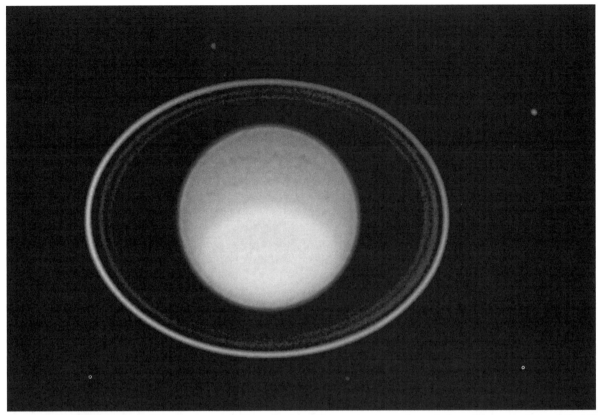

In 1995, the Hubble Space Telescope revealed much about Uranus's atmosphere: Infrared images showed that it is composed of hydrogen and traces of methane; the inner atmosphere is clear, an intermediate yellow layer is hazy, and a very thin outer layer is red. In addition, Uranus's rings are bright in the infrared. (NASA/Erich Karkoschka, University of Arizona)

somewhere between twenty and thirty times that of solar abundance.

One thing that makes Uranus particularly curious is the fact that its rotational axis is tilted 97.8° from the perpendicular to the ecliptic plane. This tilt is why many refer to Uranus as the planet that rotates on its side. There is no universally accepted explanation for this high degree of tilt, but many believe the planet was knocked on its side by a collision with a large body early in the Uranian system's development. This curious tilt means that for roughly half of each orbit the north pole receives solar radiation, and for roughly half of the rest of the orbit the south pole is in sunlight. This makes for unusual seasons and atmospheric dynamics. Although the planet's interior rotates once every 17 hours 14 minutes, the atmosphere rotates differentially. Features in the upper atmo-

sphere have been clocked at as much as 0.25 kilometer per second and thus may experience a full rotation in less than 14 hours.

KNOWLEDGE GAINED

Earth-based telescopic studies of Uranus revealed it to have a bizarre orientation of its rotational axis, several relatively small satellites, and an orbital period of 84.3 years. In 1977, observations made from an aircraft-based telescope as Uranus occulted a star revealed the presence of dark rings around the mysterious seventh planet from the Sun. That same year the Voyager 2 spacecraft launched on an approved mission to fly by both Jupiter and Saturn. The National Aeronautics and Space Administration (NASA) originally proposed sending an armada of sophisticated spacecraft on what had been termed the "Grand Tour."

This "tour" referred to the fact that every 176 years, planetary alignments are such that gravitational slingshot maneuvers in the outer solar system can be used to send spacecraft to investigate all the outer planets from Jupiter to Pluto. Unfortunately, that ambitious plan was not funded, but NASA was given authorization to build two modest Voyager spacecraft for exhaustive investigations of Jupiter and Saturn. When Voyager 1 was successful at both gas giants, the Voyager 2 spacecraft was targeted through the Saturn system in such a way as to make possible a flyby of Uranus and Neptune.

Atmospheric structure is often discussed in terms of either pressure levels or temperature or both. If one defines Uranus's "surface" as the site where the pressure is 1 bar (1 Earth atmosphere, or 10^5 pascals), then that atmosphere can be described as follows. Uranus has a troposphere found from −300 to 50 kilometers above the surface, where the pressure varies from 100 to 0.1 bar, respectively. A stratosphere exists between 50 and 400 kilometers, where the pressure varies from 0.1 to 10^{-10} bar. Then, from 400 kilometers out to as much as two planet radii, or roughly 50,000 kilometers, is the thermosphere and corona, where the pressure dwindles down to near vacuum from the upper stratospheric level of 10^{-10} bar.

One might think that because its poles receive more solar illumination than the equatorial region, Uranus would be warmer at the pole presently facing toward the Sun, but that is not the case. Near the equator is the planet's only portion to experience fairly rapid day-night variation due to the excessive tilt of Uranus. Near the equator the warmest temperatures are recorded. Upper atmospheric temperatures near the equator can rise to 57 kelvins. Why the equatorial region is warmer than the illuminated polar region is currently unknown.

The Hubble Space Telescope routinely was used by planetary scientists to examine Uranus for features and changes in those features within the planet's atmosphere. In 1998 an image credited to NASA and Erich Karkoschka of the University of Arizona revealed on the order of twenty clouds in Uranus's atmosphere. That was rather remarkable; prior to that time, in the entire history of Uranus observations, there

had been fewer than that number of clouds seen in the planet's usually unremarkable-looking atmosphere. This Hubble image was taken in infrared and clearly showed the planet's rings and many of its known satellites. In addition to the clouds, it revealed a bright band circling the planet. Wind speeds of clouds near the band were determined to be in excess of 500 kilometers per hour. One of the clouds seen in this infrared image was the brightest Uranian cloud ever observed.

In 2006, Hubble images, in concert with near-infrared observations made using the ground-based Keck telescope, revealed a new, more dynamic picture of Uranus. The planet was seen to have a great dark spot and some degree of banding. These observations were summarized by astronomer Heidi Hammel at a Hubble science overview briefing held in 2002, about a month in advance of what was then proposed to be the final shuttle servicing mission to the Hubble. Hammel used her appearance at that briefing as an opportunity to stress how much Hubble had already done to change the picture of an inactive Uranian atmosphere as presented by Voyager 2 in 1986. Hammel and other astronomers had detected an increase in the number and scope of clouds in the ice giant's atmosphere, in addition to finding that great dark spot. She expressed enthusiasm about extended Hubble operations advancing understanding of Uranus, perhaps the planet about which the least is known. This increased activity appears related to seasonal changes as Uranus orbits the Sun.

CONTEXT

Uranus was the first planet for which clear records for discovery exist. Although Herschel was not the first to note Uranus in astronomical records, he was the first to identify it correctly as a planet and not a comet or unidentified star, as others had done previously. From the time of its discovery to the dawn of the space age, little could be learned about Uranus from ground-based telescopes. However, observations of Uranus's orbit about the Sun led to the recognition that there was good reason to believe that it was not the last planet to be discovered in the solar system. Based on gravitational perturbations in the orbit of Uranus, Neptune was discovered by and large by mathematical analysis. Observations verified the correctness of those calculations. Uranus and Neptune were believed to be very similar. Both displayed a bluish-green tint in telescopic views. Spectroscopic analysis indicated that both planets had atmospheres different from those of Jupiter and Saturn. Like their larger gas giant cousins, Uranus and Neptune were known to have atmospheres rich in hydrogen and helium, but their bluish-green color was identified as due to extensive absorption of red light by methane.

Voyager 2 provided the greatest portion of current understanding about the Uranian system. Planetary scientists interested in Uranus await a return mission, most likely an orbiter, perhaps with a lander probe for one of the icy satellites and an atmospheric probe to ram through the upper atmosphere of Uranus and conduct measurements until it is crushed. Ground-based observations and imaging by the Hubble Space Telescope continue in the meantime. The biggest question about detection and observation of clouds in Uranus's atmosphere centers on the source of energy driving those storms, since Uranus's internal heat flow appears to be insufficient to cause such airflow.

David G. Fisher

FURTHER READING

Elkins-Tanton, Linda T. *Uranus, Neptune, Pluto, and the Outer Solar System.* New York: Chelsea House, 2006. This book explores the Sun's relationship with the three outer planets and their moons, considering these planets as recorders of the formation of the solar system. Aimed at a general or high school audience. Illustrations, bibliography, index.

Encrenaz, Thérèse, et al. *The Solar System.* New York: Springer, 2004. A thorough exploration of the solar system from early telescopic observations through the space missions that had investigated all planets by the publication date. Takes an astrophysical approach to place our solar system in a broad context as just one member of similar systems throughout the universe.

Freedman, Roger A., and William J. Kaufmann III. *Universe.* 8th ed. New York: W. H. Freeman, 2008. A college text on astronomy, somewhat more advanced than many introductory texts but with a wealth of detail and excellent diagrams. Chapters 6 through 16 describe the solar system, including Uranus and what is know about its atmosphere. Comes with a CD-ROM.

Irwin, Patrick G. J. *Giant Planets of Our Solar System: An Introduction.* 2d ed. New York: Springer, 2006. Focuses on Jupiter, Saturn, Uranus, and Neptune and their atmospheres. Suitable as a textbook for upper level college courses in planetary science. Filled with figures and photographs. Available to the serious general audience.

McBride, Neil, and Iain Gilmour, eds. *An Introduction to the Solar System.* Cambridge, England: Cambridge University Press, 2004. A complete description of solar-system astronomy suitable for an introductory college course. Accessible to nonspecialists as well. Filled with supplemental learning aids and solved student exercises.

Miller, Ron. *Uranus and Neptune.* Brookfield, Conn.: Twenty-First Century Books, 2003. Considers Uranus and its satellites in comparison with other gas giants, especially Neptune, including their atmospheres.

Morrison, David, and Tobias Owen. *The Planetary System.* 3d ed. San Francisco: Pearson/Addison-Wesley, 2003. Geared for the undergraduate college student. Planetary atmospheres are treated as important physical features of the various members of the Sun's family. They are discussed individually in the context of what is known about each planet's characteristics and with regard to theories about their evolution and the evolution of the entire solar system. Comprehensive for the average reader.

Tocci, Salvadore. *A Look at Uranus.* New York: Franklin Watts, 2003. As part of the Out of This World series, this book covers all aspects of the planet Uranus from its discovery through 2002. Includes several photographs. Suitable for all readers.

See also: Auroras; Earth's Atmosphere; Earth's Composition; Eclipses; Infrared Astron-

Uranus's Interior

Categories: Planets and Planetology; The Uranian System

Uranus, the seventh planet from the Sun, has much in common with the other Jovian gas giants, but its interior is different from the interiors of the larger planets Jupiter and Saturn and more like that of Neptune.

OVERVIEW

Uranus is largely composed of hydrogen and helium and is considered to be a Jovian planet, or gas giant. The interior of Uranus differs significantly from that of Jupiter and Saturn, however. It shares much more in common with Neptune. Like the interior of Jupiter and Saturn, Uranus's interior cannot be directly sampled. Models of the interiors of the Jovian planets are inferred by external observation. For example, a magnetic field tells much about the nature and physical characteristics of a given planet's interior.

Accurate determinations of orbital motions of Uranus's satellites led to a precise value of the planet's mass. With its size slightly in excess of Neptune, Uranus is the third largest planet in the solar system, but the second least dense. Uranus has a mass 14.5 times that of Earth. With a mean radius four times the Earth's, its overall density is 1.27 grams per cubic centimeter. Of course, the different structures of Uranus, from the core to the upper atmosphere, have specific characteristics of their own. Nevertheless, in gross terms, only Saturn is less dense than Uranus as a planet. Uranus is composed, like Jupiter and Saturn, primarily of hydrogen and helium, although the percentages of

both are different in Uranus from the percentages of Jupiter and of Saturn. Uranus contains methane, water vapor, ammonia, carbon monoxide, and carbon dioxide. Hydrocarbons such as ethane, acetylene, methyl acetylene, and diacetylene are also present, presumably created by photolysis of methane in the atmosphere under illumination of solar ultraviolet light.

Like Jupiter and Saturn, hydrogen in Uranus is found in gaseous form in the atmosphere, but deep inside the planet at greater pressure hydrogen may exist in more exotic forms. Whether helium is found in forms other than as a simple gas is a matter of debate. Because of their size, temperature, pressure, compositional, and interior structural differences from Jupiter and Saturn, Uranus and Neptune both are often referred to as ice giants.

Several models exist for Uranus's interior. They agree on major features and differ in only minor ways. A great deal of the interior is composed of water, ammonia, and methane, with the percentage of water included in each model varying. That water percentage ranges between 9.3 to 13.5 Earth masses, depending on which model a scientist supports. With only 0.5 to 1.5 Earth masses of hydrogen and helium in the atmosphere and interior, that leaves between 0.5 to 3.7 Earth masses for what is considered rocky material.

Under the atmosphere of hydrogen, helium, and methane are two inner layers. First, an icy mantle actually incorporates the majority of the planet's mass. Under that is the rocky core. The core represents less than 20 percent of the planet's radius, and the relatively thin upper atmosphere of gases represents another 20 percent of the planet's radius. This leaves the mantle as 60 percent of the planet, with as much as 13.5 Earth masses. Obviously the densities of the atmosphere, mantle, and core are different, just as pressures and temperatures in these three distinctly different regions vary.

The mantle is icy in the sense that it is a hot, dense, electrically conducting mixture of water, ammonia, and less abundant volatile substances. In the mantle, under the great pressures present, molecules dissolved in the water become ionized and therefore create the high electrical conductivity displayed by the mantle. Because of its combined fluid and highly conductive nature, this layer is often referred to as a water-ammonia ocean. The latter quality of the mantle is responsible for generating the planet's complex magnetic field.

The core is far denser, at approximately 9 grams per cubic centimeter. The planet's central pressure and temperature are believed to reach 8 million bars and 5,000 kelvins, respectively. High internal heat flows out to the atmosphere. Materials that have high electrical conductivities usually also have high thermal conductivities. In the case of Uranus, however, it is obvious, based on atmospheric quiescence, that heat flow from the planet's core to the atmosphere is less than that of Neptune. Whereas Neptune radiates more energy than it intercepts by solar irradiance, Uranus is barely 6 percent greater in the infrared than the solar energy absorbed by Uranus's atmosphere. All models agree that Uranus's heat flow from the interior is only 0.042 watt per square meter. The heat flow from the much less massive and smaller Earth, by comparison, is 0.075 watt per square meter. Uranus's low heat flow and hence cold atmosphere (portions of the troposphere have been recorded at a mere 49 kelvins) could be tied to the planet's bizarre rotational configuration. Most planetary scientists believe that Uranus suffered a catastrophic event early in its history, presumably a collision with a large planet-sized body, in order to be left rotating on its side. During such an impact Uranus lost a great deal of its primordial interior heat. Such heat is left over from the gravitational collapse that created the planet. Not all scientists subscribe to that explanation. Some propose that instead the planet's mantle could be layered by composition in such a way that heat flow toward the atmosphere could be diminished by convective action.

METHODS OF STUDY

The Voyager 2 spacecraft has been the only spacecraft to encounter the Uranus system. Originally intended only to fly by Jupiter and Saturn, the Voyager 2 spacecraft was specially targeted to pass through the Uranian system in early 1986. While Voyager 2 cruised to Uranus,

a number of astronomers such as Heidi Hammel trained some of the best earthbound telescopes suitable for planetary studies at Uranus to get a feel for what Voyager would encounter. Those studies provided additional information about Uranus's atmosphere but also hinted that Uranus had some internal heat sufficient to drive cloud dynamics. Earthbound telescopic studies did little, however, to enhance understanding of Uranus's interior. A particularly important discovery during the Uranus encounter that aided in describing Uranus's interior was detection of a planetary magnetic field. The nature of Uranus's magnetic field helped planetary scientists devise models for the planet's interior, models quite different from those for the interiors of both Jupiter and Saturn. Then, in 1989, Voyager 2 fulfilled a similar task at Neptune. Planetary scientists realized that the relative similarities of Uranus and Neptune extended to their interiors.

It took the Voyager 2 spacecraft nearly five years to cross the gulf from the Saturn system to the Uranus system. The encounter phase began on November 4, 1985. A great many fundamental questions were about to be answered. At first only ever-increasingly revealing images were produced by Voyager 2. Scientists eagerly awaited detection of a magnetosphere which would indicate Uranus possessed a magnetic field. That would also tell much about the nature of the planet's interior.

When Voyager 2 indeed picked up radio signals indicating the spacecraft crossed the magnetosphere, it clearly revealed Uranus has a magnetic field. After a jam-packed investigation, Voyager 2 ended the Uranus encounter on January 25, 1986, snapping a farewell image of the crescent planet.

CONTEXT

The planets Mercury through Saturn were known to the ancients. No individual can be credited with their discovery. Uranus, however, is the first planet for which definite records exist indicating when the planet was observed and charted. Herschel was not the first to pay special attention to Uranus. Others had misidentified it as an unnamed star or an unknown comet. Herschel recognized Uranus as a newly discovered planet orbiting the Sun. From that time forward to the dawn of the space age, even as Earth-based telescopes grew in size and resolution, little could be learned substantively about Uranus other than its mass, size, mean distance from the Sun, rotation rate, and that it has a very unusual axial tilt relative to the ecliptic plane. It was realized that its atmosphere was quite different from that of both Jupiter and Saturn, suggesting the Uranus's interior differs from those of its two larger gas giant cousins due to its smaller mass. The interior was suspected to be more akin to that of Neptune, which was the next planet in the outer solar system to be discovered after Uranus.

Indeed, models of the two ice giants, generated based on Voyager 2 results and observations, are quite similar. Both planets are believed to have an outer envelope of molecular hydrogen, helium, and methane. Underneath that both planets have mantles that contain water, methane, and ammonia under conditions of high pressures and temperatures. Beneath that is an icy and rocky core. However, the difference between Uranus's and Neptune's interiors is that Uranus's is less active: The planet does not have as great a heat flow from the interior to drive atmospheric dynamics.

David G. Fisher

FURTHER READING

Burgess, Eric. *Uranus and Neptune: The Distant Giants.* New York: Columbia University Press, 1988. Covers the Voyager 2 spacecraft's mission, technical difficulties, and its encounters with Jupiter, Saturn, and Uranus. Focuses on data collected by Voyager 2 about Uranus. Includes several illustrations and tables. Well written, suitable for the general audience.

Elkins-Tanton, Linda T. *Uranus, Neptune, Pluto, and the Outer Solar System.* New York: Chelsea House, 2006. This book explores the Sun's relationship with the three outer planets and their moons. It looks at these planets as recorders of the formation of the solar system. Aimed at a general or high school audience. Illustrations, bibliography, index.

Encrenaz, Thérèse, et al. *The Solar System.* New York: Springer, 2004. A thorough exploration of the solar system from early telescopic observations through the space missions that had investigated all planets by the publication date. Takes an astrophysical approach to place our solar system in a wider context as just one member of similar systems throughout the universe.

Hunt, Garry E., and Patrick Moore. *Atlas of Uranus.* New York: Cambridge University Press, 1988. This was the first volume after the 1986 Voyager encounter to offer a comprehensive history of Uranus: its discovery, satellites, rings, and the data returned by Voyager, including photographs.

Irwin, Patrick G. J. *Giant Planets of Our Solar System: An Introduction.* 2d ed. New York: Springer, 2006. Suitable as a textbook for upper-level college courses in planetary science. Focuses on Jupiter, Saturn, Uranus, and Neptune and their satellites, rings, and magnetic fields. Filled with figures and photographs.

Loewen, Nancy. *The Sideways Planet: Uranus.* Mankato, Minn.: Picture Window Books, 2008. An educational children's book devoted to the planet Uranus. Covers Uranus's rings, moons, and tilted axis.

Miner, Ellis. *Uranus: The Planet, Rings, and Satellites.* New York: Ellis Horwood, 1990. The author thoroughly covers the topics of both the Uranian system and the Voyager mission. Illustrations, bibliography, index.

Schmude, Richard W. *Uranus, Neptune, and Pluto and How to Observe Them.* New York: Springer, 2008. Ideal for backyard or amateur astronomers who are interested in observing the outer planets. Also includes up-to-date information about the planets.

Tocci, Salvadore. *A Look at Uranus.* New York: Franklin Watts, 2003. As part of the Out of This World series, this book covers all aspects of the planet Uranus from its discovery through 2002. Includes several photographs. Suitable for all readers.

See also: Auroras; Earth's Atmosphere; Earth's Composition; Eclipses; Infrared Astronomy; Jovian Planets; Planetary Atmospheres;

Uranus's Magnetic Field

Categories: Planets and Planetology; The Uranian System

Uranus is the seventh planet outward from the Sun. Although a Jovian or gas giant planet, it has more in common with Neptune than Jupiter or Saturn. Uranus's magnetic field is believed to be produced in a manner similar to that which generates Neptune's magnetic field.

OVERVIEW

Uranus was the first planet to be discovered through telescopic observations. In March, 1781, while searching for binary stars using a 2-meter telescope, William Herschel noted an object he initially thought was either a comet or a nebula. Recognizing its observed orbital motion to be that of a planet, Herschel first proposed the object be named in honor of King George III of England. However, the planet was finally named after the Roman god of the heavens.

Observations of Uranus continued over the next two centuries, but by and large Uranus remained an enticing mystery. Five satellites were discovered between 1787 and 1948. The rotational axis of the planet proved extremely surprising. Uranus is a world rotating virtually on its side. The axis is inclined 97.8° relative to the ecliptic plane of the solar system. Since Uranus's atmosphere—as seen from Earth-based telescopes between the time of Herschel and the dawn of the space age—did not reveal significant features to observe over time, it was not until the second half of the twentieth century that this planet's rotational period was accurately determined.

Knowledge of the orientation of Uranus's rotational axis, its interior structure, and its rotational rate is key to determining the nature of Uranus's magnetic field. Indeed, until the Voyager 2 spacecraft encountered the planet close up in January, 1986, there was no definitive proof that Uranus even possessed a magnetic field, although one was strongly suspected based on a contemporary model of the planet that shared similarities with gas giants Jupiter and Saturn.

Voyager 2 carried a magnetometer and a radio astronomy experiment used in concert to sample the magnetic environment of each planet (Jupiter, Saturn, Uranus, Neptune) that it encountered on its historic "Grand Tour." Basically, a sophisticated magnetometer operates on the very fundamental principal of an induced voltage being produced in a coil of wire when it intercepts a time-varying magnetic flux. That flux is directly proportional to the instantaneous strength of the magnetic field. Spacecraft such as Voyager 2 detect and investigate magnetic environments in space by picking up radiation in radio wavelengths produced by oscillating charged particles, such as those in the solar wind or ions trapped in planetary magnetic fields.

Uranus and Neptune were considered to be gas giants like Jupiter and Saturn until it was realized, based largely upon computer models and Voyager 2 data, that being smaller in mass and having significant atmospheric differences from Jupiter and Saturn, Uranus and Neptune were better classified as ice giants. Finding out about Uranus's interior would be key to determining the means whereby a magnetic field could be produced by the planet. However, the converse is true as well. Directly measuring the magnetic field of the planet would help to develop a model of Uranus's interior. What was known about Uranus's composition in the atmosphere prior to the Voyager 2 encounter was that the planet is composed primarily of hydrogen (83 percent) and helium (15 percent). However, Uranus has a pale blue-green color due to the presence of methane (2 percent), which selectively absorbs red wavelengths of light. Uranus also contains ices such as water, ammonia, and an assortment of hydrocarbons.

KNOWLEDGE GAINED

Voyager 2 determined that the magnetic field generated by Uranus has quite unusual characteristics. The planet's dipole moment is approximately fifty times that of Earth. The value of Uranus's dipole moment is 3.8×10^{17} telsa meters cubed. By comparison to giant Jupiter, this is a mere 0.26 percent of the dipole moment of the largest planet in the solar system. Uranus's average magnetic field strength, the maximum value or amplitude of the field, at the plant's "surface" was measured to be 23 micro-teslas. However, magnetic field strength varied with latitude. At the "surface" in the southern hemisphere, the field strength was seen to dip as low as 10 micro-teslas. Then again, at the "surface" in the northern hemisphere the field was found to be as strong at 110 micro-tesla. This situation contrasts greatly with Earth, where the magnetic field is nearly as intense at both poles. Earth's field is centered close to the planet's physical center. However, that is not the case with Uranus. The center of Uranus's magnetic field actually is displaced from the physical center of the planet by about a third of its radius; the magnetic center is closer to the south rotational pole. Further complicating the field is the fact that the magnetic axis, the line from the south to the north pole through the planet, is tilted 59° relative to the line running between the north and south rotational poles.

Uranus's magnetosphere thus displays a highly unusual tilt. That tilt and the rotational tilt of the planet give the dynamic behavior of Uranus's magnetosphere a twisting structure. To make the magnetosphere's character even stranger, it appears that the ring system around Uranus actually streams ions in the magnetosphere down into Uranus's atmosphere. Auroras are produced, but they differ somewhat from the familiar auroral displays seen in Earth's polar regions.

By detecting variations in radio waves produced by the planet's magnetic field, astronomers have detected the rotational rate of the planet more accurately than by following the differential rotation of those atmospheric features that could be found. That rotation value is 17.233 hours.

Despite the unusual orientation and subsequent twisting behavior of the field lines as the planet rotates, Uranus's magnetosphere does share some things with other planetary magnetic fields in our solar system. Its magnetosphere is affected by the solar wind, forming a bow shock ahead of the planet, a magnetopause, and a magnetotail. The bow shock was crossed at a distance equivalent to 23 Uranus radii, whereas the magnetopause was determined by Voyager 2 to be located at 18 Uranus radii. The magnetotail appears as a corkscrew structure due to the twisting of the planet's magnetic field lines. This magnetic field structure also has given Uranus radiation belts of trapped charged particles. Those particles consist primarily of protons and electrons but have a minor component of molecular hydrogen ions. Uranian satellites create gaps in the radiation belts by "sweeping up" charged particles as they revolve about that planet. Ring particles and the surfaces of the planet's satellites struck by this ionizing radiation are darkened by that exposure.

What generates such an unusual planetary magnetic field? Whereas Earth's field is created deep in the planet by a dynamo effect involving electrical currents generated by its molten core, Uranus's magnetic field is speculated to be produced by the ice giant's mantle. Between the atmosphere and the planet's core, Uranus has a mantle layer believed to be composed of a highly pressurized water, ammonia, and other ices that become ionized under that tremendous pressure and in the presence of temperatures in excess of 1,000 kelvins. Therefore, currents flow through the mantle. Some scientists do not accept this explanation, but at present there is no way, short of direct investigation of the planet's interior, to validate the mantle "ocean" hypothesis or show it to be incorrect.

Why does Uranus's magnetic field have such an unusual orientation relative to the planet's rotational axis? Here there are two reasonable hypotheses. One suggests that the planet is in the process of reversing its magnetic field. (Rocks on Earth present a record that Earth's magnetic field has reversed itself many times over geologic time.) The second hypothesis is that the disruption of the magnetic field alignment resulted from a collision between Uranus and one or more large bodies. Further study will be needed to decide between these two theories or replace them with a better explanation.

The Hubble Space Telescope in concert with the Keck telescope has imaged new storms in Uranus's atmosphere. If Hubble received another servicing mission to extend its life sufficiently, Hubble and Keck would, in addition to searching for atmospheric dynamics, produce data that could assist in developing a better understanding of Uranus's interior. Such data could help explain Uranus's complex magnetic field without getting direct measurements of the field characteristics, as would be possible only by sending another spacecraft to sample the field close to the planet.

CONTEXT

Every 176 years, planetary alignments are such that it is possible through ingenuous use of gravity-assist (or "slingshot") maneuvers to send a pair of spacecraft to visit all the outer planets from Jupiter to Pluto. Prior to the authorization for Mariner Jupiter-Saturn, which later was named Voyager, the National Aeronautics and Space Administration (NASA) had originally proposed sending a pair of sophisticated spacecraft on a journey that had been called the Grand Tour, to take advantage of this rare opportunity to reach four planets. However, that ambitious plan was not funded. The end result was that Voyager 2 would be the only spacecraft to visit Uranus and Neptune in the twentieth century, and most likely for many decades to come after that initial spacecraft encounter. As such, Voyager 2 has provided the greatest share of data about the Uranian system.

The interpretation of those data led to our current understanding about this unique and in many ways bizarre planet. Despite the tremendous insights provided by the Voyager 2 data, many questions remain unanswered. Among those are important questions concerning the internal structure of the planet, the production of the planet's magnetic field, and the nature of the relatively minor amount of internal heating in the planet.

Just as the Galileo probe orbited Jupiter for a prolonged period of time and the Cassini spacecraft did the same at Saturn, the next logical step for Uranus studies would be to dispatch a dedicated orbiter, a spacecraft outfitted with a wide-ranging suite of scientific instruments to allow focused investigations of the planet's atmosphere, internal structure, magnetic field, ring system, and collection of satellites. However, nearly three decades after the Voyager 2 encounter no such program was on the horizon and no proposal was considered likely to be funded unless a nuclear propulsion capability was developed to lessen the travel time to a planet as far away from Earth as Uranus. In the meantime, Uranus will continue to be studied using the Hubble Space Telescope and ground-based facilities (such as the Keck telescope) in attempts to gain further insight into the many unanswered questions remaining from the Voyager encounter.

David G. Fisher

FURTHER READING

Bredeson, Carmen. *NASA Planetary Spacecraft: Galileo, Magellan, Pathfinder, and Voyager.* New York: Enslow, 2000. A part of Enslow's Countdown to Space series, this volume provides an overview of NASA planetary exploration during the last two decades of the twentieth century. Designed for younger readers, but suitable for all audiences.

Burgess, Eric. *Uranus and Neptune: The Distant Giants.* New York: Columbia University Press, 1988. Covers the Voyager 2 spacecraft's mission, technical difficulties, and encounters with Jupiter, Saturn, Uranus, and Neptune. Describes data collected by Voyager 2 about Uranus. Includes several illustrations and tables. Well written and suitable for a general audience.

Elkins-Tanton, Linda T. *Uranus, Neptune, Pluto, and the Outer Solar System.* New York: Chelsea House, 2006. This book explores the Sun's relationship with the three outer planets and their moons. It looks at these planets as recorders of the formation of the solar system. Aimed at a general or high school audience. Illustrations, bibliography, index.

Encrenaz, Thérèse, et al. *The Solar System.* New York: Springer, 2004. A thorough exploration of the solar system from early telescopic observations through the space missions that had investigated all planets by the publication date. The astrophysical approach gives our solar system a wider context as just one member of similar systems throughout the universe.

Freedman, Roger A., and William J. Kaufmann III. *Universe*. 8th ed. New York: W. H. Freeman, 2008. A college text on astronomy, somewhat more advanced than many introductory texts, but with a wealth of detail and excellent diagrams. Chapters 6 through 16 describe the solar system. Comes with a CD-ROM.

Hunt, Garry E., and Patrick Moore. *Atlas of Uranus*. New York: Cambridge University Press, 1988. This was the first volume after the 1986 Voyager encounter to offer a comprehensive history of Uranus: its discovery, satellites, rings, and the data returned by Voyager, including photographs.

Irwin, Patrick G. J. *Giant Planets of Our Solar System: An Introduction*. 2d ed. New York: Springer, 2006. Suitable as a textbook for upper-level college courses in planetary science. Focuses on Jupiter, Saturn, Uranus, and Neptune and their satellites, rings, and magnetic fields. Filled with figures and photographs. Accessible to the serious general audience.

Loewen, Nancy. *The Sideways Planet: Uranus*. Mankato, Minn.: Picture Window Books, 2008. An educational children's book devoted to the planet Uranus. Covers the planet's rings, moons, and tilted axis.

McBride, Neil, and Iain Gilmour, eds. *An Introduction to the Solar System*. Cambridge, England: Cambridge University Press, 2004. A complete description of solar system astronomy suitable for an introductory college course but useful to interested laypersons as well. Filled with supplemental learning aids and solved student exercises. A Web site is available for educator support.

Miner, Ellis. *Uranus: The Planet, Rings, and Satellites*. New York: Ellis Horwood, 1990. The author thoroughly covers the topics of both the Uranian system and the Voyager mission. Illustrations, bibliography, index.

Schmude, Richard W. *Uranus, Neptune, and Pluto and How to Observe Them*. New York: Springer, 2008. Ideal for backyard or amateur astronomers who are interested in observing the outer planets. Includes up-to-date information about the planets.

Tocci, Salvadore. *A Look at Uranus*. New York: Franklin Watts, 2003. As part of the Out of This World series, this book covers all aspects of the planet Uranus from its discovery through 2002. Includes several photographs. Suitable for all readers.

See also: Auroras; Earth's Atmosphere; Earth's Composition; Eclipses; Infrared Astronomy; Jovian Planets; Planetary Atmospheres; Planetary Formation; Planetary Interiors; Planetary Magnetospheres; Planetary Ring Systems; Planetary Rotation; Planetary Satellites; Planetology: Comparative; Telescopes: Ground-Based; Uranus's Atmosphere; Uranus's Interior; Uranus's Rings; Uranus's Satellites; Uranus's Tilt.

Uranus's Rings

Categories: Planets and Planetology; The Uranian System

Uranus has thirteen known rings, eleven inner and two outer ones. These rings are mostly very narrow and faint. Uranus's ring system is less complex than Saturn's but more so than that of Jupiter.

OVERVIEW

When Uranus was first observed (perhaps as early as 1690), many astronomers considered it to be a star rather than a planet. William Herschel studied Uranus in 1781, when he thought he had discovered a new comet. After two years of further study, astronomers agreed that Uranus was in fact a planet.

Herschel appears to have observed the rings of Uranus in February of 1789. He sketched an image of Uranus in his journal, making a note that the planet had rings of a faint reddish hue. Rings around Uranus remained an open issue for a long time. The existence of Uranus's ring system was finally confirmed, albeit accidentally, in 1977. Astronomers James Elliot, Ed-

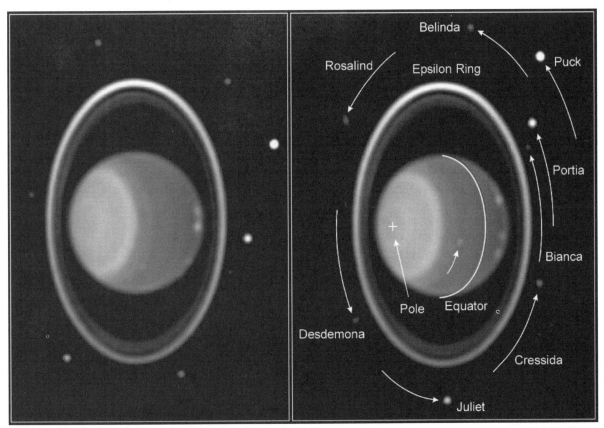

This 1997 image from the Near Infrared Camera and Multi-Object Spectrometer (NICMOS) on the Hubble Space Telescope shows clouds on Uranus (left) and clearly displays the planet's rings, satellites, and rotation (right). (NASA)

ward Dunham, and Douglas Mink set out to study Uranus's atmosphere by observing the occultation of the star SAO 158687. They noticed that this star briefly disappeared from view before and after being eclipsed by the planet. Each of the five occultations that they observed yielded the same results. When the group published their work, they referenced these occultations and resulting rings using the Greek letters Alpha, Beta, Gamma, Delta, and Epsilon. In 1978 another group of scientists found four additional rings. The Eta ring was found between the Beta and Gamma rings. The other three were discovered inside the orbit of the Alpha ring and were named Six, Five, and Four (in that order).

The National Aeronautics and Space Administration's (NASA's) robotic Voyager 2 probe flew by Uranus in 1986. The spacecraft took the first photographs of the Uranian ring system. Voyager 2 also discovered two more faint rings, Lambda and 1986U2R/Zeta, bringing the total number of known rings around the planet to eleven. In 2003, the Hubble Space Telescope discovered, and in 2005 confirmed, the existence of an additional pair of rings. They form an outer ring system that is separate from the other eleven rings. These two outer rings have an orbital radius of more than 100,000 kilometers from Uranus's center—double that of the inner rings. In addition to the twelfth and thirteenth rings, Hubble found two satellites. Mab, which is only 24 kilometers in diameter, shares an orbit with the outermost ring. Every time a meteoroid impacts the small satellite, dust particles and other debris ejected become part of the Mu ring. The Nu ring lies between Uranus's small satellites Rosalind and Portia.

Rings of Uranus

	Radius (km)	Radius/ Eq. Radius	Optical Depth	Albedo ($\times 10^{-3}$)	Width (km)	Eccentricity
Uranus equator	25,559	1.000	—	—	—	—
6	41,837	1.637	~0.3	~15	1.5	0.0010
5	42,234	1.652	~0.5	~15	~2	0.0019
4	42,571	1.666	~0.3	~15	~2	0.0011
Alpha	44,718	1.750	~0.4	~15	4-10	0.0008
Beta	45,661	1.786	~0.3	~15	5-11	0.0004
Eta	47,176	1.834	~0.4-	~15	1.6	—
Gamma	47,627	1.863	~1.3+	~15	1-4	0.0011
Delta	48,300	1.900	~0.5	~15	3-7	0.00004
Lambda	50,024	1.957	~0.1	~15	~2	0.
Epsilon	51,149	2.006	0.5-2.3	~18	20-96	0.0079

Source: Data are from the National Aeronautics and Space Administration/Goddard Space Flight Center, National Space Science Data Center.

The Mu and Nu rings are very different from the inner rings. With widths of 17,000 kilometers and 30,000 kilometers, respectively, the Mu and Nu rings are much broader. These two rings are also much fainter then the others, but they can be seen on the Voyager 2 photographs.

Many similarities exist between Uranus's outer rings and Saturn's E and G rings. The E ring includes Enceladus, which contributes dust to it the same way Mab is believed to contribute to the Mu ring. The Nu ring, like Saturn's G ring, contains no embedded "shepherding" satellites and is composed of dust and larger particles. Scientists working with the Keck telescopes in Hawaii studied the rings at near-infrared wavelengths. The Nu ring was visible, meaning it has a reddish hue. This possibly gives some credit to Herschel's original claim about observing Uranus's rings, despite critics' claims that the rings are too faint for him to have seen. The Mu ring was not visible, meaning that its small dust particles appear blue in color. Red is a typical color for planetary rings. Blue however, is not. Saturn's E ring is the only other ring known to have the unusual blue hue.

The inner ring system contains two types of rings: narrow and dusty. The closest ring to Uranus is 1986U2R. It was discovered in 1986 by Voyager 2. This ring is only about 12,000 ki-lometers above the cloud tops of Uranus. The 1986U2R (or Zeta) ring was observed in 2003 and 2004 using the Keck telescopes. Scientists found the ring to be broad, very faint, and composed of dust grains.

The next set of rings is Six, Five, and Four, which were named for the occultations that led to their discoveries. They are the faintest of Uranus's narrow main rings. These three lie outside Uranus's equatorial plane by 0.06°, 0.05°, and 0.03°, respectively. Six, Five, and Four do not contain dust and are the thinnest of Uranus's narrow rings.

After the Epsilon ring, the Alpha and Beta rings are the brightest of Uranus's rings. Alpha and Beta are narrowest and faintest at their closest points to Uranus. At their farthest, the two rings are their broadest and brightest. Like all of Uranus's rings, Alpha and Beta are composed of extremely dark material. They are much darker than Uranus's inner satellites, meaning that the rings cannot be composed of pure water ice. The composition of the rings is thus unknown, but astronomers think it is probably a mixture of dark materials and ices.

The seventh ring outward from Uranus's core is Eta, at 47,176 kilometers. Eta has both a narrow part and a broader, dustier section. When Voyager 2 photographed the ring in forward scattered light, it appeared very bright, indicating a large amount of dust. Eta has an inclination and eccentricity of zero, meaning that the ring lies in the planet's equatorial plane, and the ring particles execute circular orbits. In 2007 Uranus's rings were viewed edge-on for the first time. The Eta ring appeared to be the second brightest, which was a significant increase. This finding has led planetary scientists to believe that, while the ring is optically narrow, it is geometrically thick. The next chance for astronomers to view the rings in that unique geometry will occur in 2049.

The Gamma ring is narrow, with an inclination close to zero. Gamma's width varies from 3.6 to 4.7 kilometers. The ring was not visible during the 2007 ring plane crossing. This means that Gamma is both optically and geometrically narrow. The ring also does not contain any dust. Scientists are uncertain about what holds this small ring together.

Like the Eta ring, Delta has both a narrow and a broad component. The thinner part varies from 4.1 to 6.1 kilometers wide, while the thicker part ranges from 10 to 12 kilometers. The wide section is composed of dust, unlike the narrow part. In 2007, only the broad area of Delta was visible. At the outer edge of the Delta ring, a small satellite named Cordelia orbits Uranus.

The Lambda ring lies between Cordelia (a shepherd satellite) and the Epsilon ring. Lambda is faint and narrow even when backlit. The dusty ring was first detected by Voyager 2 during stellar occultation observations, but only at ultraviolet wavelengths. The Lambda ring is composed of micrometer-sized dust, which was confirmed in 2007, when it appeared very bright.

The brightest and densest of Uranus's rings is Epsilon.

It is the outermost ring of the inner system. Epsilon reflects two-thirds of all the light visible from Uranus's rings. It is only one of two rings that Voyager 2 was able to photograph clearly. Epsilon is the most eccentric of the rings but has a near-zero orbital inclination. The dense ring has particles ranging in size from 0.2 to 20 meters in diameter. In 2007, the ring was not observable, because of its lack of dust. Epsilon contains many dense, narrow ringlets, and possibly partial arcs. The ring may stay so compact because of its shepherd satellites: Cordelia on the inner side and Ophelia on the outer.

KNOWLEDGE GAINED

The first rings of Uranus were officially discovered by accident in 1977. Elliot, Dunham, and Mink were using the Kuiper Airborne Observatory (KAO) to study Uranus's atmosphere during five stellar occultations. The KAO is an

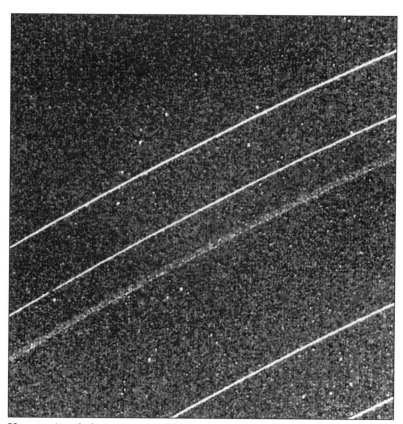

Voyager 2 took these images of Uranus's rings in January, 1986, from a distance of 1.12 million kilometers. From the top, the Delta, Gamma, Eta, Beta, and Alpha rings are visible. (NASA/JPL)

airplane with a 36-inch (91.5-centimeter) telescope mounted on the side. With the KAO, scientists can conduct research while flying 14 kilometers above the Earth's surface, thereby making infrared observations readily. At that elevation there is significantly less atmospheric water vapor, which blocks infrared wavelengths from reaching the surface of the Earth. The KAO therefore combines many benefits of a space telescope with the accessibility of a ground-based telescope.

Launched in 1977, Voyager 2 is the only spacecraft to have visited Uranus. It came within 81,500 kilometers of the planet on January 24, 1986. The spacecraft had several instruments on board, including cameras, magnetometers, and spectroscopes. At the time, Uranus's south pole was pointed toward the Sun. Voyager discovered ten satellites as well as the Lambda and Zeta (1986U2R) rings. The Mu and Nu rings have since been located on Voyager 2 photographs.

The Hubble Space Telescope was studying Uranus in 2003 when it discovered the Mu and Nu rings. Scientists were able to confirm the finding in 2005. When the Keck telescope studied the two rings at near-infrared wavelengths, only the Nu ring was visible. This means that the Nu ring has a reddish color. The Mu ring therefore has a bluish tint, because it was not visible.

In 2007, astronomers were able to view Uranus's rings edge-on. Teams of scientists used the Keck II telescope in Hawaii, the Hubble Space Telescope, and the European Southern Observatory's Very Large Telescope in Chile to study the event. Images taken with the Keck telescope show that the rings have changed since Voyager 2 visited the planet more than two decades ago. The broad, dusty inner Zeta ring appears very different. If it is the same ring discovered by Voyager, Zeta has moved several thousand kilometers away from Uranus. Similar shifts have been detected in the ring systems of Saturn and Neptune.

CONTEXT

All of the Jovian planets in our solar system have ring systems. Each set of rings is unique. Neptune's rings are simpler than Uranus's, containing only five rings and partial arcs. Saturn's ring system on the other hand is more complex. Jupiter has but two faint rings.

The thirteen rings of Uranus are mostly faint and narrow. They are in two groupings—an inner system of eleven rings and an outer set of two. In 2007, the Nu ring was determined to be red in color, and the Mu ring was found to be blue. Red seems to be a typical color for planetary rings, like Saturn's G ring. The blue color of the Mu ring, however, is not common. The only other example in the solar system is Saturn's E ring. What caused this odd blue color is still a mystery to scientists. The rings around Uranus are believed to be made of debris from collisions between Uranus's satellites.

Scientists can learn more about the formation and evolution of the solar system by investigating planetary ring systems. As ground-based and space telescopes improve, astronomers could unlock the secrets of Uranus's rings and the solar system itself.

Jennifer L. Campbell

FURTHER READING

Chaisson, Eric, and Steve McMillan. *Astronomy Today*. 6th ed. New York: Addison-Wesley, 2008. A well-written college-level text for introductory astronomy courses. Has a chapter on Uranus and Neptune that covers the ring systems.

Elkins-Tanton, Linda T. *Uranus, Neptune, Pluto, and the Outer Solar System*. New York: Chelsea House, 2006. Explores the Sun's relationship with the three outer planets and their moons, considering these planets as recorders of the formation of the solar system. Aimed at a general or high school audience. Illustrations, bibliography, index.

Esposito, Larry. *Planetary Rings*. New York: Cambridge University Press, 2006. A synopsis of current knowledge of the outer planets' ring systems. Includes information from the Cassini mission on Uranus's rings, and on ring ages and evolution. Geared toward scientists and college students.

Fridman, Alexei M., and Nikolai N. Gorkavyi. *Physics of Planetary Rings: Celestial Mechanics of Continuous Media*. New York: Springer, 1999. Compares the ring systems of Jupiter, Saturn, Uranus, and Neptune using observational and mathematical data. Designed for scientists, astronomy and physics students, and amateur astronomers wishing to know more about the rings of the outer planets.

Hunt, Garry E., and Patrick Moore. *Atlas of Uranus*. New York: Cambridge University Press, 1988. This was the first volume after the 1986 Voyager encounter to offer a comprehensive history of Uranus: its discovery, satellites, rings, and the data returned by Voyager, including photographs.

Loewen, Nancy. *The Sideways Planet: Uranus*. Mankato, Minn.: Picture Window Books, 2008. An educational children's book devoted to the planet Uranus. Covers Uranus's rings, moons, and tilted axis.

Miner, Ellis. *Uranus: The Planet, Rings, and Satellites*. New York: Ellis Horwood, 1990. The author thoroughly covers the topics of both the Uranian system and the Voyager mission. Illustrations, bibliography, index.

Miner, Ellis D., Randii R. Wessen, and Jeffrey N. Cuzzi. *Planetary Ring Systems*. New York: Springer Praxis, 2006. Looks at the ring systems of each gas giant. Covers recent research in the field, as well as the many questions that remain unanswered.

Schmude, Richard W. *Uranus, Neptune, and Pluto and How to Observe Them*. New York: Springer, 2008. Ideal for backyard or amateur astronomers who are interested in observing the outer planets. Includes up-to-date information about the planets.

Tocci, Salvadore. *A Look at Uranus*. New York: Franklin Watts, 2003. As part of the Out of This World series, this book covers all aspects of the planet Uranus, including the planet's discovery and research to 2002. Includes several photographs. Suitable for all readers.

See also: Auroras; Earth's Atmosphere; Earth's Composition; Eclipses; Infrared Astronomy; Jovian Planets; Planetary Atmospheres; Planetary Formation; Planetary Interiors; Planetary Magnetospheres; Planetary Ring Systems; Planetary Rotation; Planetary Satellites; Planetology: Comparative; Telescopes: Ground-Based; Uranus's Atmosphere; Uranus's Interior; Uranus's Magnetic Field; Uranus's Satellites; Uranus's Tilt.

Uranus's Satellites

Categories: Natural Planetary Satellites; Planets and Planetology; The Uranian System

Uranus's natural satellites form a miniature solar system with distinctive properties that teach us the complexity and diversity of planetary and satellite formation. The peculiar surface features of some of the satellites and their unusual orbital characteristics suggest an earlier epic in the solar system with violent collisions among its members.

OVERVIEW

Uranus holds its place in the solar system as a member of the subgroup of planets that are called Jovian, after the largest planet in the group, Jupiter. These planets are also referred to as "gas giants" in that their atmospheres comprise the greatest portion of the planet's structure. It is the physical nature and orbital properties of the satellites that set apart one Jovian planet from another, and Uranus is no exception. Surveying the most interesting properties of this planetary system and highlighting their respective features can provide insights about the origin and evolution of the solar system itself.

Uranus has twenty-seven satellites that have been identified. Their names follow a theme that is distinctive in the solar system in that the satellites are not named for mythological figures, like those of the other planets, but instead take their names from characters in plays by William Shakespeare and poems by Alexander Pope. Oberon, Titania, and Puck, for example, were named for characters from Shakespeare's *A Midsummer Night's Dream*; Ariel, Umbriel, and Belinda are named for characters in Pope's *The Rape of the Lock*. Oberon and Titania were first discovered by William Herschel in 1781. Ariel and Umbriel were discovered in 1851 by William Lassell. In 1948, Gerard Kuiper discovered the last moon of any significant size, Miranda.

The only spacecraft to visit Uranus to date has been Voyager 2, which flew by the planet in

1986. Despite the briefness of its visit, Voyager 2 discovered ten small satellites: Juliet, Puck, Cordelia, Ophelia, Bianca, Desdemona, Portia, Rosalind, Cressida, and Belinda. (Perdita was also imaged by Voyager 2, but its discovery was not confirmed until 1999.) The number of satellites has swelled to the current known number of twenty-seven through observation with both orbiting telescopes (Hubble Space Telescope) and ground-based observatories.

A way to organize the Uranian systems of satellites is to think of them as distributed in three divisions. The first division consists of the inner thiteen, relatively small, circular satellites starting just outside the ring system. The second division comprises the five midsized satellites that were discovered prior to the Voyager 2 mission. Finally, the last division comprises nine irregular satellites discovered more recently.

None of the natural satellites of Uranus can be considered on the scale of the largest satellites of the other planets of the solar system, such as Titan (Saturn), Triton (Neptune), the four Galilean satellites of Jupiter (Io, Europa, Callisto, and Ganymede), or even Earth's own Moon. Nor does any of these satellites contain an atmosphere, as does as Titan, or show active volcanoes, as seen on Io. However, the five largest Uranian satellites in Division 2—Miranda, Ariel, Umbriel, Titania, and Oberon (in order outward from the planet)—all have enough mass to be spherical. Hence, if they were not orbiting Uranus and were free from any debris, they would qualify as dwarf planets, like Pluto. The composition of these satellites is mostly ice, with mixtures of ammonia and methane. They all exhibit synchronous rotation, rotating exactly at the same rate they revolve around Uranus, always showing the same side facing the planet.

The large satellites of Uranus all show extensive cratering. The consensus of scientific opinion is that the larger craters were formed from collisions with planet-sized objects during the formation of the solar system, and the smaller craters were produced afterward from impacts from comets and meteoroids. Ariel, Umbriel, and Miranda have unusual surface features. Ariel, the lightest-colored, has craters, valleys, and canyons. It has a diameter of about 1,300 kilometers with grooves and crevices that extend over its entire surface, suggesting recent lava flow that has cooled and solidified. Umbriel is the darkest-colored satellite; its surface is old and cratered. Although Umbriel (with a diameter of 1,110 kilometers) shows the most uniform cratering, with little evidence of geological activity, it has a large, bright ring at the top edge of its southern hemisphere; this ring's origin is unknown, but it has unofficially been named the "fluorescent Cheerio." Miranda (about 500 kilometers diameter),

A near-infrared image of the moons of Uranus from the European Southern Observatory, 2002. (European Southern Observatory)

the smallest of these five large satellites, has the most distinctive surface features. There are sharp grooves and ridges along its surface. One feature resembles a large chevron, and another looks like a carved racetrack. Titania and Oberon are both larger than the other satellites (with diameters of approximately 1,500-1,600 kilometers), and they are also several times farther away from Uranus. Both Titania and Oberon show large craters, but the presence of circular regions on Oberon suggests that it has experienced more geological activity than Titania. Oberon's surface is frozen and has a mountain 6 kilometers high. Titania, the biggest satellite among those of Uranus, has impact basins, craters, and rifts.

Inside Miranda's orbit there are the thirteen Division 1 satellites, all with enough mass to be spherical and with diameters ranging from about 25 to 170 kilometers. They are (in order out from the planet) Cordelia, Ophelia, Bianca, Cressida, Desdemona, Juliet, Portia, Rosalind, Cupid, Belinda, Puck, and Mab. Their composition appears to be about half water and half rocklike materials. Their orbital distribution presents a more crowded arrangement than that of the large satellites, and it appears that these thirteen bodies interact gravitationally with each other, crossing paths and periodically colliding with one another. Some of these inner satellites may serve as shepherds for Uranus's narrow rings, which orbit closest to the planet.

Outside the orbit of Oberon are the small irregular satellites in Division 3: Francisco, Caliban, Stephano, Trinculo, Sycroax, Margaret, Prospero, Setebos, and Ferdinand. They are considered irregular, though, because they orbit in odd directions far from Uranus. Their respective sizes and compositions have not been measured accurately. Scientists have estimated that their diameters range from 18 to 150 kilometers. They have very eccentric (elliptical) orbits, and all orbit retrograde, that is, in the direction opposite to the general revolution of other bodies in the solar system, which is counterclockwise around the north polar axis of the Sun.

Major Uranian Satellites		
Satellite	Diameter (km)	Distance from Planet (km)
Miranda	500 ± 220	130,000
Ariel	1,300 ± 130	192,000
Umbriel	1,110 ± 100	267,000
Titania	1,600 ± 120	438,000
Oberon	1,630 ± 140	596,000

Source: Data are from Jet Propulsion Laboratory, California Institute of Technology. *Voyager at Uranus: 1986.* JPL 400-268. Pasadena, Calif.: Author, 1985, p. 5.

KNOWLEDGE GAINED

The study of Uranus's system of natural satellites, including their orbital and physical properties, makes it possible to understand more fully the processes not only of planet formation but of the evolution of the solar system itself. Satellites can be formed at the time of the formation of the planet they orbit, but they can also be captured by the planet's gravitational pull at a later stage. It appears that both processes are at work in the Uranian system. The group of satellites that are in Division 2 (Miranda, Ariel, Umbriel, Titania and Oberon), as well as the thirteen satellites in Division 1, most likely formed at the same time Uranus condensed. However, events have transpired to alter these satellites' orbits and surface features. For example, Miranda has a scarred surface that is presumed to be the result of a collision with another object and was fractured into pieces. These pieces then came back together unevenly and show the rough terrain and scarps that characterize the satellite. An alternative hypothesis is that Miranda, being too small an object to complete its internal mixing, froze midway through the process of separating its structure into layers.

Several of the Division 1 moons are in such close orbits with each other that they conceivably collide and switch orbital positions around Uranus. In addition, Cordelia and Ophelia serve as shepherding moons for the Epsilon ring system.

The irregular satellites in Division 3 appear to be satellites that were previously solar system bodies, such as comets and asteroids, that were captured by Uranus's gravitational pull and orbit around the planet in elliptical and retrograde orbits.

This accumulated knowledge points to a solar system that evolved sequentially over vast periods of time with many diverse objects that are still changing their orbital shapes and constitutions.

CONTEXT

There were only five known satellites of Uranus until Voyager 2 visited this planetary system and made its closet approach on January 24, 1986. Voyager 2 was launched in 1977 and visited Jupiter in 1979, Saturn in 1981, and, after its flyby of Uranus, Neptune in 1989.

As Voyager 2 approached the Uranus system, its onboard computers were reprogrammed by scientists and engineers back on Earth at the Jet Propulsion Lab to enable the cameras to produce high-quality photographs in the reduced light and at the high speeds at which the spacecraft would be traveling on its flyby. Most of the photographs were taken in a six-hour period in and around the time of closest approach (9:59 A.M. PST) on January 24. On the way to the rendezvous with Uranus, Voyager 2 obtained clear, high-resolution images of each of the five large Uranian satellites (Miranda, Ariel, Umbriel, Titania, and Oberon). It was its discovery of eleven new satellites that added to our knowledge base of this planet. During processing of images of the outer ring (Epsilon ring) of Uranus, it was discovered that two small satellites, Cordelia and Ophelia, were shepherding or keeping this thin ring in orbit around the planet. Also sighted were Bianca, Cressida, Desdemona, Juliet, Portia, Rosalind, Cupid, Belinda, and Perdita which belong to what has been called the Portia Group of satellites, These satellites have similar orbits and light-reflecting properties. The closeness of their respective orbits leads to the hypothesis that this group interacts with each other and may at times collide.

Until 1997, the Uranian system was distinct from the other Jovian planets in that there were no identified irregular satellites. However, the discovery on September 6, 1997, of two irregular satellites, Caliban and Sycroax, by Brett J. Gladman, Philip D. Nicholson, Joseph A. Burns, J. J. Kavelaars, Brian G. Marsden, Gareth V. Williams, and Warren B. Offutt, using the 200-inch Hale telescope (at Palomar Observatory in Southern California) removed that distinction. Subsequently, Stephano,

Light-colored Ariel, seen in this 1986 image from Voyager 2, has grooves and crevices that extend over its entire surface, suggesting recent lava flow that has cooled and solidified. (NASA/JPL)

Prospero, and Setebos were discovered by Matthew J. Holman, Kavelaars, Gladman, Jean-Marc Petit, and Hans Scholl on July 18, 1999. Trinculo, Margaret, and Ferdinand were discovered by Holman, Kavelaars, and Dan Milisavljevic on August 13, 2001.

Joseph Di Rienzi

FURTHER READING

Bennett, Jeffery, Megan Donahue, Nicholas Schneider, and Mark Voit. *The Cosmic Perspective*. 3d ed. San Francisco: Pearson Addison Wesley, 2004. This textbook provides a thematically organized overview of the universe. Chapter 12 discusses the Jovian systems and contains a subsection on the medium-sized satellites of Uranus.

Burgess, Eric. *Uranus and Neptune: The Distant Giants*. New York: Columbia University Press, 1988. Covers the Voyager 2 spacecraft's mission, technical difficulties, and its encounters with Jupiter, Saturn, and Uranus. Describes the data collected by Voyager 2 about Uranus. Includes several illustrations and tables. Well written and suitable for the general audience.

Carroll, Bradley W., and Dale A. Ostlie. *An Introduction to Modern Astrophysics*. San Francisco: Pearson Addison Wesley, 2007. This is an encyclopedic textbook that covers all of modern astronomy and astrophysics. Although much of the book is for the advanced student, the chapters on the solar system are very descriptive. Chapter 21, "The Realms of the Giant Planets," includes a section on their satellites and discusses Miranda in particular.

Elkins-Tanton, Linda T. *Uranus, Neptune, Pluto, and the Outer Solar System*. New York: Chelsea House, 2006. This book explores the Sun's relationship with the three outer planets and their moons. It looks at these planets as recorders of the formation of the solar system. Aimed at a general or high school audience. Illustrations, bibliography, index.

Encrenaz, Thérèse, et al., eds. *The Outer Planets and Their Moons: Comparative Studies of the Outer Planets Prior to the Exploration of the Saturn System by Cassini-Huygens*. New York: Springer, 2005. An in-depth look at the current understanding of the solar system's outer planets. Focuses on the studies of their formation, evolution, magnetospheres, satellites, and ring structures. For scientists, first-year graduate students, and advanced undergraduates.

Hunt, Garry E., and Patrick Moore. *Atlas of Uranus*. New York: Cambridge University Press, 1988. This was the first volume after the 1986 Voyager encounter to offer a comprehensive history of Uranus: its discovery, satellites, rings, and the data returned by Voyager, including photographs.

Loewen, Nancy. *The Sideways Planet: Uranus*. Mankato, Minn.: Picture Window Books, 2008. An educational children's book devoted to the planet Uranus. Covers Uranus's rings, moons, and tilted axis.

Miner, Ellis. *Uranus: The Planet, Rings, and Satellites*. New York: Ellis Horwood, 1990. The author thoroughly covers the topics of both the Uranian system and the Voyager mission. Miranda is featured as a remarkable satellite. Illustrations, bibliography, index.

Schmude, Richard W. *Uranus, Neptune, and Pluto and How to Observe Them*. New York: Springer, 2008. Ideal for backyard or amateur astronomers who are interested in observing the outer planets. Also includes up-to-date information about the planets.

Tocci, Salvadore. *A Look at Uranus*. New York: Franklin Watts, 2003. As part of the Out of This World series, this book covers all aspects of the planet Uranus, from its discovery through 2002. Includes several photographs. Suitable for all readers.

See also: Jupiter's Satellites; Miranda; Neptune's Satellites; Planetary Ring Systems; Planetary Satellites; Saturn's Satellites; Telescopes: Ground-Based; Titan; Triton; Uranus's Magnetic Field; Uranus's Rings.

Uranus's Tilt

Categories: Planets and Planetology; The
Uranian System

*All planets in our solar system rotate on an
axis tilted in relation to the ecliptic plane
(the plane carved out by their orbit around
the Sun). However, Uranus's axis is tilted
at such an extreme angle that the planet rotates while virtually lying on its side. Despite several theories, scientists do not fully
understand what causes Uranus's tilt.*

OVERVIEW

Uranus was observed as early as 1690, but
astronomers thought it was a star. Using a telescope he built himself, Sir William Herschel observed Uranus over a series of nights in 1781.
Herschel initially reported to the Royal Society
that he had discovered a new comet. After tracking the "comet" for two years, astronomers finally agreed that Uranus was actually the seventh planet in the solar system.

In 1829, astronomers determined that the rotation of Uranus was unique. All of the planets
rotate on an axis that is tilted with respect to
the orbital plane of the solar system. The orbital plane is the imaginary surface on which
the planets orbit and almost lies on the Sun's
equator (the plane is tilted at a 7° angle with
respect to the Sun's equator). Axial tilt is calculated by drawing a line perpendicular to the
orbital plane. The rotational axis of the planet
is compared to the perpendicular line. For example, the rotational axis of the Earth has a
tilt of 23.5°. Mars is tilted at 25.19°, and Saturn's axis is tilted at 26.73°. Uranus, on the
other hand, has an axial tilt of 97.8°. Because
of this, Uranus is often referred to as the
"sideways planet." Either the north or the
south pole of Uranus is usually pointed toward
the Sun. Uranus's equator experiences day
and night the same way as the Earth's polar ice
caps do. The poles of Uranus each experience
forty-two years of sunlight, followed by forty-
two years of complete darkness. Only around
equinoxes is the Sun facing Uranus's equator,
causing "normal" Earth day-night conditions.
Its last equinox occurred on December 7, 2007,

and the the next will not happen until the year
2049.

Due to its unusual orientation, scientists have
conflicting methods for determining which pole
is "north" and which is "south." The International Astronomical Union (IAU) refers to which-
ever pole lies above the orbital plane as the
north pole. Most scientists use this designation.
Others use the right-hand rule from physics and
the direction the planet is spinning to designate
the poles north or south. This method contradicts the IAU's determination, instead naming
the pole below the orbital plane as "north."

The only spacecraft to date that has visited
Uranus is Voyager 2. Launched in 1977, the
probe reached Uranus in 1986. Voyager 2 came
within 81,500 kilometers of the planet. It discovered and photographed ten new satellites
and nine rings orbiting Uranus. The spacecraft
also helped scientists determine more precisely
the axial tilt of Uranus.

There are two main competing theories to explain why Uranus is tilted on its side. No one
knows who proposed the popular "collision" theory, which posits that Uranus formed and then
a large Earth-sized object crashed into it with
such force that it left the planet on its side. The
current accepted theory of planetary formation
is the idea of nebular condensation, developed
in the seventeenth century by French philosopher René Descartes. As a massive cloud of interstellar dust and debris condensed, it would
collapse and start to spin. Planets slowly would
begin to form from clumps of matter joining together. The bigger the planets grew, the faster
they would be able to attract more material
through a process known as accretion. The debris cloud that the planets formed from is called
the accretion disk, which became the orbital
plane. The collision knocking Uranus on its side
would have had to happen early in its formation. Possibly an object struck the planet's core
before Uranus's satellites had condensed from
the debris cloud surrounding it. Another theory
is that the impact left behind debris that later
became Uranus's satellites. However, there are
several questions that remain unanswered with
this scenario.

Why does Uranus have a nearly circular obit,
like the other planets? Would not a large impact

have affected Uranus's orbit? If Uranus's satellites had formed before the collision, why were their orbits not changed? The satellites orbit Uranus's equator, just like its ring system. Two very small captured satellites, however, have been found orbiting Uranus's poles. The nebular theory also fails to explain other oddities of the solar system, such as why Venus has a retrograde rotation (rotates backward), why Mercury and Pluto have elliptical orbits, and why Uranus and Neptune have tilted magnetic fields.

In 1997, Argentinean scientists Adrian Brunini and Mirta Parisi published a paper giving plausible ways that Uranus became tilted. They believed that if a collision had taken place, it had to be when Uranus was a more solid core surrounded by a planetary envelope. The impacting object would have hit the proto-Uranus from the opposite direction as it traveled around the Sun. The two scientists thought that studying Uranus's satellites was the key to figuring out its odd axial tilt. They concluded that either the satellites of Uranus were created by the collision itself, or no collision happened. Brunini and Parisi's study found that Uranus's satellite Prospero (S/1999 U3) set a number of constraints on any possible conclusion. Therefore, they believe, it is possible that a new theory of solar-system formation is needed to explain the tilt of Uranus. A number of scientists seem to be shifting toward the second explanation: that Uranus formed tilted on its side, and that the nebular theory for the formation of the solar system fails to explain how this could have happened. Researchers have been working on finding a simulation that solves this and other oddities of the solar system.

In 2006, Brunini published a new theory of the formation of the solar system in *Nature* magazine. His mathematical model is based on the idea that Jupiter and Saturn once had a 1:2 orbital resonance. This means that in the time it took Saturn to orbit the Sun once, Jupiter went around twice. The gravitational effect of Jupiter and Saturn gradually changed the orbits of Uranus and Neptune. Brunini's simulation shows that his model would take about a million years for the outer planets to reach the orbital positions we now observe. He argues that during the

close encounter of Saturn and Uranus, the angular momentum of the planets shifted, which over time caused their axial tilts to change. This scenario, Brunini argues, can explain the orbits of Uranus's rings and satellites, which would have slowly changed their orientation along with Uranus. Unlike a collision, Brunini's scenario would have taken hundreds of thousands of years to play out.

No definitive answer has been found for what caused Uranus's unique tilt. Only Voyager 2 has visited Uranus; new spacecraft would be able to provide more data but cannot be sent until either the planets are again aligned for a "slingshot" (gravity-assist) approach (more than a century away) or the necessary nuclear propulsion systems are developed. Until then, researchers are left making mathematical and computer models in their efforts to solve the mysteries of Uranus's axial tilt.

METHODS OF STUDY

Uranus can be observed from Earth with telescopes, and on dark, clear nights can be viewed with the unaided eye. Scientists have also taken photographs of the Uranus system using the Hubble Space Telescope. In late 2002, astronomers in Chile were able to image Uranus, its rings, and some of its satellites. The pictures were taken with the Very Large Telescope (VLT) at the European Southern Observatory (ESO) Paranal Observatory. The rings that are normally unable to be viewed from Earth, along with seven satellites, appeared in the image because it was taken at near-infrared wavelengths.

The Voyager program is the only spacecraft that has visited Uranus. Launched in 1977, Voyager 2 came within 81,500 kilometers of Uranus on January 24, 1986. Voyager 2 was equipped with more than a dozen scientific instruments, including cameras, television cameras, magnetometer, and spectroscopes. Voyager 2 viewed Uranus's "south" pole (located south of the orbital plane), which was pointed toward the Sun. At Uranus, Voyager 2 discovered ten satellites and two rings. The spacecraft also studied the planet's five largest moons (Oberon, Umbriel, Titania, Ariel, and Miranda), taking the first close-up photographs of them.

Voyager 2 provided the first close-up photographs of Uranus and detailed information about its magnetic field, ring system, weather, and unusual axial tilt. By the early nineteenth century, scientists knew that the planet was tilted, but it was not until Voyager 2 arrived at Uranus that astronomers knew precisely how tilted it was.

Computer modeling can be used to explore the dynamics of complex systems over time, where geologic time is essentially replaced by computation time. Modern computers allow a tremendous amount of computational power, and the magnitude of that computational power is continuously increasing. Basically, a computer modeling effort such as that used by Brunini and similar researchers seeks to begin with certain basic assumptions about initial conditions of a complex system such as Uranus in its interaction with larger bodies such as Jupiter and Saturn, and then introduce the gravitational interactions between all of these bodies and allow the computational cycle to mimic the passage of time as each of these bodies orbits the Sun and continues to interact with the others. This sort of thing cannot be easily done by hand. Sir Isaac Newton, in presenting his development of mechanics in *Philosophiae Naturalis Principia Mathematica* (1687; commonly known as *The Principia*), provided a means of quantifying the gravitational interaction between two bodies. That relatively simple problem can be solved in closed form in both spatial and temporal coordinates. However, the three-body problem requires numerical analysis, which is largely done at present by computer programming or software packages, since it cannot be solved in closed form. The more bodies are involved in a calculation, the more computing power is required.

CONTEXT

Scientists may never know the real reason for Uranus's axial tilt. Further study of the planet Uranus by spacecraft, and even possibly by humans, could lead to the answer. Computer simulations and mathematical models can help scientists speculate what might have happened. Maybe the accepted nebular theory for how the solar system formed is incorrect. Maybe the gravitational effects of Jupiter and Saturn slowly caused Uranus to lean to its side. Maybe Adrian Brunini and his colleagues are correct, and the only way to make this determination is to study Uranus's satellites. The quest to explain Uranus's extreme axial tilt could lead to a new view of how Earth and the solar system formed.

Jennifer L. Campbell

FURTHER READING

Burgess, Eric. *Uranus and Neptune: The Distant Giants*. New York: Columbia University Press, 1988. Covers the Voyager 2 spacecraft's mission, technical difficulties, and its encounters with Jupiter, Saturn, and Uranus. Focuses on data collected by Voyager 2 about Uranus. Includes several illustrations and tables. Well written, suitable for the general audience.

Chaisson, Eric, and Steve McMillan. *Astronomy Today*. 6th ed. New York: Addison-Wesley, 2008. A well-written college textbook for introductory astronomy courses. Includes a chapter on Uranus and Neptune.

Elkins-Tanton, Linda T. *Uranus, Neptune, Pluto, and the Outer Solar System*. New York: Chelsea House, 2006. Explores the Sun's relationship with the three outer planets and their moons. Looks at these planets as recorders of the formation of the solar system. Aimed at a general or high school audience. Illustrations, bibliography, index.

Encrenaz, Thérèse, et al., eds. *The Outer Planets and Their Moons: Comparative Studies of the Outer Planets Prior to the Exploration of the Saturn System by Cassini-Huygens*. New York: Springer, 2005. An in-depth look at the current understanding of the solar system's outer planets. Focuses on the studies of their formation, evolution, magnetospheres, satellites, and ring structures. For scientists, first-year graduate students, and advanced undergraduates.

Fraknoi, Andrew, David Morrison, and Sidney Wolff. *Voyages to the Stars and Galaxies*. Belmont, Calif.: Brooks/Cole-Thomson Learning, 2006. An introductory college text that gives students easy-to-understand analogies to help them with more complex theories.

Well written and easy to read. Includes a CD-ROM featuring InfoTrac software.

Freedman, Roger A., and William J. Kaufmann III. *Universe*. 8th ed. New York: W. H. Freeman, 2008. A thorough and well-written introductory college astronomy textbook. Covers all aspects of Uranus.

Hunt, Garry E., and Patrick Moore. *Atlas of Uranus*. New York: Cambridge University Press, 1988. This was the first volume after the 1986 Voyager encounter to offer a comprehensive history of Uranus: its discovery, satellites, rings, and the data returned by Voyager, including photographs.

Loewen, Nancy. *The Sideways Planet: Uranus*. Mankato, Minn.: Picture Window Books, 2008. An educational children's book devoted to the planet Uranus. Covers Uranus's rings, moons, and tilted axis.

Miner, Ellis. *Uranus: The Planet, Rings, and Satellites*. New York: Ellis Horwood, 1990. The author thoroughly covers the topics of both the Uranian system and the Voyager mission. Illustrations, bibliography, index.

Schmude, Richard W. *Uranus, Neptune, and Pluto and How to Observe Them*. New York: Springer, 2008. Ideal for backyard or amateur astronomers who are interested in observing the outer planets. Also includes up-to-date information about the planets.

Tocci, Salvadore. *A Look at Uranus*. New York: Franklin Watts, 2003. As part of the Out of This World series, this book covers all aspects of the planet Uranus, from its discovery through 2002. Includes several photographs. Suitable for all readers.

See also: Auroras; Earth's Atmosphere; Earth's Composition; Eclipses; Infrared Astronomy; Jovian Planets; Planetary Atmospheres; Planetary Formation; Planetary Interiors; Planetary Magnetospheres; Planetary Ring Systems; Planetary Rotation; Planetary Satellites; Planetology: Comparative; Telescopes: Ground-Based; Uranus's Atmosphere; Uranus's Interior; Uranus's Magnetic Field; Uranus's Rings; Uranus's Satellites.

Van Allen Radiation Belts

Category: Earth

The Van Allen radiation belts are concentrated rings of ionized particles in Earth's magnetosphere. Detailed study of the radiation belts led to an understanding of certain phenomena occurring in the ionosphere and the determination of the physical properties of the exosphere.

OVERVIEW

The Van Allen radiation belts are concentrated, torus-shaped regions of charged particles within Earth's magnetosphere. These particles, made up of protons, electrons, and other ions, spiral about in great numbers between Earth's magnetic poles. The magnetic and charged particles within the Van Allen belts can be divided into four regions: the Van Allen geomagnetically trapped radiation region, the auroral region, the magnetosheath, and interplanetary space. The inner and outer belts are part of the Van Allen geomagnetically trapped radiation region. In discussions of the Van Allen belts, the magnetic storm is often referred to as a third radiation belt.

The inner zone stretches from about 1,000 to more than 5,000 kilometers above Earth. It is mainly independent of time. Its composition is nearly consistent with that expected for the decay products of cosmic-ray-produced neutrons in the atmosphere (a neutron is an elementary, neutral particle of mass); this zone is of cosmic-ray origin. The radiation in the middle of the inner zone is composed of electrons with energies exceeding 40 kilo-electron volts (keV) and protons with energies greater than 40 million electron volts (MeV). (Electrons are elementary particles with a negative charge; protons are positively charged elementary particles.) In the inner belt, many of the high-energy protons are capable of penetrating several inches of lead. At the edge of the inner zone, in the region of geomagnetic latitudes 35° to 40°, low-energy electrons are found. The decay of light-scattering neutrons gives rise to high-energy protons. Beyond Earth's magnetic field, the mean ionizing capacity is 2.5 times higher than the minimum ionizing capacity. Particles in the inner zone are stable and exist for a long period of time.

The outer zone stretches about 15,000 to 25,000 kilometers above Earth. This zone undergoes very large temporal fluctuations appearing to be caused by solar activity and auroras, atmospheric heating, and magnetic storms. The outer belt contains soft particles; it is of solar origin. The outer zone contains electrons with more than 40 keV in energy and protons with more than 60 MeV in energy. The outer zone has greater geophysical significance than the inner zone. According to the comparison of E. V. Gorchakov, the boundaries of the outer zone coincide with isochasms (lines of equal probability of auroras). Trapped particles introduce magnetic effects in the outer radiation belt. This effect was measured by Luna 1. The increase in ionization of the outer zone is unstable. Particles exist for a short period of time compared with those of the inner belt.

A third radiation belt is produced by magnetic storms. Protons are transported from the Sun in a corpuscular stream and injected by magnetic field perturbations into Earth's field. The charge exchange with neutral hydrogen in Earth's exosphere is the fastest mechanism of removal. This is about a hundred times faster than scattering from ions in the exosphere. With the exception of trapped radiation, the entire region in the magnetic cavity is known as the auroral region. Auroral particles, the islands or pulses in the long tail and spikes at high latitudes of 1,000 kilometers, are phenomena that occur in the auroral region. Electrons of uniform angular distribution have a roughly constant intensity between 100 and 180 kilometers in altitude.

The magnetosheath lies between the shock

front formed by the solar wind and the magnetic cavity. Islands of electrons have been observed in the magnetosheath. At its widest, the magnetosheath is about four times the radius of Earth. It contains a compressed, seemingly chaotic interplanetary magnetic field. The interplanetary field connected to the Sun is predominantly in the ecliptic plane. The field terminates when the solar wind undergoes a shock transition to subsonic flow.

The lifetime of trapped particles decreases with distance from Earth. The lifetime of electrons with energies greater than 1 MeV at a distance of 1.2 to 1.5 times Earth's radius is about a year. The lifetime of the same electrons is reduced to days and months at a distance of 1.5 to 2.5 times Earth's radius. At even greater distances, the lifetime of the particles is measured in minutes. Because Earth is strongly influ-

enced by the Sun's magnetic field, Earth's geomagnetic field does not decrease indefinitely with increasing distance. The solar wind pushes Earth's magnetic field and is deflected by it. At about 10 Earth radii, the radiation belt ends abruptly.

Particles of trapped radiation may be lost in two ways. During a magnetic storm, the magnetosphere may lose or gain particles. This occurs at distances of 1.0 to 1.5 times Earth's radius. The other mechanism occurs at distances greater than 8 times Earth's radius. Small, rapid variation in the magnetic field at such distances scatters trapped particles, dumping them into the atmosphere. In a similar fashion, it is seen that charged particles in Uranus's magnetosphere are swept down into the planet's upper atmosphere by collisions with particles in its ring system.

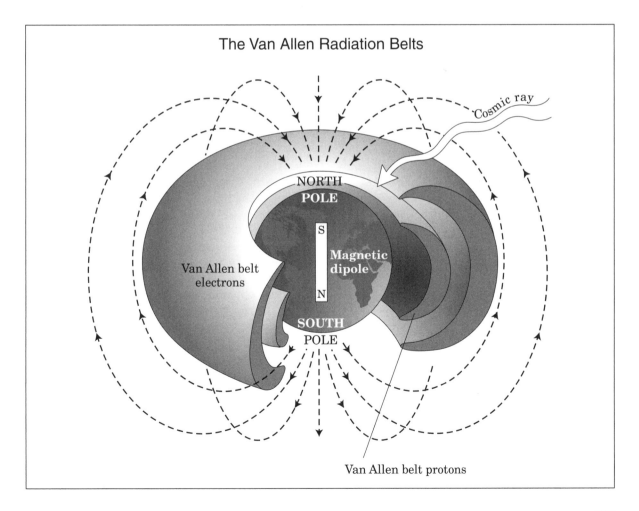

The Van Allen Radiation Belts

Cosmic ray

NORTH POLE

S

Magnetic dipole

N

Van Allen belt electrons

SOUTH POLE

Van Allen belt protons

Beautiful auroral displays occur when the charged particles are dumped into Earth's upper atmosphere. Solar flares eject into space streams of high-energy protons and electrons. When these beams of high-energy particles are directed toward Earth, Earth's magnetic field is partially disrupted. Particles trapped within the field lines can escape downward toward Earth at the lower ends of the radiation belts. High-energy particles, reinforced with particles from the Sun, energize the upper atmosphere, causing luminous and often colorful auroras.

KNOWLEDGE GAINED

The years 1957 and 1958 were designated the International Geophysical Year (IGY), an international scientific tour de force to advance understanding of Earth sciences. As contributions to IGY, the Soviet Union and the United States both pledged to place a satellite in orbit about the Earth. Russia's Sputnik 1 beat the American effort to orbit. However, the American effort was the first to gather useful scientific information. With the data returned by Explorer 1, America's first artificial satellite, a high-energy radiation belt was detected by James A. Van Allen and his assistants, George H. Ludwig, Carl E. McIlwain, and Ernest C. Ray. The same observations were made by Explorer 3, launched by the U.S. Army on March 26, 1958, and Sputnik 3, launched by the Soviet Union on May 15, 1958.

Later, a satellite was launched as part of Project Argus, which studied the location, height, and yield of electron blasts. This project was carried out by the Advanced Research Projects Agency. The belt of electrons produced by the Argus nuclear explosions developed at a distance of twice Earth's radius. Explorer 4, launched on July 26, 1958, carried four Geiger counters to handle high levels of radiation. One of these Geiger counters was shielded with a thin layer of lead to keep out most of the radiation. The satellite reached a height of 2,200 kilometers and registered an intensity of high-energy radiation. From the data returned, scientists concluded that Earth is surrounded by belts of high-energy radiation consisting of particles originating from the Sun and trapped in the lines of force of Earth's magnetic field.

These were named the "Van Allen radiation belts."

Explorer 4 obtained a kidney-shaped intensity contour of Earth's inner belt. Data from early Pioneer spacecraft suggested a solar origin of soft particles populating the outer zone. Three Pioneer probes and Luna 1 discovered the crescent-shaped intensity contours of the outer belt. Sputnik 3 data helped identify the bulk of the outer belt particles as low-energy electrons (10 to 50 keV).

Several more human-made belts were produced in 1962. The Starfish project, an American venture, created a belt much wider than the Argus belt. Decay of some of the particles took several years in low altitudes. In the same year, the Soviets created at least three similar belts. More sophisticated versions of the instrumentation used in these early probes of Earth were incorporated into spacecraft sent to other planets in the solar system. Probes to Mercury, Jupiter, Saturn, Uranus, and Neptune have discovered radiation belts similar to Earth's Van Allen belts. A planet needs a magnetic field to trap charged particles into a radiation belt or system of radiation belts. For that very reason, there are no significant belts of trapped charged particles at Venus or Mars.

CONTEXT

In the earliest days of space exploration, gauging the intensity of Earth's radiation belts with uncrewed spacecraft was crucial as a first step toward sending humans into space. Both the United States and the Soviet Union had a vested interest in the results of early investigations of the magnetosphere.

The Van Allen belts, while lifesaving in that they keep dangerous radiation from reaching the surface of the planet, are potentially hazardous to Earth-orbiting spacecraft. They threaten electronics systems and instrumentation and can interfere with radio transmissions. In the late 1950's, it was not known just how hazardous the radiation surrounding Earth would prove to humans. While it would not be wise to base a space station within the radiation belts, the belts themselves pose little threat to humans, who quickly punch through on voyages of exploration beyond the Earth; that was the case back in the Apollo program and will be true of the National Aeronautics and Space Administration's (NASA's) planned Constellation program flights to the Moon and later to Mars. However, leaving the safety of orbit beneath the Van Allen belts does expose astronauts to the potential hazards of ionizing radiation streaming outward into the solar system from the Sun. Therefore, special protection must be provided to humans on expeditions beyond low-Earth orbit.

The relationship between auroras and the Van Allen belts has been studied for decades, but although the overall phenomenon has been well characterized, all is not completely understood. Scientists do know that most bright auroras are produced by electrons dumped into Earth's atmosphere by solar flares. The auroral particles are the electrons escaping from the outer Van Allen radiation belt. The average kinetic energy of the electrons is 32 keV. The leakage of corpuscular radiation into the auroral zones is the most important loss of corpuscular radiation from the outer Van Allen belt.

Satya Pal

FURTHER READING

Bone, Neil. *The Aurora: Sun-Earth Interaction*. New York: John Wiley, 1996. One volume in the Ellis Horwood Library of Space Science and Space Technology. Devoted to describing the electrodynamics of the Sun-Earth environment that produce auroral displays.

Bothmer, Volker, and Ioannis A. Daglis. *Space Weather: Physics and Effects*. New York: Springer Praxis, 2006. A selection from the publisher's excellent Environmental Sciences series, this is an overview of the Sun-Earth relationship and provides a historical and technological survey of the subject. Projects the future of space weather research through 2015 and includes contemporary spacecraft information.

Foerstner, Abigail. *James Van Allen: The First Eight Billion Miles*. Iowa City: University of Iowa Press, 2007. An engaging portrait of the legendary physicist, discussing his contributions to the World War II effort as well as to the advancement of studies of Earth's geomagnetic environment, his early efforts to study cosmic rays using balloon-launched rockets, the Explorer 1 story, and Van Allen's continuing participation in studying space physics until his passing in 2006.

Gregory, Stephen A. *Introductory Astronomy and Astrophysics*. 4th ed. San Francisco: Brooks/Cole, 1997. Suitable as a textbook for introductory college courses or advanced high school courses in general astronomy. Covers all topics from solar-system bodies to cosmology. Some errors and issues with mathematical presentations.

Kallmann-Bijl, Hildegaard, ed. *Space Research: Proceedings of the First International Space Science Symposium*. New York: Interscience, 1960. A detailed explanation of the theory behind the nature, origin, and composition of the inner, outer, and third radiation belts is provided. Dated and technical, but provides a useful historical perspective.

Milone, Eugene F., and William Wilson. *Solar System Astrophysics: Background Science on the Inner Solar System*. New York: Springer, 2008. Rigorous and highly mathematical presentation involving geophysics, atmospheric physics, and mineralogy covering all aspects of planetary science. Includes results from the Mars Exploration Rovers and Cassini spacecraft.

Moldwin, Mark. *An Introduction to Space Weather*. Cambridge, England: Cambridge University Press, 2008. This text introduces space weather, the influence the Sun has on Earth's space environment, to the nonscience reader. Discusses both the scientific aspects of space weather and issues of technological and societal import.

Savage, Candace. *Aurora: The Mysterious Northern Lights*. New York: Firefly Books, 2001. Heavily illustrated with photographs of auroral displays, this book provides a history of scientific investigation of auroral phenomena.

Sullivan, Walter. *Assault on the Unknown*. New York: McGraw-Hill, 1961. Sullivan describes the discovery of the Van Allen belts in great detail.

Van Allen, James A. *The Magnetospheres of Eight Planets and the Moon*. Oslo: Norwegian Academy of Science and Letters, 1990. A technical summary of all major magnetic structures in the solar system, written by the prolific researcher after whom the Van Allen belts are named.

_____. "Radiation Belts Around the Earth." *Scientific American* 200 (March, 1959): 39-47. A seminal article by the discoverer of the Van Allen belts. Well illustrated. Provides insight into the nature of the original discovery and the early days of the space race.

See also: Auroras; Earth's Atmosphere; Earth's Magnetic Field: Origins; Earth's Magnetic Field: Secular Variation; Earth's Magnetic Field at Present; Earth's Magnetosphere; Earth-Sun Relations; Jupiter's Magnetic Field and Radiation Belts; Planetary Magnetospheres; Solar Flares; Solar Wind.

Venus's Atmosphere

Categories: Planets and Planetology; Venus

The atmosphere of Venus has a surface temperature of about 743 kelvins and a surface pressure of about 90 Earth atmospheres. Its clouds consist largely of carbon dioxide, and droplets of sulfuric acid rain down to the surface.

OVERVIEW

The second planet from the Sun and Earth's immediate inner neighbor, Venus is often called Earth's twin, because the masses and radii of the two planets are very similar. Venus's mass is 82 percent that of Earth. Its radius is only 5 percent less than Earth's. Under ordinary circumstances, no objects in the sky other than the Sun and the Moon surpass Venus in brightness. Viewing Venus with one or more of the other planets also visible in the sky shortly before sunrise or briefly after sunset can be an awe-inspiring sight. In ancient times, since Venus could at times be seen in the morning skies and at other times in the evening skies, the planet was actually thought to be two different objects; they were given the names Phosphoros and Hesperus for the morning and evening star, respectively.

Venus revolves once around the Sun every Earth 224.7 days. It rotates on its axis once every 243.01 days. The inclination of its orbit is about 3.5° with respect to the ecliptic plane. Its orbit, like those of all the other planets, is an ellipse, but it is very close to being a perfect circle. The tilt of Venus's axis of rotation is about 17.8°, as compared with the Earth's 23.5°. As a result, any seasonal changes of weather on Venus would be less extreme than on Earth. Remarkably, its rotation is retrograde, backward, as compared with the direction of revolution, or clockwise, as seen from above the ecliptic plane. It is not known with certainty why Venus has a retrograde rotation. Virtually all other objects in the solar system rotate and revolve prograde, or counterclockwise as seen from above the ecliptic.

At each inferior conjunction (each time Venus and the Sun are aligned in the sky with Venus closer to Earth than to the Sun), the same face of Venus points toward the Earth. This phenomenon may mean that Venus's rotation is influenced by the Earth's gravitational pull; however, some information indicates that the alignment at inferior conjunction is not exactly perfect and therefore may be coincidental.

As seen from Earth, Venus's maximum disk size, or angular diameter on the sky, is about 0.02°, and its minimum angular diameter is about 0.003°. The maximum or minimum size on the sky corresponds to the closest and farthest distances from Earth. In contrast, the Sun and the Moon have an apparent angular diameter of about 0.5°. Given these observational circumstances, little can be learned about the planet's atmosphere from telescopic observations. The planet's average density is 5.25 grams per cubic centimeter, and the surface acceleration caused by gravity is 0.903 times that of Earth. Escape velocity from Venus's surface is 10.3 kilometers per second. That means Venus can retain its atmosphere virtually indefinitely, since most of the molecules at the top of the atmosphere will travel at speeds well under 10.3 kilometers per second.

Through a telescope, an image of Venus is somewhat disappointing; only the planet's phases are obvious. Exceptional atmospheric and viewing conditions at good astronomical sites have allowed scientists to see and photograph subtle variations in shading on bright cloud tops of the planet's lit side. However, no part of the surface of Venus is visible through telescopes, because of its thick cloud cover.

Sulfur is probably released into the atmosphere of Venus by outgassing volcanic processes. Sulfur rises and bonds with water and oxygen to produce sulfuric acid. It appears as a fairly thick haze and is more strongly concentrated than the acid found in the battery of an automobile. These clouds are less murky than fog, with visibilities within them of perhaps several hundred meters. Top layers of these clouds are about 80 kilometers above the planet's surface. The tops of the sulfuric acid-laden clouds have winds that move faster than 300 kilometers per hour, a speed comparable to that of the Earth's jet streams. These high winds swirl

This image of one complete Venusian hemisphere is a composite of high-resolution Magellan images whose gaps are filled by images from the Arecibo radio telescope in Peru. (NASA/JPL/USGS)

around the planet in about four days. Circulating motions cause gas to rise near the equator and descend near the poles, probably a direct consequence of excess solar heating in the equatorial region.

The atmosphere of Venus is mostly carbon dioxide. Venus has undergone a spectacular greenhouse effect, caused by particles in the clouds trapping or absorbing infrared radiation. Most sunlight is reflected by clouds back into space. The albedo of Venus is about 0.8; that is, 80 percent of incident radiation striking the cloud tops is reflected back into space. The 20 percent that penetrates the clouds warms the surface sufficiently to heat the surface rocks and terrain. These surface structures radiate, essentially at infrared wavelengths, and that radiation cannot penetrate the clouds and gets trapped. Heat builds, and the high temperature releases gases (mainly carbon dioxide) from rocky minerals into the atmosphere, which in

turn drives the greenhouse effect further. A cycle develops: The carbon dioxide traps the radiation, which heats the surface and atmosphere, which then triggers the release of more carbon dioxide. This continues until the entire atmosphere reaches a sufficiently high temperature (740 kelvins) to radiate as much energy back into space as it receives.

High above the clouds, the atmospheric layers called the exosphere consist mostly of hydrogen and helium. These strata are affected by intense incoming ultraviolet solar radiation. The radiation ionizes atoms, making them electrically charged. The uppermost layers form the ionosphere. Venus's ionosphere is not as intensely ionized as Earth's.

Venus's surface is quite flat compared with that of the Earth; however, there are at least three major elevated regions, and they exhibit features that influence the atmosphere. The largest, Ishtar Terra, is similar in size to Australia. The others are about the size of the largest islands in Indonesia. Many less elevated regions are also present. On Ishtar Terra, a mountain called Maxwell Montes is apparently higher than Mount Everest with respect to the surrounding flat terrain. Large volcanoes are present on each of the three plateaus. Volcanoes imply gas release from the planetary interior; therefore, Venus's atmosphere is attributable in part to volcanic outgassing. Although the atmosphere is composed mostly of carbon dioxide, there is some nitrogen and sulfur dioxide, very little water vapor, and trace amounts of various other gases. Sulfur dioxide is outgassed by volcanoes on the Earth but is quickly diluted by rain and moisture. In contrast, sulfur dioxide outgassed in the dry atmosphere of Venus is very stable, accounting for the efficient production of sulfuric acid in the clouds and elsewhere.

Continent-building processes caused by plate tectonics on the Earth may have occurred on Venus, though to a far lesser extent. This idea is based mostly on the fact that only a few elevated regions exist. Therefore, the outgassing brought on by a variety of volcanic actions related to plate tectonics is probably slower and less effective on Venus, compared with the heat-induced gases, such as carbon dioxide, from rocks.

One other feature of Earth's atmosphere that probably does not exist to any extent on Venus is production of high-level auroras (or, as they are known on Earth, the northern and southern lights), because Venus has little or no magnetic field. Since its rotation rate is very slow, one would expect a weak but nevertheless measurable magnetic field. Several theories have been put forth to explain this lack. One theory is that, like the Earth, Venus undergoes polarity changes of its overall dipolar magnetic field. The Earth's magnetic field is explained by the dynamo hypothesis. The rotating core produces loop currents in the heated molten regions, which in turn produce a magnetic field. Geological and paleontological evidence suggests that the Earth's magnetic field undergoes reversals of polarity at irregular intervals. During the reversal periods, little or no field is present. It is possible that Venus could be undergoing such a magnetic reversal phase. In fact, either it is in such a phase or the dynamo hypothesis is incorrect.

METHODS OF STUDY

The first attempt to send an interplanetary probe to Venus did not fare well. Mariner 1 launched on July 22, 1962. A software error caused its booster to veer dangerously off course while low in the atmosphere, and it was destroyed on purpose by the range safety personnel at Cape Canaveral.

Mariner 2, a sister spacecraft to the failed first American Venus probe, launched successfully on August 27, 1962. Fortunately, this probe was able to fly within

Venus's Atmosphere Compared with Earth's

	Venus	Earth
Surface pressure (bars)	92	1.014
Surface density (kg/m^3)	~65	1.217
Avg. temperature (kelvin)	737	288
Scale height (kilometers)	15.9	8.5
Wind speeds (meters/second)	0.3-1.0	up to 100
Composition		
Argon	70 ppm	9,430 ppm
Carbon dioxide	96.5%	350 ppm
Carbon monoxide	17 ppm	—
Helium	12 ppm	5.24 ppm
Hydrogen	—	0.55 ppm
Hydrogen chloride	tr	—
Hydrogen fluoride	tr	—
Krypton	—	1.14 ppm
Neon	7 ppm	18.18 ppm
Nitrogen	3.5%	78.084%
Oxygen	—	20.946
Sulfur dioxide	150 ppm	—
Water	20 ppm	1%
Xenon	—	0.08 ppm

Note: Composition: % = percent; ppm = parts per million; tr = trace amounts.

Source: Data are from the National Space Science Data Center, NASA/Goddard Space Flight Center.

34,833 kilometers of the Venusian surface on December 14, 1962. Although it carried no photographic equipment, Mariner 2 provided a treasure trove of new information about the shrouded planet. The spacecraft was outfitted with Geiger tubes, an ion chamber, a cosmic dust detector, a microwave radiometer, and a magnetometer experiment. Mariner 2 determined that the Venusian surface temperature was more than 670 kelvins. It detected neither a planetary magnetic field nor any Van Allen-like radiation belts about the planet. It continued to collect data about particles and fields in interplanetary space until contact was lost on January 3, 1963, at a distance of 87 million kilometers from Earth.

Mariner 5 was launched on June 14, 1967, and was sent to fly by Venus. This spacecraft also did not include photographic or television cameras. It was equipped with radio science and ultraviolet experiments as well as particle and magnetic field detectors. Mariner 5 encountered Venus on October 19, 1967, coming within 4,000 kilometers. The spacecraft investigated Venus's cloud tops and the solar wind interacting with interplanetary magnetic fields.

Early images taken by the space probes Mariner 10 and Pioneer Venus in reflected solar ultraviolet light revealed considerable variation in shading on Venus. Variation is caused by radiation coming from different levels of the clouds. The study of the motion of these clouds has indicated that the top layers can rotate at very high speeds, approaching perhaps 100 kilometers per hour. These high-strata, rapid wind velocities are in part caused by the hot, sunlit clouds transferring heat to the colder dark side. The cloud structure shows three distinct strata: a high, thick layer; a medium-high haze layer; and a lower, medium-thick layer. From this lower layer downward it is essentially clear all the way to the surface.

The Pioneer Venus atmospheric probes, sent to the surface via a combination of small retrorockets and parachutes, and Soviet Venera landers measured a decrease in the wind velocity at lower altitudes. On the surface, winds are essentially gentle breezes. Heat transfer and exchange in the atmosphere are very dependent on the density and pressure of various layers.

Lower levels of the atmosphere are under superhigh pressures. Soviet Venera landers and the Pioneer space probes measured pressures of about 90 atmospheres at the surface. Heat transfer is so efficient that there is no large-scale difference between daytime and nighttime temperatures at the surface. Mariner 10 and Pioneer Venus ultraviolet images indicated an overall circulation pattern: Atmospheric gas rises at the equator and descends at the poles. With slow rotation of the planet, this circulation pattern is highly stable.

Four Venera landers managed to set down on the surface, perform experiments, and obtain electronic images of the surroundings using a fish-eye lens or wormlike view of the terrain to the horizon. At both landing sites, rocks and the horizon in the clear atmosphere are visible. Rocks are clearly of volcanic origin. In some cases their sharp edges indicate little or no erosion, suggesting recent volcanic origin. One would expect that erosion, under the high pressure and intense heat, would quickly deform and erode the rocks' edges. Pioneer Venus detected the existence of sulfuric acid droplets, which at the high temperatures and pressures present is very corrosive.

Pioneer Venus was equipped with a radar ranger. Scientists could send radar beams to the surface of Venus from the spacecraft and measure the time interval from the emission of the radar to the subsequent receiving of the reflected echo. This procedure allowed accurate measurement of distances between the spacecraft and the surface. After a compilation of such observations, the Pioneer Venus mission team was able to provide a detailed map of the surface terrain for the first time. Better maps would have to await a more sophisticated radar system placed in orbit about Venus.

Vega 1 and 2 were ambitious missions involving identical carrier spacecraft that each delivered both a lander based on the Venera design and an instrumented balloon to Venus before both carrier spacecraft were then redirected to join an international group of spacecraft intercepting and studying Halley's comet near its 1985/1986 perihelion passage. The carrier craft were outfitted with an imaging system, an infrared spectrometer, and a spectrometer capable of ultraviolet through infrared observations, detectors of dust and micrometeoroids, a plasma energy analyzer, a magnetometer, wave and plasma analyzers, a neutral gas mass spectrometer, and an energetic particle analyzer. The Vega 1 and 2 carrier craft encountered Venus on June 11 and 15, 1985, respectively, having several days earlier ejected their lander and balloon payloads. Neither carrier craft provided deep new insights into the nature of the Venusian atmosphere, but they did set the stage for the unique balloon payload and their results.

Venus's atmosphere apparently provided a particularly strong wind gust when the Vega 1 lander was still 20 kilometers above the planet's surface. This even activated the surface experiments early, and no results were produced after touchdown. Vega 2 operated properly and on June 15, 1985, safely touched down in the eastern Aphrodite Terra region. This lander determined the local atmospheric pressure to be 91 atmospheres, with a surface temperature of 736 kelvins. It endured the extreme environment for just under an hour, but, before it failed, the lander determined a rock sample to be a variety of anorthesite.

The Vega 1 and 2 balloons floated in the planet's atmosphere, providing data for about forty-six hours at an altitude of approximately 54 kilometers. The balloons were small in mass and size (25 kilograms, about 55 pounds, and 3.4 meters, or 11 feet, in diameter) but were able to dangle a gondola assembly filled with instruments to sample and measure the Venusian atmosphere. The balloons began their mission after being deposited on Venus's dark side. They sank to a depth of 50 kilometers before rising again to an altitude of 54 kilometers where they determined the pressure and temperature to be similar to those conditions on Earth. However, wind speed was nearly that of hurricane status, and at this altitude the carbon dioxide-rich atmosphere had a strong concentration of sulfuric acid with far less hydrofluoric and hydrochloric acid. Before losing electrical power, the balloons registered a variable vertical component to the atmospheric winds upon which they floated. Also they survived long enough to move from the dark side to the planet's illuminated side.

The National Aeronautics and Space Admin-istration's (NASA's) next probe to Venus was named after the great Portuguese explorer Ferdinand Magellan. Its goal was no less daunting than to use an imaging radar to map at least 98 percent of Venus's surface to a resolution of 100 meters or less. Magellan was deployed from the space shuttle *Atlantis* on the STS-30 mission on May 5, 1989, and sent on its way toward the inner solar system. The spacecraft arrived in Venus orbit on August 10, 1990. Its synthetic aperture radar system was able to peer through the thick atmosphere. After four years of mapping, radar altimetry, and gravitational field measurements, NASA intentionally drove Magellan through the planet's atmosphere on October 12, 1994. In one final experiment, information was inferred about the atmosphere as the spacecraft heated up, and contact was eventually lost when Magellan was destroyed.

The European Space Agency (ESA) launched the Venus Express spacecraft on November 9, 2005. Largely a twin of ESA's successful Mars Express spacecraft, but modified to study Venus, Venus Express entered a nine-day-period polar orbit about Venus on April 11, 2006. Then for science operations to commence, that orbit was altered to have a twenty-four-hour period. Equipped with a penetrating radar, Venus Express began generating a surface map of the planet at resolutions even better than Magellan had achieved. However, science objectives of Venus Express also included detailed atmospheric studies. Aboard the spacecraft were infrared, visible-spectrum, and ultraviolet instruments to observe Venusian atmospheric characteristics and determine temperature profiles as a function of altitude.

Venus Express discovered a rather unexpected double vortex feature located around the south pole of the shrouded planet. This remarkable find occurred on the spacecraft's very first highly elongated orbit about Venus. Thus Venus Express was able to examine the planet's atmospheric patterns in ultraviolet and infrared from a global perspective (when far from Venus) and at close range (as it approached its low point in orbit). A vortex feature had been seen previously over the planet's north pole by earlier spacecraft, but a double vortex with a stable structure was quite unusual. Invoking the high wind speed of the upper atmosphere and the convection of rising hot air was insufficient to explain this double vortex. Using infrared sensors, Venus Express was able to map out windows in the atmosphere through which thermal radiation could escape to space. That modeling assisted scientists in determining cloud structures as a function of altitude above the tremendously hot planetary surface.

Some earlier probes had provided circumstantial evidence that lightning was present in Venus's atmosphere, but others produced data strongly suggesting there was a total lack of lightning. In 2006, the magnetometer aboard Venus Express provided definitive data that lightning does occur in Venus's atmosphere.

Venus Express made another important discovery in 2008, detecting hydroxyl molecules. This was the first time on a planet other than Earth that hydroxyl molecules had been detected; hydroxyl is a molecular ion consisting of one oxygen and one hydrogen atom bonded together covalently. Hydroxyl was detected at an altitude of 100 kilometers above the Venusian surface, which Venus Express accomplished by means of the Visible and Infrared Thermal Imaging Spectrometer, picking up the faint infrared light emitted by these molecules in a very narrow band of Venus's atmosphere. That band appears to be only 10 kilometers thick.

Hydroxyl has been found around comets. However, planetary atmospheres produce the molecule in a very different manner from that involved in comets. Hydroxyl on Earth is associated with the abundance of ozone in the upper atmosphere. Thus, detection of hydroxyl in Venus's atmosphere suggests that Venus still retains some Earth-like aspects. Absorption of ultraviolet light by hydroxyl molecules is important to the heating balance of any planetary atmosphere. The hydroxyl data from Venus Express would greatly assist planetary scientists in fine-tuning their models of the Venusian atmosphere.

Finally, in late 2008 Venus Express for the first time detected water being lost from Venus's daylight side. The previous year, this spacecraft's Analyzer of Space Plasma and Energetic Atoms detected the signature of hydrogen being stripped away from the planet's nightside. The orbiter's magnetometer was used to find hydrogen dissociated from water coming off the daylight side to be lost into space. Solar wind particles penetrate Venus's atmosphere, since the planet lacks a protective magnetic field. Scientists believe that solar wind particles break water molecules into two parts hydrogen and one part oxygen. Oxygen and hydrogen have been found escaping the nightside in the right proportion, but oxygen escaping from the daylight side was not seen in the 1:2 ratio required if the hydrogen seen comes from water. In any event, the solar wind mechanism was believed by many to be the means whereby, over time, Venus lost much of its original water.

CONTEXT

Earth's atmosphere has the potential to become more like that of Venus. There are two ways that such a situation could develop. If Earth moved closer to the Sun, increased solar heat would release more carbon dioxide into the atmosphere. Limestone rock (calcium carbonate) and dissolved carbon dioxide in the oceans provide a tremendous store of trapped carbon dioxide. Under higher temperatures, the rocks and seashells would chemically release carbon dioxide, and the greenhouse mechanism would raise the atmosphere's temperature. As a result, new carbon dioxide would be released and would speed the greenhouse mechanism. The second way that Earth's atmosphere could become more like that of Venus involves pollution of the atmosphere to the extent that enough carbon dioxide accelerates the existing greenhouse effect.

More generally, study of Venus's atmosphere helps scientists to better understand terrestrial weather and climate. Earth has an atmosphere composed mostly of nitrogen. The weather, however, is influenced primarily by water molecules and carbon dioxide molecules. These substances are found in Earth's atmosphere only in trace amounts; nevertheless, they are responsible for most of the heat transfer around the globe. On Venus, the weather is controlled by the atmosphere's main constituent, carbon dioxide. The contrasts between weather processes on Venus and those on the Earth have led to a more complete understanding of the latter. In an even larger context, comparative planetology studies of Venus, Earth, and Mars contribute to a better understanding of Earth's complex weather system and atmospheric physics. Moreover, such study helps us learn why three planets, all within the Sun's habitable zone, could evolve so differently. To understand Earth's evolution fully, it is necessary to know why the Venusian atmosphere became thick in carbon dioxide at great pressure (so that a planetary greenhouse effect led to runaway temperatures), and also to understand why the Martian atmosphere became thin in carbon dioxide at low pressure and low temperature. Earth, on the other hand, developed a nitrogen-oxygen atmosphere with traces of carbon dioxide; this led to reasonable temperatures and pressures and the development of a complex biosphere and interactive oceanic-atmospheric processes to maintain a dynamic equilibrium.

James C. LoPresto and David G. Fisher

FURTHER READING

Beatty, J. Kelly, Carolyn Collins Petersen, and Andrew Chaikin, eds. *The New Solar System*. 4th ed. Cambridge, Mass.: Sky, 1999. Amply illustrated with color images, diagrams, and informative tables, this book is aimed at a popular audience, but it can also be useful to specialists. Contains an appendix with planetary data tables, a bibliography for each chapter, planetary maps, and an index.

Cattermole, Peter John. *Venus: The Geological Story*. Baltimore: Johns Hopkins University Press, 1996. Provides a comprehensive presentation of the latest understanding of Venus based on Magellan data.

Elkins-Tanton, Linda T. *The Sun, Mercury, and Venus*. New York: Chelsea House, 2006. Examines the innermost portion of the solar system and the star, our Sun, which plays such a prominent role in the evolution of both planets. For the general audience with an interest in science.

Esposito, Larry W., Ellen R. Stofan, and Thomas E. Cravens, eds. *Exploring Venus as a Terrestrial Planet*. New York: American Geophysical Union, 2007. A collection of articles covering all major areas of planetary research on Venus. Technical.

Fimmel, Richard O., Lawrence Colin, and Eric Burgess. *Pioneering Venus: A Planet Unveiled*. Washington, D.C.: National Aeronautics and Space Administration, 1995. A complete summary of the findings of Pioneer Venus as of 1983. Pioneer Venus orbited Venus and sent several probes into the atmosphere of the planet. It also mapped the planet using a radar-ranging device.

Freedman, Roger A., and William J. Kaufmann III. *Universe*. 8th ed. New York: W. H. Freeman, 2008. A college text on astronomy, somewhat more advanced than many introductory texts, but with a wealth of detail and excellent diagrams. Comes with a CD-ROM. The chapter on Venus is lucid and filled with spectacular diagrams and photographs.

Grinspoon, David Harry. *Venus Revealed: A New Look Below the Clouds of Our Mysterious Twin Planet*. New York: Basic Books, 1998. A thorough examination of the geology of Venus, incorporating Magellan mapping and other data. Explains the Venusian greenhouse effect. A must for the planetary science enthusiast who wishes to read an integrated approach to science and history. Includes speculation about Venus's past.

Hartmann, William K. *Moons and Planets*. 5th ed. Belmont, Calif.: Thomson Brooks/Cole, 2005. An updated version of a classic text that covers all aspects of planetary science. A comparative planetology approach is used rather than presenting just one chapter on all characteristics of Venus.

Marov, Mikhail Ya, and David Grinspoon. *The Planet Venus*. New Haven, Conn.: Yale University Press, 1998. Marov was Soviet Venera mission chief scientist, Grinspoon a NASA-funded scientist studying Venus. Together they provide a coordinated description of American and Soviet attempts to learn the secrets of Venus, a planet shrouded in mystery. For both general readers and specialists.

Morrison, David, and Tobias Owen. *The Planetary System*. 3d ed. San Francisco: Pearson/Addison-Wesley, 2003. A textbook at the beginning college level, introducing the scientific knowledge of the solar system as of 1988. The chapter on Venus goes into detail about the greenhouse effect and the contrasting atmospheres of Venus and the Earth.

Spanenburg, Ray, and Kit Moser. *A Look at Venus*. New York: Franklin Watts, 2002. A look beneath the thick clouds of Venus written for a younger audience.

See also: Earth's Atmosphere; Earth's Composition; Earth's Magnetic Field: Origins; Earth-Sun Relations; Extraterrestrial Life in the Solar System; Greenhouse Effect; Mars's Atmosphere; Planetary Atmospheres; Planetary Interiors; Planetary Magnetospheres; Planetary Rotation; Planetary Tectonics; Planetology: Comparative; Planetology: Venus, Earth, and Mars; Terrestrial Planets; Venus's Craters; Venus's Surface Experiments; Venus's Surface Features; Venus's Volcanoes.

Venus's Craters

Categories: Planets and Planetology; Venus

Impact craters are the most numerous and most easily recognized surface features in the solar system. Because of their pristine nature, Venusian impact craters provide a unique opportunity for astrogeologists to study the effects of atmospheric variabilities and gravity in the formation of planetary surfaces.

OVERVIEW

One of the earliest and most crucial phases in the early development of the solar system was the Great Bombardment. During this phase, planetary surfaces were under intense bombardment by cosmic debris. For the Earth and Moon, the Great Bombardment began about 4.6 billion years ago and declined about 3. 8 billion years ago. Because the Moon has no active tectonism or atmospheric processes, impact craters there are mostly scars left from this time of intense bombardment. Because Earth, by contrast, is tectonically active and continuously subjected to weathering processes, craters of such ancient ages are not visible on Earth's surface and have been identified only on the

cratons of continents, which undergo relatively little resurfacing from weathering, tectonism, volcanism, or other processes. The majority of impact craters identified on Earth, therefore, are geologically young and do not represent impacts from the Great Bombardment. Venus, Earth's closest planetary neighbor, presumably was also subjected to this intense cosmic bombardment, but it appears that Venus totally lacks the ancient, heavily cratered surface that occurs on Mercury, Mars, the Moon, and the rocky moons of other planets.

The detailed morphology of the surface of Venus is known through research done on remote-sensing data obtained from successful missions mounted by the National Aeronautics and Space Administration (NASA)—the Mariner (1962-1975), Pioneer (1978), Galileo (1990), Magellan (1990-1994), Cassini-Huygens (1998-1999), and MESSENGER (2004) uncrewed spacecraft missions—as well as the Soviet Union's Vega (1985) and Venera missions (1967-1984) and the European Space Agency (ESA) Venus Express (2005) mission. Other data have been accumulated through study of high-resolution images from Earth-based radar. The Magellan spacecraft, inserted into Venus orbit in 1990, yielded radar images with a resolution of a few hundred meters covering nearly 98 percent of the planet. In addition, the Soviet Venera missions mapped nearly 25 percent of Venus with additional radar imaging. These data have resulted in the mapping of nearly 100 percent of Venus and suggest that Venus lacks the ancient, heavily cratered surface occurring on other terrestrial planets and rocky satellites. The Venus Express mission has provided valuable data on the atmosphere of Venus and evidence suggesting that past oceans may have existed on the Venusian surface. A new Venus mission, Planet-C is planned for 2010 by the Japan Aerospace Exploration Agency (JAXA).

Astrogeologists use the density of impact craters to determine the age of planetary surfaces. The older the surface, the more impact craters it will have accumulated over time. On Venus this dating technique is problematic, because there are relatively few impact craters. Based on the density of Venus's impact craters larger than 30 kilometers in diameter, esti-

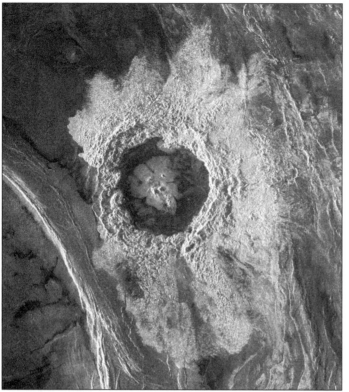

Venus's Dickinson Crater, 69 kilometers in diameter, from the Magellan spacecraft. (NASA/JPL)

mates of cratering rates scaled for other terrestrial planets and rocky satellites, and the known population of asteroids crossing Venus's orbit, the planet's average surface age is estimated at between 450 million and 250 million years old, the younger age being more likely. This suggests the surface terrain of Venus may be less than 5 percent of the age of the solar system. However, these ages are average estimates, and based on superposition some Venusian impact craters, volcanic structures, and tectonic terrains are thought to be as young as 50 million years old. Average age estimates aside, for the last 700 million years Venus has been subjected to significant surface volcanism, which probably is the reason that so few impact craters are visible on the planet's surface: Older impact craters have been covered over with lava flows or destroyed during episodes of catastrophic volcanic eruptions.

Venus has the densest atmosphere of any terrestrial planet. The Venusian surface pressure

is equivalent to 94 bars—more than ninety times the pressure humans feel from Earth's atmosphere (90 bars is approximately the weight of water at 1 kilometer below the surface of Earth's oceans). In addition, Venus's atmosphere is composed of 96 percent carbon dioxide and trace amounts of nitrogen, water vapor, argon, carbon monoxide, and other gases. These clouds seal in the Sun's heat, creating a perpetual greenhouse effect that boosts surface temperatures on Venus to around 753 kelvins. Clouds within the Venusian atmosphere are composed mainly of sulfuric acid and small amounts of hydrochloric and hydrofluoric acid. The presence of such a dense atmosphere effectively filters the numbers of potential impactors by severely decreasing their kinetic energy during transit and preventing all but the largest incoming objects from impacting the Venusian surface. Craters smaller than 1.5 kilometers appear not to exist on Venus. Many craters of this size are distinctly noncircular and form groups or clusters of craters. This phenomenon is attributed to the impactor's becoming fragmented as it passes through the dense Venusian atmosphere and hitting the surface like a shotgun blast rather than like an artillery shell.

The variety of morphologies seen in Venus's impact craters tends to depend on their size. As the diameter of Venusian craters increases, changes in crater morphology take place and appear to correlate directly with Venus's surface gravity and dense atmosphere. Much of the morphology of Venusian impact craters is unique. Craters larger than 11 kilometers in diameter exhibit morphological characteristics similar to comparable complex craters on other planets: a circular shape, surrounding ejecta blankets, well-defined rims, terraced walls, central ring structures or central peak complexes, and, in the largest craters, multiple-ring basins. However, smaller Venusian craters tend to display a wide variation in shape and structural complexity—the opposite of the cratering patterns seen on other terrestrial planets and rocky satellites.

The morphological divergence is most directly attributed to the greater atmospheric density of Venus. Large multiring basins on Venus display at least two, and sometimes three or more, rings and near-pristine morphology; are surrounded by blocky ejecta distributed in lobes or raylike patterns; and in some cases produce lavalike flows of ejecta traveling several radii from the crater. The ejecta patterns are attributed to the dense Venusian atmosphere's slowing the travel path and speed of debris exiting the crater during impact. Many of the largest Venusian impact craters appear to have little to no topographic relief. Shallowness of these craters may be linked to Venus's lower gravity, producing slower impact speeds, and the planet's high surface and crustal temperatures, producing a large volume of impact-generated melt that remains in a near-molten state, allowing it to flow over long periods of time and eventually fill in the crater.

It is also suggested large Venusian impacts could trigger the subsequent volcanic or tectonic activity that disguises, or eventually erases, them within the landscape. One of the more difficult aspects of studying Venusian craters is distinguishing impact craters from circular volcanic calderas. High-resolution radar images help in defining the morphology of impact craters versus volcanic features by distinguishing ejecta deposits from lava flows. Unfortunately, lavas generated from impact-triggered volcanism can complicate discerning these structures because the lava flows may infill the impact craters, making them look like calderas.

One of the most unusual phenomena associated with Venusian impact craters is parabolic halos. These halos surround about 10 percent of the youngest craters and usually expand westward. The halos are attributed to the formation of a pre-impact bow-shock wave created by the impactor's

producing strong turbulence as it travels through the dense Venusian atmosphere. The turbulence lifts surface dust high into the air, and then prevailing easterly winds resettle the dust after the impact. Because the halos appear unaffected by volcanic, tectonic, or atmospheric processes, the haloed craters may be no more than 50 million years old, making them useful dating horizons.

KNOWLEDGE GAINED

Identifiable impact craters on Venus are rare—slightly less than one thousand, or approximately 1 crater per million square kilometers—and large craters and basins are uncommon. Impact craters on Venus are randomly distributed and range in size from 1.5 to 270 kilometers in diameter. Venusian impact craters are unusual in that, almost without exception, they appear to be fresh, characterized by sharp rims and well-preserved ejecta deposits. This morphology suggests that the craters have not

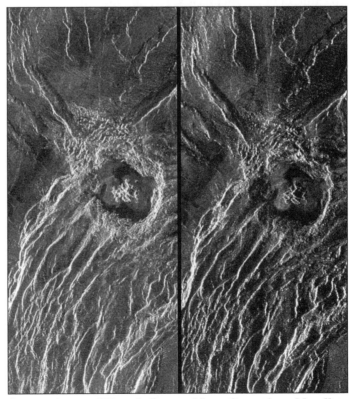

A stereo image of Venus's Geopert-Meyer Crater from Magellan. (NASA/JPL)

been subjected to significant erosional, volcanic, or tectonic activity. Only about 40 percent of Venusian craters appear slightly modified, 5 percent appear embayed by volcanic deposits, and 35 percent appear modified by tectonic activity. Venus's dense atmosphere filters out small meteors, so there is a lack of small impact craters to chip away at larger craters. This situation favors the preservation of existing large craters. Furthermore, while there is currently no hydrogeologic cycle on Venus, there is evidence to suggest that there may once have been liquid oceans of some kind on the surface.

The pristine appearance of craters on Venus makes it appear the surface is both geologically young and of a relatively uniform age. This observation has significant implications for the geologic history of Venus. While resurfacing processes have most likely removed Venusian craters older that 450 million years, the morphology of existing craters is not what is expected for a steady balance between crater formation and crater loss caused by tectonic, volcanic, or erosional process. The unique observation is that Venusian craters of all ages look "fresh," suggesting that most of the present surface characteristics of Venus date from the end of a global resurfacing event that ceased about 450 million years ago.

Because of their lack of weathering, Venusian craters provide a unique opportunity for scientists to study the effects atmospheric variabilities and gravity have in forming planetary surfaces. While the total number of impact craters on Venus are not comparable to those on Mars, Mercury, and the rocky satellites, they do fall into morphological and age classifications similar to those of impact craters on Earth. Crater density and morphology suggest that cratering records of Venus and Earth are similar. Because the cratering data from these two planets are complementary, they provide interpretive guidelines for researching the roles that volcanism, tectonics, and erosional processes play in planetary resurfacing.

CONTEXT

The high temperatue and dense atmosphere of Venus slow incoming projectiles, destroying the smaller, high-velocity objects. This shielding effect influences the size of Venusian craters. Smaller craters appear to be absent on Venus because only large impactors can penetrate the Venusian atmosphere to reach the surface. It is estimated that as many as 98 percent of the craters between 1.5 and 35 kilometers in diameter that could have formed on Venus did not as a result of its dense atmosphere. Venus's high temperature and dense atmosphere also impeded the emplacement of ejecta during cratering by limiting flight distance and decelerating fragments, resulting in lobate ejecta blankets that are sharply defined and make up coarse blocks. Because of Venus's high surface temperature, rocks tend to be softer, less solid, and somewhat viscous. During an impact, these viscous rocks produce large amounts of impact melt, which works to fill the craters and make them topographically low. It is also suggested that large Venusian impacts could trigger regional tectonic or volcanic events by transferring their tremendous heat and shock energies into the planet's thin crust.

Randall L. Milstein

FURTHER READING

Esposito, L. W., E. R. Stofan, and T. E. Cravens, eds. *Exploring Venus as a Terrestrial Planet: Geophysical Monograph 176*. Washington, D.C.: American Geophysical Union, 2007. Addresses the open questions regarding Venus's geology, atmosphere, surface evolution, and future exploration. Includes results from the Venus Express mission.

Lopes, R. M., and T. K. P. Gregg. *Volcanic Worlds: Exploring the Solar System's Volcanoes*. New York: Springer, 2004. A general review of volcanic activity throughout the solar system. Comparisons are made between volcanic activity on Earth and other planets, showing how data from one planet can aid in the understanding of physical processes on another.

Spudis, P. D. *The Geology of Multi-ring Impact Basins: The Moon and Other Planets*. New York: Cambridge University Press, 1993. Although this is a technical book, the chapter on Venus is well illustrated and easy to read, with good comparisons to similar Earth structures.

Trefil, James S. *Other Worlds: Images of the Cosmos from Earth and Space.* Washington, D.C.: National Geographic Society, 1999. A richly illustrated book with exploration mission images of Venus and computer-generated three-dimensional perspectives of Venus's surface.

Uchupi, E., and K. Emery. *Morphology of the Rocky Members of the Solar System.* New York: Springer, 1993. Focuses on the morphology of planets and their satellites and the reasons for the differences and similarities between them. The book's theme is that the solar system should be approached as a single entity, not as a group of individual planets.

See also: Earth's Atmosphere; Earth's Composition; Earth's Magnetic Field: Origins; Earth-Sun Relations; Extraterrestrial Life in the Solar System; Greenhouse Effect; Mars's Atmosphere; Planetary Atmospheres; Planetary Interiors; Planetary Magnetospheres; Planetary Rotation; Planetary Tectonics; Planetology: Comparative; Planetology: Venus, Earth, and Mars; Terrestrial Planets; Venus's Atmosphere; Venus's Surface Experiments; Venus's Surface Features; Venus's Volcanoes.

Venus's Surface Experiments

Categories: Planets and Planetology; Venus

An understanding of the geology of the other planets in the solar system is important for understanding the geologic past and future of Earth. Venus holds many clues to this understanding, including its surface geology, which appears to be mostly igneous and basaltic in nature.

OVERVIEW

The planet Venus is considered one of the terrestrial or Earth-like planets because of its position in the solar system, its planetary diameter, its geology, and other characteristics. Despite being called Earth's twin because of those similarities, Venus is actually very different. The study of the planet Venus has been at best difficult because of its heavy cloud cover.

The best information on the Venusian surface and its soils came early from the Soviets, who focused on exploring the planet, successfully landing six spacecraft on the surface. Even though these craft operated for only limited amounts of time because of the planet's extreme temperatures and the pressure of its atmosphere, the data provided from them have given astronomers and geologists important clues to the soils on Venus. Much of the information obtained by the Soviets can be compared with that known about Earth, and to the data obtained from firsthand examination of the lunar rocks and soils. For example, photographs can be useful in examining the appearances of the soil, rocks, and their distribution. Images taken by Venusian landers can be compared with photographs of similar materials found on Earth and the Moon.

The first of the Soviet landers to provide clues to the Venusian soils was Venera 8, which made the first soft landing on Venus, on July 22, 1972, in a region generally thought to be like the rolling plains of Earth. The probe analyzed the surface and soils directly underneath it with a gamma-ray spectrometer designed to determine the chemical composition of surface material. Results showed that the soils under the Venera 8 were igneous, or volcanic, in origin. The layer was found to be approximately 4 percent potassium, approximately 200 parts per million uranium, and approximately 650 parts per million thorium. The layer was also determined to have a density of approximately 1.5 grams per cubic centimeter (in comparison, water has a density of 1 gram per cubic centimeter at a temperature of 277 kelvins). From these data, astronomers and geologists were able to ascertain not only that the soils under Venera 8 were igneous but also that they were probably similar to the granites or basalts found on Earth.

In 1975, Veneras 9 and 10 provided an even better look at the Venusian soils. Each lander transmitted an image that showed the soil and rocks surrounding it. These photographs re-

Venus Compared with Earth		
Parameter	*Venus*	*Earth*
Mass (10^{24} kg)	4.8685	5.9742
Volume (10^{10} km^3)	92.843	108.321
Equatorial radius (km)	6,051.8	6,378.1
Ellipticity (oblateness)	0.000	0.00335
Mean density (kg/m^3)	5,243	5,515
Surface gravity (m/s^2)	8.87	9.80
Surface temperature (Celsius)	+450 to +480	−88 to +48
Satellites	0	1
Mean distance from Sun		
millions of km (miles)	108 (67)	150 (93)
Rotational period (hrs)*	−5,832.5	23.93
Orbital period	224.7 days	365.25 days

*The minus sign signifies a retrograde rotational period.
Source: National Space Science Data Center, NASA/Goddard Space
  Flight Center.

vealed rocks that were on the average 20 centimeters wide, about 50 to 60 centimeters long, and slablike in appearance. A few of the rocks showed evidence of volcanic origin. Many of the rocks had jagged edges, which demonstrates little erosion, although some did show signs of weathering. This relative lack of erosion surprised many astronomers and geologists. They had believed that, because of the planet's extremes in temperature, atmospheric pressure, wind velocity, and chemical composition, the photographs would show well-eroded landscapes. Astronomer Carl Sagan, among others, hypothesized that low wind velocities at the surface levels of Venus produce little effect on the rock. Apparently, the Venusian surface temperature stays fairly constant and thus does not create much wind. Chemical analysis of the rocks again showed the elements potassium, uranium, and thorium. Nevertheless, the sites differed in the type of rock material. At one Venera lander site, the rocks were basaltic in appearance, similar to those lining Earth's oceans. At the other site, the rocks were more like granite, similar to that found in Earth's mountains. The rocks appear to be relatively young in age. This would indicate that the planet has been geologically active in the geologically recent past. The Venusian soil in the areas observed photographically appeared to be loose, coarse-grained dirt. It was also evident from the photographs that Venus (or at least parts of it) is a dry and dusty planet. Radar images from other Venera missions, as well as Pioneer Venus and the Magellan spacecraft, verified this for the rest of the planet.

The Soviets continued their studies of the Venusian surface with two additional spacecraft, Veneras 13 and 14. These two spacecraft performed similar examinations, but in a much more complex manner. Rather than single images, near-panoramic views of the landing sites were produced. Photographs showed rocks somewhat similar to those found at the Venera 9 and 10 landing sites. Rocks also showed evidence, however, of what appears to be thin layering, ripple marks, and fracturing, especially around Venera 14. Some rocks showed evidence of erosion. On Earth, rocks that show layering—such as sandstone and limestone—are usually sedimentary. Based on the photographs and measurements made by the spacecraft, several Soviet scientists suggested that the Venusian rocks might be sedimentary, but that has not been confirmed.

The possible cause or causes of the erosion remain unknown. In the absence of water, several possibilities have been suggested. These include chemical weathering or erosion caused by nearby volcanism and its resulting ash, dust, and lava. Chemical weathering seems the most likely explanation. Venus's thick atmosphere has cloud layers laced with sulfuric acid, which rains down as a caustic, corrosive agent on the surface.

Both spacecraft collected a cubic centimeter of Venusian soil for analysis. The probes utilized an X-ray source to stimulate emissions from the collected soil samples. This chemical analysis revealed that the samples were similar to basalt in composition, although the basalts differed at the two sites. Near the Venera 13 landing site, the type of basalt found is referred to as leucitic high-potassium basalt, while near the Venera 14 landing site, a tholeiitic basalt, similar to that found on the ocean floors on Earth, was found. The soil itself appeared fine-grained, and the photographs revealed many small rocks. It has been speculated that this also indicates that weathering processes of some type are at work, breaking down larger rocks into smaller ones, eventually reducing them to soil.

Another pair of Soviet probes, the Vega 1 and 2 spacecraft, landed on Venus in June, 1985. Vega 2 results revealed a Venusian soil and surface that are again similar to basalt. Nevertheless, the new data also revealed a surface rich in the element sulfur, which is usually associated with volcanism. This presence has provided another clue to the surface and geology of Venus.

Venera 8 landed about 5,000 kilometers east of an area referred to as the Phoebe region. Veneras 9, 10, 13, and 14 all landed between 900 and 3,000 kilometers east of the raised areas known as Beta Regio and Phoebe Regio. Even though the craft landed on and took samples from an area that could be of the same or similar geologic makeup, they have given astronomers and geologists a good idea of the planet's surface composition.

The probes produced mostly photographic data, although some chemical analysis was conducted on site. Thus, any discussion of soil samples is based on the evidence reported by these spacecraft, since no samples have ever been returned to Earth for detailed study. Spacecraft data have enabled astronomers and geologists to begin to understand not only the surface of the planet Venus and its chemical makeup but also the planet's evolutionary path.

KNOWLEDGE GAINED

It appears that Venus may still be an active planet geologically, which scientists inferred from the discovery of high concentrations of sulfur in Venus's atmosphere. Thus, its soils, for the most part, must be considered with that fact in mind.

Analysis of Venusian rocks around the landers provided scientists with interesting but sometimes confusing data. For example, the fact that most of the rocks appear to lack signs of erosion at first seemed puzzling. An understanding of the weather patterns on Venus and the planet's atmospheric chemistry, however, has led to the development of theories relating the small-scale erosion to a low wind velocity at the surface because of its virtually uniform temperature. The rocks themselves appeared to be mostly igneous in nature. Most igneous samples appeared to be similar to basalt, much like those rocks and materials that line Earth's ocean floors. Some of the rocks resembled granite, like those that form Earth's mountains. However, despite apparent volcanic origins for Venus's crust, some specimens appeared to be sedimentary. This led to further questions which remain unanswered. Although the sedimentation process on Earth is usually accomplished by water, present-day Venus has no water, nor is there any evidence of water in its near past. The origins of this phenomenon remain unknown.

Analyzed samples varied slightly from site to site, as was expected by geologists, since samples on Earth also differ. In fact, variation of Earth samples is greater than that of the limited Venusian ones. Nevertheless, potassium—a key element in igneous and especially basaltic materials—was detected, as were uranium and thorium. Geologically, the rocks are relatively young, presenting additional evidence that Venus is a planet that may be experiencing continuous changes. Fine-grained soils were found at some sites, while coarser soils appeared at others. At one site, at least, smaller rocks led scientists to theorize that erosion does occur on the planet, thus producing soil.

CONTEXT

When the planets of the solar system are categorized, one usually finds two major groupings: the terrestrial or Earth-like planets; and the Jovian or Jupiter-like planets (sometimes also called the "gas giants"). Venus, because of its relative size, atmosphere, position within the solar system, and surface, is naturally among the set of terrestrial planets: Mercury, Venus, Earth, and Mars.

An understanding of the nature of the terrestrial planets, their atmospheres, planetary geologies, and soils can give astronomers and geologists clues to the pasts not only of these worlds but also of our own—revealing how these planets were formed, what geological changes they have undergone, and how they might be related. Venus holds many clues to the formation of the solar system. Unfortunately, observations of the Venusian surface are nearly impossible because of the dense atmosphere that surrounds the planet. Orbiting spacecraft provide information regarding the general geologic contours on the surface—the planet's mountains, valleys, craters, and other surface features—but hard evidence of the nature of the surface, particularly its soil, can come only from the surface of the planet. Prior to the landing of Soviet probes, no information about the Venusian surface existed.

Materials sampled and photographed in the vicinities of the Soviet landers proved to be mostly igneous in nature. Additional on-site chemical analysis showed these materials—both rocks and soils—to be similar to granite or basalt. Basalt-type materials are not unique to the second planet from the Sun. These materials have been found on Mars in the vicinities of the American Viking landers, in samples brought back from the Moon by the American Apollo crews, and by uncrewed Soviet spacecraft. As Soviet spacecraft became more sophisticated and knowledge of the harsh Venusian environment grew, landers were able to provide data on the surface of Venus, among other things. Additional information may provide scientists with clues to the past of the terrestrial planets, part of which is hidden in the Venusian surface and soil. Perhaps more important, Venusian soil information may provide clues to Earth's future, particularly regarding our planet's fragile environment.

Comparative planetology is essential for achieving a more complete understanding of our own planet. As Venus, Earth, and Mars started out relatively similar in the early solar system, and all three are in the habitable zone, why is it then that Venus is devoid of water and hot with a thick atmosphere of carbon dioxide, Earth is capable of supporting life, and Mars has no liquid surface water and is cold, with a thin atmosphere of carbon dioxide? Only when that question is answered will scientists have a clear idea of Earth's complex planetary environment.

Work performed along the way to that understanding has included spacecraft dispatched to the veiled planet Venus by both the Soviet Union and the United States. American spacecraft have only flown by or Venus or studied it from orbit. The National Aeronautics and Space Administration (NASA) has had more interest in exploring Mars than Venus. However, because the Soviets have had only bad luck when it comes to Martian exploration, they have emphasized the study of Venus with flyby craft, landers, orbiters, and even balloons temporarily floating within its hellish atmosphere. At the dawn of the twenty-first century, however, neither NASA nor the Russians had any plans to return to Venus for at least two decades. In the meantime, the European Space Agency's Venus Express began orbiting Venus on April 11, 2006, conducting mapping operations and other scientific investigations of Venus's surface and atmosphere.

Mike D. Reynolds

FURTHER READING

Cattermole, Peter John. *Venus: The Geological Story*. Baltimore: Johns Hopkins University Press, 1996. A comprehensive presentation of the latest findings on Venus, based on Magellan data.

Corliss, William R., ed. *The Moon and the Planets*. Glen Arm, Md.: Sourcebook Project, 1985. A discussion of many solar-system phenomena that cannot be easily explained by prevailing scientific theories. Each anomaly is defined, substantiating data are presented, and the challenge the anomaly presents to astronomers is explained. Examples and references are also listed.

Esposito, Larry W., Ellen R. Stofan, and Thomas E. Cravens, eds. *Exploring Venus as a Terrestrial Planet*. New York: American Geophysical Union, 2007. A collection of articles covering all major areas of planetary research on Venus. Technical.

Fimmel, Richard O., Lawrence Colin, and Eric Burgess. *Pioneering Venus: A Planet Unveiled*. Washington, D.C.: National Aeronautics and Space Administration, 1995. A profusely illustrated scientific and technical publication from NASA that includes the Pioneer Venus data as well as a good deal of information from the Russian spacecraft dispatched to investigate Venus. Illustrations, bibliographic references, index.

Frazier, Kendrick. *Solar Systems*. Rev. ed. Alexandria, Va.: Time-Life Books, 1985. This text contains outstanding color photographs, diagrams, and coverage of the planets of the solar system.

Grinspoon, David Harry. *Venus Revealed: A New Look Below the Clouds of Our Mysterious Twin Planet*. New York: Basic Books, 1998. A thorough examination of the geology of Venus. Incorporates Magellan mapping and other data. Explains the Venusian greenhouse effect. Includes speculation about Venus's past. A must for the planetary science enthusiast who wants an integrated approach to science and history.

Hartmann, William K. *Moons and Planets*. 5th ed. Belmont, Calif.: Thomson Brooks/Cole, 2005. An updated version of a classic text that covers all areas of planetary science. The chapter on Venus covers all fundamental knowledge about the planet and spacecraft exploration of it.

Morrison, David, and Tobias Owen. *The Planetary System*. 3d ed. San Francisco: Pearson/Addison-Wesley, 2003. The authors provide a detailed description of each of the planets and other bodies of our solar system. Some coverage of general astronomy and chemistry is included in introductory chapters. College level.

Snow, Theodore P. *The Dynamic Universe*. Rev. ed. St. Paul, Minn.: West, 1991. A general introductory text on astronomy. Covers historical astronomy, equipment used in astronomy, the solar system, stellar astronomy, galactic astronomy, cosmology, and life in the universe. The book features special inserts, guest editorials, and a list of additional readings at the end of each chapter. College level.

See also: Earth's Atmosphere; Earth's Composition; Earth's Magnetic Field: Origins; Earth-Sun Relations; Extraterrestrial Life in the Solar System; Greenhouse Effect; Mars's Atmosphere; Planetary Atmospheres; Planetary Interiors; Planetary Magnetospheres; Planetary Rotation; Planetary Tectonics; Planetology: Comparative; Planetology: Venus, Earth, and Mars; Terrestrial Planets; Venus's Atmosphere; Venus's Craters; Venus's Surface Features; Venus's Volcanoes.

Venus's Surface Features

Categories: Planets and Planetology; Venus

Enormous strides have been made in understanding the nature of the surfaces of the solid bodies in the inner solar system and the processes that have shaped them. Venus, however, has been a particularly difficult planet to study. The picture that is emerging suggests that Venus may be the only member of the four terrestrial planets, besides Earth, that remains geologically active.

OVERVIEW

Of all the terrestrial planets, Venus is the most similar to Earth in size and geologic composition. At 12,258 kilometers in diameter, it is only slightly smaller (by 511 kilometers) than Earth, and its density is within 2 percent of being identical. It lacks the polar flattening, equatorial bulge, and planetary magnetic field that Earth exhibits.

Geologic study of Venus is exceedingly difficult because of the fact that the planet is perpetually shrouded from view by thick clouds. Its surface has never been photographed from Earth-based telescopes or from spacecraft in orbit above the planet. However, in 1990 the Magellan spacecraft began generating high-resolution radar images of Venus's surface and thereby began producing detailed maps with a resolution of approximately 100 meters. A few panoramic photographs taken by several Soviet spacecraft have revealed the barren, rocky character of Venus's surface in the proximity of their landing sites. However, a very high surface temperature averaging 750 kelvins has limited the operating life spans of spacecraft that have landed on Venus to a maximum of about two hours. Still, Venus has been the subject of very persistent research by space scientists and has yielded enough data about its topography and composition to permit informed speculation about the processes responsible for creating its surface. Scientists' knowledge of the planet's geologic features rests primarily on techniques involving radar imaging, while preliminary impressions of the chemical and structural nature of the surface have been provided through experiments conducted by Soviet spacecraft at several different landing sites.

The Venusian surface is generally smoother than that of any other terrestrial planet. Sixty percent of it lies within 500 meters of Venus's mean radius of 6,051 kilometers. Because Venus has no equivalent to sea level, the mean radius is used as the baseline elevation for topographic measurements. Despite this prevailing uniformity, Venus does have some high mountains and deep valleys. The total range between highest and lowest points on the planet is nearly 14 kilometers, a value that is similar to Earth's.

Planetologists divide Venusian topography into several distinct types of terrain. Rolling plains dominate the globe and form an irregular, planet-girdling area covering more than 70 percent of the surface. About 16 percent of the surface lies below the level of the rolling plains. The remainder is divided among upland plains (0.5-2.0 kilometers above mean radius) and several types of true highlands. Among the latter are the regios, also called domed uplands. These are large, roughly circular areas that rise gently toward their centers, where they achieve

heights of between 3 and 5 kilometers above the mean radius. They are thought to be situated over interior "hot spots," which have caused the surface to bubble outward on a gigantic scale. Huge shield volcanoes sit atop many of the domed uplands, a fact that adds credibility to the theory that these landforms are similar to volcanic domes on Earth. Alpha Regio and Beta Regio, the first two surface features identified by Earth-based radar studies, are examples. Surfaces of the domed uplands are generally smooth, like those of the rolling plains, and appear to be the same age. Unlike plains, however, they seem to be crisscrossed by fault lines indicating crustal stresses. Two continent-sized highland areas lie within the mapped region of the surface, but together they account for only 8 percent of the planet's surface. Ishtar Terra and Aphrodite Terra, about the size of Australia and Africa respectively, exhibit a rich variety of landscapes, including two types of mountainous terrain as well as areas of flat and complex plains.

The most common highland topography is one of ridges and valleys that intersect in chevron-shaped or chaotic patterns. This terrain is called tessera terrain, from the Greek word for "mosaic tile," and resembles the deformation patterns that occur on the top of a moving glacier; there are no glaciers on Venus, however. In contrast to the tesserae are more dramatic but less common mountain systems that thrust their peaks 4 to 12 kilometers above the mean radius. The Maxwell Montes region of Ishtar Terra, which includes the highest known point, 11,800 meters above mean radius, is an example

In 1975, the Venera 10 lander survived for sixty-five minutes on the Venusian surface and took this picture of a volcanic-looking surface and part of the spacecraft before it lost function. (NASA)

of the latter. It consists of a series of parallel ridges and valleys 15 to 20 kilometers apart. The Maxwell system appears much like the ridge and valley province of the Appalachian Mountains, although it is substantially higher. Its features, and those of at least three other mountain chains on Ishtar Terra, closely resemble those produced when plates of the Earth's crust are thrust together by tectonic forces.

Contrasting with the mountain ranges are great rifts that cleave the surface to depths of up to 2 kilometers. One huge rift system stretches from east to west for more than 20,000 kilometers and can be traced as a series of chasms along the entire southern edge of Aphrodite Terra. From there, it continues across the rolling plains to link with Beta Regio. Another rift splits Beta Regio and continues to Phoebe Regio. This complex system consists of many related but distinct chasms, the largest of which is 3,500 kilometers long and 100 kilometers wide, with its deepest point lying 2.1 kilometers below mean radius.

Among the most interesting surface features thus far discovered are Venus's large volcanoes. Two excellent examples are Thea and Rhea Mons, which, along with several other volcanoes, rise from the Beta Regio dome. They appear to be situated on a fault that forms one edge of the great rift. The mass of material that has issued from them is greater than the total output of the volcanic mountains that have formed the Hawaiian Island chain. Both are shield volcanoes like Olympus Mons on Mars, formed by chronic, nonexplosive eruptions of lava that flow long distances before solidifying. Radar images show what may be geologically recent lava flows from both Thea and Rhea, and there is intriguing but very controversial evidence that these volcanoes may have been in eruption in the 1950's and again in the 1970's. Since that time no direct evidence of volcanism has been found by even the long-lived Magellan orbiter.

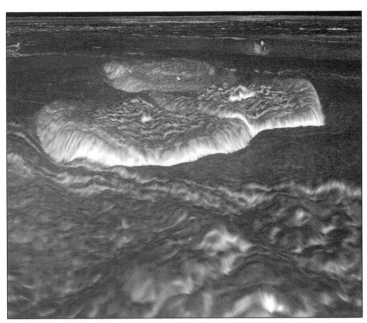

Part of Alpha Regio, a mountainous region with seven domed hills, likely built by outflows of lava. (NASA/JPL)

Impact craters, so characteristic of the surfaces of the Moon and Mercury and even fairly common on Mars, are apparently not nearly so plentiful on Venus. More than one hundred have been observed in radar images, ranging in diameter up to a maximum of 144 kilometers. However, the number of craters discovered is considered low, and the largest crater is only modest in size. These facts are considered to be important evidence that the present Venusian surface is not an ancient one. An old surface should bear numerous scars of past encounters with meteors, comets, and asteroids, as is the case with the Moon and Mercury.

Another class of circular geologic feature bearing superficial similarity to impact craters is apparently an unrelated phenomenon. This group comprises the so-called coronae, of which at least eighty have been found. They average 500-800 kilometers in diameter, but their depth is only 200-700 meters, a fact that is inconsistent with an impact origin. Many researchers interpret the coronae to be collapsed bubbles in the crust, caused by localized heating from hot spots in the mantle beneath.

Many Soviet spacecraft failed, but four successfully took panoramic photographs of their

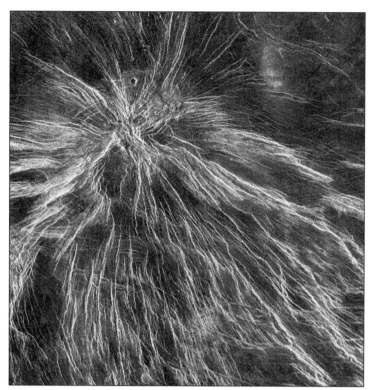

Graben—long, linear depressions usually found between parallel faults—are visible in Venus's Themis Regio. These particular graben form a nova, or a series of graben radiating from a central area; about fifty such novae have been found on the Venusian surface. (NASA)

landing sites. These pictures are remarkably similar in showing barren landscapes dominated by flat-topped rocks with relatively little loose, fine-grained material that might be described as soil. Closer inspection of images of the rocks shows that they seem to be partially exposed outcroppings of a horizontally layered rock mass that exhibits a marked tendency to break into platelike slabs. Additional measurements indicate that the rocks are of low density (1.5 grams per cubic centimeter) and high porosity. They have a bearing strength of only a few kilograms per square centimeter, meaning that they can be broken rather easily. These findings are regarded as surprising, for they are characteristic of sedimentary rocks on Earth. Under close inspection, the panoramic photographs seem to support this conclusion, showing what appear to be striations, indicating ripple marks and crossbedding, two common features of sedimentary deposits. Even the electrically nonconductive properties of the rocks, as revealed by their radar reflectivity, agrees well with the behavior of sedimentary rock. If the surface rocks are indeed of sedimentary origin, they presumably formed from deposits of windblown sand.

The velocity of surface winds is low by terrestrial standards, not exceeding 1.3 meters per second. However, under Venus's dense atmosphere, which is ninety times heavier than Earth's, this velocity is more than sufficient to move fine-grained materials and raise dust. At a number of sites on the rolling plains, researchers have detected depressions that seem to be filled with volcanic ash, which may become lithified over time by yet unknown processes.

The mass of the planet, the density of its surface materials, and the relative abundance of certain elements present in its rocks all point to the likelihood that Venus, like Earth, experienced a planet-wide "meltdown" early in its history. The result was differentiation, a process in which the lighter elements migrated to the surface and the heavier elements settled toward the center. Escape of the residual heat from that meltdown is presumed to have been the major architect of the surface features observed on Venus, just as it has been on Earth. Whether the interior remains molten has not been determined. Venus unquestionably lacks a planetary magnetic field; because such a field is thought to be generated by planetary rotation around a molten iron core, this lack suggests that the interior of Venus has cooled and solidified. However, the evidence that volcanism has occurred on the surface in recent geologic time contradicts this view. It may be that Venus, which rotates 243 times more slowly than does Earth, simply does not spin fast enough to create the dynamo effect that gives rise to a magnetic field.

METHODS OF STUDY

Study of Venus by optical telescope has not been productive for elucidating the nature of the planet's surface. The first successful attempts to penetrate Venusian clouds were made in 1961. Teams of American, British, and Soviet scientists were able to use large radio telescope antennas to beam radar waves at the planet and receive faint return echoes. Earth-based radar studies of Venus have continued but are seriously limited by the fact that good results can be achieved only when the planet passes near the Earth, at which time Venus always presents the same "face."

To gain a global picture of the Venusian terrain, the National Aeronautics and Space Administration's (NASA's) Pioneer Venus orbiter and the Soviet Venera 15 and 16 spacecraft carried radar-imaging instruments into orbit around Venus. In principle, the American and Soviet spacecraft operated similarly, combining synthetic aperture radar (SAR) imaging with radar altimeter measurements. The SAR images from Pioneer Venus cover 70 percent of the surface but are only about as detailed as those shown by a desktop physical relief globe. Veneras 15 and 16 reached Venus in late 1983. Equipped with larger radar antennas, they were capable of resolving surface details as small as one to two kilometers in size and could measure elevations to within 50 meters. Venera radar images look remarkably like high-altitude black-and-white photographs. Unfortunately, this imaging was obtained only for the northern quarter of the planet (from 90° to 30° north latitude).

Beginning in late 1970, a series of soft landings on Venus were made by Soviet Venera craft equipped to conduct experiments to detect the presence of certain rock-forming minerals. One approach used a gamma-ray spectrometer to detect gamma radiation emitted by radioactive uranium, thorium, and potassium. Venera landers of the 1980's employed an automated drill that bored into the surface to obtain samples not contaminated by chemicals from the atmosphere or from the lander itself. The sample material was transferred to an automated laboratory inside the craft, where it was subjected to X-ray fluorescence. Results showed that the rocks at most of the landing sites appear to resemble basalt, an igneous rock enriched with iron and magnesium. The exact composition, however, varied from site to site and, while not identical to that of Earth basalts, was chemically closer to Earth rocks than to Moon rocks.

At two locations, the experiments detected minerals more characteristic of granite, another common igneous rock. These findings are not seen to be in conflict with evidence that Ve-

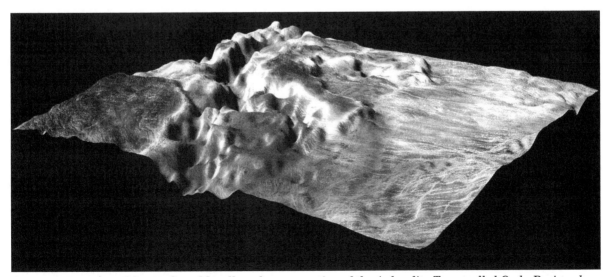

A computer-generated image from Magellan shows a portion of the Aphrodite Terra called Ovda Regio, where plains meet highlands. (NASA/JPL/USGS)

nusian surface rocks may be sedimentary in nature, for the experiments detected the presence and ratio of identifying minerals but not the type of matrix that contained them.

There is no water on the surface of Venus at present, but two discoveries suggest that such has not always been the case. If water was ever plentiful, it must have boiled away, so that the water molecules were dissociated into oxygen and hydrogen. Investigators have sought evidence for the "missing" oxygen and hydrogen, and some believe that they may have found both. Deuterium, a hydrogen isotope, has been detected to be one hundred times more abundant in the Venusian atmosphere than it is in Earth's. Meanwhile, an experiment has shown that oxidized terrestrial basalts, when heated to the Venusian surface temperature, appear identical in visible and micrometer wavelength imagery to the surface rocks of Venus. The likeliest source for the deuterium and the oxygen to oxidize the basalt is dissociated water molecules.

An intriguing possibility exists that Venus may harbor active volcanoes—perhaps the largest in the solar system (the large Martian volcanoes are definitely extinct). Spacecraft and Earth-based observations have detected large amounts of sulfur dioxide, a common volcanic effluent, in the Venusian atmosphere. Moreover, sulfur dioxide content increased dramatically in the 1950's and again in the 1970's.

The first color pictures taken on the surface by the Venera 13 lander seemed to show that the landscape had an orange or amber tint, which proved to be an effect of sunlight filtered through the heavy overcast. Computer processing of the photographs has since shown that, in normal white light, the rocks are a uniform, colorless gray.

Previous spacecraft had performed preliminary radar investigations of Venus's surface, identifying the major types of features on that surface and identifying prominent examples of each. However, high-resolution maps of the entire surface were lacking. The goal of NASA's Magellan spacecraft was to use a synthetic aperture radar in prolonged orbit about Venus to produce a global map at a resolution even in excess of the best contemporary maps of Earth's

surface. Detailed geological interpretations and altimetry data were obtained in the process.

Magellan launched aboard the space shuttle Atlantis on May 4, 1989, and was the primary payload of the STS-30 mission. The spacecraft was deployed from the shuttle's cargo bay, and was dispatched on an trajectory that concluded with orbital insertion about Venus on August 10, 1990. Magellan entered a highly elliptical orbit, often ranging from as little as 300 to as much as 8,500 kilometers above the surface. With Magellan in a polar orbit, the planet Venus rotated underneath it, thereby allowing the spacecraft to image a different ground track on each low pass. The spacecraft turned toward Earth as it climbed toward its highest orbital point and then transmitted the radar imagery it had collected during its low pass over Venus.

Science activities were slightly varied with each of Magellan's six cycles, with the spacecraft's orbit occasionally being altered for different research requirements. During the first cycle, Magellan concentrated on global radar mapping and imaged 84 percent of Venus. Later cycles filled in gaps and concentrated on specific features of interest. In a lower orbital altitude late in its operational mission, Magellan was able to collect precise gravitational data as Venus slightly altered the spacecraft's orbital parameters. Magellan was used to test aerobraking techniques by having the spacecraft fly through the upper portions of Venus's atmosphere; its large solar panels experienced a retarding torque due to atmospheric drag. How the spacecraft responded to the atmosphere indirectly informed scientists about Venus's atmospheric particle density as a function of altitude. With its primary and extended missions completed, flight controllers decided to send Magellan plunging into the upper atmosphere to remove it from orbit. Maneuvers were conducted to force the spacecraft's orbit to decay due to orbital drag. On October 11, 1994, the final spacecraft maneuver was conducted. Controllers lost contact with Magellan the following day. Then, on October 14, Magellan was destroyed in the atmosphere. Although it could not be verified, many believed pieces of descending debris survived long enough to impact the surface.

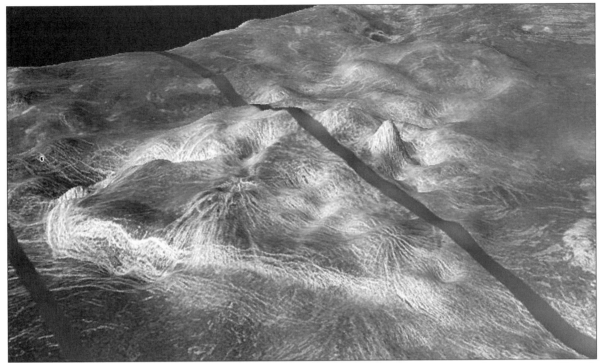

Yavine Corona, one of many coronae, circular regions averaging 500-800 kilometers in diameter, possibly collapsed crustal bubbles caused by localized heating from hot spots in the mantle beneath. Yavine Corona contains two novae. (NASA/JPL/USGS)

Perhaps the most important results of Magellan's intense investigation of Venus was determining a total lack of plate tectonics based on the two primary processes observed on Earth. Instead of continental drift and basin floor spreading, Venus's global rift zones and coronae move as a result of upwelling and subsidence of magma in the planet's mantle. That suggested that Venus's surface is indeed quite young geologically speaking, perhaps less than 800 million years old. The enormous data set from Magellan was made available to interested researchers and individuals on compact disc.

Despite the extensive research with Magellan, many questions remained to be investigated. The European Space Agency (ESA) dispatched its Venus Express spacecraft to Venus in order to examine the planet's atmosphere at infrared wavelengths. Venus Express launched on November 9, 2005, and was inserted successfully into Venus orbit on April 11, 2006. Its Visible and Infrared Thermal Imaging Spectrometer (VIRTIS) was used to identify the amount of sulfur dioxide in the atmosphere between 35 and 40 kilometers and to monitor that constituent for changes in concentration over time that would indicate active volcanism. Venus Express's Spectroscope for Investigation of Characteristics of the Atmosphere of Venus (SPICAV) used stellar occultation methods to determine the identity of atoms and molecules in the upper atmosphere at an altitude between 70 and 90 kilometers. SPICAV saw rapid drops in the amount of sulfur dioxide in the upper atmosphere, strongly indicating that Venus has active volcanoes. VIRTIS was then used to identify hot spots on the surface.

CONTEXT

In order to understand a system as complex as a terrestrial planet, it is necessary to have more than one example of how such planets function and evolve. Hence, the study of geologic processes on another world, far from being simply an esoteric and impractical inquiry, holds promise for improving human understanding of the forces that have acted on Earth and still continue to shape its surface. For this reason, the goal of Venusian geological studies is to discover the relationships and sequences of events that have resulted in the landforms that can be imaged. "Looking at Venus is like running the experiment that produced the Earth a second time," according to Robert Kunzig, senior editor of *Discover* magazine. Indeed, scientists appreciate the opportunity to "run the experiment" again under slightly different conditions in order to see whether their set of explanatory theories can accommodate any observed deviations in the results. Venus presents a marvelous opportunity to test the plate tectonics theory on a planet that has many fundamental similarities to Earth but also exhibits numerous significant differences.

It is generally believed that Earth's loss of internal heat is primarily a result of convection and occurs mainly along the 75,000-kilometer-long mid-ocean ridge. Seafloor spreading that results is responsible for producing a large expanse of young and renewable crust. Even the older continental masses are invigorated by the tectonic activity that is driven by this convective heat loss. Venus seems to have experienced similar horizontal crustal movements in the past and may still be experiencing them. The driving force, however, might be quite different. Most authorities interpret the present surface as having been formed through the release of heat at localized hot spots.

Another fundamental question is whether rocks making up Venus's vast, rolling plains differ significantly in composition from those of the higher terrain. This issue is of interest because it bears on where and how crustal materials originated. It is not inconceivable that Venus could reveal hitherto unknown relationships that may have acted on Earth when the original continental rocks were first solidifying some 3.8 billion years ago.

The mystery of whether Venus once had oceans is of particular interest for two reasons. First, most scientists now accept as true evidence that Earth's own atmosphere is beginning to warm as a result of increases in carbon dioxide content. There is growing concern that atmospheric pollution may trigger a runaway "greenhouse effect," similar to the process that appears to have happened on Venus. If Venus retains any "memory" of conditions before it became so hot, it will only be in the record of the rocks themselves. Second, scientists are still uncertain about how Earth got its abundant water in the first place. If Venus had oceans at a former time, that fact would have significant implications for theories of the origin of planetary water.

A large amount of the Venusian surface was mapped by Pioneer Venus's radar, but its images were good enough only to show gross features of the surface and could not address cause-and-effect relationships. Better imaging has been obtained by the Arecibo radio telescope and the Soviet Venera 15 and 16 spacecraft, but the total area covered was too small to permit generalizations to be drawn. Together, all these data allowed planetary scientists the luxury of asking better questions, which might then be answered by the higher-resolution Magellan spacecraft's synthetic aperture radar imaging system. Magellan produced a spectacular increase in knowledge of Venusian topography. Magellan imagery provided planetary scientists with the information needed to correlate the roles of volcanic activity, tectonic motion, and impact events in the formation and evolution of Venus's surface features. Magellan established that some surface features resulted from tectonics and Venusian volcanoes have been active in recent geologic time, but some key questions are likely to remain unanswered until more complicated surface experiments can be conducted.

Richard S. Knapp

FURTHER READING

Bazilevskiy, Aleksandr T. "The Planet Next Door." *Sky and Telescope* 77 (April, 1989): 360-368. The author, a senior member of the Soviet Venera science team, provides a comprehensive overview of the surface of Venus in clear and nontechnical terms. The article is particularly valuable for its discussion of Venus's medium- and small-scale surface features and for its fair-minded discussion of issues about which there is significant debate or uncertainty.

Bredeson, Carmen. *NASA Planetary Spacecraft: Galileo, Magellan, Pathfinder, and Voyager*. New York: Enslow, 2000. This book, part of Enslow's Countdown to Space series, provides an overview of NASA planetary exploration during the last two decades of the twentieth century. Suitable for all audiences.

Burgess, Eric. *Venus: An Errant Twin*. New York: Columbia University Press, 1985. Probably the general reader's most complete single source of information about Venus and how present knowledge has been obtained. Chapters on the Veneras, Pioneer Venus, and the relationship of Venus's geological history to that of Earth and Mars round out the detailed discussion of the surface and atmosphere. Nearly one hundred well-chosen photographs and diagrams illustrate the text.

Cattermole, Peter John. *Venus: The Geological Story*. Baltimore: Johns Hopkins University Press, 1996. Provides a comprehensive presentation of the latest understanding of Venus, based on Magellan data.

De Pater, Imke, and Jack J. Lissauer. *Planetary Sciences*. New York: Cambridge University Press, 2001. A challenging and thorough text for students of planetary geology, this volume offers an excellent reference for the most serious reader with a strong science background. Provides an in-depth contemporary explanation of solar-system formation and evolution.

Faure, Gunter, and Teresa M. Mensing. *Introduction to Planetary Science: The Geological Perspective*. New York: Springer, 2007. Designed for college students majoring in Earth sciences, this textbook provides an application of general principles and subject material to bodies throughout the solar system. Excellent on comparative planetology.

Fimmel, Richard O., Lawrence Colin, and Eric Burgess. *Pioneering Venus: A Planet Unveiled*. Washington, D.C.: National Aeronautics and Space Administration, 1995. A profusely illustrated scientific and technical publication from NASA that includes the Pioneer Venus data as well as a good deal of information from the Russian spacecraft dispatched to investigate Venus.

Grinspoon, David Harry. *Venus Revealed: A New Look Below the Clouds of Our Mysterious Twin Planet*. New York: Basic Books, 1998. Incorporates Magellan mapping and other data in its coverage of Venusian geology; explains the Venusian greenhouse effect; speculates about Venus's past. A must for the planetary science enthusiast who wants an integrated approach to science and history.

Harvey, Brian. *Russian Planetary Exploration: History, Development, Legacy, and Prospects*. New York: Springer, 2007. Early Russian space programs attempted a large number of Moon, Venus, and Mars investigations. Many were successful, many not. These robust programs are often overlooked. This is their story in one illuminating book about the engineering, development, flight operations, and science returns.

Kerr, Richard. "Venusian Geology Coming into Focus." *Science* 224 (May 18, 1984): 702-703. A brief, easily understood summary of the major geologic features revealed by radar mapping. Interpretation of the Venera 15 and 16 results by leading American planetologists forms the basis of the article. *Science* frequently publishes articles dealing with research on Venus. Many are suitable for lay readers and those with a general science background.

Marov, Mikhail Ya, and David Grinspoon. *The Planet Venus*. New Haven, Conn.: Yale University Press, 1998. Marov was Soviet Venera mission chief scientist, Grinspoon a NASA-funded scientist studying Venus. Together they provide a coordinated description of American and Soviet attempts to learn the secrets of Venus, a planet shrouded in mystery. For both general readers and specialists.

Morrison, David, and Tobias Owen. *The Planetary System*. 3d ed. San Francisco: Pearson/Addison-Wesley, 2003. A full chapter is devoted to the geologic and atmospheric processes on Venus. An additional chapter covers the origin of the solar system, useful to those not familiar with current theories on planetary formation. Each of the other terrestrial bodies also receives a chapter of discussion.

Spanenburg, Ray, and Kit Moser. *A Look at Ve-*

nus. New York: Franklin Watts, 2002. A look beneath the thick clouds of Venus. Written for a younger audience.

Young, Carolynn, ed. *The Magellan Venus Explorers' Guide*. Pasadena, Calif.: Jet Propulsion Laboratory, California Institute of Technology, National Aeronautics and Space Administration, 1990. Prepared as a field and educational guide to the Magellan mission. Published prior to mission launch, this volume contains no spacecraft results, but it does describe the expected research and its value to Venus studies.

See also: Earth's Atmosphere; Earth's Composition; Earth's Magnetic Field: Origins; Earth-Sun Relations; Extraterrestrial Life in the Solar System; Greenhouse Effect; Mars's Atmosphere; Planetary Atmospheres; Planetary Interiors; Planetary Magnetospheres; Planetary Rotation; Planetary Tectonics; Planetology: Comparative; Planetology: Venus, Earth, and Mars; Terrestrial Planets; Venus's Atmosphere; Venus's Craters; Venus's Surface Experiments; Venus's Volcanoes.

Venus's Volcanoes

Categories: Planets and Planetology; Venus

The planet Venus has at least sixteen hundred major volcanoes and many more minor ones, which is more volcanoes than any other planet in the solar system. Most are shield volcanoes, but Venus also has pancake domes and other volcanic features. About 80 percent of Venus's surface has been shaped by some type of volcanic activity. Venus does not have volcanic chains like those formed on Earth from plate tectonics. Comparing volcanic features on Venus and Earth helps us better understand volcanic processes on both planets.

OVERVIEW

Venus is sometimes referred to as Earth's twin sister because its size, mass, and density are similar to those of Earth. Venus's surface conditions, however, definitely make Venus Earth's "evil" twin sister. Owing to a runaway greenhouse effect from the carbon dioxide atmosphere, Venus has a surface temperature hot enough to melt lead. The surface atmospheric pressure is nearly one hundred times what it is on Earth. Thick layers of sulfuric acid clouds veil the surface of Venus. Landers on Mars can last for years, but on Venus they are destroyed by the harsh surface conditions within about an hour. These atmospheric conditions make it impossible to study the surface of Venus using direct optical means or long-term robotic landers. Astronomers must use radar maps rather than optical photographs to study the planet's surface features. Radar maps from both Earth and spacecraft have, however, unveiled the surface of our mysterious twin sister.

Radar maps show that volcanic activity has played a major role in shaping the surface of Venus. Volcanic activity includes not only erupting volcanoes but also lava flows and other activity whereby solid, liquid, or gaseous material escapes from the planet's interior. Volcanic activity is often caused by tectonic activity but can occur independently of tectonic activity. There are more than sixteen hundred large volcanic features on the surface of Venus and possibly as many as hundreds of thousands of smaller volcanic features. In addition, about 80 percent of the planet's surface is covered with flat plains that are probably solidified lava flows. These lava plains formed when lava flooded areas covering thousands of square kilometers and then solidified.

Most of the volcanoes on Venus are shield volcanoes. Shield volcanoes derive their name from their resemblance to ancient warriors' shields lying on the ground pointing upward. Shield volcanoes are often very large, but they have fairly gentle, rather than very steep, slopes. Shield volcanoes form when lava flows out from a single central vent. Rather than forming on the boundaries of tectonic plates, shield volcanoes usually form over a volcanic hotspot. These hotspots are places in the planet's crust where lava wells up from the planet's mantle. When the lava breaks through the surface, it erupts to form a shield volcano. Successive eruptions can form very large shield volcanoes.

Two of the larger known shield volcanoes on Venus are Sif Mons and Gula Mons. They have peak altitudes of about 4 kilometers above the surrounding surface, which compares to Mauna Loa, Earth's largest shield volcano, which rises about 8 to 9 kilometers above the Pacific ocean floor. (The largest shield volcano in the solar system is Mars's Olympus Mons, which towers about 25 kilometers above the Martian surface.) Often the top of a shield volcano will collapse to form a crater, known as a caldera. This collapse occurs when the lava flow retreats back to the planet's mantle, leaving nothing to support the top of the volcano. The calderas formed on Sif Mons and Gula Mons are about 100 kilometers across. Calderas are fairly common on the surfaces of both Venus and Earth. Calderas, however, are not the only types of craters found on Venus. Venus has many large impact craters that formed from meteorite impacts rather than volcanic activity.

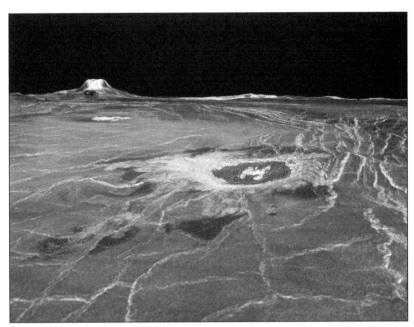

Gula Mons, seen on the Venusian horizon (left), rises to an altitude of about 3 kilometers and is one of the larger shield volcanoes on Venus. Cunitz Crater can be seen in the center middleground. This image was returned by the Magellan spacecraft. (NASA/JPL)

The largest volcanic features found on Venus are coronae. These features are not found on the other terrestrial planets. Coronae, which are approximately circular in shape (hence their name, from the Latin for "crown"), form from an uplifting process. Hot mantle material swells and pushes the crust upward. Coronae usually have associated volcanoes and lava flows. Aine is a large corona on Venus that is about 300 kilometers in diameter. On a larger scale, Lakshmi Planum, which is part of Ishtar Terra—one of Venus's two large continental sized features—is about 1,500 kilometers at its widest point. Lakshmi Planum likely formed from the same process that formed the coronae, but on a larger scale.

Another common type of volcano found on Venus is the lava dome or pancake dome. Venus's lava domes are much smaller than its shield volcanoes, being typically tens of kilometers in diameter or less. They are usually circular and relatively flat—hence the name "pancake dome." They form when lava slowly flows out onto the surface and then flows back. However, a thin crust solidifies on the surface of the lava. When the lava subsides, the crust stays and cracks because it lacks support. Pancake domes are often found near coronae.

On Earth, volcanoes often form at the boundaries of the tectonic plates. Examples are the volcanoes on the western coasts of North and South America and the Mid-Atlantic Ridge, which runs along the Atlantic Ocean's floor between the North American and Eurasian plates. Such volcanoes are not found on Venus. Venus apparently does not have tectonic plates on its crust. Venus does not have plate tectonics similar to Earth's, but it does have tectonic activity. On Earth, plate tectonics is caused by convection currents in Earth's mantle slowly moving the crustal plates horizontally. On Venus, the crust is not divided into plates. Convection currents in the mantle cause vertical rather than horizontal crustal movement on Venus. Coronae are a good example of volcanic features formed on Venus from the crust's vertical tectonic motion.

Are the volcanoes on Venus still active, as on Earth, or are they extinct, as on Mars? Planetary scientists do not yet know the answer to this question. Despite being volcanically active, at any given time few of Earth's many volcanoes are actively erupting. The same would be true on Venus. Hence scientists would not expect to see volcanoes continually erupting on Venus, even if it is still volcanically active. Partly because of the planet's thick cloud layer, no one has observed a volcanic eruption on Venus, but there is some indirect circumstantial evidence to suggest that volcanoes on Venus are still active. Volcanic eruptions emit sulfur dioxide gas. Scientists observe frequent variations in the amount of sulfur dioxide in Venus's upper atmosphere. These variations could be caused by occasional volcanic eruptions spitting sulfur dioxide into the atmosphere. Space probes to Venus have also detected radio outbursts from Venus that are similar to those produced by lightning discharges from erupting volcanoes on Earth. These observations are evidence, but not proof, that Venus's volcanoes are still active. If planetary scientists were to observe a volcano on Venus in the act of erupting, then Venus would join Earth and Jupiter's satellite Io as the worlds in the solar system with still-active volcanoes.

KNOWLEDGE GAINED

Because thick clouds veil the surface of Venus, astronomers for a long time could only speculate about the planet's surface. Prior to the space age, speculations varied: The surface was envisioned as hot and steamy by some, as a hot and dry desert by others. With the coming of the space age, astronomers were finally able to gather data on the surface of Venus. They did not, however, suspect just how hot Venus really was.

The first radar maps of Venus from Earth were made using the Arecibo radio telescope in Puerto Rico, beginning in the late 1970's. Because of the distance to the planet, these images had a relatively low resolution, on the order of a few kilometers. These Earth-based radar maps did, however, allow planetary scientists to observe the large-scale surface features of Venus.

Earth-based radar maps of Venus can reveal only part of the Venusian surface, because during Venus's closest approach to Earth only one side faces Earth. Orbiting spacecraft have therefore been sent to map the surface of Venus. Pioneer Venus 1 went into orbit around Venus in late 1978. This mission was the first orbital mission to use radar to map much of the surface of Venus, with a resolution of about 7 kilometers. In 1990, the Magellan mission went into orbit around Venus. The Magellan orbiter made much more extensive and detailed radar maps of Venus. Because it used a polar rather than an equatorial orbit, Magellan was able to map essentially the entire surface, including the polar regions, which were hidden to previous missions. The best resolution of the Magellan radar maps is about 100 meters.

The fact that Venus has volcanic activity, including both volcanic mountains and lava plains, in-

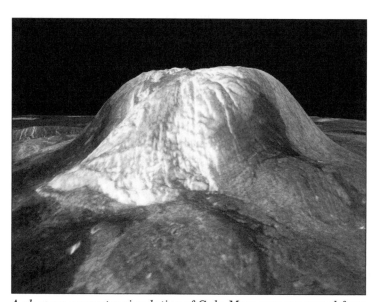

A close-up computer simulation of Gula Mons was generated from data returned by the Magellan spacecraft. It rises 3 kilometers above the area known as Eistla Regio. (NASA/JPL)

dicates that, geologically speaking, the surface of Venus is very young. The surface, not the planet itself, is probably less than a half a billion years old.

CONTEXT

The volcanoes on Venus contribute to its very harsh surface environment. Volcanoes on Earth outgas significant amounts of carbon dioxide gas. Those on Venus and Mars probably do the same. On Earth, biological activity, such as plant respiration, uses the carbon dioxide. Earth therefore has a very small percentage of carbon dioxide in its atmosphere. On Venus, however, the carbon dioxide is still in the atmosphere; 97 percent of Venus's atmosphere is carbon dioxide. All this carbon dioxide produces a runaway greenhouse effect and surface temperatures greater than 700 kelvins.

Jupiter's moon Io is also volcanically active. However the volcanoes on Io differ from the volcanoes on Venus and other terrestrial planets. In addition to rock, Io's composition includes significant amounts of ice.

Venus, Mars, and Earth all have large volcanoes. However, the volcanic and tectonic activity is different on each of these three planets. The differences arise from differences in size and internal heating of the planets. Mars has had the least amount of volcanic activity. Earth's crust is divided into several tectonic plates. Movement of these plates is an important force in shaping the volcanic activity on Earth. Venus does not have a crust broken into several plates. Hence, Venus has much volcanic activity, but it does not have the types of features formed by crustal plate movement. Understanding how volcanic and tectonic activity differs among the various planets is one of the frontiers of planetary science.

Paul A. Heckert

FURTHER READING

Chaisson, Eric, and Steve McMillan. *Astronomy Today*. 6th ed. New York: Addison-Wesley, 2008. Chapter 9 of this very readable introductory astronomy textbook covers the planet Venus. There is a good section on the planet's surface volcanism.

Freedman, Roger A., and William J. Kaufmann III. *Universe*. 8th ed. New York: W. H. Freeman, 2008. Chapter 11 of this introductory astronomy textbook is a complete and readable overview of Mercury, Venus, and Mars, including volcanic activity.

Hartmann, William K. *Moons and Planets*. 5th ed. Belmont, Calif.: Thomson Brooks/Cole, 2005. This textbook on the planets and satellites of the solar system summarizes our understanding of volcanic and other tectonic processes on Venus as well as other planets and moons.

Hester, Jeff, et al. *Twenty-first Century Astronomy*. New York: W. W. Norton, 2007. Chapters 6 and 7 of this well-illustrated astronomy textbook are about the terrestrial planets. Volcanic and tectonic processes are well covered, and the comparison of these processes on different planets is very good.

Morrison, David, Sidney Wolf, and Andrew Fraknoi. *Abell's Exploration of the Universe*. 7th ed. Philadelphia: Saunders College Publishing, 1995. Venus is covered in chapter 15 of this classic astronomy textbook.

Zeilik, Michael. *Astronomy: The Evolving Universe*. 9th ed. New York: Cambridge University Press, 2002. Chapter 9 of this astronomy textbook provides an overview of the terrestrial planets.

Zeilik, Michael, and Stephen A. Gregory. *Introductory Astronomy and Astrophysics*. 4th ed. Fort Worth, Tex.: Saunders College Publishing, 1998. Pitched at undergraduate physics or astronomy majors, this more advanced textbook goes into greater mathematical depth than most introductory astronomy texts. Chapters 4 and 5 cover the basic principles of the terrestrial planets, including volcanic processes.

See also: Earth's Atmosphere; Earth's Composition; Earth's Magnetic Field: Origins; Earth-Sun Relations; Extraterrestrial Life in the Solar System; Greenhouse Effect; Mars's Atmosphere; Planetary Atmospheres; Planetary Interiors; Planetary Magnetospheres; Planetary Rotation; Planetary Tectonics; Planetology: Comparative; Planetology: Venus, Earth, and Mars; Terrestrial Planets; Venus's Atmosphere; Venus's Craters; Venus's Surface Experiments; Venus's Surface Features.

White and Black Dwarfs

Category: The Stellar Context

White dwarfs are stars of about one solar mass in the last stage of their lives. They have no more ways to generate energy, so they shine only because they are very hot. As they radiate their energy away, they cool and fade, becoming cold, dark black dwarfs.

OVERVIEW

White dwarfs are a unique class of stars. They have high surface temperatures, at least initially, and low luminosities. This combination means that they are very small, about the size of Earth and thus much smaller than stars in the energy-generating stages of their lives. With a mass approximately the same as the Sun but contained in a sphere about the size of the Earth, they have average densities in the neighborhood of a billion kilograms per cubic meter.

The first white dwarf was discovered in 1862 by the American astronomer and telescope maker Alvan Graham Clark, while testing the 18.5-inch refracting telescope he made for Dearborn Observatory. He found that Sirius (Alpha Canis Majoris), the brightest appearing star in the night sky, had a very faint companion, given the name Sirius B. In 1896, a similar but even fainter companion was discovered orbiting the star Procyon (Alpha Canis Minoris) and was given the name Procyon B. Then Harvard astronomers discovered that the star 40 Eridani had a faint companion similar to Sirius B, which was named 40 Eridani B. Shortly thereafter, Adriaan van Maanen found a similar but still fainter single star (not a part of the binary system), which was subsequently named after him, Van Maanen's star. During the middle of the twentieth century, Willem Luyten of the University of Minnesota found several hundred of these white dwarf stars, as they had come to be called, by looking for faint blue stars with large proper motions. (The large proper motions showed that the stars were not very far away, so their faint appearance meant they really were intrinsically faint. The blue color showed they had hot surfaces, which, together with their faint luminosities, meant they were very small.) Today even more white dwarfs have been found with improved detectors and observing techniques.

By the early 1900's, it was known that these stars were intrinsically very faint, very small, and incredibly dense. Most, but not all, had a bluish-white color, indicating a hot surface. Modern observations have refined these early findings. Sirius B, the first white dwarf to be discovered, has a surface temperature of about 24,000 kelvins, a luminosity about 0.020 times the Sun's, a radius about 0.008 times the Sun's (about 5,600 kilometers, or about 0.9 times the Earth's radius), a mass about 1.1 times the Sun's, and an average density of about 3 billion kilograms per cubic meter (about 3 million times the density of water). The white dwarf 40 Eridani B has a surface temperature of about 12,000 kelvins, a luminosity about 0.004 times the Sun's, a radius about 0.014 times the Sun's (about 9,800 kilometers, or about 1.5 times the Earth's radius), a mass about 0.43 times the Sun's, and an average density of about 200 million kilograms per cubic meter (about 200,000 times the density of water).

The unusual properties of white dwarfs, especially their great densities, initially made many astronomers question whether such stars could really exist or whether there was something wrong with the observations or the analysis of them. The English astrophysicist Sir Arthur Stanley Eddington summed up the scientific community's reactions this way:

The message of the companion of Sirius, when decoded, ran: "I am composed of material three thousand times denser than anything you've ever come across. A ton of my material would be a little nugget you could put in a matchbox."

What reply could one make to something like that? Well, the reply most of us made in 1914 was, "Shut up; don't talk nonsense."

The theoretical work to try to figure out this "nonsense" began with Eddington himself, who in 1924 suggested that the ultrahigh density of white dwarfs might be caused by their extremely high temperatures, which would totally ionize of the matter in them. The resulting bare nuclei and free electrons could be forced into a much smaller volume of space than un-ionized atoms surrounded by orbiting electrons could be packed.

In 1927, Ralph Howard Fowler used the newly developed Fermi-Dirac statistics (named for Enrico Fermi and Paul Adrien Maurice Dirac), and the Pauli exclusion principle (named for Wolfgang Pauli) from quantum mechanics to show that at the extreme densities within white dwarf stars, electrons are completely degenerate. This means that the electrons fill all the "cells" in combined momentum-position phase space up to some maximum momentum, with two and only two electrons in each "cell." Unlike an ideal gas, which releases gravitational energy when it contracts, a sphere of degenerate electrons cannot shrink without an input of energy to increase the momentum of the electrons; shrinking reduces the position part of phase space, and this requires an increase in the momentum part of phase space. Consequently, without an input

An artist's rendition of a comet (right) as it is broken apart by the gravitational field of white dwarf G29 38 (upper left). (NASA/JPL-Caltech/T. Pyle, SSC)

of energy, degenerate electrons behave collectively like an incompressible fluid.

With high enough densities, the upper momentum states of degenerate electrons are relativistic; that is, the fastest electrons move at a speed that is a significant fraction of the speed of light. In the early 1930's, Subrahmanyan Chandrasekhar applied relativity theory to derive an equation of state for relativistically degenerate electrons. The pressure of the degenerate electrons far exceeds the pressure of the nuclei, which still behave like an ideal gas. Thus it is degenerate electron pressure that balances gravity to maintain hydrostactic equilibrium in white dwarfs. Employing the degenerate electron equation of state and the condition of hydrostatic equilibrium, Chandrasekhar was able to compute models for the interior structure of white dwarfs of various masses. He found that the greater is the mass of a white dwarf, the smaller its size is and the greater its density is. This is shown in the data for Sirius B and 40 Eridani B: Sirius B is more massive, but smaller and denser, than than 40 Eridani B. Chndrasekhar's model implied that there is an upper mass limit for white dwarf stars, now called the Chandrasekhar limit, of about 1.4 solar masses. Although only a few white dwarfs in binary systems have had their masses measured directly, they are below and thus consistent with the Chandrasekhar upper mass limit. For his wide-ranging contributions to astrophysics throughout his career, of which his work on the physics of white dwarfs was just the beginning, Chandrasekhar was awarded the Nobel Prize in Physics in 1983.

The spectra of white dwarf stars are difficult to interpret. Their high surface gravity leads to pressure broadening of the absorption lines in their spectra, and in some cases (such as the white dwarf Wolf 489) the lines become so broad and shallow that they almost disappear into the continuous background spectrum. Some white-dwarf spectra display prominent hydrogen lines, others show strong helium lines, and still others have lines representing a mixture of hydrogen, magnesium, potassium, calcium, iron, or a combination of these elements. White dwarf spectra also show a very large Stark effect (named for Johannes Stark), caused by the presence of strong electrostatic fields that result from high densities and the accompanying degeneracy. The spectra of some white dwarfs indicate the presence of magnetic fields up to about 10 million times stronger than that of Earth.

APPLICATIONS

Eddington realized that the large density and high surface gravity of white dwarfs provided a testing ground for a prediction of Albert Einstein's general theory of relativity. According to general relativity, light emitted in the presence of a strong gravitational field (such as exists at the surface of a white dwarf) is redshifted; basically, photons of light lose energy climbing out of the strong gravitational field, thus lengthening their associated wavelengths. At Eddington's request, Walter Adams at Mount Wilson Observatory attempted to measure this gravitational redshift in white dwarf spectra. His measurements were not accurate enough for a definitive confirmation, though they were consistent with the prediction of general relativity.

In 1954, using better instrumentation, D. M. Popper measured the wavelengths of the lines in the spectrum of 40 Eridani B with greater accuracy. He found that, after allowing for the Doppler shift of the spectral lines due to the star's radial velocity, there was a residual wavelength increase of 0.0070 percent. Based on the mass and radius of this star, general relativity predicts its spectral lines should have their wavelengths increased by 0.0057 percent. (These percentage shifts mean that for a spectral line with a "normal" wavelength of 500 nanometers, general relativity predicts its wavelength should be increased by 0.029 nanometer, while Popper's measurements corresponded to an increase of 0.035 nanometer— very good agreement considering the difficulty in accurately measuring the wavelengths of white dwarf spectral lines.) The confirmation of a prediction of general relativity through observation of a gravitational redshift in a white dwarf spectrum was a major achievement.

CONTEXT

All stars spend most of their energy-producing lives as main sequence stars, fusing hydrogen into helium in their cores. White dwarfs and black dwarfs now are understood as stars in the last stages in their lives, with less than about 8 solar masses when first formed. When stars with initial masses below about 8 solar masses exhaust the hydrogen in their cores, they expand to become red giants, eventually fusing helium into carbon and maybe oxygen in their cores. However, they are not massive enough to generate energy by any other nuclear fusion reactions. Strong stellar winds and thermal pulsations in their bloated atmospheres puff off their outer layers as expanding shells of gas called planetary nebulae (a term that has nothing to do with planets but, rather, originated in the 1800's when, with the telescopes available then, these objects looked round, like planets, and fuzzy, like nebulae). The stars remaining at the centers of planetary nebulae are the former cores of red giants, exposed to view as the planetary nebulae expand and dissipate. These central stars of planetary nebulae are progenitors of white dwarfs.

To become white dwarfs, stars must lose enough of their original mass during the last stages of their lives—whether by strong stellar winds, planetary nebulae, or some other mechanism—that their final mass is less than the Chandrasekhar limit of 1.4 solar masses. Stars with initial masses below about 0.25 to 0.5 of a solar mass are not massive enough to ignite helium fusion in their cores and probably will never become red giants or planetary nebulae; instead they may progress slowly from the main sequence directly to the white dwarf stage, taking perhaps hundreds of billions to trillions of years to do so.

White dwarf stars can no longer generate energy through any nuclear fusion process. They cannot release gravitational energy by contracting because their electrons are degenerate. They shine only because they are very hot, with central temperatures perhaps as high as 100 million kelvins initially. As they shine, they radiate their energy away, slowly cooling and fading, like an ember plucked from a fire. At the initial high temperatures, the atomic nuclei in white dwarfs behave like an ideal gas, but as the stars cool, the nuclei "freeze" into a regular lattice-like pattern, similar to a giant crystal,

through which the degenerate electrons move freely. The white dwarfs, now essentially solid, continue to cool and grow fainter, eventually becoming cold, dark black dwarfs. This is the fate that awaits our Sun several billion years in the future.

However, this may not be the end for all white dwarfs. If a white dwarf is a member of a close binary system with a red giant companion, gas (mostly hydrogen) can be transferred from the red giant onto the white dwarf. The hydrogen that accumulates on the white dwarf's surface may be heated sufficiently to fuse explosively into helium. A shell of hot gas is blasted into space, becoming as much as 100,000 times brighter than the white dwarf itself. This explosive outburst is called a nova. Since the process can repeat over and over again, novae can recur for decades or even centuries.

If the white dwarf in a close binary system is already almost at the Chandrasekhar limit (the maximum mass a white dwarf can have), any additional matter transferred from the companion can push the white dwarf over the mass limit. If this happens, the white dwarf collapses on itself and heats up to about a billion kelvins. The high temperature initiates a series of nuclear fusion reactions that blow the star apart as a Type Ia supernova. (A Type II supernova, on the other hand, is produced by a massive supergiant that explodes once it develops an iron core that collapses.) Because Type Ia supernovae all are produced in the same way by essentially identical objects (white dwarfs gaining enough mass to exceed the Chandrasekhar limit), they reach approximately the same peak luminosity, nearly 10 billion solar luminosities, making them reliable "standard candles." Since their peak luminosity is so high, they can be used as standard candles to determine the distances of galaxies billions of light-years away.

V. L. Madhyastha

FURTHER READING

Chaisson, Eric, and Steve McMillan. *Astronomy Today*. 6th ed. New York: Addison-Wesley, 2008. Very well-written college-level textbook for introductory astronomy courses. Several chapters deal with the later stages of stellar evolution, including white dwarfs, black dwarfs, recurrent novae, and Type Ia supernovae.

Chandrasekhar, Subrahmanyan. "On Stars, Their Evolution, and Their Stability." *Reviews of Modern Physics* 56 (1984): 137-147. Although this is a technical article, it is seminal: the Nobel lecture delivered by Chandrasekhar in 1983. The general reader may easily skip over the mathematical equations and concentrate on the text, which is highly informative and easy to comprehend. Includes references to several important works on the subject.

Fraknoi, Andrew, David Morrison, and Sidney Wolff. *Voyages to the Stars and Galaxies.* Belmont, Calif.: Brooks/Cole-Thomson Learning, 2006. A well-written, thorough college textbook for introductory astronomy courses. Covers not only white and black dwarfs but also the lifetimes of stars generally, including recurrent novae, and Type Ia supernovae.

Freedman, Roger A., and William J. Kaufmann III. *Universe.* 8th ed. New York: W. H. Freeman, 2008. College-level introductory astronomy textbook. Addresses the various paths of star evolution, including white dwarfs.

Kafatos, Minas C., and Andrew G. Michalitsianos. "Symbiotic Stars." *Scientific American* 251 (July, 1984): 89-94. An informative article that describes some of the intriguing aspects of binary systems in which matter is transferred from a red giant to a hot white dwarf.

Luyten, Willem J. "White Dwarfs." In *Advances in Astronomy and Astrophysics.* Vol. 2. New York: Academic Press, 1963. Luyten succinctly introduces all the observed properties of white dwarfs and major problems facing

astronomers. Having spent his career identifying and studying the strange and unusual features of these stars, he presents a readable introduction with a list of some thirty-seven white dwarfs and relevant references.

Schneider, Stephen E., and Thomas T. Arny. *Pathways to Astronomy.* 2d ed. New York: McGraw-Hill, 2008. Very thorough college textbook for introductory astronomy courses. Divided many short sections on specific topics, with several that address the later stages of stellar evolution, including white dwarfs, black dwarfs, recurrent novae, and Type Ia supernovae.

Van Horn, Hugh M. "Physics of White Dwarfs." *Physics Today* 32, no. 1 (1979): 23. An elementary introduction to the mathematically involved subject of what occurs inside a white dwarf. Van Horn provides a list of useful references.

Weidemann, V. "Masses and Evolutionary Status of White Dwarfs and Their Progenitors." *Annual Review of Astronomy and Astrophysics* 28 (September, 1990): 103-137. A lengthy article with numerous references describing developments in observing and modeling the evolution of white dwarfs. The same issue contains an article on the cooling of white dwarfs by Francesca D'Antona and I. Mazzitelli. The articles, although technical, are readable by nonspecialists.

See also: Brown Dwarfs; Gamma-Ray Bursters; Gravity Measurement; Hertzsprung-Russell Diagram; Main Sequence Stars; Novae, Bursters, and X-ray Sources; Nuclear Synthesis in Stars; Protostars; Pulsars; Red Dwarf Stars; Red Giant Stars; Solar Evolution; Stellar Evolution; Supernovae; Thermonuclear Reactions in Stars.

X-Ray and Gamma-Ray Astronomy

Category: Scientific Methods

X-ray and gamma-ray astronomy involves the observation of events in the universe that occur at energies far greater than what is normally shown by visible light or other forms of astronomy. This branch of astronomy promises great advances in our understanding the formation and fate of matter in the universe.

OVERVIEW

Like visible light, both X rays and gamma rays are electromagnetic radiation, but at much shorter wavelengths (and correspondingly higher frequencies). Because of this extreme difference, it is much more difficult to collect large quantities of X rays and gamma rays and to detect a source. In general, the field is called high-energy astrophysics, a name that also refers to cosmic rays (atomic and subatomic particles spewed by various nuclear reactions).

Because of the extremely short wavelengths involved, X rays and gamma rays are both measured in terms of their energy, from thousands to billions of electron volts (eV). Although there is no firm demarcation, X rays span the range of 1,000 to 100,000 eV and gamma rays from 100,000 eV upward (visible light is around 1-2 eV). X rays are emitted by the innermost electrons of an atom releasing energy or through Bremsstrahlung (braking) radiation. Gamma rays, by definition, are emitted by reactions in the nuclei of atoms or by subatomic reactions. There is some overlap between the two spectral bands, and it may be impossible to tell if an emission was from an energetic electron or a nuclear reaction.

Since X rays and gamma rays both pass through solid matter, it is difficult to focus them onto a detector and difficult to make them interact with the matter in the detector itself. A common method of "focusing" high-energy radiation is to use collimators to exclude all radiation except that coming from a particular direction. The basic technique can be imitated by looking through a cluster of straws held at arm's length: The view is narrow and restricted. Essentially, all light that is nearly "on axis" (that is, parallel with the centerline of the straws) passes through, while light that is "off axis" strikes the sides of the straw. The longer the straws, the narrower the field of view. A variety of complex collimation techniques have been developed to allow only radiation from a desired source to fall directly on a detector. While simple, this method generally yields images of the sky that lack the finer resolution of optical telescopes and thus leaves much uncertainty about the location of energy sources.

The most direct solution to this problem is the use of grazing-incidence mirrors to focus "soft" (low energy) X rays. Grazing incidence occurs when radiation strikes a surface at an extremely shallow angle and is reflected rather than being absorbed or scattered. The effect is readily seen when visible light strikes a windshield or pond at less than the critical angle and is reflected to cause glare. Because of the energy of X rays, the angle of incidence must be extremely shallow, typically less than 1°, and the surface must be exceptionally smooth to allow an image to form. X-ray telescopes generally use two reflectors—a parabolic primary and a hyperbolic secondary—to focus the radiation into an image without aberration. To provide the shallow angle of incidence, the mirrors resemble tubes (the segment of the paraboloid or hyperboloid surface is more distant from the focus than with conventional telescopes). This also requires that the secondary mirror be mounted directly behind and precisely aligned with the primary. Such an arrangement is known as a Wolter Type 1 telescope. A number of variations are available. Because only a small region of the radiation is intercepted by the mirrors, modern X-ray telescopes typically will use a nesting scheme in which up to six complete telescopes are built within each other.

Another type of X-ray telescope is the Kirkpatrick-Baez, which uses curved plates of glass in banks one behind the other. This arrangement will focus the light in one axis, then the other. While lacking the fine resolution of Wolter telescopes, the Kirkpatrick-Baez arrangement is well suited to all-sky surveys.

Grazing-incidence telescopes were developed in the 1960's and 1970's. The 1980's saw the development of a radically new approach: normal-incidence X-ray mirrors. In Bragg crystals, the internal structure of a crystal can refract X rays almost as effectively as glass refracts light. Normal-incidence mirrors are built up in layers of microscopic crystals that intercept the X rays and reverse their direction. When laid upon parabolic or hyperbolic surfaces, one can build an X-ray telescope that resembles a conventional reflector telescope and provide comparable resolution for low- to medium-energy X rays.

At higher energies, though, the efficiency of reflectors decreases until virtually no image is being focused. At very high energies, a variation on the pinhole camera technique may be used. In this approach, the front end of the telescope is a plate punctured by holes in a quasi-random pattern. Each pinhole allows an X-ray image to be projected onto the detectors behind the plate. However, because so many images are projected at once, the scene must be mathematically deconvolved (decoded) to produce a true picture of the scene.

There is no way to focus gamma rays. Their energies are simply so high that they would pass unhindered through the mirrors. Instead, gamma-ray "telescopes" constitute instruments that are designed to detect interactions of gamma rays with matter.

X rays and gamma rays, because of the energies involved, are detected less directly than light; specifically, the energy yielded by interaction with some intermediate object is detected. The first and simplest X-ray detector is photographic film. Exposure to X rays will cause the film to darken (that is, produce a negative image). Film is used only in X-ray telescopes that can be recovered, such as crewed satellites and sounding rockets;

The Chandra X-Ray Observatory undergoing inspection at TRW Space and Electronics Group, Redondo Beach, California, in 1998. (NASA/TRW)

various electronic detectors are used in uncrewed satellites. The two are complementary in use. Film provides the highest spatial resolution and reveals fine details of the source and its position relative to other objects, but over a broad energy range. Electronic detectors can measure energy levels with a fair degree of accuracy but at the expense of spatial resolution. In both cases, the energy resolution can be enhanced by placing filters in front of the detector, but this generally establishes a low-energy cutoff and higher-energy radiation will still pass through.

Most development work has gone into electronic detectors that measure the energies yielded when an X-ray or gamma-ray photon strikes matter and releases enough electrons to generate a current, or stimulates electrons to release lower-energy photons as visible light, which can then be detected by conventional devices.

Proportional counters are similar to Geiger-tube counters. A high-energy photon enters a gas-filled chamber, intercepts a gas atom or molecule, and generates an ion and an electron. These are attracted to the electrodes (anode and cathode) in the tube and cause an increase in the electrical current that is measured by the instrument's electronics. In its simplest form, the counter simply registers the arrival of a photon regardless of direction. Most proportional counters, though, are actually gangs of many small counters mounted behind collimators to narrow the field of view. They also use shielding to reduce the chance of photons entering from behind the detector and giving a false signal. In some cases, the collimators move so as to make sources wink on and off and thus improve the precision of its location and even to deduce the structure of extended sources.

To increase the chances of detecting an X ray, detectors sometimes are electrically charged so that a single X ray will produce a shower of electrons that is readily detected (comparable to electric eyes that generate an electric current when struck by light). The electrically charged device is a microchannel plate, which is like a microscopic collimator with its channels as a slant. This ensures that an X ray will strike the wall of the channel, which has an electric charge

not quite high enough to cause an electric arc. The incoming X ray provides the needed extra jolt and releases a cascade of electrons that exit the back of the plate to be measured by the wires of a grid behind the detector.

In the late 1980's, progress was made on developing X-ray charge-coupled devices (CCDs) that could provide a more efficient means of creating X-ray images. CCDs are electronic analogs of the retina and comprise thousands of small photosensors that are read individually by computer. Unlike conventional CCDs, an X-ray CCD will be illuminated from behind, again so that the incident X rays will generate light or an electrical charge that is detected by the CCD.

Spectral measurements in X rays use proportional counters, too, but require a grating or crystal to spread the X rays. The familiar visible-light analog is a wedge prism that breaks white light into the color spectrum. With X rays, a similar effect is desired since the intensity of the X rays at different energies is a measure of the activity at the source. Gratings are surfaces that are microscopically ribbed. The degree to which radiation scatters depends on the energy of the photon. A proportional counter can then be moved along an arc and radiation can be measured. Spectral and imaging instruments can be combined by placing a transmission grating or crystal in the telescope's line of sight. The focused radiation then is spread across the imager and its intensity measured. Normally, this technique works only for strong X-ray sources.

In scintillation detectors, the incoming radiation is trapped in a crystal and causes a flash of light, which is measured by a photomultiplier tube. The most sensitive of these instruments require that the crystal be supercooled by liquefied or solid gas so that the body temperature of the crystal does not cause a false reading. Such crystals can be used in spectrometers. A variation of the effect is Compton scattering, in which a high-energy X ray strikes matter and is scattered at lower energy and with release of an electron whose energy corresponds to the original energy. At higher energies, however, the efficiency of this effect declines. Gamma rays can also produce electron-positron (antielectron) pairs that can be detected in spark chambers, which cause sparks between electrified plates

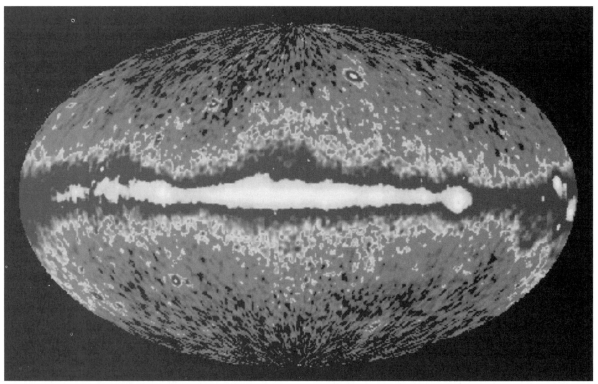

This image, made possible by the Compton Gamma Ray Observatory, shows the entire sky in the gamma range above 100 million electron volts. Without an orbiting observatory such as Compton, such an image would not be possible to acquire from Earth, since the atmosphere protects the planet's surface, and hence terrestrial life, from the penetration of gamma radiation. The bright middle band is the gamma radiation from the Milky Way. (NASA, Compton Gamma Ray Observatory)

or wires. These are then registered by photo-multiplier tubes.

Because gamma rays cannot be focused, different techniques are used to determine the shape of a source. The instrument can be tilted back and forth and the flux change measured to determine its origin. Or, the instrument can be built as a pair, one behind the other, and the two sets of signals collated to determine the origin of each gamma ray.

X-ray and gamma-ray detectors, in general, have anticoincidence detectors that are designed to detect cosmic rays and thus allow the signals they generate to be subtracted from the instrument signal, somewhat like filtering the noise in a radio.

METHODS OF STUDY

X-ray and gamma-ray astronomy is conducted above the atmosphere of the Earth, since the atmosphere absorbs all X rays (even the thin layer above 36,000 meters absorbs soft X rays). While both fields are best conducted from satellites, both started on (and still use) suborbital platforms: X-ray astronomy from suborbital rockets and gamma-ray astronomy from balloons because the lower fluxes required larger, heavier detectors. Suborbital rockets (also called sounding rockets) can expose a payload to the space environment for several minutes, depending on the weight of the instrument and the power of the rocket. Suborbital rockets were a major tool during much of early X-ray astronomy and continue to serve the same purpose as new telescopes are developed and tested. They also have filled the gap between flights of major X-ray satellites. Gamma-ray instruments, though, have to be larger and spend more time at altitude to intercept a sufficient flux from gamma-ray sources, so uncrewed balloons have been their preferred suborbital platform. In either case, satellites are the best means of operation, since they provide essentially indefinite observing time for medium to heavy instruments.

Several X-ray astronomy satellites have been launched, the most notable being the High-Energy Astrophysical Observatories (HEAO), the European Space Agency's X-ray satellite EXOSAT, and the Roetgen X-Ray Satellite (ROSAT). Three of these satellites—HEAO 2, EXOSAT, and ROSAT—carried Wolter Type 1 X-ray telescopes to produce images and spectra of the heavens.

Between 1967 and 1969 Orbiting Solar Observatory (OSO) 3 instrumentation registered 621 gamma-ray photons, confirming a diffuse gamma-ray background, but it must be noted that bursts as discovered by the military Vela satellites remained classified until 1973 rather than immediately becoming a subject of scientific investigation.

In 1972, Small Astronomy Satellite (SAS) 2 discovered an unexpected point source of gamma rays later identified as the neutron star Geminga. As a result of the Cosmic Origins Spectrograph (COS) B observations, the number of gamma-ray point sources was raised to twenty-five between 1975 and 1981. The first extragalactic gamma-ray source was identified as the quasar 3C 273. The study of gamma-ray astronomy greatly expanded during the two years (1979-1981) of HEAO 3 observations. One of its most important detections was 511-kilo-electron-volt radiation coming from the annihilation of electrons and positrons in the center of the Milky Way. The long-lived (1980-1989) Solar Maximum Mission observatory discovered soft gamma rays originating from solar flares.

Gamma-ray astronomy in the 1990's was conducted primarily by the National Aeronautics and Space Administration's (NASA's) Compton Gamma Ray Observatory and the European Space Agency's (ESA's) BeppoSax. However, ground-based Atmospheric Cherenkov Telescopes (ACTs) detected hard gamma radiation coming from blazars; emissions varied on timescales ranging from a few minutes to several hours.

NASA's High Energy Transient Experiment (HETE) 2 and the Reuven Ramaty High Energy Solar Spectroscopic Imager (RHESSI) satellites and ESA's International Gamma-Ray Astrophysics Laboratory (INTEGRAL) continued to expand the nature of high-energy astrophysics research until NASA launched the Swift satellite in 2004, which began finding as many as one hundred gamma-ray bursters annually, and doing so in such a way that the source of the intense gamma-ray emission could be identified quickly enough for other observatories, such as Chandra, to begin rapidly recording the afterglow of the bursters. Swift data provided evidence for mergers of neutron stars or the collapse of a neutron star into a black hole. When NASA launched the Gamma-Ray Large Area Space Telescope on June 11, 2008, gamma-ray astronomers anticipated another tremendous advance in their discipline.

The principal X-ray observatory for the twenty-first century is the Advanced X-Ray Astrophysics Facility (AXAF), conceived as an X-ray complement to the optical Hubble Space Telescope. Once in orbit after delivery by the space shuttle *Columbia* mission STS-93 in July, 1999, AXAF was renamed the Chandra X-Ray Observatory, or CXO. Chandra has a set of six nested Wolter Type 1 mirror pairs (each pair equivalent to a single telescope) to produce X-ray images with resolutions of 0.1 arc second or better. It carries advanced instrumentation derived from the HEAO 2 suite of instruments to produce images and spectra of X-ray objects. Also included is a unique X-ray CCD camera that takes advantage of the latest in solid-state detectors.

The Compton Gamma Ray Observatory (GRO) was launched on the space shuttle *Atlantis* in 1991. It carried four different but complementary gamma-ray instruments. Because gamma rays cannot be focused, they are not arranged in a common focal plane, like detectors in the Hubble Space Telescope, but stand alone on the satellite bus. GRO's instruments spanned a broad energy range, from 10,000 to 10 billion eV. The Burst and Transient Source Experiment (BATSE) recorded gamma-ray "flashes" that have puzzled scientists since 1967. These flashes may be caused by matter falling into neutron stars or black holes, but their unpredictability has hampered efforts to locate them. The burst experiment involved eight detectors that surveyed the entire sky to record and locate bursts so that optical telescopes and other types of telescopes could identify candidates for the burster. The Oriented Scintillation Spectrometer Experiment (OSSE) measured the spectra of objects when gamma rays caused scintillations in crystals. The four scintillation units were tilted to observe "empty" sky next to a source and subtract its background noise in order to obtain the signal of the object being observed. The Imaging Compton Telescope (COMPTEL) was actually two detector arrays using the Compton effect. Light was scattered and recorded as the gamma-ray encounters a scintillation liquid in the first detector, and the gamma ray was reemitted at lower energy to be detected in the same manner at the next level. The energy of the two emitted photons revealed the energy of the original gamma ray, and the locations of the two "hits" pointed back toward the source.

GRO's Energetic Gamma Ray Experiment Telescope (EGRET) was a large instrument designed to detect even low fluxes of the highest-energy gamma rays. Gamma rays entering the telescope struck a tantalum sheet and created an electron-positron pair that then traveled through two spark chambers and finally into a crystal scintillator. A complex anticoincidence system discounted sparks caused by cosmic rays. The energy of the gamma ray was recorded and its direction was revealed by the paths of the two particles it creates.

GRO succeeded in detecting gamma-ray bursters. It has also found antimatter clouds above the Milky Way and uncovered a bizarre new type of star, a "bursting pulsar," that flashes gamma rays. GRO's mission concluded on June 4, 2000, when the massive observatory was directed to plunge through the upper atmosphere in such a way that debris was not likely to fall on populated areas. There were safety concerns about large pieces of debris that might well survive reentry if GRO fell out of orbit in an uncontrolled fashion; however, there were no confirmed reports of debris making it to the deserted South Pacific Ocean area chosen for impact.

CONTEXT

X-ray and gamma-ray astrophysics has become one of the most revealing disciplines in astrophysics since its development following World War II. Extraterrestrial cosmic rays and X rays were detected in the 1800's by balloon crews who carried electrostatic instruments and cloud chambers aloft. Even when their true meaning became known, it was not appreciated that the sources might be caused by stars. As laboratory physics developed an understanding of nuclear fusion and how matter decays, it became obvious that X rays and gamma rays were generated by stars. Nevertheless, it was doubted that the flux (or total energy flow) would be great enough to be measured, at least for any star other than Earth's Sun.

With the availability after World War II of captured V-2 rockets to carry instruments aloft in tests, however, some rudimentary instruments were flown, followed by larger instruments aboard uncrewed balloons starting in the 1950's. It was discovered that as sensitivity increased, what could be seen became richer and more detailed. In the early 1960's, it appeared that the sky was suffused with a strong background glow of X rays. The Uhuru (Swahili for "freedom") satellite carried an all-sky survey detector that had sufficient resolution to detect more than three hundred discrete sources among the background glow. This led to the development and launch of a series of three High-Energy Astrophysical Observatories. HEAO 1 and 3 carried detectors that mapped the entire sky in X rays through gamma rays (HEAO 3 also carried cosmic-ray detectors). HEAO 2 carried the first stellar X-ray telescope and discovered that the X-ray background was composed largely of point sources that could not be distinguished at lower resolutions. This led to the discovery of the X-ray components of known visible objects and of previously unknown objects. In many cases, what was seen in X rays matched very nicely with the visible and radio components. In other cases, it seemed as though two different objects were being viewed.

Comparable work has been done in solar physics. The Orbiting Solar Observatories (OSOs) carried a number of X-ray and gamma-ray instruments in the 1960's and 1970's.

The Skylab space station in 1973-1974 included X-ray spectrometers and imaging telescopes in its array of eight solar telescopes. The Solar Maximum Mission (SMM) satellite (1980-1989) carried X-ray burst detectors and gamma-ray spectrometers to measure the output of the Sun.

In general, astronomers have found that the universe is far more energetic than previously believed. The Crab nebula—the remnant of a star that exploded in 1054—was found to pulse in X rays at the same rate as its visible pulsar, thus suggesting a very compact object. X rays have been found coming from the cores of quasars and most "normal" galaxies. The heart of the Milky Way seems to be the site of matter-antimatter annihilation. Instruments have measured X rays being emitted at an energy of 511,000 eV (or 5.11 keV), which corresponds to the conversion of electrons and positrons (anti-electrons) into energy.

Gamma-ray astronomy provides a different view of the universe. Specifically, it reveals the creation and destruction of matter (properly, its conversion from energy to matter and back) in supernovae, neutron stars, pulsars, black holes, quasars, active galaxies, and other objects. Gamma-ray astronomy has confirmed that the heavier elements are created in the blast furnace of a supernova when a star self-destructs. The famous Supernova 1987A produced gamma-ray lines indicating that cobalt 56 was decaying into iron. Cobalt 56 is unstable and had to have been created shortly before the observation—that is, when the star exploded.

The most important result from high-energy astronomy—as well as from radio and infrared—is the increasing awareness that objects must be studied in terms of their total output rather than as emitters in different spectral bands. What puzzles scientists in one band may be solved in another, or at least a new line of investigation can be illuminated.

Dave Dooling

FURTHER READING

Arny, Thomas T. *Explorations: An Introduction to Astronomy.* 3d ed. New York: McGraw-Hill, 2003. A general astronomy text for the nonscience reader. Includes an interactive CD-ROM and is updated with a Web site. Comets are covered.

Fabian, A. C., K. A. Pounds, and R. D. Blandford. *Frontiers of X-Ray Astronomy.* New York: Cambridge University Press, 2004. For the most serious astronomy reader or students of astrophysics. Covers contemporary research with space-based X-ray telescopes.

Freedman, Roger A., and William J. Kaufmann III. *Universe.* 8th ed. New York: W. H. Freeman, 2008. College-level introductory text covering the field of astronomy. Contains descriptions of astrophysical questions and their relationships. Informative.

Hirsh, Richard F. *Glimpsing an Invisible Universe: The Emergence of X-Ray Astronomy.* New York: Cambridge University Press, 1983. History of X-ray astronomy through the High Energy Astrophysical Observatories. Written for the well-informed reader.

Maccarone, Thomas J. *From X-Ray Binaries to Quasars: Black Holes on All Mass Scales.* New York: Kindle, 2006. Provides descriptions of high-energy processes that produce X-ray emissions. Covers the cosmological significance of quasars, black holes, and other high-energy objects.

McLean, Ian S. *Electronic and Computer-Aided Astronomy: From Eyes to Electronic Sensors.* Chichester, England: Horwood, 1989. A survey of the history and applications of electronics and astronomy. While much of the book is a technical survey, several chapters provide a general history and explanation of CCDs.

Schlegel, Eric M. *The Restless Universe: Understanding X-Ray Astronomy in the Age of Chandra and Newton.* New York: Oxford University Press, 2002. Covers X-ray astronomy for early rocket-launched studies to the Chandra and XMM-Newton observatories. Explains the cosmological implications of current X-ray astronomy research. For the serious layperson.

Trumper, Joachin, and Gunther Hasinger. *The Universe in X Rays.* New York: Springer, 2008. Reviews the development of X-ray astronomy and its impact upon advancements in astrophysics and cosmology. Covers

ROSAT, RXTE, BeppoSax, Chandra, and XMM-Newton results.

Tucker, Wallace. *The Star Splitters: The High Energy Astronomy Observatories*. NASA SP-466. Washington, D.C.: Government Printing Office, 1984. History of the HEAO satellite program and an overview of its results. Includes descriptions of X-ray optics and detectors.

Verschuur, Gerrit L. *The Invisible Universe: The Story of Radio Astronomy*. New York: Springer Praxis, 2006. Provides a history of developments in radio astronomy, and along the way describes the discovery of pulsars, quasars, and radio galaxies. Suitable for a general science course in college as well as for astronomy majors as background information.

Weeks, T. C. *Very High Energy Gamma Ray Astronomy*. New York: Taylor & Francis, 2003. Covers gamma-ray astronomy through results from the Compton Gamma Ray Observatory. For students of either theoretical or experimental high-energy astrophysics.

See also: Archaeoastronomy; Coordinate Systems; Earth System Science; Gamma-Ray Bursters; Gravity Measurement; Hertzsprung-Russell Diagram; Infrared Astronomy; Neutrino Astronomy; Optical Astronomy; Radio Astronomy; Telescopes: Ground-Based; Telescopes: Space-Based; Ultraviolet Astronomy.

Appendixes

Glossary

Absolute magnitude: The brightness of a star or other celestial body measured at a standard distance of 10 parsecs. *See also* Apparent magnitude, Luminosity, Parsec.

Absolute space, absolute time: An absolute idea that there is an overlying stationary structure in space and time which never changes, and against which all events and objects can be measured. The idea now generally is considered outdated in terms of relativity theory.

Absolute temperature scale: A temperature scale which sets the lowest possible temperature (absolute zero, or the temperature at which molecular and atomic translational motion stops) at zero. *See also* Kelvin.

Absolute zero: The temperature at which all translational motion of atoms and molecules ceases.

Absorption spectrum: An electromagnetic spectrum that shows dark lines which result from the passage of the electromagnetic radiation through an absorbing medium, such as the gases found in a star's atmosphere. The resulting *absorption lines* are characteristic of certain chemical elements and reveal much about the composition of the star's atmosphere. *See also* Electromagnetic spectrum, Emission spectrum, Spectrum.

Acceleration: The change in the velocity of an object divided by the time required for the change to occur; commonly expressed in units of meters per second squared.

Acceleration of gravity: The average acceleration of an object which is in free-fall near the Earth's surface, ignoring air resistance; this approximately constant acceleration has a value of 9.8 meters (32.15 feet) per second squared.

Accretion: The accumulation of matter that can result eventually in the formation of a planet or smaller-sized body. *See also* Condensation.

Accretion disk: A disk of material that spirals in toward a black hole or other compact object.

Achondrite: A stony meteorite that contains mostly silicate minerals and a small amount of metal formed from the cooling of molten rock.

Acquisition: The detection and tracking of an object, signal, satellite, or probe to obtain data or control the path of a spacecraft. *See also* Star tracker.

Active galaxy: A galaxy that contains a compact, highly energetic nucleus.

Aerodynamics: The study of the behavior of solid bodies, such as an airplane, moving through gases, such as Earth's atmosphere.

Aerography: The branch of meteorology that collects atmospheric data for the production of weather charts.

Aeronautics: The study of aircraft and aerodynamic spacecraft and the flight of these human-made objects in the Earth's atmosphere.

Aeronomy: The study of the physics and chemistry of the atmospheres of Earth and other planets.

Aerospace: The space extending from Earth's surface outward and beyond the atmosphere; also refers to the engineering discipline that designs and builds craft capable of operating in both the atmosphere and vacuum of space.

Airglow: A faint glow emitted by Earth which results from interaction between solar radiation and gases in the ionosphere, perceived from space as a halo around the planet. Airglow is known for interfering with Earth-based astronomical observations, making space-based telescopes desirable.

Albedo: The fraction of incident light that is reflected from planets, moons, and asteroids.

Alpha particle: A helium nucleus emitted during the radioactive decay of uranium, thorium, or other unstable nuclei.

Altitude: The distance of an object directly above a surface. Also, the arc or angular distance of a celestial object above or below the horizon.

Andesite: A type of volcanic igneous rock intermediate in composition and density between granite and basalt.

Anemometer: An instrument for measuring wind speed.

Angstrom (Å): One ten-thousand-millionth of a meter; a unit used to measure electromagnetic wavelengths.

Angular momentum: A property of a rotating body or a system of rotating bodies. For a discrete body, it is calculated as the product of the mass and the square of distance of the body from the axis about which rotation takes place. For a body with continuous distribution of mass, this calculation involves an integration over the body of distance squared times differential mass. In an analogy to translational motion where force results in a change in linear momentum, in rotational motion a torque results in a change in angular momentum. However, if a system has no net torque, then conservation of angular momentum applies, as it does for bodies orbiting the Sun or the planets. Angular momentum can also be expressed as the product of *moment of inertia*, a body's inertial resistance to rotational acceleration under the application of a torque, and *angular velocity*. When a planet is closest to the Sun, a point in its orbit when its distance is smallest, then its angular velocity is greatest. When a planet is farthest from the Sun, a point in its orbit when its distance is greatest, then its angular velocity is smallest. The latter two sentences describe Kepler's second law of planetary motion: The radius vector to a planet sweeps out equal areas in equal times.

Annular solar eclipse: An eclipse occurring when the apparent size of the Moon is slightly less than that of the Sun, so that even at points where the Moon is seen to move directly across the center of the Sun, the eclipse is not total, since a ring or annulus of the Sun can always be seen.

Anorthosite: A low-density igneous rock, consisting mostly of plagioclase feldspar, that comprises most of the outer crust of the Moon.

Anthropic principle: The philosophical viewpoint that the universe is structured so that galactic and stellar evolution inevitably will lead to the evolution of life, including intelligent life.

Antimatter: Matter composed of antiparticles: positrons (like electrons but with a positive charge), antiprotons (like protons but with a negative charge), antineutrons, and so forth. Small amounts of anti-hydrogen have been created for very brief times. The result of an encounter between a matter particle and its antimatter equivalent is mutual annihilation, with their mass converted into a burst of energy.

Antiparticles: The counterpart of elementary particles, having the same mass and spin as their corresponding particle, but with opposite electric charge and magnetic moment. *See also* Elementary particles.

Aphelion: The point in its orbit around the Sun at which an object, traveling an elliptical path, is farthest from the Sun.

Apoapsis: The point in one object's orbit around another at which the orbiting object is farthest away from the object being orbited.

Apocynthion: The point in an object's orbit around the Moon at which it is farthest away from the Moon.

Apogee: The point in an object's orbit around Earth at which it is farthest away from Earth.

Apolune: Apocynthion.

Apparent magnitude: The brightness of a star or other celestial body as seen from a point, such as Earth. The brightness is apparent because stars vary in their distance from Earth. *See also* Absolute magnitude, Luminosity.

Apparent motion: The path of movement of a body relative to a fixed point of observation.

Apparent size: The perceived or angular size of an object as viewed from a specific perspective, regardless of its true linear size.

Archean eon: The older of a two-part division of the Precambrian, the earliest era of geologic history. Also known as the Archeozoic.

Artificial satellite: A human-made satellite or object sent into orbit around a celestial body.

Asteroid: A small, solid body (also known as a minor planet) orbiting the Sun and ranging in size from about 1,000 kilometers in diameter down to a minimum of between 10 and 100 meters. The solar system contains many thousands of these objects, most of them having orbits in between the orbits of Mars and Jupiter, a region known as the asteroid belt. The Trojan asteroids are located 60° ahead of and behind Jupiter in its orbit. Some asteroids cross inside the orbit of the Earth.

Asteroid belt: The region between the orbits of Mars and Jupiter containing the majority of asteroids.

Asthenosphere: A region of Earth's upper mantle that has less rigid and probably plastic rock material that is near to but below its melting temperature.

Astrobleme: The remnant of a large impact crater on Earth; erosion will have altered the superficial appearance, but confirmation can be made from deeper structural damage and the presence of characteristically shattered and shocked rock.

Astronautics: The science and technology of spaceflight, including all aspects of aerodynamics, ballistics, celestial mechanics, physics, and other disciplines as they affect or relate to spaceflight. *See also* Aerodynamics, Aeronautics, Celestial mechanics.

Astronomical unit (AU): The mean distance between the centers of Earth and the Sun: 149,597,870 kilometers (92,955,630 miles). Used for measuring distance within the solar system.

Astronomy: The study of all celestial bodies and phenomena within the universe.

Astrophysics: The branch of astronomy dealing with the chemical and physical properties and behaviors of celestial matter and their interactions.

Ataxites: Iron meteorites that contain a very high nickel content.

Atmosphere: Any gaseous envelope surrounding a planet or star. Earth's atmosphere consists of five layers: the troposphere, stratosphere, mesosphere, thermosphere (which roughly coincides with the ionosphere), and exosphere.

Atmospheric pressure: The pressure exerted by a planet's atmosphere, decreasing with height and increasing with depth. Earth's atmospheric pressure at sea level is approxi-

mately 14.7 pounds per square inch, or 101,325 newtons per square meter, or 1.01325 bars. This pressure often is referred to as "one atmosphere" and can be used to describe the atmospheric pressure of other planets; for example, the atmospheric pressure at the surface of Venus is about 90 atmospheres, meaning 90 times Earth's atmospheric pressure at sea level.

Atom: The smallest particle of an element that can exist alone. Atoms consist of electrons (negatively charged particles) orbiting a nucleus made of protons (positively charged particles) and neutrons (particles with no net charge). The combinations of these particles determine the identity of the atom as a particular chemical element or isotope of that element. The number of protons (same as the number of electrons in a neutral atom) is the atomic number and therefore determines the chemical element as found in the periodic table; the total number of protons and neutrons in the nucleus is the atomic mass number and therefore identifies the particular *isotope* of that element.

Atomic number: The number of protons, or positively charged nuclear particles, in the nucleus of an atom.

Attitude: The orientation of a spacecraft or other body in space relative to a point of reference.

AU. *See* Astronomical unit.

Aurora: The colored lights appearing in the sky near the poles when charged particles issuing from the Sun become trapped in Earth's magnetic field. The arching, spiraling glows result from these particles interacting with atmospheric gases as they follow Earth's magnetic force lines.

Aurora australis: The aurora occurring near Earth's South Pole.

Aurora borealis: The aurora occurring near Earth's North Pole.

Avionics: The electronic devices used on board a spacecraft, or the development, production, or study of those devices.

Axis: The imaginary line around which a celestial body or human-made satellite rotates.

Axis tilt: A tilt in the pole-to-pole line about which a planet rotates, relative to the plane of the ecliptic. For example, Earth's axis tilt is 23.5°.

Azimuth: The arc, or angular distance, measured horizontally and moving clockwise, between a fixed point (usually true north) and a celestial object. *See also* Altitude.

Ballistics: The study of the motion of projectiles in flight, including their trajectories—especially important in the launching and course-correction maneuvers of spacecraft.

Band. *See* Frequency, Hertz.

Bar: A unit of pressure; one bar is defined as 100,000 Newtons per square meter, which is nearly the pressure of the Earth's atmosphere at sea level. *See also* Atmospheric pressure.

Barycenter: The center of mass of a system of two or more bodies.

Baryons: A class of particles that includes protons, neutrons, and the unstable hyperons; all baryons are composed of a combination of three quarks.

Basalt: A fine-grained, dark igneous rock composed chiefly of pyroxenes and feldspars, typically found at or near the surface of differentiated planets and moons.

Basin: A large, typically flat-bottomed crater formed by the impact of a very large body.

Beta decay: A radioactive decay process in which an electron and a neutrino are emitted from an atomic nucleus, transforming a neutron into a proton.

Big bang theory: The cosmological theory that the universe originated from a primordial ultra-hot, ultra-dense state, about thirteen to fifteen billion years ago. The big bang created space and time, matter and energy. Space rapidly expanded, and as it did so, energy and matter rapidly cooled, leading to the expanding universe we observe today. (The name "big bang" was first employed by an early opponent of this theory, Sir Fred Hoyle, who used it in a derogatory sense—but the name stuck.) *See also* Steady state theory.

Big crunch: The eventual recompression of all matter in the universe that may occur if the universe is closed.

Binary star: A star system composed of two stars orbiting their combined center of mass.

Binary stars are termed "visual" if both components can be seen with a telescope, "spectroscopic" if their spectral lines are Doppler shifted alternately toward shorter and longer wavelengths by the orbital motion of the stars toward and away from us, and "eclipsing" if one star passes directly between us and the second star blocking at least some of its light.

Biosatellite: An artificial satellite carrying life-forms for the purpose of discovering their reaction to conditions imposed in the space environment.

Biotelemetry: The remote measurement and monitoring of the life functions (such as heart rate) of living beings in space, and the transmission of such data to the monitoring location, such as Earth.

Black dwarf star: A star that has cooled to the point that it no longer emits visible radiation; the end state of a white dwarf star. *See also* White dwarf star.

Black hole: A celestial body, predicted by Albert Einstein's general theory of relativity, in which gravity is so strong that nothing, not even light, can escape from it. Two types of black holes have been detected observationally. Stellar-mass black holes (or simply stellar black holes) have masses at least several times the Sun's mass and form from massive stars that collapse at the end of their energy-producing lives. Supermassive black holes have masses millions to billions of times the Sun's mass and occur at the centers of many galaxies, including our Milky Way. A possible third type of black hole, mini black holes, having about the mass of a mountain or less, may have formed in the very early universe in the aftermath of the big bang; mini black holes have not yet been detected.

Blueshift: An apparent shortening of electromagnetic wavelengths emitted from a star or other celestial object, indicating movement toward the observer. *See also* Doppler effect.

Bolometric magnitude: The brightness of a star as detected from above Earth's atmosphere and recorded in all wavelengths.

Bow shock: A shock wave (analogous to the shock wave preceding a supersonic aircraft) that is formed at the point where the solar wind (the stream of ionized gases flowing outward from the Sun) encounters a planet's magnetosphere.

Breccia: Rock composed of a random mixture of angular, broken fragments of other rocks and minerals.

Brown dwarf star: An intermediate-mass object, between a planet and a star, about ten to eighty times the mass of Jupiter, which formed like a star, but without enough mass to sustain nuclear fusion as a source of energy.

Caldera: A very large crater formed by the collapse of the central part of a volcano.

Caloris Basin: The largest known structure on Mercury; it is similar to the Moon's Imbrium Basin and was formed by a large impact.

Canopus: The brightest star in the sky after Sirius, visible south of 37° latitude. Canopus is often the target of a spacecraft's star tracker, which uses it as a reference point in steering a course toward the spacecraft's destination.

Carbon dioxide: A molecule consisting of one carbon atom and two oxygen atoms; it is a common constituent of planetary atmospheres.

Carbonaceous asteroid: An asteroid made up principally of carbon-based materials.

Carbonaceous chondrites: A class of stony meteorites found to contain large amounts of carbon in conjunction with other elements; used to date the solar system and to provide clues to the chemical composition of the early solar nebula.

Cassegrain telescope: A type of reflecting telescope, named for its inventor, Guillaume Cassegrain (1672). The telescope contains two mirrors: a concave mirror near its base, which reflects light from the sky onto a convex mirror above it; the convex mirror, in turn, reflects the light back down through a hole in the middle of the concave mirror to the focal point. The Hubble Space Telescope is a variant of the Cassegrain design called a Ritchy-Cretian system that uses aspheric mirrors.

Cassini division: The largest gap in Saturn's rings, directly viewable from Earth-based telescopes.

Catastrophism: The theory that the large-scale features of Earth were created suddenly by catastrophes in the past; the opposite of uniformitarianism. *See also* Uniformitarianism.

Celestial equator: A great circle on the celestial sphere 90° from the celestial poles, separating the northern and southern hemispheres of the sky.

Celestial mechanics: The branch of physics concerned with those laws which govern the motion (especially the orbits) of celestial bodies, both artificial and natural.

Celestial poles: Imaginary points around which the celestial sphere appears to rotate.

Celestial sphere: An imaginary sphere surrounding an observer at a fixed point in space (the sphere's center), with a radius extending to infinity, a *celestial equator* (a "belt" cutting the sphere into two even halves), and *celestial poles* (north and south). By reference to these points on the celestial sphere, the observer can describe the position of an object in space.

Celsius scale: A temperature scale, named for its inventor, Anders Celsius (1701-1744), which sets the freezing point of water at 0° and the boiling point at 100°. Also referred to as the centigrade scale, its increments correspond directly to kelvins. To convert kelvins to degrees Celsius, subtract 273.15. *See also* Kelvin.

Centrifugal force: The pseudo-force tending to impel a body outward from a center of rotation, equal and opposite to the *centripetal force* caused by the inertia of the body.

Centrifuge: A device for whirling objects or human beings at high speeds around a vertical axis, exerting centrifugal force to test spacecraft hardware or train astronauts to withstand the forces of launch.

Cepheid variable: A massive star that has passed its main sequence phase (the greater part of its lifetime) and has entered a transitional phase in its evolution, during which the star expands and contracts, pulsating in brightness. By measuring the star's period of pulsation and estimating from that its absolute magnitude, then comparing the absolute magnitude to the apparent magnitude, astronomers find the distance of the star.

Chandrasekhar limit: The maximum possible mass for a white dwarf star, calculated by Subramanyan Chandrasekhar in 1931 as approximately 1.4 solar masses (later modified upward for rapidly rotating white dwarf stars). When the star's mass exceeds the Chandrasekhar limit, gravity compresses it into a neutron star.

Charge: A property of matter defined by the excess or deficiency of electrons in comparison to protons. Negative charge results from excess electrons; positive charge, from a deficiency of electrons.

Chemical evolution: The synthesis of amino acids and other complex organic molecules—the precursors of living systems—by the action of atmospheric lightning, solar ultraviolet radiation on atmospheric gases, and other sources of energy.

Cherenkov light: Light emitted by a particle that is exceeding the speed of light in the medium through which it is traveling.

Chert: A hard rock of minutely crystalline, and often partly hydrous, silica.

Chondrite: A stony meteorite containing glassy spherical inclusions called chondrules, which are usually composed of iron, aluminum, or magnesium silicates.

Chromosphere: The layer of the solar atmosphere between the photosphere and the corona, several thousands of kilometers thick, which is visible only when the photosphere is obscured, as during a solar eclipse or by using special filters. The term also applies to corresponding regions of other stars.

Circle of illumination: The circle on Earth's surface that bisects Earth and separates the sunlit half from the shadowed half.

Circular orbit: An orbit described by an orbiting body that maintains a constant distance about the body around which it orbits. The orbit's eccentricity is zero as the orbit's pair of foci are identical.

Climate: The sum total of the prevailing long-term weather conditions of an area, determined by such factors as latitude, altitude, and location.

Closed universe: A universe in which all matter will eventually recompress into a tiny volume of space.

Coesite: A high-density type of quartz formed under the pressures and temperatures involved in impact cratering.

Coma: The gaseous envelope surrounding the head of the comet and consisting of evaporated gases from the comet's nucleus.

Comet: A luminous celestial object orbiting the Sun, consisting of a nucleus of water ice and other ices mixed with solid matter, and, as the comet approaches the Sun, a growing coma and tail. The coma is a collection of gases and dust particles that evaporate from the nucleus and form a glowing ball around it; the tail forms as these materials are swept away from the nucleus. Comets appear periodically, depending on the parameters of their solar orbits.

Comet period: The time required for a comet orbiting the Sun to complete a single orbit.

Condensation: A condition in the early solar nebula when hot gases cooled to form solids; a preliminary stage to accretion. *See also* Accretion.

Condensation temperature: The temperature at which gases of the primitive solar nebula condensed into solid particles.

Conjunction: The alignment of two planets or other celestial bodies so that their longitudes on the celestial sphere are the same. *Inferior conjunction* occurs when Mercury or Venus passes between the Earth and the Sun; *superior conjunction* occurs when Mercury or Venus passes behind the Sun as seen from Earth.

Conservation law: A rule of physics that states that the total value of some quantity does not change.

Conservation of angular momentum: The principle that the total angular momentum of a body or group of objects in a system remains constant in the absence of any external torque.

Constellation: A collection of stars which form a pattern as seen from Earth. The stars in these groupings are often quite distant from one another, their main common characteristic being the illusory picture they form (such as the Big Dipper, or Ursa Major) against the backdrop of the night sky. Constellations provide points of reference for astronomers and other stargazers.

Convection: A flow of material resulting from temperature differences that cause warm, light material to rise and cool, dense material to sink.

Convective equilibrium: A state of stellar stability characterized by a fluid, convective transfer of energy from hotter inner regions to cooler outer layers.

Coorbital satellites: Bodies that share the same orbit; these bodies are in a 1:1 orbital resonance.

Core: The central portion of any celestial body, such as planets and stars. Earth's core is located 2,900 kilometers below the planet's surface.

Core-diffracted phases: Those elastic waves incident at the Earth's outer core at a grazing angle that are diffracted and arrive within the shadow zones for direct waves.

Core-reflected phases: Elastic waves that are reflected from the Earth's core-mantle boundary.

Core sample: A sample of rock and soil taken from Earth, the Moon, or another terrestrial planet by pressing a hollow cylinder down into the planet's surface.

Core-transmitted phases: Elastic waves that travel through the Earth's core.

Coriolis force: A pseudo-force that deflects moving objects to the right (Northern Hemisphere) or to the left (Southern Hemisphere) because of the Earth's rotation. The term describes an inertial movement in a noninertial reference frame.

Corona: The outermost portion of the Sun's atmosphere, extending like a halo outward from the Sun's photosphere. The corona consists of extremely hot ionized gases that eventually escape as solar wind. The term is also used to refer to the corresponding region of any star's atmosphere.

Coronagraph: A device for viewing the solar (or another star's) corona, consisting of a solar telescope outfitted with an occulting mechanism to obscure the photosphere, as during a solar eclipse, so that the corona is more easily perceived.

Cosmic background radiation: Relic radiation produced a few hundred thousand years after the big bang, when the early expanding universe had cooled sufficiently (down to a few thousand kelvins) for electrons to join with protons to form neutral hydrogen atoms. Before that time, the free electrons made the early universe opaque to electromagnetic radiation. Once the free electrons were incorporated into neutral hydrogen atoms, the universe became transparent to electromagnetic radiation, and photons were free to fly throughout the expanding universe. When the universe first became transparent, the photons had a distribution of energies and wavelengths characteristic of a thermal, blackbody source at a temperature of several thousand kelvins. As the universe expanded, the photon energies decreased and the wavelengths increased. Today the radiation is observed greatly redshifted (by about a factor of 1,000) into the microwave portion of the spectrum, with wavelengths of approximately 1 million nanometers or 1 millimeter, appearing like thermal, blackbody radiation from a source with a temperature of approximately 3 kelvins.

Cosmic dust: Tiny solid particles found throughout the universe, thought to have originated from the primordial universe, the disintegration of comets, the condensation of stellar gases, and other sources. Also known as interstellar dust.

Cosmic microwave background radiation: Microwave radiation from the glowing of the hot early universe, now cooled by the expansion of the universe to a temperature of about 2.7 kelvins.

Cosmic radiation: Atomic particles, also known as cosmic rays, that are the most energetic known, consisting mainly of protons, along with electrons, positrons, neutrinos, gamma-ray photons, and various atomic nuclei. These particles emanate from a number of sources, both within and beyond the Milky Way, and they bombard atoms in Earth's atmosphere to produce showers of secondary particles such as pions, muons, electrons, and nucleons. If a primary cosmic-ray particle is sufficiently energetic when it hits an atmospheric atom, the secondary particles can reach Earth's surface.

Cosmic string: A hypothetical early concentration of energy that initiated formation of galaxy filaments.

Cosmogony: The study of the origin and nature of the solar system.

Cosmology: The study of the large-scale structure of the universe and its movements, origin, evolution, and ultimate fate.

Coulomb repulsion: A repulsive electrical force experienced between charged particles of similar sign, such as nuclei of elements as a result of their similar positive charges, and that must be overcome for fusion to take place. The magnitude of this force is directly proportional to the product of the charges of the two interacting objects and inversely proportional to the distance between the centers of these two charged bodies or particles.

Covariant: Interdependent according to a particular mathematical rule, such as that which connects the space coordinates and time of an event in relativity.

Crab nebula: An important supernova remnant that contains a pulsar at its center; also known as M1.

Crater: A depression in the surface of a planet or moon caused by the force of a meteorite's fall. Also, the depression that forms at the mouth of a volcano.

Crater morphology: The structure or form of craters and the related processes that produced them.

Crust: The outermost layer of a planet, generally composed of materials removed from the interior by chemical and physical processes.

Crustal differentiation: The process resulting in the origin of continental and oceanic crust through remelting of original, heavier crust.

Crystal: A solid made up of a regular periodic arrangement of atoms or molecules; its form and physical properties express the repeat units of the structure.

Crystallization: The process by which minerals are formed at various temperatures and pressures, resulting in an orderly arrangement of their atoms.

Cumulate: An igneous rock composed chiefly of crystals which accumulated by sinking or floating from a magma.

Cumulus crystals: Dense minerals within liquid magma that accumulate by gravity upon the floor of a magma chamber.

Curie temperature: The temperature above which a permanently magnetized material loses its magnetization.

Curved space: A space which does not obey the rules of Euclidean geometry; may be positively curved or negatively curved.

Cyclotron radiation: The radiation produced by charged particles as they spiral around magnetic field lines at extremely high speeds.

Dark cloud: An interstellar cloud of dust with sufficient density to block the passage of starlight.

Dark energy: The name given to the hypothetical energy that is causing the expansion of the universe to accelerate. According to modern cosmological models, it should amount to approximately 70 percent of all the energy and matter in the universe.

Dark matter: Mass in the universe that does not give off any form of electromagnetic radiation and is thus invisible but is known by its gravitational influence.

Day: The interval of time between two successive passages of the Sun over the meridian of a planet.

Declination: The angular distance north or south from the celestial equator measured along a circle passing through the celestial poles.

Deep space: Regions of space beyond the Earth-Moon system.

Deep space probe: A device launched beyond the Earth-Moon system that is designed to investigate other parts of the solar system or beyond. Sometimes called an *interplanetary space probe* in reference to spacecraft investigating the planets and the space between them.

Degenerate electron gas: An assembly of electrons occupying all the quantum mechanically allowed space, occurring within white dwarf stars; the pressure in an electron degenerate gas does not depend on temperature but only on the density.

Degradation: Erosion from all processes, including wind, rain, and other mechanisms.

Density: The amount of matter that is contained within a given volume of space; mass per unit volume (grams per cubic centimeter or kilograms per cubic meter).

Differentiation: A process that separates materials by some criteria, such as the heating and melting of a terrestrial protoplanet in which denser material sinks toward the center to become the core and less dense material rises toward the surface to become the crust.

Digital imaging. *See* Imaging.

Dipole field: Electrically, the field shape produced prototypically by a pair of electrically charged particles of equal magnitude but opposite sign separated by a small distance relative to the distance at which the dipole is being observed. Magnetically, the field created by the north and south poles which cannot be found alone, but must always occur in pairs.

Dirty snowball model: The model of a comet's nucleus as a dirty snowball, consisting of various frozen ices containing sand and dust grains.

Discontinuity: An abrupt change in some property at a boundary, such as the change in speed of seismic waves at the boundary between the Earth's crust and mantle (called the Mohorovičić discontinuity).

Dissociation: The breaking up of a compound into simpler components, such as the separation of a molecule into its constituent atoms.

Diurnal: Occurring daily. The diurnal motion of a planet or other celestial body is its daily path across the sky as seen from a fixed point such as Earth.

Doppler effect: The effect of an object's motion toward or away from an observer on the wavelength and frequency of electromagnetic or sound waves that it emits. If the object is moving away from the observer, the observed wavelength is longer and the observed frequency is lower compared to what the object emitted. If the object is moving toward the observer, the observed wavelength is shorter and the observed frequency is higher compared to what the object emitted. The faster the speed of the object, the more the wavelength and frequency are shifted away from their emitted values. Since red light is at the long wavelength end of the visible spectrum, and blue and violet light are at the short wavelength end of the visible spectrum, an object moving away from the observer will appear redder (called a redshift) and an object moving toward the observer will appear bluer (called a blueshift). A source of sound will be heard at a higher pitch as it approaches and at a lower pitch as it moves away.

Double star. *See* Binary star.

Draconic month: The period from one crossing of the ecliptic plane by the Moon to the next (in the same direction), about 27.21 days.

Dynamo effect: The movement of an electrical conductor through a magnetic field, producing an electrical current that in turn generates a magnetic field.

Earth day: Twenty-four hours, or the time required for Earth to complete one rotation on its axis.

Earth-emitted radiation: The portion of the electromagnetic spectrum, from about 4 to 80 microns, in which Earth emits about 99 percent of its radiation.

Earth-orbital probe: An uncrewed spacecraft carrying instruments for obtaining information about the near-Earth environment.

Earth tide: The slight deformation of Earth resulting from the same forces that cause ocean tides, those that are exerted by the Moon and the Sun.

Eccentricity: The degree to which an ellipse (or orbital path) departs from circularity. Eccentricity is characterized as "high" when the ellipse is very elongated. For conic sections, eccentricity is zero for a circle, is between zero and one for an ellipse, is unity for a parabola, and is greater than one for a hyperbola.

Eclipse: The obscuring of one celestial body by another. In a lunar eclipse, Earth's shadow obscures the Moon when Earth is situated directly between the Sun and Moon. In a solar eclipse, the Moon is situated between the Sun and Earth in such a way that part or all of the Sun's light is blocked; the total blockage of sunlight (with the exception of the Sun's corona) is called a total eclipse of the Sun.

Ecliptic: The apparent annual path of the Sun on the celestial sphere.

Ecliptic plane: The plane in which Earth orbits the Sun. From Earth, the ecliptic plane is perceived as the Sun's yearly path through the sky.

Ecology: The relationship between organisms and their environment.

Effective temperature: The temperature of a blackbody (an ideal thermal radiator) that emits the same total flux of electromagnetic radiation as the object under consideration.

Einstein cross: A collection of four images of a distant object formed by a massive object acting as a gravitational lens.

Einstein ring: The images of a distant extended object (such as a galaxy) elongated into arcs when viewed past a mass that bend the light rays.

Ejecta: Material thrown out of a volcano during eruption. Also, the material ejected from the crater made by a meteoric impact.

Electrolysis: A process whereby water is broken down into oxygen and hydrogen.

Electromagnetic radiation: A phenomenon processing both wave and particle characteristics (referred to as wave-particle duality). In some situations, it behaves as waves of electric and magnetic fields oscillating perpendicular to each other and the direction of propagation, and traveling at the speed of light. In other situations it behaves as particles called photons that carry energy and momentum, but have no rest mass. The energy and momentum of a photon are directly proportional to the frequency and inversely proportional to the wavelength of the electric and magnetic field oscillations. *See also* Electromagnetic spectrum.

Electromagnetic spectrum: The continuum of all possible electromagnetic wavelengths, from the longest, radio waves (longer than 0.3 meter), to the shortest, gamma rays (shorter than 0.01 nanometer). The shorter the wavelength, the higher the frequency

and the greater the energy. Within the electromagnetic spectrum is a range of wavelengths that can be detected by the human eye, visible light. Its wavelengths correspond to colors: Red light emits the longest-wavelength visible radiation; violet light, the shortest-wavelength radiation. None of these types of electromagnetic radiation is discrete; each blends into the surrounding forms. Detection of nonvisible radiation by special instruments (used in such branches of astronomy as infrared astronomy and X-ray astronomy) reveals much about the behavior of celestial bodies and the origins of the universe.

Electron: An elementary particle which carries a negative charge and a mass about one eighteen-thousandth of a proton. The number of electrons in an atom in its neutral state is determined by its atomic number and therefore is the same as the number of protons in its nucleus. If an atom accepts or loses one or more electrons then it becomes ionized and has a net negative or positive charge, respectively.

Electron volt: The amount of energy gained by an electron when it is accelerated through an electrical potential difference of one volt; one electron volt (symbol eV) equals 1.6×10^{-19} joules.

Electrophoresis: A process for separating cells using a weak electric field, more easily accomplished in space than on Earth.

Element: The simplest chemical substance, made up of atoms of identical atomic number; every element has a unique atomic number.

Elementary particles: The smallest units of matter, characterized by electrical charge, mass, and angular momentum. Among elementary particles are electrons, neutrons, protons, neutrinos, the various mesons, and their corresponding antiparticles (which form antimatter). Photons, the smallest units of electromagnetic radiation, are also considered as elementary particles.

Ellipse: An oval-shaped geometric figure traced by a point moving so that the sum of its distances from two other points (called the foci) remains constant. The eccentricity of an ellipse varies between zero and one. A circle is a special case of an ellipse in which the two foci are the same point, i.e. the center of the circle.

Elliptical orbit: An orbit which departs from circularity, as most orbits do. A *highly elliptical orbit* is one whose apoapsis is much greater than its periapsis, resulting in an orbit that traces out an elongated ellipse.

Emission nebula: A cloud of gaseous material hot enough to be observed by its own emitted light.

Emission spectrum: Also known as emission line spectrum or bright line spectrum, a spectrum consisting of a series of bright, colored lines at certain specific wavelengths, emitted by a diffuse glowing gas. Each chemical element in gaseous form emits its own characteristic set of bright, colored lines at a particular set of wavelengths.

Enantiomer: A particular version of the same kind of asymmetric chemical compound, such as sugars and amino acids, that may be left- or right-handed and so polarize light in a clockwise (D enantiomer) or counterclockwise (L enantiomer) direction, respectively.

Energy: The ability to do work.

Enstatite chondrites: A rare group of meteorites composed of recrystallized agglomerates whose textural features and mineralogy represent conditions of thermal metamorphism under reducing conditions.

Eolian erosion: A mechanism of erosion or crater degradation caused by wind.

Ephemerides: Calculations showing predicted positions of celestial objects, which can include the Sun, Moon, and planets, and times of eclipses, sunrise, and sunset; ephemerides may also plot a particular object, such as a comet or minor planet.

Epicenter: The region at a planet's surface directly above the focus, or hypocenter, of a seismic quake.

Equatorial orbit: An orbit that directly overlies a planet's equator.

Erg: The amount of kinetic energy of a mass of 2 grams moving at 1 centimeter per second; a mosquito in flight possesses about one erg of energy.

Escape velocity: The speed at which an object must travel to escape the gravitational attraction of a celestial body. In order for a spacecraft to leave Earth orbit, for example, its engines must exert enough in-orbit thrust to achieve escape velocity.

Event: A fundamental "point" of space-time, specified not only by a place but also by a time of occurrence.

Event horizon: The boundary beyond which an observer cannot see. Also, the boundary beyond which nothing can escape from a black hole, where escape velocity equals the speed of light and thus nothing, not even light, can escape. Therefore, the event horizon is theoretically the spherical delineation of a black hole. *See also* Escape velocity.

Exclusion principle. *See* Pauli exclusion principle.

Exobiology: The study of the conditions for and potential existence of life-forms beyond Earth.

Exosphere: The outermost region of the atmosphere.

Extraterrestrial life. *See* Exobiology.

F region. *See* Thermosphere.

Faculae (*sing.* facula): Bright spots or streaks on the solar photosphere associated with the magnetic field of the Sun.

False-color image: An image resembling a photograph, created from data collected by instruments (such as an infrared sensor) aboard a spacecraft and deliberately assigned unnatural colors in order to make nonvisible radiation visible or to highlight distinctions. *See also* Imaging.

Faraday cup: A probe flown on spacecraft to determine the energy (and therefore the velocity) of the plasma in the vicinity of the probe.

Fireball: A very large and bright meteor that often explodes with fragments falling to the ground as meteorites; sometimes called a bolide.

Fission (atomic): The splitting of an atomic nucleus into less massive parts, resulting in a great release of energy.

Fission tracks: Regions of damage to a crystal along the path taken by a moving ion, usually a fragment resulting from fission decay or a cosmic ray.

Flat space: A space of any number of dimensions which obeys the rules of Euclidean geometry.

Flight path: The trajectory of an airborne or spaceborne object relative to a fixed point such as Earth.

Fluid mechanics: The study of the behavior of fluids (gases and liquids) under various conditions, including that of microgravity in spaceflight. Understanding fluid mechanics in space is important to the technology of spaceflight and may have applications on Earth as well.

Fluorescence: Re-emission of visible light due to absorption of electromagnetic radiation of a higher energy (usually ultraviolet light) from an external source.

Flyby: A close approach to a planet or other celestial object, usually made by a probe for the purpose of gathering data; the maneuver does not include orbit or landing. Also used to refer to a mission which undertakes a flyby.

Focal ratio (f-number): The ratio of (1) the distance between the center of a lens or mirror and its point of focus (focal-length) and (2) the aperture, or diameter, of the lens.

Focus (*pl.* foci): (1) The region within Earth from which earthquake waves emanate; also called its hypocenter. (2) The two points that help define an ellipse or hyperbola geometrically. (3) The point at which the light refracted by a lens or reflected by a curved mirror forms an image.

Force: A physical phenomenon capable of changing the momentum of an object; the four *fundamental forces* in the universe are gravity, electromagnetism, the strong nuclear force, and the weak nuclear force.

Frame of reference: A particular position, moving or stationary, from which objects and events are observed.

Fraunhofer lines: Prominent absorption lines in the Sun's spectrum, first observed by Joseph von Fraunhofer in 1814, indicating the presence of certain elements in the Sun's atmosphere. Also used to refer to such absorption lines in other stars' spectra.

Free return trajectory: An orbital flight path which allows a disabled spacecraft to reenter Earth's atmosphere without assistance.

Frequency: The number of times an event recurs within a specific period of time. Frequency characterizes all wave phenomena (such as sound, seismic, and electromagnetic waves) and is determined by dividing the wave speed by its wavelength. Frequency is measured in hertz or multiples of hertz.

Fusion: A thermonuclear reaction in which the nuclei of light elements are joined to form heavier atomic nuclei, releasing energy. It is the process whereby stars form the elements with atomic numbers up to that of iron. It is also the process whereby the stars formed the elements with atomic numbers up to iron.

Gabbro: An igneous rock consisting mostly of pyroxene, feldspar, and often olivine, and containing less than 55 percent silica.

Gain: The increase in power of a transmitted signal as it is picked up by an antenna.

Galactic cluster: An archaic, obsolete, ambiguous designation that could refer to a cluster of several hundred to several thousand stars in the main disk of a spiral galaxy, or to a cluster of galaxies. The star clusters now are referred to as *open clusters* or *open star clusters* to distinguish them from globular star clusters. The clusters of galaxies now are referred to as *galaxy clusters*. In both cases, the clusters are held together by the mutual gravitational attraction of their members.

Galaxy: A collection of millions to trillions of stars, gas, and dust gravitationally bound together.

Gamma radiation: The most energetic form of electromagnetic radiation, with wavelengths less than 0.01 nanometer. The ability of gamma rays to penetrate the interstellar matter of the universe makes them especially valuable to astronomers.

Gamma-ray astronomy: The branch of astronomy that investigates gamma radiation and its sources. Gamma-ray observatories sent into orbit have included the Orbiting Solar Observatory, SAS 2, COS-B, the Compton Gamma Ray Observeratory, and Fermi.

Gaps: Spaces in planetary rings caused by gravitational interactions between the planet, its moons, and the ring particles.

Gegenschein: A patch of faint light about 20°

across, visible from the nightside of Earth at a point opposite the Sun in the ecliptic plane, and caused by the back-scattering of sunlight from interplanetary dust grains in the ecliptic plane. *See also* Zodiacal light.

Geiger counter: A device that detects high-energy radiation (in the form of particles and photons) by means of a tube containing gas and an electric current. The radiation causes the gas to ionize, which is transmitted to the current and detected as a sound or a needle jump.

General relativity: Albert Einstein's theory of gravitation, which extends Sir Isaac Newton's theory by stating that matter curves space and time; gravity is explained as matter moving along shortest paths, called geodesics, in the curved space-time.

Geocentric orbit: An orbit with Earth as the object orbited.

Geochemical sinks: The means by which elements and compounds are removed from the crust, atmosphere, and oceans to be recycled in active chemical cycles.

Geochronology: The study of the time scale of Earth; it attempts to develop methods that allow the scientist to reconstruct the past by dating events such as the formation of rocks.

Geodesy: The science concerned with the size and shape of Earth and its gravitational field.

Geodynamo theories: Theories that explain the cause of Earth's magnetic field and its secular variation in terms of electric charges in the Earth's molten metallic outer core.

Geomagnetic elements: Measurements that describe the direction and intensity of Earth's magnetic field.

Geomagnetism: The external magnetic field generated by forces within Earth; this force attracts magnetic materials, inducing them to line up their magnetic moments along the Earth's magnetic field lines.

Geometry: A set of rules which describes the structure of a region of space; traditional geometries described space as flat, while relativistic geometries describe it as curved.

Geomorphology: The study of landforms on planetary surfaces and the processes responsible for their origin.

Geostationary orbit: A type of geosynchronous orbit which is circular and lies in Earth's equatorial plane, at an altitude of approximately 36,000 kilometers (22,320 miles). As a result, a satellite in geostationary orbit appears to hover over a fixed point on Earth's surface. *See also* Geosynchronous orbit.

Geosynchronous orbit: A geocentric orbit with a period of 23 hours, 56 minutes, 4.1 seconds, equal to Earth's rotational period. Such an orbit is also geostationary if it lies in Earth's equatorial plane and is circular. If inclined to the equator, a geosynchronous orbit will appear to trace out a figure eight daily; the size of the figure eight will depend on the angle of inclination. These orbits are used for satellites whose purpose it is to gather data on a particular area of Earth's surface or to transmit signals from one point to another. Communications satellites are geosynchronous.

Geothermal: Pertaining to the heat of the interior of a planet.

Gigahertz. *See* Hertz.

Glass: A solid consisting of a disordered pattern of atoms, which represents a rapid cooling from a molten state; in meteorites, it is found in chondrules and within the matrix as fragments.

Globular clusters: Spherically shaped congregations of tens to hundreds of thousands of stars that occur throughout the universe, although more often near elliptical galaxies than spiral galaxies such as the Milky Way. It is believed that globular clusters contain the oldest stars.

Granite: A silica-rich igneous rock light in color, composed primarily of the mineral compounds quartz and potassium- and sodium-rich feldspars.

Gravitation *or* Gravity: The force of attraction which exists between two bodies, such as Earth and the Moon. In 1687, Sir Isaac Newton described this force as proportional to the product of the masses of the two bodies and as inversely proportional to the distance between them squared. Although gravitation is the weakest of the naturally occurring forces (the others being electromagnetic and nuclear in nature), it has the broadest range and is responsible for much celestial movement, including orbital dynamics.

Gravitational constant, *G*: The constant (6.67×10^{-11} newton meters^2/kilograms^2) in Newton's law of gravity that determines the gravitational attraction between two bodies.

Gravitational contraction: The shrinking of an object, such as a protostar or protoplanet, due to the gravitational attraction of every bit of matter in the object for every other bit of matter in the object.

Gravitational differentiation: The separation of minerals, elements, or both as a result of the influence of a gravitational field wherein heavy phases sink or light phases rise through a melt.

Gravitational field: The acceleration field (force per unit mass) created by the mass of a celestial body. It assigns to every point in space around the body both an acceleration and a direction of motion that would be experienced by another body placed in the vicinity of the gravitating body that sets up this field.

Gravitational force: An attraction that acts on all masses, causing weight and orbital motion.

Gravitational lens: A large mass that bends the light from distant objects.

Gravitational mass: That property of an object that produces its gravitational attraction for other objects; it is the mass that appears in Newton's formula for the gravitational force between two masses $F = Gm_1m_2/r^2$ (force equals Newton's universal gravitation constant G times the product of the masses of the two gravitationally interacting bodies divided by the square of the distance between the centers of these two masses). In contrast, inertial mass is the property of an object that resists any change in its motion. In all tests, the gravitational mass and inertial mass of an object are individual.

Gravitational potential energy: The energy a body has by virtue of its location in a gravitational field; in a star, the extended outer atmosphere stores a large amount of gravitational potential energy, which can be released during contraction.

Gravity assist: A technique, first used with the Mariner 10 probe to Mercury, whereby a spacecraft uses the gravitational and orbital energy of a planet to gain energy to achieve a trajectory toward a second destination or to return to Earth.

Grazing incidence: Reflection where the incoming radiation strikes at an extremely shallow angle (typically less than 1°).

Great Red Spot: A vast, oval-shaped cloud system occurring at 22° south latitude in Jupiter's atmosphere, rotating counterclockwise. Its red color comes from an unknown substance that convection pulls to the surface; the substance absorbs violet and ultraviolet radiation and consequently delivers a red hue to the Spot. The Great Red Spot has been observed for more than three centuries.

Greenhouse effect: The heating of a planet's surface and lower atmosphere as a result of trapped infrared radiation. Such radiation becomes trapped when there is an excess of certain gases such as carbon dioxide in the atmosphere, which absorbs and reemits infrared radiation rather than allowing it to escape. As a result, the atmosphere acts like a greenhouse, heating the planet.

Groundwater: The water that occurs in the subsurface of a planet; it particularly applies to the subsurface zone that is saturated with such water.

Guest star: Another term, in ancient Chinese astronomical records, for a nova, or "new" star.

Gyroscope: A device which uses a rapidly spinning rotor to assist in stabilization and navigation.

H I region: A region in which the element hydrogen exists primarily in the form of neutral atoms.

H II region: A region in which the element hydrogen is ionized, existing as separate protons and electrons, necessarily at a higher temperature than an H I region.

Hadron: Any particle that participates in the strong interaction; hadrons are divided into baryons, which obey the Pauli exclusion principle, and mesons, which do not.

Heat budget: The balance between incoming short-wavelength solar radiation and outgoing long-wavelength infrared terrestrial or planetary radiation.

Heat death: The eventual loss of all usable energy in the universe that will occur if the universe is open (that is, if it will expand forever).

Heliocentric orbit: An orbit with the Sun at its center or at one of its two foci.

Heliopause: The border between the solar system and the surrounding interstellar space, where the solar wind gives way to interstellar matter and winds.

Heliosphere: The region of interplanetary space extending outward from the Sun in which the solar magnetic field controls the behavior of charged particles.

Hertz: An SI unit of frequency, equaling one cycle per second. Multiples include kilohertz (10^3 hertz), megahertz (10^6 hertz), and gigahertz (10^9 hertz).

Hertzsprung-Russell diagram: A graph widely used in astronomy that depicts stellar properties such as luminosity or absolute magnitude versus spectral type or effective photospheric temperature. Named after Ejnar Hertzsprung and Henry Norris Russell, who independently devised the graph in the early twentieth century.

Hexahedrites: Iron meteorites that contain less than 6 percent nickel content; they usually consist of large single crystals of kamacite and may show Neumann bands when polished.

High-Earth orbit: Any Earth orbit at a relatively great distance from Earth, such as the geosynchronous orbits of telecommunications satellites.

High-gain antenna: A single-axis, strongly directional antenna that is able to receive or transmit signals at great distances.

Highlands: Densely cratered regions on the lunar surface, which when seen with the naked eye take on a pale color; they are primarily anorthositic breccia.

Homogeneity problem: The difficulty of reconciling observations of the extreme uniformity of the cosmic microwave background radiation with the early inhomogeneity required to account for large-scale structures in the universe, such as galaxy clusters and voids.

Homogeneous accretion theory: One of two major theories on Earth's differentiation: that differentiation occurred after the Earth had formed through accretion of debris. *See also* Inhomogeneous accretion theory.

Horizon: The line formed where land meets sky, from the perspective of an observer. In astronomy, the horizon also means the circle on the celestial sphere that is formed by the intersection of the observer's horizontal plane with the sphere. The *particle horizon* is the theoretical limit beyond which particles cannot yet have traveled.

Horizontal branch: A stage of helium fusion in the cores of stars; in star clusters, this appears as a nearly horizontal grouping in the Hertzsprung-Russell diagram.

Hour circle: A great circle on the celestial sphere passing through the celestial poles.

Hubble constant: The constant of proportionality between the recessional velocity of galaxies and their distances in Hubble's Law.

Hubble's law: The principle, articulated in 1929 by Edwin Hubble, that the galaxies are moving away from one another at speeds proportional to their distance, that is, uniformly across time. Hubble deduced this principle from observations of the redshifts in galactic spectra. Along with the discovery of the cosmic microwave background radiation by Arno Penzias and Robert Wilson in 1965, Hubble's law forms the basis for the big bang theory of the expanding universe. *See also* Big bang theory, Doppler effect, Redshift.

Hydrocarbons: Molecules containing hydrogen, carbon, and oxygen.

Hydrostatic equilibrium: A state of stellar stability characterized by a balance between gravitational contraction and thermal pressure expansion.

Igneous rock: A rock formed by the cooling of molten material.

Imaging: The process of creating a likeness of an object by electronic means.

Impact basin: A large circular basin produced by a meteorite impact.

Impact breccia: Angular, fragmental rock produced by meteorite impact.

Impact crater: A generally circular depression

formed on the surface of a planet by the impact of a high-velocity projectile such as a meteoroid, asteroid, or comet.

Inclination. *See* Orbital inclination.

Inclusions: In meteorites, rounded or irregular shapes that have textures and mineralogies suggestive of unmelted aggregates of solid particles, indicative of a primitive origin.

Index of refraction: A physical parameter of matter that describes the degree to which a ray of light will bend upon entering that medium. Specifically the index of refraction n is given by the speed of light in vacuum divided by the speed of light within the medium; since the latter is always slower than the former, the index of refraction of a medium is always greater than one, being equal to unity only for vacuum.

Inertia: The tendency of a body to stay at rest or in uniform motion unless acted upon by an unbalanced force.

Inertial mass: That property of an object that resists any change in its motion; it is the mass that appears in Newton's law of motion $F = ma$ (force equals mass times acceleration). In contrast, gravitational mass is the property of an object that makes it attract other masses. In all tests, the inertial mass and gravitational mass of an object are indistinguishable.

Inflation: The quantum cosmological scenario, which postulates that, almost immediately after the big bang, the universe underwent an enormous expansion.

Infrared: A component of the electromagnetic spectrum that is found just beyond the red part of the visible spectrum.

Infrared astronomy: The branch of astronomy which examines the infrared emissions of stars and other celestial phenomena. Studying the infrared emissions tells astronomers much about the composition and dynamics of their sources. Because infrared radiation cannot readily penetrate most of Earth's atmosphere, infrared astronomy is done at special observatories located at great altitude, such as at Mauna Kea in Hawaii or with orbiting infrared telescopes such as the Spitzer Space Telescope.

Infrared radiation: Electromagnetic radiation of wavelengths from 1 to 1,000 microns, wavelengths that occur beyond the red end of the visible portion of the electromagnetic spectrum.

Infrared spectrometer: A spectrometer that takes spectra of infrared radiation.

Inhomogeneous accretion theory: One of two major theories on Earth's differentiation: that differentiation occurred while the Earth was accreting debris because denser debris formed first. *See also* Homogeneous accretion theory.

Intercrater plain: Terrain consisting of gently rolling plains littered with small secondary craters; the crater density is higher than in a smooth plain.

Interferometer: An instrument that uses interference of electromagnetic waves to produce images with better resolution (sharper detail).

Interferometry: A data acquisition technique which uses more than one signal receiver (such as a series of radio telescopes). Signals are combined to form one highly detailed image. *See also* Interferometer, Very long baseline interferometry.

Intergalactic medium: Matter that exists between galaxies.

International Geophysical Year (IGY): The eighteen-month period from July, 1957, to December, 1958, during which many countries cooperated in the study of Earth and the Sun's effect on it. During this time, the space age can be said to have begun with the launch of Sputnik 1 on October 4, 1957.

Interplanetary space probe. *See* Deep space probe.

Interstellar dust. *See* Cosmic dust.

Interstellar medium: The material that lies between the stars; it consists mainly of grains of dust and gas, mostly hydrogen, along with heavier elements released by supernova explosions.

Interstellar wind. *See* Solar wind, Stellar wind.

Interval: A measure of the separation in space-time between two events; intervals may be timelike, spacelike, or null (lightlike).

Invariant: Unchanged by a transformation of

coordinates, such as the interval between two events in space-time.

Inverse beta decay: The process by which a neutrino interacts with a proton to produce a neutron and a positron.

Ion: An atom that is not electrically balanced but rather has either more electrons than protons or more protons than electrons. These therefore bear net electrical charge.

Ionization: The process whereby atoms are made into ions, by removal or addition of electrons. Such a process often occurs as a result of excitation of atoms into an energy state from which they easily lose electrons.

Ionosphere: The ionized layer of gases in Earth's atmosphere, occurring between the thermosphere (below) and the exosphere (above), between about 50 and 500 kilometers (31 to 310 miles) above the planet's surface. Within the ionosphere, ionized gases are maintained by the Sun's ultraviolet radiation. The resulting free electrons reflect long radio waves, making long-distance radio communication possible. Other planets are known to have ionospheres, including Jupiter, Mars, and Venus.

Iridium: A highly dense metallic element that is more abundant in materials of extraterrestrial origin and the Earth's core than at the Earth's surface.

Isomagnetic charts: Maps on which are traced curves, all the points of which have the same value of some magnetic property.

Isostasy: The concept that Earth's crust is in, or is trying to achieve, flotational equilibrium by buoyantly floating on denser mantle rocks beneath.

Isotope: Atoms with the same number of protons in the nucleus but with differing numbers of neutrons; a particular element will generally have several different isotopes occurring naturally.

Isotropic: Having properties that are the same in all directions; the opposite is anisotropic—having properties that vary with direction.

Jupiter-like planets: Giant gaseous planets with about the same composition as that of Jupiter, ranging from about one-tenth to ten times its mass (thirty to three thousand Earth masses).

K-band: A radio frequency range of about 11 to 15 gigahertz. *See also* Hertz.

Kamacite: A form of ferritic iron containing up to 7.5 percent nickel in solid solution.

Kelvin: A unit of temperature on the kelvin temperature scale, which begins at absolute zero (−273.15° Celsius). One unit kelvin is equal to 1° Celsius. The kelvin scale is particularly suited to scientific (especially astronomical) measurement. *See also* Absolute temperature scale.

Kepler's laws of motion: Three laws of motion discovered by Johannes Kepler and published by him in 1618-1619: (1) Each planet moves in an ellipse around the Sun, with the Sun at one of the two foci of that ellipse. (2) A line from the Sun to the planet sweeps out equal areas in equal times. (3) The square of the period of a planet's orbit is proportional to the cube of its mean distance from the Sun. *See also* Angular momentum.

Kilogram: A metric unit of mass, the equivalent of 1,000 grams. On Earth, a kilogram of mass weighs about 2.205 pounds.

Kilometer: A metric unit of distance, the equivalent of 1,000 meters or approximately 0.62 mile.

Kinetic energy: The energy of motion.

Kinetic pressure: The average force per unit area produced by movement of atoms and molecules.

KREEP-norite: A rock formation occupying small regions of the lunar surface, caused by the release of lava from liquid pockets of deep lunar crust as a result of meteoritic impacts; called KREEP because it is enriched in potassium (K), rare-Earth elements (REE), and phosphorus (P).

Kuiper Belt: A disk of icy, rocky objects that lies beyond Neptune, approximately 30 to 1,000 astronomical units from the Sun.

Lagrangian points: Stable points in the orbit of an intermediate body around a larger body where small particles may accumulate; these are sometimes referred to as L4 and L5.

Lapse rate: The rate at which temperature changes with altitude.

Laser: Originally an acronym for "light amplification by stimulated emission of radiation." A beam of infrared, visible, ultraviolet, or shorter-wavelength radiation produced by using electromagnetic radiation to excite the electrons in a suitable material to a higher energy level around their atomic nuclei. These electrons are then stimulated in such a fashion that they jump back down to their normal energy levels. When they do, they emit a stream of "coherent" radiation: photons with the same wavelength, direction, and phase as the originating radiation. This results in a narrow, intense beam of light (or nonvisible radiation), which bounces off a reflector and directly back to the propagating material, where the process is repeated and thus the laser is maintained. Laser technology has a vast range of applications in telecommunications, medicine, and astronomical measurements.

Laser ranging: A technique whereby scientists at two different Earth stations can determine, very precisely, their distance from each other by bouncing a laser beam off a satellite retroreflector. The time it takes to receive an "echo" allows each scientist to calculate his or her distance from the satellite; knowing both distances allows calculation of the distance between the two points on Earth. Over time, these measurements are repeated; changes in the distance between the two Earth locations are noted, providing much information about crustal movements and the likelihood of earthquakes.

Laser reflector: An optical device off which a laser beam can "bounce" or reflect; used to measure (usually great) distances by determining time of flight of the laser beam from original emission to final reception after reflection from the laser reflector on a body whose distance is to be determined in this way, recognizing that the laser beam travels at the speed of light (3×10^8 meters per second).

Latitude: The angular distance from a specified plane of reference. On Earth, the angular distance north or south of the equatorial plane; in the solar system, the angular distance of a celestial body above or below the ecliptic plane. *See also* Longitude.

Launch site: A location housing a facility designed to handle preparations for launch as well as the launch itself.

Lava: Molten magma from the interior of a planet extruded through cracks or holes in the planet's surface.

Leptons: A class of elementary particles, including electrons, neutrinos, and their antiparticles, which are not affected by the strong nuclear force that binds protons and neutrons.

Light, speed of. *See* Velocity of light.

Light curve: A graph that indicates how the brightness of a star changes with time.

Light-year: A unit of distance equal to the distance that light travels in a vacuum in one year. At a speed of approximately 300,000 kilometers (186,000 miles) per second, a light-year is about 10 trillion kilometers.

Limb: The outer edge of the visible disk of the Sun, Moon, a planet, or another celestial body.

Liquid metallic hydrogen: Hydrogen that behaves as a metal under pressure of about 5 million Earth atmospheres.

Lithification: The conversion of loose mineral materials into rock.

Lithium: Element 3 in the Periodic Table of Elements following hydrogen and helium, produced within protostars but destroyed by nuclear reactions in stars.

Lithosphere: A rigid layer consisting of the Earth's crust and the top part of the underlying mantle, about 80 kilometers (49.6 miles) thick. The term can be used in reference to the rigid outer part of any planet.

Longitude: The angular distance from a specified plane perpendicular to the latitude reference plane. On Earth, the angular distance east or west of the prime meridian, which runs from pole to pole through the Royal Observatory at Greenwhich, England.

Look angle: Angular limits of vision.

Low-Earth orbit: Generally, any orbit at an altitude of about 300 kilometers (186 miles) or less. Such an orbit has a period (time required to complete one orbit around Earth) of 90 minutes or less.

Lowell bands: Dark areas on the periphery of the Martian polar caps in summer.

Luminosity: The rate at which a star radiates electromagnetic energy, usually expressed in joules per second or in watts; the *intrinsic brightness* or light output of a star, as distinct from its *apparent brightness*.

Lunar day: The time it takes the Moon to complete one rotation on its axis. Relative to distant stars, it takes approximately 27.33 Earth-days. Relative to the Sun, it takes approximately 29.53 Earth-days; this is the time from one sunrise to the next, as seen from the Moon.

Lunar (synodic) month: The period from one new moon to the next, about 29.53 days.

Mach: The ratio of the speed of a moving object to the speed of sound in the surrounding medium. At Mach 1, the speed of an aircraft equals the speed of sound.

Mafic *and* ultramafic: Rock-forming magmas that are high in dense, refractory elements such as iron and magnesium; oceanic basalts are examples of mafic rocks.

Magellanic Clouds: Two small, irregular galaxies, the two nearest galaxies outside the Milky Way, visible from the Southern Hemisphere as the Large Cloud and the Small Cloud, respectively 160,000 and 185,000 light-years away. The Magellanic Clouds have been instrumental in establishing an extragalactic distance scale.

Magma: Molten rock material formed in a planet's interior; when extruded at the surface, it becomes known as lava.

Magnetic anomalies: Distortions in the magnetic field, produced by an object such as an iron ore body.

Magnetic field: A field created by magnetic properties of an object. When an object of mass m and charge q having a speed v enters an external magnetic field B created by an external influence, it experiences a magnetic sideways deflecting force given in magnitude by the product of q, v, B, and the sine of the angle between the direction of the velocity vector and the magnetic field. The direction of that sideways deflecting force is perpendicular to both the velocity and the magnetic field vectors; positive and negative charges are deflected in opposite directions from one another when entering a magnetic field. *See also* Gravitation.

Magnetic pole: The location on Earth's surface where Earth's magnetic field is perpendicular to the surface.

Magnetic storm: Rapid changes in Earth's magnetic field as a result of the bombardment of Earth by electrically charged particles from the Sun.

Magnetic surveys: Measurements of the magnetic field at many points, on or above Earth's surface, carried out by field teams, airborne magnetometers, ships at sea, or satellites.

Magnetometer: A scientific instrument used to measure disturbances in Earth's magnetic field.

Magnetopause: The outer limit of a planet's magnetic field and the boundary of its magnetosphere.

Magnetosheath: A region of magnetic turbulence between the bow shock and the magnetopause.

Magnetosphere: The domain around a planet in which the behavior of charged particles is controlled by the planet's magnetic field and not by the Sun's.

Magnetotail: A "tail" of nearly parallel lines of magnetic force extending from a planet in the direction away from the Sun.

Magnitude: The brightness of a celestial body expressed numerically; the lower the number, the brighter the body. *See also* Absolute magnitude, Apparent magnitude.

Main sequence star: A star, such as the Sun, which produces energy mainly by a hydrogen-to-helium fusion reaction. Most stars spend the greater part of their lifetimes in this state.

Mantle: A layer of dense silicate rock that lies at depths of approximately 34 to 2,885 kilometers (21 to 1,789 miles) between the crust and outer core and comprises the majority of Earth's volume.

Mare (*pl.* maria): A large, flat area on the Moon or Mars, so named (after the Latin for "sea") because these areas appear dark, thus sealike, to the Earth observer.

Maritime satellite: A satellite designed for

telecommunications by and for shipping industries. These satellites occupy geostationary orbits over oceans to transmit ship-to-shore communications and data.

Mascon: One of several concentrations of mass located beneath lunar maria, which causes a distortion in the orbit of a spacecraft around the Moon.

Maser: An acronym for "microwave amplification by stimulated emission of radiation." A device similar to a laser in which energy is generated as in a laser, but at microwave levels. A maser can exist in nature as a celestial object. Artificial masers are used to amplify weak radio signals. *See also* Laser.

Mass: The amount of matter contained within a body, which determines the amount of gravitational force it exerts. Mass is measured in units such as kilograms; it differs from weight, which is the force exerted on a mass by gravity and which therefore would be measured in units such as newtons or pounds.

Mass spectrometer: An instrument that identifies the chemical composition of a substance by separating ions by mass and charge.

Matrix: In meteorities, the fine-grained material that surrounds both chondrules and inclusions; it consists of hydrous silicate minerals, troilite, magnetite, and other lower-temperature phases.

Matter: A substance that has mass and occupies space, which along with energy is responsible for all observable phenomena.

Maunder minimum: Named for E. W. Maunder, who in 1890 discovered a period in the three-hundred-year history of sunspot observations when few sunspots were recorded. Confirmed independently in 1976 by evidence from tree rings, the Maunder minimum covers the years 1645 to 1715, a period also known as the Northern Hemisphere's "Little Ice Age."

Megahertz. *See* Hertz.

Megaparsec: A unit of measurement equaling 3.26 million light-years.

Mesosphere: The layer of Earth's atmosphere occurring above the stratosphere and below the thermosphere, from about 40 kilometers to 85 kilometers (24.8 to 52.7 miles) above sea level. This is the coldest layer of the atmosphere.

Messier number: The number of an object listed in the catalog of 103 nebulae and star clusters prepared by Charles Messier in 1784; an object is referred to as M followed by the catalog number, such as the globular star cluster M13 in Hercules.

Metabolic process: A chemical process in a living organism which provides it with the energy to function.

Metamorphism: A process by which heat and pressure applied to a rock cause it to change without causing it to melt.

Meteor: A bright streak of light in the sky, sometimes called a shooting star, produced by a meteoroid entering Earth's atmosphere at high speed and heating the air column along its path to incandescence.

Meteor shower: A large number of meteors resulting from the passage of Earth through a meteoroid stream or swarm believed to be the debris left in the orbit of a comet.

Meteorite: A metallic or stony meteoroid (or combination) that survives its passage through the atmosphere as a meteor and falls to the surface of the Earth.

Meteoritics: The study of the naturally occurring masses of matter that have fallen to the Earth's surface from outer space.

Meteoroid: A small solar system body, probably a fragment from a comet or asteroid, which produces a meteor when it enters Earth's atmosphere.

Meter: The metric unit of length, equivalent to approximately 39.37 inches, or a little more than 1 yard.

Metric system: The decimal system of weights and measures, which forms part of the Système International d'Unités. *See also* SI units.

Metric ton (or tonne): A unit of mass equal to 1,000 kilograms. On Earth, it is equivalent to 2,205 pounds, which is close to an ordinary ton, or short ton, of weight. *See also* Newton, Pound.

Microgravity: Nearly zero gravity. Microgravity exists in a space vehicle because of the minute gravitational forces exerted by objects on one another. The microgravity environment is of great importance as an ideal environment for certain types of materials processing.

Micrometeorite: A micrometeroid that has reached Earth's surface.

Micrometeoroid: A meteoroid with a diameter of less than 0.1 millimeter. Because of their size, micrometeoroids rarely burn up but reach Earth's surface instead, as spherules or as cosmic dust particles.

Micron: A unit of measure convenient for measuring wavelengths of infrared radiation; a micron is equal to one millionth of a meter.

Micropaleontology: The study of microscopic fossils, of potential importance in exobiology as well as life sciences on Earth.

Microwaves: A form of electromagnetic radiation with wavelengths ranging between 1 millimeter and 30 centimeters, located between infrared and long-wave radio on the electromagnetic spectrum.

Milky Way: The galaxy in which our solar system is located. The Milky Way, a spiral or barred spiral galaxy, contains about 10^{11} stars and is about 100,000 light-years across. *See also* Galaxy.

Miller-Urey synthesis: The production of amino acids by repeatedly passing an electrical spark through a mixture of methane, ammonia, water vapor, and hydrogen.

Millibar: A pressure of 100 newtons per square meter.

Missing mass of the universe. *See* Dark matter.

Model: A simulation of a phenomenon that is difficult to observe or specify by direct means; models abstract from phenomena under study those qualities that the investigator perceives to be essential for understanding.

Molecular clouds: Massive, very large clouds of various molecules and dust; the breeding ground of stars in deep space.

Molecule: The smallest unit of a substance, formed by a characteristic complex of atoms joined together. The smallest unit of the substance water, for example, is a molecule formed by two hydrogen atoms and one oxygen atom.

Moon: Any natural satellite orbiting a planet, especially Earth's Moon.

Nanometer: One-thousand-millionth (one-billionth) of a meter; a unit used to express electromagnetic wavelengths.

Near-Earth space: Roughly defined as the space environment from the outer reaches of Earth's atmosphere to the path of the Moon's orbit, the area beyond which is known as deep space.

Near-infrared: A portion of the electromagnetic spectrum that lies beyond the red end of the visible spectrum and has a photon energy that is approximately the same as that typical of many chemical bonds.

Nebula: A celestial body composed of aggregated gas and dust, which may be either luminous, reflecting or emitting light under the influence of nearby stars (a reflection nebular or an emission nebula), or dark, obscuring the light of distant stars and appearing as a silhouette (an absorption nebula).

Nebular hypothesis: The concept that the solar system and all of its parts are the result of the contraction of a cloud of gas and dust.

Neumann bands: A textural pattern that is common to iron meteorites with less than 6 percent nickel content; they reflect deformational twinning paralled to trapezohedral planes in kamacite.

Neutral gas analyzer: An instrument that determines the chemical composition of an atmosphere.

Neutrino: An elementary particle of enormous penetrating power as a result of its lack of electric charge and its nearly total lack of mass. Traveling directly out from the cores of stars as a by-product of nuclear reactions, neutrinos have enormous potential as a source of information on the stars and other astrophysical phenomena.

Neutron: An elementary particle composed of quarks in such a way that it has no net electric charge. Found in atomic nuclei and is attracted to other nucleons (other neutrons and protons) by the strong nuclear force. Its mass is approximately equal to that of a proton.

Neutron stars: The smallest stars known, with diameters of about 20 kilometers (12.4 miles) and masses about two times that of the Sun, consisting of a thin iron shell enclosing a "liquid" sea of degenerate neutrons.

Although astrophysicists lack a full understanding of these objects, neutron stars are thought to originate from stars much more massive than the Sun which explode as supernovae. Much mass is lost in the process, and what remains as the neutron star may spin rapidly and be observable as pulsars.

New astronomy: A term used collectively to refer to the areas of astronomy (such as gamma-ray astronomy, infrared astronomy, and X-ray astronomy) investigating electromagnetic emissions by celestial phenomena observable from satellites in orbit around the Earth or Sun.

New General Catalog: A catalog—in full, the *New General Catalog of Nebulae and Clusters of Stars*—compiled in 1888 by Johan Ludwig Emil Dreyer, which lists many nebulae, star clusters, and galaxies. Frequently the objects listed in the catalog are identified by their NGC numbers; for example, the globular star cluster Omega Centauri is NGC 5139.

Newton: The basic metric or SI (Système Internationale) unit of measure for force or weight; also used to measure thrust. Named after Sir Isaac Newton.

North point: That intersection of the celestial meridian and astronomical horizon lying due north.

Northern lights. *See* Aurora.

Nova: A star that emits a sudden burst of visible light and other forms of electromagnetic radiation, and quickly (over months or years) returns to its former brightness. *See also* Supernova.

Nuclear burning: The process by which the nuclei of light elements, under conditions of extreme temperature and density, are converted into the nuclei of heavier elements through fusion.

Nuclear energy: Energy that is released as a result of interactions between elementary particles and atomic nuclei.

Nuclear reaction: A very high-energy process in which nuclei break apart (or fission), if very massive, or join together (or fusion) if very light, causing the emission of heat and light.

Nucleosynthesis: The process by which heavier elements are produced from hydrogen and helium in stars.

Nucleus: The central part of something. In galaxies, the central region or core; in spiral galaxies, sometimes called the central bulge; in comets, the solid body of the comet; in atoms, the central massive part consisting of protons and neutrons about which the electrons orbit.

Oblate spheroid: A nearly spherical shape that is flattened at the poles and bulges at the equator.

Occultation: An eclipse of any astronomical object other than the Sun or the Moon caused by the Moon or any planet, satellite, or asteroid.

Octahedrites: Iron meteorites that usually contain between 6 and 16 percent nickel.

Olivine: A silicate mineral of magnesium and iron that is common in some igneous chondritic meteorites.

Oort Cloud: A vast region of billions to trillions of comets that surrounds the solar system and is located tens of thousands of astronomical units from the Sun.

Open universe: A universe that expands forever and will eventually lose all usable energy.

Opposition: The alignment of Sun, Earth, and a superior planet (one whose orbit is farther from the Sun than Earth's) in a straight line; that is, the superior planet appears in the sky at 180° celestial longitude from the Sun. In this position, the planet is closest to Earth and therefore most easily observed by ground-based instruments.

Optical radiation: Visible light; a range of wavelengths that form a small section of a larger range of wavelengths called the electromagnetic spectrum.

Optical window: The region of the electromagnetic spectrum (295 to 1,100 nanometers) that is passed by the atmosphere and that is easily manipulated by lenses and mirrors; "visible" light (400 to 700 nanometers) lies near the center of this window.

Orbit: The path traced out by one body as it moves around another. The distinguishing characteristics of an orbit are called its orbital parameters and include apoapsis, periapsis, inclination, eccentricity, and period. All closed orbits trace out an ellipse, of which the body orbited lies at one of the two foci. *See also* Apoapsis, Circular orbit, Eccentricity, Ellipse, Elliptical orbit, Equatorial orbit, Geocentric orbit, Geostationary orbit, Geosynchronous orbit, Heliocentric orbit, Orbital inclination, Parabolic orbit, Parking orbit, Periapsis, Period, Polar orbit, Prograde orbit, Retrograde orbit, Synchronous rotation (or orbit), Transfer orbit.

Orbital eccentricity: A measure of the elongation of an elliptical orbit, ranging from zero (for a circular orbit) to one (for an orbit that approaches an open-ended parabola).

Orbital inclination: The angle formed between the orbital plane of a satellite and the equatorial plane of the object orbited.

Orbital period: The time required for one celestial object to execute a complete revolution around another object.

Ordinary chondrites: Chondrule-bearing stony meteorites not containing carbonaceous compounds.

Organic molecules: Molecules including the elements carbon, hydrogen, and oxygen.

Oscillating universe: A theoretical type of closed universe in which a big bang and subsequent recompression of all matter periodically occur forever.

Outer core: A zone of the Earth's interior, located at depths of approximately 2,885 to 5,144 kilometers (1,789 to 3,189 miles), that is in a liquid state and consists of iron sulfides and iron oxides.

Outer space: All space beyond Earth's atmosphere. *See also* Deep space.

Outgassing: The process by which trapped gases in a planet leak out gradually over time to form the planetary atmosphere and oceans.

Ozone layer: The thin layer of Earth's atmosphere, located between 12 and 50 kilometers (7.44 to 31 miles) above Earth's surface (in the stratosphere), in which ozone (O_3) is found in its greatest concentrations. This layer, which absorbs most of the ultraviolet radiation entering the atmosphere, forms a protective blanket around the planet, shielding it from excess radiation.

P waves. *See* Primary (P) waves.

Paleomagnetism: The study of the record of remanent or fossil magnetism in rocks, indicative of past states of Earth's magnetic field and very useful in determining secular variation, sea-floor spreading, and past locations of continents.

Panspermia: A theory proposed by chemist Svante Arrhenius in 1906, and later modified by Sir Fred Hoyle, which holds that organic molecules (hence the beginnings of life) were transported to Earth by means of comets.

Parabolic orbit: An orbit that describes a parabola around the object orbited and hence escapes from the gravitational field of that object.

Parallax: The apparent angular displacement of a star as seen from opposite sides of the Earth's orbit around the Sun. Measuring this angle (and knowing the size of the Earth's orbit) allows the distance to the star to be calculated.

Parking orbit: An interim orbit around a celestial body between launch and injection into another orbit or into a trajectory toward another destination.

Parsec: A unit for measuring astronomical distances equivalent to 3.26 light-years; the distance at which the stellar parallax is 1 second of arc.

Partial lunar eclipse: An eclipse in which the Moon passes completely or partially into the penumbra of the Earth but does not enter (or entirely enter) the umbra.

Partial melting: Melting of some minerals in a rock but not others, resulting in a magma concentrated in some elements and depleted in others as compared to the original unmelted rock.

Partial solar eclipse: An eclipse in which part but not all of the Sun is covered by the Moon.

Particle. *See* Elementary particles.

Path of the Sun: The apparent motion of the Sun as it tracks across the sky.

Pauli exclusion principle: The principle that no two particles of the same type can occupy precisely the same quantum state; it is obeyed by baryons, leptons, and quarks, but not by photons or mesons.

Pedestal crater: A crater that has assumed

the shape of a pedestal as a result of the wind's unique shaping processes.

Penumbra: A region of partial shadow where some but not all parts of the source of illumination are obscured.

Perfect spheroid: A three-dimensional body that has the same circumference regardless of the direction by which it is measured; that is, it is perfectly "round."

Periapsis: The point in one object's orbit around another at which the orbiting object is closest to the object being orbited.

Pericynthion: The point in an object's orbit around the Moon at which it is closest to the Moon.

Peridotite: A silicate igneous rock consisting largely of the mineral olivine.

Perigee: The point in an object's orbit around Earth at which it is closest to Earth.

Perihelion: The point in a solar orbit at which the orbiting object is closest to the Sun.

Perilune: Pericynthion of an artificial satellite.

Period: The time span between repetitions of a cyclic event. An *orbital period* is the time required for a satellite or moon to make one complete orbit around a planet, a moon, the Sun, or another celestial body.

Permafrost: Permanently frozen soil which is laced with water ice.

Permeability: The property or capacity of porous geological materials to transmit fluids; it indicates the relative ease of fluid flow through a medium.

Perturbation: The act of altering the orbital course (direction or speed) of satellite or planetary orbits usually initiated by gravitational effects from or collision with another object.

Petrography: The description and systematic classification of rocks.

Photochemistry: Chemical reactions caused by the action of strong ultraviolet light, which excites or dissociates some compounds and leads to the formation of others.

Photometer: An instrument that measures the brightness of a light source.

Photometry: The technique of measuring the brightness of light sources.

Photomultiplier: An instrument for increasing the apparent brightness or strength of a source of light by means of secondary excitation of electrons; effectively, a light (or other radiation) amplifier.

Photon: A particle of electromagnetic radiation possesing energy and momentum but neither rest mass nor charge.

Photopolarimeter: An instrument used to measure the brightness and polarization of light.

Photosphere: The region of the Sun that separates its exterior (the chromosphere and corona) from its interior, forming the boundary between the transparent and opaque gases. The photosphere appears as the bright central disk from Earth, and it is the source of most of the Sun's light.

Photosynthesis: The utilization of carbon dioxide and water by chlorophyll-containing organisms in the presence of sunlight to metabolically produce carbohydrates used by the plant for food; oxygen is a by-product in the photosynthesis process.

Photovoltaic cell: A solid-state energy device that converts sunlight into electricity.

Pitch, roll, and yaw: Movements that a spacecraft undergoes as a result of launch or other stresses. Pitch is up-down movement; roll is longitudinal rotation; yaw is side-to-side movement.

Pixel: A small unit arranged with others in a two-dimensional array which contains a discrete portion of an image (as on a television screen) or an electrical charge (as on a charge coupled device). Together, these pixels form an image or other meaningful information.

Planck length, *l*: The Planck length l_p is defined by $[Gh/(2pc^3)]^{0.5}$ and is approximately equal to 1.6×10^{-35} meters. Combining gravitation's G with quantum mechanics' h, it marks the distance at which quantum effects dominate.

Planck's constant, *h*: Named after the German physicist Max Planck, h (6.63×10^{-34} joule-seconds) first arose in modern physics in an attempt to explain blackbody radiation characteristics. It is a quantum of "action," having units of angular momentum. Among its many fundamental aspects in quantum physics, it is the constant of proportionality between the energy and frequency of a photon of light.

Plane of the ecliptic: The plane of the Earth's orbit around the Sun.

Planet: A nonluminous natural celestial body that orbits the Sun (or another star) and is not categorized as an asteroid or comet. There are eight known planets in the solar system: Mercury, Venus, Earth, Mars, Jupiter, Saturn, Uranus, and Neptune in increasing distance from the Sun. Pluto lost it planetary status as a result of a redefinition of "planet" adopted by the International Astronomical Union (IAU) in 2006.

Planetesimal: A small solid body (from less than a millimeter up to hundreds of kilometers) that accreted in the solar nebula during the formation of the solar system. Many of these bodies grew into protoplanets and planets, while others remain as meteoroids and asteroids.

Plasma: Ionized gas, consisting of roughly equal numbers of free electrons and positive ions. Plasma occurs in stars, nebulae, and interplanetary, interstellar, and intergalactic space. It is considered the fourth state of matter, along with solid, liquid, and gas.

Plasma sheath: The definite outer boundary of Earth's ionosphere.

Plate tectonics: The theory that the crust and upper mantle of the Earth are divided into a number of moving plates about 100 kilometers (62 miles) thick that converge at trench sites and diverge at oceanic ridges.

Polar orbit: An orbit in which a satellite passes over a planet's or moon's poles.

Polarimeter: An instrument for measuring the degree to which electromagnetic radiation is polarized.

Pole: One of two points on the surface of a planet where it is intersected by its axis of rotation. In a magnetic field, one of two or more points of concentration of the lines of magnetic force.

Polygonal ground: The distinctive geological formation caused by the repetitive freezing and thawing of permafrost.

Polymorphism: The characteristic of a mineral to crystallize into more than one form.

Porphyritic chondrules: Generally small spherules (1 to 5 millimeters), often with crystals of the minerals olivine and pyroxene set into a glass matrix.

Posigrade: Moving in the direction of travel.

Positron: The antiparticle of an electron.

Potential energy: Energy due to an object's location in some sort of field, such as gravitational potential energy due to a gravitational field or electrical potential energy due to an electric field.

Pound: An Imperial unit used to measure force, weight, and thrust in the United States and some other English-speaking nations.

Precession: A type of motion that occurs in a rotating body in response to torque: A planet or other rotating body orbiting around the Sun will slowly turn its rotation axis in such a way that, over a long period of time, the rotating each of the body's two poles describes a circle in space. The fact that the Earth precesses means that adjustments must be made in the direction in which astronomers look to observe stars and other celestial phenomena. Earth's period of precession is approximately 25,800 years.

Primary minerals: Those minerals formed when magma crystallizes.

Primary (P) waves: The fastest elastic wave generated by an earthquake or artificial energy source; basically an acoustic or shock wave that compresses and stretches material in its path.

Primordial solar nebula: An interstellar cloud of gas and dust that condensed under gravity to form the Sun, the planets and their satellites, asteroids, comets, and meteoroids some 4.6 billion years ago.

Principle of equivalence: The rule that, in a limited region of space-time, the effects of the acceleration of a given frame of reference are not distinguishable from those of a gravitational field.

Probe. *See* Deep space probe.

Prograde orbit: An orbit that moves in the same direction as the rotation of the body orbited.

Prominences: Arches of glowing gases above the Sun's photosphere often seen in the vicinity of sunspots.

Proper motion: The motion of a star across the line of sight.

Protein: A high-molecular-weight compound

that is a long chain or aggregate of amino acids joined by hydrogen bonds.

Proto solar system (solar nebula): The cloud of gas and dust that separated from a larger cloud and eventually collapsed to form the Sun in the center and the planets in the outer portions of the cloud.

Proton: The constituent particle in the nucleus made up of quarks in such a way that it has a net positive electric charge. It interacts with other protons and neutrons inside a nucleus via the strong nuclear force, interacts with electrons and protons (at distances greater than nuclear dimensions) via the Coulomb electric force, and decays according to the weak nuclear force. Chemically speaking, the number of protons in the nucleus of an atom determines which chemical element it is.

Protoplanet: A stage of planet formation in which a large precursor of a planet gravitationally contracts, perhaps attracting more material or maybe losing part of a gaseous envelope.

Protostar: The stage just before actual star formation, when a vast cloud of matter has coalesced to about 1 percent of the cloud's dispersed volume, causing nuclear reactions to begin.

Protosun: A stage of star formation in which gravitational contraction shrinks and heats the developing star, before the interior becomes hot enough to initiate nuclear fusion reactions.

Pulsar: A rapidly spinning neutron star that emits a narrow beam of electromagnetic radiation in the form of visible light, radio waves, X rays, and gamma rays.

Pyroxene: A silicate mineral of magnesium, iron, and sometimes calcium which contains more silicon than is present in olivine.

Quantum theory: An important theory in physics that associates waves with particles; particle momentum is h divided by the particle's wavelength, while energy is h times the particle's frequency; light has frequency and wavelength (thus, energy and momentum), and so acts like a particle and a wave.

Quarks: Subatomic particles hypothesized to form electrons, protons, neutrons, and their antiparticles, characterized by electric charge, "flavor," and "color." The forces required to break elementary particles into their component quarks is so great that quarks do not exist as free particles in nature, although it is thought that neutron stars may consist of a "quark soup" within a solid iron shell.

Quasar: An acronym for "quasi-stellar radio source" or "quasi-stellar object." An object continuously releasing a tremendous amount of energy, equivalent to the output of between one million and 100 trillion suns, which includes virtually all kinds of electromagnetic radiation (gamma, X, ultraviolet, optical, infrared, microwave, and radio) from a very small volume of space about the size of the solar system. As far as is known, all objects satisfying these criteria are located in the nuclei of galaxies; it is thought that they may be associated with supermassive black holes at the centers of galaxies. Discovered in 1963, the first quasar caused much excitement among astronomers, and these phenomena continue to be among the most fascinating and mysterious in the universe.

Radar: An acronym for "radio detection and ranging." A means of locating and determining the distance of objects by bouncing radio waves off them and measuring the time required to receive the echo.

Radial velocity: Movement in the line of sight, toward or away from the observer.

Radiant: The point in the sky from which a meteor shower seems to emanate, whose associated constellation provides the name for a given shower.

Radiating pyroxene chondrules: Generally small spherules (1 to 5 millimeters) composed of excentroradial pyroxene crystals, often resembling a fanlike growth pattern.

Radiation. *See* Electromagnetic radiation.

Radiation (of stars): Stars radiate energy in all portions of the electromagnetic spectrum, whereas one band of wavelengths predominates according to the star's photospheric temperature.

Radiative equilibrium: When radiation becomes the predominant, balanced mechanism for carrying heat away from the interior of a star.

Radio astronomy: The branch of astronomy that examines the radio emissions of celestial objects. Because radio radiation, along with visible radiation, can penetrate Earth's atmosphere, radio receivers have provided much of the data detectable by ground-based, as well as space-based, instruments. Radio emissions also form a significant portion of certain celestial phenomena, such as radio galaxies, quasars, and pulsars.

Radio noise: Any body that generates radio waves or oscillations of a random nature over the radio frequencies, from millimeters to several thousand meters; often heard as a hissing sound in radio receivers.

Radio spectrum: Those frequencies generally ranging from 1 centimeter to about 30,000 centimeters.

Radio telescope: A telescope designed to gather radio waves from extraterrestrial sources.

Radioactive decay: The conversion of one element into another by the emission of charged particles from an atom's nucleus.

Radioactivity: The process by which an unstable atomic nucleus spontaneously emits a particle (or particles) and changes into the nucleus of another atom.

Radiogenic heating: Heating caused by the decay of radioactive materials in a planetary body; energy released during the decay produces the heat.

Rampart crater: A type of crater found most often on Mars and produced by some subsurface shaping mechanism that causes a unique, rampart-type wall formation.

Real time: Referring to the transmission of signals or other data at the same time that they are used.

Reconnaissance satellite: A satellite that gathers information about enemy military installations.

Red dwarf star: A relatively small, cool star having low luminosity.

Red giants: Stars with surface temperatures less than 4,700 kelvins and between 10 and 1,000 times the diameter of the Sun, which emit mostly red and infrared light, and have huge surface areas.

Red Planet: Mars, so named because of its color as seen through Earth-based telescopes.

Redshift: The apparent lengthening of electromagnetic wavelengths issuing from an object as a result of the object's movement away from the observer. As a result, the spectral lines in the spectra of such an object will shift toward (or beyond) the red end of the visible spectrum. *See also* Doppler effect.

Reentry: The return of a spacecraft into Earth's atmosphere.

Reflecting telescope: An optical telescope that uses a mirror or mirrors to gather and focus light from the object observed. These telescopes are widely used for Earth-based as well as space-based optical astronomy. *See also* Cassegrain telescope, Refracting telescope.

Reflection: The "bounce" of wave energy off a boundary that marks a change in the material; it must be noted that at an interface between two different media, wave energy can be reflected to a certain degree back into the original medium, and what is not reflected off the interface is refracted across the interface into the other medium.

Reflection nebula: A cloud of dust, usually bluish in color, visible by virtue of light from nearby stars scattered by the dust grains.

Reflectivity: The amount of light reflected from a body.

Refracting telescope: An optical telescope that uses a lens to magnify and focus light from the object observed. Refracting telescopes were used by the earliest astronomers. When reflecting telescopes were perfected in the twentieth century, refracting telescopes became less important in astronomy, although they are still widely used for guided and amateur observations. *See also* Reflecting telescope.

Refraction: The change in direction of a wave path upon crossing a boundary, resulting from a change in the material light enters into and the different velocity of the wave on either side of the boundary.

Refractory: Refers to substances that melt or boil at relatively high temperatures and, conversely, condense from liquids or gas at low temperatures.

Refractory (siderophile) elements: Elements least likely to be driven off by heating; the last elements to be melted as a rock is heated to form magma.

Regolith: The layer of soil and rock fragments just above the solid planetary crust.

Relativity: The physical law, first proposed by Albert Einstein, that states that measurements of time and space are dependent upon the frame of reference in which they are measured. The general theory of relativity applies this law to gravity and mass; the special theory of relativity applies it to the propagation of electric and magnetic phenomena in space and time.

Remote sensing: Acquiring data at a distance by electronic or mechanical means.

Resolution: The degree to which a photographic or other imaging system, or the image produced, clearly distinguishes objects of a certain size. In a photograph with a resolution of 200 meters, for example, the smallest distinguishable objects are 200 meters across.

Retrograde orbit: An orbit that moves opposite to the rotational direction of the body orbited.

Revolution: One complete orbit of one body around another body, such as a planet around the Sun or a natural or artificial satellite around a planet.

Right ascension: A coordinate for measuring the east-west positions of celestial bodies; the angle measured eastward along the celestial equator from the vernal equinox to the hour circle passing through a body.

Rille: A long, narrow valley on a moon or planet.

Ring resonance: The gravitational interaction between a ringed planet, its moons, and the particles in the rings; the principal effect of the resonance is formation of discrete rings and gaps.

Robotics: The development, construction, and use of computerized machines to assist or replace humans in a variety of tasks requiring precise "hand-eye" coordination.

Roche limit: Named for Édouard Roche, who discovered it in 1848, the minimum distance from a planet at which a natural satellite can form by accretion: roughly 2.44 times the planet's radius. Within this limit an existing satellite will be torn apart by gravitational stresses. Saturn's rings, which lie within the planet's Roche limit, may be the remnants of a former moon.

Roll. *See* Pitch, roll, and yaw.

Rotate: To spin around an axis.

RR Lyrae stars: A class of regular pulsating variable stars, having a period of about half a day and an average luminosity of about one hundred times that of the Sun.

Saros cycle: A period of 223 lunar months, after which the pattern of eclipses repeats.

Satellite: Any body that orbits another of larger mass, usually a planet. Satellites include moons, the small bodies that form planetary rings, and human-made robotic spacecraft. *See also* Artificial satellite.

Scarp: A vertical or near-vertical cliff, often extending for many kilometers.

Schwarzschild radius: The radius of a mass, such as a collapsing star, at which it becomes a black hole—that is, at which its gravitational field will not allow light to escape. The length of this radius depends on the body's mass, and the formula for calculating it was established by Karl Schwarzschild in 1916.

Scintillation: Light emitted when high-energy radiation is absorbed by matter then re-emitted at lower energies.

Scintillation counter: A device that detects atomic particles and generates an electrical current proportional to the energy of the particle.

Secondary crater: A crater resulting from impact of material thrown out of a primary impact crater.

Secondary (S) wave: A transverse type of earthquake wave, slower than a primary wave, which will not travel in a liquid.

Secular variation: A change in the magnetic pole position on the Earth's surface over hundreds of years.

Sedimentary rocks: Rocks that are formed by the deposition of layers of sediment (the weathered and eroded debris from preexisting rocks).

Seismic activity: Any movement in the outer layer of a planet or moon.

Seismic waves: Elastic oscillatory disturbances spreading outward from an earthquake or human-made explosion; they provide the most important data about Earth's interior.

Seismometer: A sensitive instrument that measures movements in the outer layer of a planet or moon. The graphs produced by this instrument can be interpreted to determine the magnitude and intensity of seismic activity.

Selenography: The study of lunar surface features; the counterpart of geography on Earth.

Selenology: The study of the Moon, analogous to geology on Earth.

Self-exciting dynamo: Also called a self-sustaining dynamo, a model of the earth's outer core in which the magnetic field produced by convection is in the same direction as the field through which the motion occurs.

Seyfert galaxy: A spiral galaxy with an active core that is similar to a quasi-stellar object.

Shear (S) waves: Seismic waves transmitted by an alternating series of sideways movements in a solid; they cannot be transmitted through liquids or gases.

Shepherd satellites: Also called shepherd moons, the tiny satellites responsible for gravitationally restraining ring particles in their defined rings.

Shock wave: A zone of compression and heating of matter traveling faster than the speed of sound in the matter.

SI units: The collective units of measurement used in the Système International d'Unités, the system of measurement most widely used by scientists. Its fundamental, or *base* units are seven: the meter (the base unit of length), kilogram (mass), second (time), ampere (electric current), kelvin (temperature), mole (amount of substance), and candela (luminosity). From these seven base units, other units are derived, which are multiples, fractions, or powers of the base units such as the kilometer (1 meter $\times$ 10^3) and the square meter (the unit of area). Further derived units are derived from combinations of the base units and have their own names: hertz (the unit of frequency, which is cycles per second), newton (force or thrust, kilogram-meters per second squared), pascal (pressure, newtons per square meter), joule (energy, newton-meter), watt (power, joules per second), coulomb (quantity of electricity, the ampere-second), volt (electric potential, watts per ampere), farad (capacitance, or the ability to store energy, coulombs per volt), and ohm (electrical resistance, volts per ampere). In the United States, the base SI units are coming into increasing use. Some measures, however, remain more familiarly rendered by English units of measure, even in scientific use: It is common, for example, to refer to rocket thrust in pounds rather than newtons, and atmospheric pressure is often measured in pounds per square inch (or bars and millibars).

Sidereal period: The time for a planet or satellite to make one complete rotation on its axis, or one complete revolution around its orbit relative to distant stars.

Siderophile elements. *See* Refractory elements.

Silicate: A class of mineral whose lattice structure includes one silicon atom surrounded by four oxygen atoms at the vertices of a tetrahedron.

Singularity: In space-time, a location at which matter and energy are compressed down to a single point in space with infinite density. The initial condition of the big-bang universe, when space-time curvature and mass-energy density were infinite.

Sinuous rille: A riverlike channel produced by lava flowing across the lunar surface.

Smokers: Undersea vents in the active rift areas, emitting large amounts of superheated water and dissolved minerals from deep inside Earth.

Smooth plain: A formation that is relatively flat, with a sparsely cratered surface.

Snell's law: A statement of the fact that refraction of waves across a boundary will occur such that the ratio of the two velocities of the material on either side of the boundary is equal to the ratio of the sines of the two angles on either side of the boundary formed by the ray path and a line perpendicular to the boundary.

Solar array: An assembly of solar cells, as on a solar panel extending from a satellite.

Solar cell: A photovoltaic device that converts solar energy directly into electricity.

Solar constant: The amount of solar energy received by a square meter in one second just above Earth's atmosphere; approximately 1,370 watts per square meter per second.

Solar cycle: A period of approximately eleven years during which the number of sunspots visible near the Sun's equator increases to a maximum and then decreases. Other solar activity follows the solar cycle. *See also* Sunspots.

Solar flare: A large eruption of charged particles and electromagnetic radiation ejected from the Sun's surface (in the low corona and upper chromosphere) and lasting from a few minutes to several hours. Solar flare activity affects radio transmission on Earth and can produce auroras in Earth's atmosphere.

Solar mass: A unit equivalent to the mass of the Sun, or 1.989×10^{30} kilograms. Masses of other stars are often expressed in terms of solar masses.

Solar nebula: A cloud of mostly hydrogen and helium gas from which the Sun and planets are believed to have formed between 4.5 and 5.0 billion years ago.

Solar radiation: The radiation emitted by the Sun.

Solar system: The Sun and everything that orbits it, including planets and their satellites, plus numerous comets, asteroids, meteoroids, and other objects. *See also* Planet.

Solar ultraviolet radiation: Electromagnetic radiation emitted by the Sun in the spectral interval between approximately 90 and 400 nanometers.

Solar wind: The hot ionized gases, or plasma, that escape the Sun's gravitational field and flow in spirals outward at about 200 to 900 kilometers (124 to 558 miles) per second. It consists primarily of free protons, electrons, and alpha particles escaping from the Sun's corona. *See also* Stellar wind.

Sonar: An acronym for sound navigation ranging. A system for bouncing sonic and supersonic waves off a submerged object in order to determine its distance.

Sounding sensor: A sonarlike device that probes an atmosphere to detect data about temperature, moisture, and other conditions.

Space age: The age of space exploration, whose beginning is generally dated from October 4, 1957, the day on which the first artificial satellite, Sputnik 1, was launched into Earth orbit.

Space telescope: Any astronomical telescope that operates in space rather than on Earth.

Space-time: A four-dimensional coordinate system consisting of the three spatial dimensions and time as a fourth crewed.

Spacecraft: Any self-contained, crewed or uncrewed, space vehicle; more specifically, a deep space probe.

Specific gravity: The ratio of the density of a substance to the density of water.

Specific heat: The number of calories of heat required to raise the temperature of one gram of a substance $1°$ Celsius.

Spectral class: A system of classifying stars based on the pattern of absorption lines appearing in their spectra. Physically it indicates the photospheric temperature of the stars; in order of decreasing temperature the sequence runs O, B, A, F, G, K, M, L, T.

Spectrograph: A type of spectrometer that splits light into its component wavelengths and records the separated wavelengths photographically or by means of a charge coupled device.

Spectrometer: An instrument for obtaining and measuring a spectrum.

Spectroscope: An instrument that spreads electromagnetic radiation into its component wavelengths.

Spectroscopy: The science of breaking up light into its various components and studying them.

Spectrum: The entire range of electromagnetic radiation from long-wavelength radio waves to short-wavelength gamma rays; also, a limited range of wavelengths in which an instrument separates the component wavelengths or frequencies.

Speed of light. *See* Velocity of light.

Spherules: Rounded glass particles, probably formed by rapid cooling of molten material.

Spin: The rotation of a body about an axis through itself.

Spin axis: The line around which a body rotates.

Spiral arm: A dense region of heavy star formation in the disk of a spiral galaxy.

Spiral galaxy: A galaxy consisting of a bulge of gas and stars at the center, around which "arms" of stars and other celestial bodies, matter, and radiation rotate in a spiral fashion.

Sputtering: The bombardment of solid surfaces by high-energy particles, such that atoms and molecular fragments are eroded from the surface and the surface chemistry is altered.

Stagnation point: The point at which the repulsive force of a planet's magnetic field balances the pressure from the solar wind.

Star: A light-emitting body composed principally of hydrogen, whose heat and light are sustained by nuclear fusion, with a mass that is about one-tenth to one hundred times that of the Sun, a surface temperature ranging from about 3,000 to 50,000 kelvins, and an interior temperature ranging from about one million to hundreds of millions of kelvins.

Star color and temperature: Light energy from cool stars is emitted mostly at the low-energy red and infrared end of the spectrum, whereas hot stars are blue-white and produce more high-energy blue and ultraviolet light.

Star tracker: An electronic device programmed to detect and lock onto a celestial body, such as the star Canopus, to provide a spacecraft with a fixed point of reference for purposes of navigation.

Steady state theory: A model of the universe proposed by Hermann Bondi, Thomas Gold, and Fred Hoyle, which posits that the density of matter in the universe remains constant in an expanding universe, with new matter spontaneously created in the space between galaxies as they move apart. The theory is now generally considered obsolete because it does not explain the presence of the cosmic background radiation. *See also* Big bang theory, Cosmic background radiation, Dark energy, Dark matter.

Stefan-Boltzmann law: A relationship between the temperature of a "blackbody" and the rate at which it radiates energy.

Stellar core: The high-temperature, high-density region at the centers of stars where the nuclear reactions that form the heavy elements occur; no more than a few percent of the entire stellar radius.

Stellar evolution: The theory that describes the changes that occur in the internal structure of stars during their lifetimes.

Stellar luminosity: Energy emitted by per second by a star, commonly measured in joules per second (watts).

Stellar photosphere: The deepest layer of a star subject to direct observation, where the absorption lines caused by individual elements form; it lies at the top of the star's interior and the bottom of its atmosphere.

Stellar populations: A differentiation between stars of different age determined from their metal abundances and distribution.

Stellar wind: The ionized gases, or plasma, that flow out from stars at high speeds, composed mainly of free protons and electrons. *See also* Solar wind.

Stratosphere: The layer of Earth's atmosphere between the troposphere and the mesosphere, extending from about 15 to 50 kilometers (9.3 to 31 miles) above Earth's surface, roughly coinciding with the ozone layer, which absorbs the Sun's ultraviolet radiation and heats the stratosphere from a low of about –60° Celsius (213 kelvins) at the bottom to about 0° Celsius (273 kelvins) at its top. There is no meteorological activity or vertical air movement in this region of the atmosphere.

Strong nuclear force: The strongest of the four known fundamental forces in the universe; it acts over a very short range, essentially only within the nucleus itself. Details of the strong nuclear force are described by the standard model.

Sublimation: The transformation of a solid directly into a gas or a gas directly into a solid, passing through no liquid stage.

Subsatellite: A satellite carried into orbit by another satellite. Also, a satellite of a moon.

Sunspots: Dark spots that appear on the Sun's surface in cycles, their numbers increasing and decreasing with the approximately eleven-year solar cycle. These dark spots are about 500 kelvins cooler than a surrounding lighter area of the spot, which in turn is another 500 kelvins cooler than the surrounding photosphere. These cooler regions result from strong magnetic fields.

Supercluster: A gravitationally bound accumulation of galaxy clusters; most galaxies belong to clusters, with clusters of galaxies belonging to superclusters.

Superfluid: A fluid that flows without viscosity, as exemplified by liquid helium below its Lambda temperature.

Supergiants: The very largest stars. Red supergiants are up to one thousand times the size of the Sun.

Superimpositions: Craters that are formed within other craters, such as those formed when a meteoroid hits inside or on the walls of an existing crater.

Supernova: A violent explosion that tears a star apart. It can occur in a massive star that collapses when it has used up all its nuclear fuel (types Ib, Ic, and II) or in a white dwarf star that gains enough mass from a nearby companion to exceed the mass limit for electron degeneracy (type Ia).

Supernova remnant: The expanding debris cloud from a star that exploded as a supernova.

Swing-by: The close approach of a spacecraft as it passes by a planet. *See also* Flyby.

Synchronous rotation (or orbit): The state of a body in which the period of rotation equals the average orbital period.

Synodic period: The time between two successive occurrences of the same configuration of three or more bodies.

Système International d'Unités. *See* SI units.

T-Tauri stage: A temporary stage of instability in the early life of a star when it experiences great mass loss and an intense stellar wind.

Taenite: A nickel-iron alloy mineral with more than 25 percent nickel in solid solution; in octahedrite meteorites, taenite forms three-dimensional sheets that appear band-shaped in cut sections.

Tail: The apparent tail of the comet that consists of volatilized, ionized materials carried away from the comet by solar wind and dust blown away from the comet by the radiation pressure of sunlight.

Tectonics: The study of the processes forming the major structural features of a planet's crust.

Tektites: Glassy pellets believed to have formed when rock and soil that were vaporized by extraterrestrial impact recondensed and fell back to Earth.

Telemetry: Real-time transmission of data from a distance via radio signals.

Teleoperations: Manipulation of an orbiter, booster, or instruments in space from Earth via remote control.

Telescope: A device that permits detailed inspection of a distant object; originally, an instrument that used lenses or mirrors to collect large quantities of light and focus it into an image from a tiny area.

Terminator: The line, on a planet or moon, between dark and light that forms the boundary between day and night.

Terrestrial planet: Any of the largely solid, rocky planets of the inner solar system: Mercury, Venus, Earth, and Mars. Sometimes Earth's Moon is included in this list even though it is not strictly a planet.

Tharsis Dome: An immense bulge in the Martian crust in the Tharsis region of the planet, which rises 11 kilometers (6.82 miles) above the Martian surface.

Thermal erosion: The erosion of water ice from a solid state to vapor.

Thermal mapping: Gathering data from which to construct maps by means of instruments capable of sensing heat-producing electromagnetic radiation.

Thermokarst: Geological formations, resembling surface depressions, caused by the melting of subsurface ice or permafrost.

Thermonuclear reaction: A reaction in which atoms of a particular element are fused under pressure and temperature to form atoms of another element, with an explosive release of energy.

Thermosphere: The highest layer of Earth's atmosphere except for the exosphere, beginning at 85 kilometers (52.7 miles) above sea level. The oxygen and nitrogen that compose the atmosphere at this level are extremely rarefied, and are heated by the Sun's ultraviolet radiation to the point of ionization; hence the ionosphere (which lies between 50 and 500 kilometers, or 31 and 310 miles, above sea level) roughly coincides with the thermosphere. This is also the region in which auroras and meteors occur.

Tidal force: The gravitational attraction exerted between two bodies that may potentially distort their shape and cause internal deformation and heating.

Time dilation: The phenomenon, predicted by Albert Einstein's special theory of relativity, whereby time appears to slow down in a system moving near the speed of light from the vantage point of an observer outside that system.

Topside observation: Electronic scanning of Earth's (or another planet's) atmosphere from above. Used mainly in reference to meteorological satellites.

Total lunar eclipse: An eclipse in which the Moon passes completely into the umbra of the Earth.

Total solar eclipse: An eclipse in which the Sun, when viewed from some parts of Earth, is entirely hidden by the Moon.

Transfer orbit: The orbit into which a spacecraft is boosted from Earth orbit on its way to orbit around another celestial body. The spacecraft does not complete a full revolution of the transfer orbit, but only part of the ellipse.

Transit: The alignment of three celestial bodies, especially when a body of small apparent size passes across the disk of larger apparent size as viewed from the third body, such as the rare transit of Mercury or Venus across the Sun as viewed from Earth.

Transponder: A device that receives radio signals and automatically responds to them using the same frequency.

Triangulation: A means of determining the position of an object by calculation from known quantities: the distance between two fixed points and the angles formed between the line described by those points and the line from each of those to a third point, whose distance is not known. Triangulation is the oldest method of determining distances, both on Earth and in space.

Troilite: An iron sulfide mineral that is common to most meteorites but is very rare on Earth.

Troposphere: The lowest level of an atmosphere; it contains the highest density of material in the atmosphere and displays turbulent winds and chemical mixing. Earth's troposphere extends upward from the surface to between 12 and 15 kilometers (7.44 and 9.3 miles); the troposphere contains about 85 percent of the total mass of the atmosphere, almost all the water vapor in the atmosphere, and most of the greenhouse gases.

Tunneling: A quantum mechanical effect in which particles with small but finite probability can pass through a barrier, impenetrable according to Newtonian or "classical" mechanics, and emerge on the other side without energy loss.

Ultraviolet astronomy: The branch of astronomy that examines the ultraviolet emissions of celestial phenomena. Ultraviolet astronomy has developed with the advent of the space age; since ultraviolet rays are unable to penetrate Earth's atmosphere, instruments must be carried aloft by satellites such as the International Ultraviolet Explorer and FUSE.

Ultraviolet radiation: Electromagnetic radiation between about 90 and 400 nanometers which is between X radiation and visible violet light.

Ultraviolet spectrometer: An instrument that measures electromagnetic wavelengths in the ultraviolet range, used on satellites such as the International Ultraviolet Explorer and the Extreme Ultraviolet Explorer.

Umbra: A region of complete darkness, where all parts of the source of illumination are obscured. For example, the darkest inner region in a sunspot or the darkest inner part of the Moon's shadow.

Uniformitarianism: The principle that processes currently operating in nature have always been operating; it suggests that the large-scale features of the Earth were developed very slowly over vast periods of time.

Vacuum: A region containing absolutely nothing. Although a perfect vacuum does not exist in nature, near-vacuum conditions exist in interplanetary, interstellar, and intergalactic space.

Van Allen radiation belts: The two layers of Earth's magnetosphere, discovered by James Van Allen in the late 1950's, in which ionized particles spiral back and forth between Earth's magnetic poles. These zones pose hazards to electronic instruments aboard spacecraft.

Variable star: A star whose brightness varies over time, as a result of several intrinsic or extrinsic factors. There are thousands of such stars, and they can be categorized into seven classes based on the causes of their variation. *See also* Cepheid variable.

Velocity of light, c: the speed (3.00×10^8 meters/second or 186,000 miles per second) of electromagnetic waves, including light, through an empty vacuum.

Vernal equinox: The point on the celestial sphere where the Sun crosses the celestial equator passing from south to north.

Vernier rocket: A small thruster rocket used in space to make fine corrections to a spacecraft's orientation or trajectory.

Very long baseline interferometry: A technique used by radio astronomers to increase perceived sharpness (resolution) of radio sources by electronically linking several radio telescopes at widely spaced locations, allowing accurate mapping of celestial objects that emit radio waves. *See also* Interferometry.

Vesicular: Containing bubble-like cavities.

Virial theorem: The principle, derived from statistical mechanics, that in a cloud contracting gravitationally, the magnitude of the gravitational potential energy must be greater than or equal to twice the total of the thermal, rotational, and magnetic energies. Also known as Jeans's theorem (for Sir James Jeans, one of its developers).

Viscosity: A measure of the ability to flow; a substance with low viscosity flows easily.

Visible spectrum: That portion of the electromagnetic spectrum to which human eyes are sensitive, between about 400 and 700 nanometers, or 0.4 to 0.7 microns.

Visual binaries: Two stars that are gravitationally bound to each other and are either far enough apart or close enough to Earth to enable the two stars to be seen individually through a telescope.

Void: The region of space hundreds of millions of light-years across, which is relatively free of galaxies.

Volatile: Having a low condensation temperature, evaporating readily; the opposite of refractory. Also used in the plural, *volatiles*, as a noun to refer to substances with these properties.

Volcanism: The dynamic process in which molten material from the interior of a planet is transferred to the planet's solid surface, issuing forth explosively from cracks or other openings.

Water of hydration: Molecular water that is bound in the crystalline structure of rocks.

Wavelength: The distance from one extremum displacement to the next one of the same phase in any sort of wave phenomenon, such as water waves, sound waves, or electromagnetic radiation.

White dwarf star: A star with a mass 1.4 times the mass of the Sun or less and a diameter approximately that of Earth, which has exhausted all its nuclear fuels and shines only because it is still hot. As it radiates its energy away, it cools and fades, becoming a black dwarf star.

White spots: Atmospheric disturbances on Jovian planets, appearing at boundaries between zones and belts.

Widmanstätten pattern: A textural pattern, common to most iron meteorites containing 6-16 percent nickel, that occurs when kamacite is oriented parallel to the octahedral planes in taenite.

Wind tunnel: A large, tubular structure through which air is forced to flow at high speeds for the purpose of testing the behavior of aircraft and other structures that travel through the atmosphere.

Work: The product of a force on an object and the component of distance traveled that is parallel to that force.

World line: The graph of the motion of an object through space-time; objects moving with constant velocities have straight world lines, while accelerated motion is represented by curved world lines.

X band: The range of radio frequencies between 5.2 and 10.9 gigahertz.

X radiation: Electromagnetic radiation with wavelengths between 0.1 and 10 nanometers, the range lying between gamma and ultraviolet radiation on the electromagnetic spectrum.

X-ray astronomy: The branch of astronomy that examines the X-ray emissions of celestial phenomena. This radiation cannot be studied from the ground, since Earth's atmosphere absorbs most X radiation, and therefore has blossomed with the advent of space-craft that can carry X-ray telescopes and other detectors into space. The examination of celestial X-ray sources has led, among other things, to the discovery of neutron stars and black holes as members of binary star systems.

Yaw. *See* Pitch, roll, and yaw.

Zeeman effect: The broadening or splitting of spectral lines of light emitted by atoms caused by the presence of magnetic fields.

Zero gravity: The condition of absolute weightlessness, which occurs in free fall and in orbit. *See also* Microgravity.

Zircons: Mineral inclusions found in granitic rocks, zircons are often the only evidence left of early crustal rocks.

Zodiacal light: The glow seen in the west after sunset and in the east before dawn, caused by sunlight reflecting off microscopic dust particles.

Zones: Atmospheric high pressure bands of yellow-white clouds that encircle Jupiter.

General Bibliography

Abell, George O., David Morrison, and Sidney C. Wolff. *Exploration of the Universe.* 7th ed. Philadelphia: Saunders College Publishing, 1995.

Adams, Fred, and Greg Laughlin. *The Five Ages of the Universe.* New York: Free Press 1999.

Ahrens, C. Donald. *Essentials of Meteorology: An Invitation to the Atmosphere.* 5th ed. Florence, Ky.: Brooks/Cole, 2007.

_____. *Meteorology Today: An Introduction to Weather, Climate, and the Environment.* 8th ed. Florence, Ky.: Brooks/Cole, 2006.

Akasofu, Syun-Ichi, ed. *Dynamics of the Magnetosphere.* Dordrecht, Netherlands: D. Reidel, 1980.

Akasofu, Syun-Ichi, and Y. Kamide, eds. *The Solar Wind and the Earth.* Boston: D. Reidel, 1987.

Alexander, Arthur Francis O'Donel. *The Planet Saturn: A History of Observation, Theory, and Discovery.* New York: Macmillan, 1962.

Allen, Oliver E. *Atmosphere.* Alexandria, Va.: Time-Life Books, 1983.

American Geophysical Union. *Scientific Results of the Viking Project.* Washington, D.C.: Author, 1978.

Anderson, Geoff. *The Telescope: Its History, Technology, and Future.* Princeton, N.J.: Princeton University Press, 2007.

Andrews, David. *An Introduction to Atmospheric Physics.* Cambridge, England: Cambridge University Press, 2000.

Arnett, David. *Supernovae and Nucleosynthesis.* Princeton, N.J.: Princeton University Press, 1996.

Arny, Thomas T., and Stephen E. Schneider. *Explorations: An Introduction to Astronomy.* 5th ed. New York: McGraw-Hill, 2007.

Asimov, Isaac. *The Exploding Suns: The Secrets of the Supernovas.* New York: E. P. Dutton, 1985.

_____. *Understanding Physics.* 1966. Reprint. New York: Barnes & Noble Books, 1993.

Asimov, Isaac, and Richard Hantula. *Jupiter.* Milwaukee, Wis.: Gareth Stevens, 2002.

Asrar, Ghassem. *EOS: Science Strategy for the Earth Observing System.* New York: American Institute of Physics, 1994.

Atreya, S. K., J. B. Pollack, and M. S. Matthus, eds. *Origins and Evolution of Planetary and Satellite Atmospheres.* Phoenix: University of Arizona Press, 1989.

Aveni, Anthony. *Stairways to the Stars: Skywatching in Three Great Ancient Cultures.* New York: Wiley, 1999.

Backus, George. *Foundations of Geomagnetism.* New York: Cambridge University Press, 1996.

Bagenal, Fran, Timothy E. Dowling, and William B. McKinnon, eds. *Jupiter: The Planet, Satellites, and Magnetosphere.* New York: Cambridge University Press, 2007.

Baker, Robert H. *Astronomy.* 7th ed. Princeton, N.J.: Van Nostrand, 1959.

Baker, Victor R. *The Channels of Mars.* Austin: University of Texas Press, 1982.

Baldwin, R. A. *The Measure of the Moon.* Chicago: University of Chicago Press, 1963.

Bally, A. W. *Seismic Expression of Structural Styles.* Tulsa, Okla.: American Association of Petroleum Geologists, 1983.

Bally, John, and Bo Reipurth. *The Birth of Stars and Planets.* Cambridge, England: Cambridge University Press, 2006.

Balogh, André, Leonid Ksanfomality, and Rudolf von Steiger. *Mercury.* New York: Springer, 2008.

Balogh, André, Louis J. Lanzerotti, and S. T. Suess, eds. *The Heliosphere Through the Solar Activity Cycle.* New York: Springer, 2007.

Barbour, Julian B., and Herbert Pfister, eds. *Mach's Principle: From Newton's Bucket to Quantum Gravity.* Boston: Birkhäuser, 1995.

Barlow, Nadine. *Mars: An Introduction to Its Interior, Surface, and Atmosphere.* Cambridge, England: Cambridge University Press, 2008.

Barnes-Svarney, Patricia. *Asteroid: Earth Destroyer or New Frontier?* New York: Basic Books, 2003.

Barrow, John D., and Joseph Silk. *The Left Hand of Creation*. New York: Basic Books, 1983.

Barstow, Martin A., and Jay B. Holberg. *Extreme Ultraviolet Astrophysics*. Cambridge, England: Cambridge University Press, 2003.

Bartusiak, Marcia. *Thursday's Universe*. New York: Times Books, 1986.

Bascom, Willard. *A Hole in the Bottom of the Sea*. Garden City, N.Y.: Doubleday, 1961.

Beattie, Donald A. *Taking Science to the Moon: Lunar Experiments and the Apollo Program*. Baltimore: Johns Hopkins University Press, 2003.

Beatty, J. Kelly, Carolyn Collins Petersen, and Andrew Chaikin, eds. *The New Solar System*. 4th ed. Cambridge, Mass.: Sky, 1999.

Beiser, Arthur. *Concepts of Modern Physics*. 6th ed. New York: McGraw-Hill, 2002.

Bell, Jim. *Postcards from Mars: The First Photographer on the Red Planet*. New York: Dutton Adult, 2006.

_____, ed. *The Martian Surface*. Cambridge, England: Cambridge University Press, 2008.

Bell, Jim, and Jacqueline Mitton, eds. *Asteroid Rendezvous: NEAR Shoemaker's Adventures at Eros*. Cambridge, England: Cambridge University Press, 2002.

Belusevic, Radoje. *Relativity, Astrophysics, and Cosmology*. Weinheim, Germany: Wiley-VCH, 2008.

Bely, Pierre, ed. *The Design and Construction of Large Optical Telescopes*. New York: Springer, 2003.

Bennett, Jeffrey, et al. *The Cosmic Perspective*. 5th ed. San Francisco: Pearson/Addison-Wesley, 2008.

_____. *Stars, Galaxies, and Cosmology: The Cosmic Perspective*. 3d ed. San Francisco: Pearson/Addison-Wesley, 2004.

Benton, Julius. *Saturn and How to Observe It*. New York: Springer, 2005.

Bernal, J. D. *The Origin of Life*. New York: World Books, 1967.

Bertin, G., and C. C. Lin. *Spiral Structure in Galaxies: A Density Wave Theory*. Cambridge, Mass.: MIT Press, 1996.

Beskin, V. S., A. V. Gurevich, and Ya N. Istomin. *Physics of the Pulsar Magnetosphere*. Cambridge, England: Cambridge University Press, 2006.

Beutler, G., M. R. Drinkwater, R. Rummel, and Rudolf von Steiger. *Earth Gravity Field from Space: From Sensors to Earth Sciences*. Boston: Kluwer Academic, 2003.

Bevan, Alex, and John De Laeter. *Meteorites: A Journey Through Space and Time*. Washington, D.C.: Smithsonian Institution Press, 2002.

_____. *Thunderstones and Shooting Stars: The Meaning of Meteorites*. Cambridge, Mass.: Harvard University Press, 1986.

Bhatnagar, Arvind, and William Livingston. *Fundamentals of Solar Astronomy*. Hackensack, N.J.: World Scientific, 2005.

Blondel, Phillippe, and John Mason, eds. *Solar System Update*. New York: Springer, 2006.

Bobrowsky, Peter T., and Hans Rickman, eds. *Comet/Asteroid Impacts and Human Society: An Interdisplinary Approach*. New York: Springer, 2007.

Bok, Bart J., and Priscilla F. Bok. *The Milky Way*. 5th ed. Cambridge, Mass.: Harvard University Press, 1981.

Bolt, Bruce A. *Earthquakes*. New York: W. H. Freeman, 1988.

_____. *Inside the Earth: Evidence from Earthquakes*. San Francisco: W. H. Freeman, 1982.

Bond, Peter. *Distant Worlds: Milestones in Planetary Exploration*. New York: Copernicus Books, 2007.

Bone, Neil. *The Aurora: Sun-Earth Interaction*. New York: John Wiley, 1996.

Bortolotti, Dan. *Exploring Saturn*. New York: Firefly Books, 2003.

Bostrom, Robert C. *Tectonic Consequences of the Earth's Rotation*. New York: Oxford University Press, 2000.

Bothmer, Volker, and Ioannis A. Daglis. *Space Weather: Physics and Effects*. New York: Springer Praxis, 2006.

Bott, M. H. P. *The Interior of the Earth*. New York: Elsevier, 1982.

Bottke, William F., Jr., Albertoi Cellino, Paolo Paolicchi, and Richard P. Binzel, eds. *Asteroids III*. Tucson: University of Arizona Press, 2002.

Boucher, C., ed. *Earth Rotation and Coordinate Reference Frames.* International Association of Geodesy Symposia 105. New York: Springer, 1990.

Boyce, Joseph M. *The Smithsonian Book of Mars.* Washington, D.C.: Smithsonian Institution Press, 2002.

Brancazio, Peter J., ed. *The Origin and Evolution of Atmospheres and Oceans.* New York: John Wiley & Sons, 1964.

Brandt, John C., and Robert D. Chapman. *Introduction to Comets.* New York: Cambridge University Press, 2004.

Bredeson, Carmen. *NASA Planetary Spacecraft: Galileo, Magellan, Pathfinder, and Voyager.* New York: Enslow, 2000.

Brewer, Bryan. *Eclipse.* 2d ed. Seattle: Earth View, 1991.

Briggs, G. A., and F. W. Taylor. *The Cambridge Photographic Atlas of the Planets.* Cambridge, England: Cambridge University Press, 1982.

Brody, Judit. *The Enigma of Sunspots: A Story of Discovery and Scientific Revolution.* Edinburgh, Scotland: FlorisSunspots, 2002.

Brown, G. C. *The Inaccessible Earth: An Integrated View to Its Structure and Composition.* 2d ed. New York: Chapman and Hall, 1993.

Brown, Peter L. *Comets, Meteorites, and Men.* New York: Taplinger, 1975.

Brunier, Serge, and Anne-Marie Lagrange. *Great Observatories of the World.* Buffalo, N.Y.: Firefly Books, 2005.

Brusche, P., and Sundermann, J. *Tidal Friction and the Earth's Rotation.* Berlin: Springer, 1978.

Brush, Stephen G. *Nebulous Earth: The Origin of the Solar System and the Core of the Earth from Laplace to Jeffreys.* Cambridge, England: Cambridge University Press, 1996.

Buchler, J. Robert, and Henry Kandrop, eds. *Astrophysical Turbulence and Convection.* New York: New York Academy of Sciences, 2000.

Buchwald, Vagn F. *Handbook of Iron Meteorites: Their History, Distribution, Composition, and Structure.* Berkeley: University of California Press, 1975.

Burgess, Eric. *Uranus and Neptune: The Distant Giants.* New York: Columbia University Press, 1988.

_____. *Venus: An Errant Twin.* New York: Columbia University Press, 1985.

Burke, Bernard F., and Francis Graham-Smith. *An Introduction to Radio Astronomy.* 2d ed. Cambridge, England: Cambridge University Press, 2002.

Burke, John G. *Cosmic Debris: Meteorites in History.* Berkeley: University of California Press, 1986.

Cairns-Smith, A. G. *Genetic Takeover and the Mineral Origins of Life.* Cambridge, England: Cambridge University Press, 1982.

Calder, Nigel. *Einstein's Universe.* New York: Viking Press, 1979.

Cameron, A. G. W., ed. *Interstellar Communication.* New York: W. A. Benjamin, 1963.

Canup, R. M., and K. Righter, eds. *Origin of the Earth and Moon.* Tucson: University of Arizona Press, 2000.

Carr, Michael H. *The Surface of Mars.* Cambridge, England: Cambridge University Press, 2006.

Carroll, Bradley W., and Dale A. Ostlie. *An Introduction to Modern Astrophysics.* 2d ed. San Francisco: Pearson/Addison-Wesley, 2007.

Casoli, Fabienne, and Thérèse Encrenaz. *The New Worlds: Extrasolar Planets.* New York: Springer Praxis, 2007.

Cattermole, Peter John. *Venus: The Geological Story.* Baltimore: Johns Hopkins University Press, 1996.

Chaikin, Andrew. *A Man on the Moon: The Voyages of the Apollo Astronauts.* New York: Penguin, 2007.

Chaisson, Eric. *Cosmic Dawn: The Origins of Matter and Life.* Boston: Little, Brown, 1981.

_____. *The Hubble Wars.* New York: HarperCollins, 1994.

_____. *Relatively Speaking.* New York: W. W. Norton, 1988.

Chaisson, Eric, and Steve McMillan. *Astronomy: A Beginner's Guide to the Universe.* 5th ed. Upper Saddle River, N.J.: Pearson/Prentice Hall, 2007.

_____. *Astronomy Today.* 6th ed. New York: Addison-Wesley, 2008.

Chamberlain, Joseph W. *Theory of Planetary*

Atmospheres: An Introduction to Their Physics and Chemistry. New York: Academic Press, 1978.

Chamberlain, Von Dei, John Carlson, and M. Jane Young. *Songs from the Sky: Indigenous Astronomical and Cosmological Traditions of the World.* West Sussex, England: Ocarina Books, 2005.

Christensen, Lars Lindberg, Robert A. Fosbury, and M. Kornmesser. *Hubble: Fifteen Years of Discovery.* New York: Springer, 2006.

Christian, James L. *Extra-terrestrial Intelligence: The First Encounter.* Buffalo, N.Y.: Prometheus, 1976.

Clark, Pamela. *Dynamic Planet: Mercury in the Context of Its Environment.* New York: Springer, 2007.

Clayton, Donald. *Handbook of Isotopes in the Cosmos: Hydrogen to Gallium.* Cambridge, England: Cambridge University Press, 2003.

Close, Frank. *Apocalypse When? Cosmic Catastrophe and the Fate of the Universe.* New York: William Morrow, 1988.

Cohen, Jack, and Ian Stewart. *Evolving the Alien: The Science of Extraterrestrial Life.* London: Ebury Press, 2002.

Cohen, Martin. *In Darkness Born: The Study of Star Formation.* New York: Cambridge University Press, 1988.

Cohen, Nathan. *Gravity's Lens: Views of the New Cosmology.* New York: John Wiley & Sons, 1988.

Cole, George H. A., and Michael M. Woolfson, eds. *Planetary Science: The Science of Planets Around Stars.* Bristol: Institute of Physics, 2002.

Cole, Michael D. *Galileo Spacecraft: Mission to Jupiter.* New York: Enslow, 1999.

Collins, Michael. *Mission to Mars.* New York: Grove Weidenfeld, 1990.

Comins, N. *What if the Moon Didn't Exist? Voyages to Earths That Might Have Been.* New York: HarperCollins, 1993.

Condie, Kent, and Robert Sloan. *Origin and Evolution of Earth: Principles of Historical Geology.* Upper Saddle River, N.J.: Prentice Hall, 1998.

Consolmagno, Guy, and Martha Schaefer. *Worlds Apart: A Textbook in Planetary Sciences.* Englewood Cliffs, N.J.: Prentice Hall, 1994.

Cook, Alex. *The Greenhouse Effect: A Legacy.* Indianapolis: Dog Ear, 2007.

Cooke, Donald A. *The Life and Death of Stars.* New York: Crown, 1985.

Corfield, Richard. *Lives of the Planets: A Natural History of the Solar System.* New York: Basic Books, 2007.

Corliss, William R., ed. *The Moon and the Planets.* Glen Arm, Md.: Sourcebook Project, 1985.

Cornell, James, and Alan P. Lightman. *Revealing the Universe.* Cambridge, Mass.: MIT Press, 1982.

Coustenis, Athena, and Fredric W. Taylor. *Titan: Exploring an Earthlike World.* Hackensack, N.J.: World Scientific, 2007.

Cox, A. N., W. C. Livingston, and M. S. Matthews, eds. *Solar Interior and Atmosphere.* Tucson: University of Arizona Press, 1991.

Cramer, John A. *How Alien Would Aliens Be?* Lincoln, Nebr.: Writers Club Press, 2001.

Crick, Francis. *Life Itself: Its Origin and Nature.* New York: Simon & Schuster, 1981.

Cross, Charles A., and Patrick Moore. *The Atlas of Mercury.* New York: Crown, 1977.

Croswell, Ken. *Ten Worlds: Everything That Orbits the Sun.* Honesdale, Pa.: Boyds Mills Press, 2007.

Cruikshank, Dale P., ed. *Neptune and Triton.* Tucson: University of Arizona Press, 1995.

Dalrymple, G. Brent. *Ancient Earth, Ancient Skies: The Age of the Earth and Its Cosmic Surroundings.* Stanford, Calif.: Stanford University Press, 2004.

Darwin, G. *The Tides and Kindred Phenomena in the Solar System.* San Francisco: W. H. Freeman, 1962.

Dasch, Pat, ed. *Icy Worlds of the Solar System.* Cambridge, England: Cambridge University Press, 2004.

Dauber, Philip M., and Richard A. Muller. *The Three Big Bangs.* Reading, Mass.: Addison-Wesley, 1996.

Davies, Ashley Gerard. *Volcanism on Io: A Companion with Earth.* Cambridge, England: Cambridge University Press, 2007.

Davies, John. *Beyond Pluto: Exploring the Outer Limits of the Solar System.* New York: Cambridge University Press, 2001.

Davies, Merton E., Stephen E. Dwornik, Donald E. Gault, and Robert G. Strom. *Atlas of Mercury*. NASA SP-423. Washington, D.C.: National Aeronautics and Space Administration, Scientific and Technical Information Office, 1978.

Davies, P. C. W. *Space and Time in the Modern Universe*. New York: Cambridge University Press, 1981.

De Bremaecker, Jean-Claude. *Geophysics: The Earth's Interior*. New York: John Wiley & Sons, 1985.

Deeg, Hans, Juan Antonio Belmonte, and Antonio Aparicio, eds. *Extrasolar Planets*. New York: Cambridge University Press, 2008.

Delobeau, Francis. *The Environment of the Earth*. New York: Springer, 1971.

Delsemme, A. H., ed. *Comets, Asteroids, Meteorites*. Toledo, Ohio: University of Toledo Press, 1977.

De Pater, Imke, and Jack J. Lissauer. *Planetary Sciences*. New York: Cambridge University Press, 2001.

Dermott, S. F., ed. *The Origin of the Solar System*. New York: John Wiley & Sons, 1978.

Devorkin, David, and Robert W. Smith. *Hubble: Imaging Space and Time*. Washington, D.C.: National Geographic Society, 2008.

Dick, Steven J. *The Biological Universe: The Twentieth-Century Extraterrestrial Life Debate and the Limits of Science*. Cambridge, England: Cambridge University Press, 1996.

Dinwiddie, Robert, et al. *Universe*. New York: DK Adult, 2005.

Dixon, Dougal. *The Practical Geologist: The Introductory Guide to the Basics of Geology and to Collecting and Identifying Rocks*. New York: Fireside, 1992.

Dodd, Robert T. *Meteorites: A Petrologic-Chemical Synthesis*. London: Cambridge University Press, 1981.

_____. *Thunderstones and Shooting Stars: The Meaning of Meteorites*. Cambridge, Mass.: Harvard University Press, 1986.

Dole, Stephen H. *Habitable Planets for Man*. 2d ed. New York: Elsevier, 1970.

Domingue, D. L., and C. T. Russell, eds. *The MESSENGER Mission to Mercury*. New York: Springer, 2008.

Drexler, Jerome. *Discovering Postmodern Cosmology: Discoveries in Dark Matter, Cosmic Web, Big Bang, Inflation, Cosmic Rays, Dark Energy, Accelerating Cosmos*. Boca Raton, Fla.: Universal, 2008.

Dudly, W. W., and D. A. Williams. *Interstellar Chemistry*. New York: Academic Press, 1984.

Duncan, Todd, and Craig Tyler. *Your Cosmic Context: An Introduction to Modern Cosmology*. San Francisco: Pearson/Addison-Wesley, 2009.

Dunne, James A., and Eric Burgess. *The Voyage of Mariner 10: Mission to Venus and Mercury*. NASA SP-424. Washington, D.C.: National Aeronautics and Space Administration, Scientific and Technical Information Office, 1978.

Dvorak, Rudolf. *Extrasolar Planets: Formation, Detection, and Dynamics*. Weinheim, Germany: Wiley-VCH, 2008.

Eddington, Arthur. *The Expanding Universe*. 1933. Reprint. Cambridge, England: Cambridge University Press, 1987.

Eddy, John A. *A New Sun: The Solar Results from Skylab*. Washington, D.C.: Government Printing Office, 1979.

Elkins-Tanton, Linda T. *Jupiter and Saturn*. New York: Chelsea House, 2006.

_____. *Mars*. New York: Chelsea House, 2006.

_____. *The Sun, Mercury, and Venus*. New York: Chelsea House, 2006.

_____. *Uranus, Neptune, Pluto, and the Outer Solar System*. New York: Chelsea House, 2006.

Elliot, James, and Richard Kerr. *Rings: Discoveries from Galileo to Voyager*. Cambridge, Mass.: MIT Press, 1984.

Ellison, Mervyn Archdall. *The Sun and Its Influence*. London: Routledge & Kegan Paul, 1955.

Emiliani, Cesare. *The Scientific Companion*. New York: John Wiley and Sons, 1988.

Encrenaz, Thérèse, Jean-Pierre Bibring, Michel Blanc, and Maria-Antonietta Barucci. *The Solar System*. New York: Springer, 2004.

Encrenaz, Thérèse, Reinald Kallenbach, T. Owen, and C. Sotin, eds. *The Outer Planets and Their Moons: Comparative Studies of the Outer Planets Prior to the Exploration of the Saturn System by Cassini-Huygens*. New York: Springer, 2005.

Erickson, Jon. *Asteroids, Comets, and Meteorites: Cosmic Invaders of the Earth*. New York: Facts On File, 2003.

Esposito, Larry. *Planetary Rings*. Cambridge, England: Cambridge University Press, 2006.

Esposito, Larry W., Ellen R. Stofan, and Thomas E. Cravens, eds. *Exploring Venus as a Terrestrial Planet: Geophysical Monograph 176*. Washington, D.C.: American Geophysical Union, 2007.

European Space Agency. *The Atmospheres of Saturn and Titan*. ESA SP-241. Paris: Author, 1985.

Ezell, Edward, and Linda Ezell. *On Mars: Explorations of the Red Planet, 1958-1978*. NASA SP-4212. Washington, D.C.: Government Printing Office, 1984.

Fabian, A. C., K. A. Pounds, and R. D. Blandford. *Frontiers of X-ray Astronomy*. Cambridge, England: Cambridge University Press, 2004.

Fairbridge, Rhodes W. *The Encyclopedia of Geochemistry and Environmental Sciences*. Stroudsburg, Pa.: Bowden, Hutchinson and Ross, 1972.

Faure, Gunter, and Teresa M. Mensing. *Introduction to Planetary Science: The Geological Perspective*. New York: Springer, 2007.

Ferguson, Kitty. *Tycho and Kepler: The Unlikely Partnership That Forever Changed Our Understanding of the Heavens*. New York: Walker, 2002.

Ferington, Esther. *The Cosmos*. Alexandria, Va.: Time-Life Books, 1988.

Fernández, Julio Angel. *Comets: Nature, Dynamics, Origin, and Their Cosmological Relevance*. Dordrecht, Netherlands: Springer, 2005.

Ferreira, Pedro G. *The State of the Universe: A Primer in Modern Cosmology*. London: Phoenix, 2007.

Field, George. *The Space Telescope*. Chicago: Contemporary Books, 1989.

Fimmel, Richard O., Lawrence Colin, and Eric Burgess. *Pioneering Venus: A Planet Unveiled*. Washington, D.C.: National Aeronautics and Space Administration, 1995.

Fimmel, Richard O., James Van Allen, and Eric Burgess. *Pioneer: First to Jupiter, Saturn and Beyond*. NASA SP-446. Washington, D.C.: Government Printing Office, 1980.

Fischer, Daniel. *Mission Jupiter: The Spectacular Journey of the Galileo Spacecraft*. New York: Copernicus Books, 2001.

Foerstner, Abigail. *James Van Allen: The First Eight Billion Miles*. Iowa City: University of Iowa Press, 2007.

Forget, Françoise, Françoise Costard, and Philippe Lognonné. *Planet Mars: Story of Another World*. Chichester, England: Praxis, 2008.

Foster, J., and J. D. Nightingale. *A Short Course in General Relativity*. 2d ed. New York: Springer, 1995.

Foukal, Peter. *Solar Astrophysics*. 2d rev. ed. Weinheim, Germany: Wiley-VCH, 2004.

Fountain, John, and Rolf Sinclair. *Current Studies in Archaeoastronomy: Conversations Across Time and Space*. Durham, N.C.: Carolina Academic Press, 2005.

Fowler, C. M. R. *The Solid Earth: An Introduction to Global Geophysics*. 2d ed. New York: Cambridge University Press, 2004.

Fowler, G. C. *The Inaccessible Earth: An Integrated View to Its Structure and Composition*. 2d ed. New York: Chapman and Hall, 1993.

Fowles, Grant R., and George L. Cassiday. *Analytic Mechanics*. 7th ed. New York: Brooks/Cole, 2004.

Frakes, L. A. *Climates Throughout Geologic Time*. New York: Elsevier, 1980.

Fraknoi, Andrew, David Morrison, and Sidney Wolff. *Voyages to the Stars and Galaxies*. Belmont, Calif.: Brooks/Cole-Thomson Learning, 2006.

Frazier, Kendrick. *Our Turbulent Sun*. Englewood Cliffs, N.J.: Prentice-Hall, 1982.

_____. *Solar Systems*. Alexandria, Va.: Time-Life Books, 1985.

Freedman, Roger A., and William J. Kaufmann III. *Universe*. 8th ed. New York: W. H. Freeman, 2007.

French, Bevan M. *The Moon Book*. New York: Penguin Books, 1977.

Fridman, Alexei M., and Nikolai N. Gorkavyi. *Physics of Planetary Rings: Celestial Mechanics of Continuous Media*. New York: Springer, 1999.

Friedlander, Michael W. *Cosmic Rays*. Cambridge, Mass.: Harvard University Press, 1989.

Friedman, Herbert. *Sun and Earth*. San Francisco: W. H. Freeman, 1986.

Fukugita, Masataka, and Tsutomu Yanagida. *Physics of Neutrinos*. New York: Springer, 2003.

Fyfe, W. S. *Geochemistry*. Oxford, England: Clarendon Press, 1974.

Gabler, Robert E., Robert J. Sager, Sheila M. Brazier, and D. L. Wise. *Essentials of Physical Geography*. 8th ed. Florence, Ky.: Brooks/Cole, 2006.

Gamow, George. *Gravity*. New York: Dover, 2003.

Garland, G. D. *Introduction to Geophysics*. 2d ed. Philadelphia: W. B. Saunders, 1979.

Garlick, Mark A. *The Story of the Solar System*. Cambridge, England: Cambridge University Press, 2002.

Garrison, Tom. *Oceanography: An Invitation To Marine Science*. Florence, Ky.: Brooks/Cole, 2007.

Gasperini, Maurizio. *The Universe Before the Big Bang: Cosmology and String Theory*. Berlin: Springer, 2008.

Gehrels, Tom, ed. *Asteroids*. Tucson: University of Arizona Press, 1979.

_____. *Hazards Due to Comets and Asteroids*. Tucson: University of Arizona Press, 1994.

_____. *Jupiter*. Tucson: University of Arizona Press, 1976.

Gehrels, Tom, and Mildred Shapley Matthews, eds. *Saturn*. Tucson: University of Arizona Press, 1984.

Genet, Russell M., Donald S. Hayes, Douglas S. Hall, and David R. Genet. *Supernova 1987a: Astronomy's Explosive Enigma*. Mesa, Ariz.: Fairborn Press, 1988.

Gibson, Edward G. *The Quiet Sun*. NASA SP-303. Washington, D.C.: Government Printing Office, 1973.

Ginzburg, V. L., and S. I. Syrovatskii. *The Origin of Cosmic Rays*. New York: Macmillan, 1964.

Giovanelli, Ronald G. *Secrets of the Sun*. New York: Cambridge University Press, 1984.

Giunti, Carlo, and Chung W. Kim. *Fundamentals of Neutrino Physics and Astrophysics*. New York: Oxford University Press, 2007.

Glasstone, Samuel. *The Book of Mars*. Washington, D.C.: National Aeronautics and Space Administration, 1968.

Glendenning, Norman K. *Compact Stars: Nuclear Physics, Particle Physics, and General Relativity*. New York: Springer, 1997.

Goldsmith, Donald. *Worlds Unnumbered: The Search for Extrasolar Planets*. Sausalito, Calif.: University Science Books, 1997.

Goldsmith, Donald, and Tobias Owen. *The Search for Life in the Universe*. 3d ed. New York: University Science Books, 2001.

Golub, Leon, and Jay M. Pasachoff. *Nearest Star: The Surprising Science of Our Sun*. Cambridge, Mass.: Harvard University Press, 2001.

Gonzalez, Guillermo, and Jay W. Richards. *The Privileged Planet: How Our Place in the Cosmos Is Designed for Discovery*. Washington, D.C.: Regnery, 2004.

Goody, R. M., and J. C. G. Walker. *Atmospheres*. Englewood Cliffs, N.J.: Prentice-Hall, 1972.

Gould, Stephen Jay. *The Flamingo's Smile*. New York: W. W. Norton, 1985.

Greely, R. *Planetary Landscapes*. 2d ed. Boston: Allen and Unwin, 1994.

Green, Simon F., Mark H. Jones, and S. Jocelyn Burnell. *An Introduction to the Sun and Stars*. New York: Cambridge University Press, 2004.

Greenberg, John L. *The Problem of the Earth's Shape from Newton to Clairaut*. New York: Cambridge University Press, 1995.

Greenberg, Richard. *Europa the Ocean Moon: Search for an Alien Biosphere*. New York: Springer, 2005.

Greene, Brian. *The Elegant Universe: Superstrings, Hidden Dimensions, and the Quest for the Ultimate Theory*. New York: W. W. Norton, 2003.

Greenstein, George. *Frozen Star*. New York: Charles Scribner's Sons, 1983.

Grego, Peter. *The Moon and How to Observe It*. New York: Kindle Books, 2005.

Gregor, C. Bryan, et al. *Chemical Cycles in the Evolution of the Earth*. New York: John Wiley & Sons, 1988.

Gregory, Stephen A. *Introductory Astronomy and Astrophysics*. 4th ed. San Francisco: Brooks/Cole, 1997.

Grewing, M., F. Praderie, and R. Reinhard, eds. *Exploration of Halley's Comet*. New York: Springer, 1989.

Gribbin, John. *Blinded by the Light: The Secret Life of the Sun*. New York: Harmony Books, 1991.

_____. *In Search of the Big Bang*. New York: Bantam Books, 1986.

_____. *Spacewarps*. New York: Delacorte Press, 1984.

_____. *Timewarps*. New York: Delacorte Press, 1980.

Gribbin, John, and Mary Gribbin. *Fire on Earth*. New York: St. Martin's Press, 1996.

_____. *Stardust: Supernovae and Life, the Cosmic Connection*. New Haven, Conn.: Yale University Press, 2000.

Grinspoon, David. *Lonely Planets: The Natural Philosophy of Alien Life*. New York: Harper-Collins, 2004.

_____. *Venus Revealed: A New Look Below the Clouds of Our Mysterious Twin Planet*. New York: Basic Books, 1998.

Gussinov, Oktay, Efe Yazgan, and Askin Ankay, eds. *Neutron Stars, Supernovae, and Supernovae Remnants*. New York: Nova Science, 2007.

Haber, Frances C. *The Age of the World: Moses to Darwin*. 1959. Reprint. Westport, Conn.: Greenwood Press, 1978.

Haigh, Joanna, et al. *The Sun, Solar Analogs, and the Climate*. New York: Springer, 2005.

Halliday, David, Robert Resnick, and Jearl Walker. *Fundamentals of Physics, Extended*. 9th ed. New York: Wiley, 2007.

Hamblin, W. Kenneth, and Eric H. Christiansen. *Exploring the Planets*. New York: Macmillan, 1990.

Hammel, H. B. *The Ice Giant Systems of Uranus and Neptune*. New York: Springer, 2006.

Hansen, Carl J., Steven D. Kawaler, and Virginia Trimble. *Stellar Interiors: Physical Principles, Structures, Evolution*. New York: Springer, 2004.

Hansen, Joel E., and T. Takahashi, eds. *Climate Processes and Climate Sensitivity*. Geophysi-

cal Monograph 29. Washington, D.C.: American Geophysical Union, 1984.

Hansson, Anders. *Mars and the Development of Life*. New York: Ellis Horwood, 1991.

Hargreaves, John K. *The Upper Atmosphere and Solar-Terrestrial Relations*. New York: Van Nostrand Reinhold, 1979.

Harland, David M. *Cassini at Saturn: Huygens Results*. New York: Springer, 2007.

_____. *Jupiter Odyssey: The Story of NASA's Gelileo Mission*. New York: Springer, 2000.

_____. *Mission to Saturn: Cassini and the Huygens Probe*. New York: Springer Praxis, 2002.

_____. *Water and the Search for Life on Mars*. New York: Springer Praxis, 2005.

Harpur, Brian, and Laurence Anslow. *The Official Halley's Comet Project Book*. London: Hodder and Stoughton, 1985.

Hartmann, William K. *Moons and Planets*. 5th ed. Belmont, Calif.: Thomson Brooks/Cole, 2005.

_____. *A Traveler's Guide to Mars: The Mysterious Landscapes of the Red Planet*. New York: Workman, 2003.

_____, ed. *Astronomy*. 5th ed. Belmont, Calif.: Wadsworth, 2004.

Hartmann, William K., and Chris Impey. *Astronomy: The Cosmic Journey*. New York: Brooks/Cole, 2001.

Hartmann, William K., and Ron Miller. *The Grand Tour: A Traveler's Guide to the Solar System*. 3d ed. New York: Workman, 2005.

Hartmann, William K., and Odell Raper. *The New Mars: The Discoveries of Mariner 9*. NASA SP-337. Washington, D.C.: Government Printing Office, 1974.

Hartmann, William K., et al. *Out of the Cradle: Exploring the Frontiers Beyond Earth*. New York: Workman, 1984.

Harvey, Brian. *Russian Planetary Exploration: History, Development, Legacy, and Prospects*. New York: Springer, 2007.

Harwit, Martin. *Cosmic Discovery*. New York: Basic Books, 1981.

Hawking, Stephen. *A Brief History of Time*. New York: Bantam Books, 1988.

_____. *An Even Briefer History of Time*. New York: Bantam, 2008.

Hawking, Stephen W., and William Israel.

Three Hundred Years of Gravitation. New York: Cambridge University Press, 1987.

Hawking, Stephen, and Roger Penrose. *The Nature of Space and Time*. Princeton, N.J.: Princeton University Press, 2000.

Hawking, Stephen, et al. *The Future of Spacetime*. New York: W. W. Norton, 2003.

Hazen, Robert. *Genesis: The Scientific Quest for Life's Origins*. Washington, D.C.: Joseph Henry Press, 2005.

Heiken, Grant, and Eric Jones. *On the Moon: The Apollo Journals*. New York: Springer, 2007.

Henbest, Nigel. *Mysteries of the Universe*. New York: Van Nostrand Reinhold, 1981.

Henbest, Nigel, and Michael Marten. *The New Astronomy*. New York: Cambridge University Press, 1983.

Henderson-Sellers, A. *The Origin and Evolution of Planetary Atmospheres*. Bristol, England: Adam Hilger, 1983.

_____, ed. *Satellite Sensing of a Cloudy Atmosphere: Observing the Third Planet*. London: Taylor and Francis, 1984.

Hester, Jeff, George Blumenthal, David Burstein, and Bradford Smith. *Twenty-first Century Astronomy*. New York: W. W. Norton, 2007.

Hey, J. S. *The Evolution of Radio Astronomy*. New York: Science History Publications, 1973.

Hilbrecht, Heinz, Klaus Reinsch, Peter Volker, and Rainer Beck. *Solar Astronomy Handbook*. New York: Willmann-Bell, 1995.

Hille, Steele, and Michael Carlowicz. *The Sun*. New York: Harry N. Abrams, 2006.

Hillier, Rodney. *Gamma-Ray Astronomy*. New York: Oxford University Press, 1984.

Hirsch, Richard F. *Glimpsing an Invisible Universe: The Emergence of X-Ray Astronomy*. New York: Cambridge University Press, 1983.

Hofmann-Wellenhof, Bernhard. *Physical Geodesy*. New York: Springer, 2006.

Holland, H. D. *The Chemical Evolution of the Atmosphere and Oceans*. Princeton, N.J.: Princeton University Press, 1984.

Horton, E., and John H. Jones, eds. *Origin of the Earth*. New York: Oxford University Press, 1990.

Hoyt, Douglas V., and Kenneth H Schatten. *The Role of the Sun in Climate Change*. Oxford: Oxford University Press, 1997.

Hoyt, William G. *Lowell and Mars*. Tucson: University of Arizona Press, 1976.

Hubbard, William B. *Planetary Interiors*. New York: Van Nostrand Reinhold, 1984.

Hunt, Garry E., and Patrick Moore. *Atlas of Neptune*. Cambridge, England: Cambridge University Press, 1994.

_____. *Atlas of Saturn*. London: Mitchell Beazley, 1982.

_____. *Atlas of Uranus*. New York: Cambridge University Press, 1988.

_____. *Jupiter*. New York: Rand McNally, 1981.

Hurley, Patrick M. *How Old Is the Earth?* Garden City, N.Y.: Doubleday, 1959.

Hutchison, Robert. *The Search for Our Beginning: An Enquiry, Based on Meteorite Research, into the Origin of Our Planet and Life*. Oxford, England: Oxford University Press, 1983.

Ince, Martin. *The Rough Guide to the Earth 1*. New York Rough Guides, 2007.

Inglis, Mike. *Observer's Guide to Stellar Evolution*. London: Springer, 2003.

Irwin, Patrick G. J. *Giant Planets of Our Solar System: An Introduction*. 2d ed. New York: Springer, 2006.

Jacobs, J. A. *The Earth's Core*. 2d ed. New York: Academic Press, 1987.

Jacobs, John A., Richard D. Russell, and J. T. Wilson. *Physics and Geology*. 2d ed. New York: McGraw-Hill, 1974.

James, David E., ed. *The Encyclopedia of Solid Earth Geophysics*. New York: Van Nostrand Reinhold, 1989.

Jastrow, Robert. *God and the Astronomers*. New York: Warner Books, 1978.

_____. *Red Giants and White Dwarfs*. New York: W. W. Norton, 1979.

Jöels, Kerry Mark. *The Mars One Crew Manual*. New York: Ballantine Books, 1985.

Johnson, Francis S., ed. *Satellite Environment Handbook*. 2d ed. Stanford, Calif.: Stanford University Press, 1965.

Jones, Barrie W. *Discovering the Solar System*. New York: John Wiley & Sons, 1999.

Jordan, Stuart, ed. *The Sun as a Star*. NASA SP-450. Washington, D.C.: Government Printing Office, 1981.

Kaler, James B. *The Hundred Greatest Stars*. New York: Copernicus Books, 2002.

Kallenbach, Reinald, Thérèse Encrenaz, J. Geiss, and Konrad Mauersberger, eds. *Solar System History from Isotopic Signatures of Volatile Elements*. New York: Springer, 2003.

Kallmann-Bijl, Hildegaard, ed. *Space Research: Proceedings of the First International Space Science Symposium*. New York: Interscience, 1960.

Kargel, Jeffrey S. *Mars: A Warmer, Wetter Planet*. New York: Springer Praxis, 2004.

Karttunen, Hannu P., Pekka Kröger, Heikku Oja, and Markku Poutanen, eds. *Fundamental Astronomy*. 5th ed. New York: Springer, 2007.

Katz, Johnathan I. *High Energy Astrophysics*. Reading, Mass.: Addison-Wesley, 1987.

Kaula, William M. *Theory of Satellite Geodesy: Applications of Satellites to Geodesy*. New York: Dover, 2000.

Kelley, David, and Eugene Milone. *Exploring Ancient Skies: An Encyclopedia Survey of Archaeoastronomy*. New York: Springer, 2004.

Kenyon, Ian. *The Light Fantastic: A Modern Introduction to Classical and Quantum Optics*. New York: Oxford University Press, 2008.

Kerridge, John F., and Mildred S. Matthews, eds. *Meteorites and the Early Solar System*. Tucson: University of Arizona Press, 1988.

Kerrod, Robin. *Uranus, Neptune, and Pluto*. New York: Lerner, 2000.

Kerrod, Robin, Carole Stott, and David S. Leckrone. *Hubble: The Mirror on the Universe*. New York: Firefly Books, 2007.

Kieffer, Hugh H., Bruce M. Jakowsky, Conway W. Snyder, and Mildred Matthews, eds. *Mars*. Tucson: University of Arizona Press, 1992.

King, Elbert A., Jr. *Space Geology: An Introduction*. New York: John Wiley and Sons, 1976.

Kippenhahn, Rudolf. *Light from the Depths of Time*. New York: Springer, 1987.

_____. *One Hundred Billion Suns: The Birth, Life, and Death of the Stars*. New York: Basic Books, 1985.

Kirby-Smith, Henry T. *U.S. Observatories: A Directory and Travel Guide*. New York: Van Nostrand Reinhold, 1976.

Kitchin, Christopher R. *Astrophysical Techniques*. 5th ed. Boca Raton, Fl.: CRC Press, 2009.

_____. *Solar Observing Techniques*. New York: Springer, 2001.

Kloeppel, James E. *Realm of the Long Eyes: A Brief History of Kitt Peak National Observatory*. San Diego: Univelt, 1983.

Knapp, Ralph E. *Geophysics*. Exeter, England: Pergamon Press, 1995.

Knauss, John. *Introduction to Physical Oceanography*. 2d ed. Long Grove, Ill.: Waveland Press, 2005.

Kolerstrom, Nicholas. *Newton's Forgotten Lunar Theory: His Contribution to the Quest for Longitude*. Santa Fe, N.Mex.: Green Lion Press, 2000.

Kosofsky, L. J., and Farouk El-Baz. *The Moon as Viewed by the Lunar Orbiter*. NASA SP-200. Washington, D.C.: Government Printing Office, 1970.

Kovalevsky, Jean. *Modern Astrometry*. New York: Springer, 1995.

Kowal, Charles T. *Asteroids: Their Nature and Utilization*. Chichester, England: Ellis Harwood, 1988.

Krisciunas, Kevin. *Astronomical Centers of the World*. Cambridge, England: Cambridge University Press, 1988.

Kroner, A., G. N. Hanson, and A. M. Goodwin, eds. *Archaean Geochemistry: The Origin and Evolution of the Archaean Continental Crust*. Berlin: Springer, 1984.

Krüger, Harald. *Jupiter's Dust Disc: An Astrophysical Laboratory*. Aachen, Germany: Shaker-Verlag, 2003.

Kuiper, Gerard P., and Barbara M. Middlehurst, eds. *Telescopes: Stars and Stellar Systems*. Chicago: University of Chicago Press, 1978.

Kump, Lee R., James Kasting, and Robert Crane. *The Earth System*. Upper Saddle River, N.J.: Prentice Hall, 2003.

Kundu, M. R., B. Woodgate, and E. J. Schmahl,

eds. *Energetic Phenomena of the Sun*. Boston: Kluwer, 1989.

Kwok, Sun. *Physics and Chemistry of the Interstellar Medium*. New York: University Science Books, 2006.

Lambeck, Kurt. *The Earth's Variable Rotation: Geophysical Causes and Consequences*. New York: Cambridge University Press, 2005.

Lang, Kenneth R. *The Cambridge Guide to the Solar System*. Cambridge, England: Cambridge University Press, 2003.

_____. *Sun, Earth, and Sky*. 2d ed. New York: Springer, 2006.

Lang, Kenneth R., and Owen Gingerich, eds. *A Source Book in Astronomy and Astrophysics, 1900-1975*. Cambridge, Mass.: Harvard University Press, 1979.

Lankford, John, ed. *History of Astronomy: An Encyclopedia*. New York: Garland, 1997.

Lapedes, D. N., ed. *McGraw-Hill Encyclopedia of Geological Sciences*. New York: McGraw-Hill, 1978.

Lauretta, Dante S., and Harry Y. McSween, eds. *Meteorites and the Early Solar System II*. Tucson: University of Arizona Press, 2006.

Lemoine, M., J. Martin, and P. Peter, eds. *Inflationary Cosmology*. New York: Springer, 2008.

Leutwyler, Kristin, and John R. Casani. *The Moons of Jupiter*. New York: W. W. Norton, 2003.

Leverington, David. *Babylon to Voyager and Beyond: A History of Planetary Astronomy*. New York: Cambridge University Press, 2003.

Levin, Harold L. *The Earth Through Time*. 5th ed. Fort Worth: Saunders College Publishing, 1996.

Levine, Joel S., ed. *The Photochemistry of Atmospheres: Earth, the Other Planets, and Comets*. Orlando, Fla.: Academic Press, 1985.

Levinson, Alfred Abraham, ed. *Apollo 11 Lunar Science Conference: Proceedings*. 3 vols. Elmsford, N.Y.: Pergamon Press, 1970.

Levinton, Jeffrey S. *Genetics, Paleontology, and Macroevolution*. 2d ed. New York: Cambridge University Press, 2001.

Levy, David H. *Clyde Tombaugh, Discoverer of Planet Pluto*. New York: Sky, 2007.

_____. *David Levy's Guide to Observing Meteor Showers*. Cambridge, England: Cambridge University Press, 2008.

_____. *Impact Jupiter: The Crash of Comet Shoemaker-Levy 9*. New York: Basic Books, 2003.

_____. *The Quest for Comets: An Explosive Trail of Beauty and Danger*. New York: Plenum Press, 1994.

Lewis, Cherry. *The Dating Game: One Man's Search for the Age of the Earth*. Cambridge, England: Cambridge University Press, 2002.

Lewis, John S. *Physics and Chemistry of the Solar System*. 2d ed. San Diego, Calif.: Academic Press, 2004.

_____. *Rain of Iron and Ice: The Very Real Threat of Comet and Asteroid Bombardment*. New York: Basic Books, 1997.

Lewis, John S., and Ronald G. Prinn. *Planets and Their Atmospheres: Origin and Evolution*. New York: Academic Press, 1983.

Liddle, Andrew, and Jon Loveday. *The Oxford Companion to Cosmology*. New York: Oxford University Press, 2008.

Littmann, Mark. *Planets Beyond: Discovering the Outer Solar System*. New York: Dover, 2004.

Lockman, F. J., F. D. Ghigo, and D. S. Balsar, eds. *But It Was Fun: The First Forty Years of Radio Astronomy at Green Bank*. Washington, D.C.: National Radio Astronomy Observatory, 2007.

Lopes, Rosaly M. C., and T. K. P. Gregg. *Volcanic Worlds: Exploring the Solar System's Volcanoes*. New York: Springer, 2004.

Lopes, Rosaly M. C., and John R. Spencer. *Io After Galileo: A New View of Jupiter's Volcanic Moon*. Heidelberg: Springer, 2007.

Lorenz, Ralph, and Jacqueline Mitton. *Titan Unveiled: Saturn's Mysterious Moon Explored*. Princeton, N.J.: Princeton University Press, 2008.

Lovett, Laura, Joan Harvath, and Jeff Cuzzi. *Saturn: A New View*. New York: Harry N. Abrams, 2006.

Lowell, Percival H. *Mars and Its Canals*. New York: Macmillan, 1906.

Luisi, Pier Luigi. *The Emergence of Life: From Chemical Origins to Synthetic Biology*. New York: Cambridge University Press, 2006.

Lunar and Planetary Institute, Houston, Texas.

Basaltic Volcanism on the Terrestrial Planets. Elmsford, N.Y.: Pergamon Press, 1981.

Lyne, Andrew G., and Francis Graham-Smith. *Pulsar Astronomy*. 3d ed. Cambridge, England: Cambridge University Press, 2006.

McAnally, John W. *Jupiter, and How to Observe It*. New York: Springer, 2008.

McBride, Neil, and Gilmour Iain, eds. *An Introduction to the Solar System*. Cambridge, England: Cambridge University Press, 2004.

Maccarone, Thomas J. *From X-ray Binaries to Quasars: Black Holes on All Mass Scales*. New York: Kindle, 2006.

McConnell, Anita. *Geomagnetic Instruments Before 1900*. London: Harriet Wynter, 1980.

McElhinny, M. W., ed. *The Earth: Its Origin, Structure, and Evolution*. New York: Academic Press, 1979.

McFadden, Lucy-Ann Adams, Paul Robest Weissman, and T. V. Johnson, eds. *Encyclopedia of the Solar System*. San Diego: Academic Press, 2007.

Mackenzie, Dana. *The Big Splat: Or, How Our Moon Came to Be*. Hoboken, N.J.: John Wiley & Sons, 2003.

MacKenzie, Fred T. *Our Changing Planet: An Introduction to Earth System Science and Global Environmental Change*. Upper Saddle River, N.J.: Prentice Hall, 2002.

McLean, Ian S. *Electronic Imaging in Astronomy: Detectors and Instrumentation*. New York: Springer, 2008.

McSween, Harry Y., Jr. *Meteorites and Their Parent Planets*. 2d ed. New York: Cambridge University Press, 1999.

_____. *Stardust to Planets*. New York: St. Martin's Griffin, 1993.

Magli, Giulio. *Mysteries and Discoveries of Archaeoastronomy: From Giza to Easter Island*. New York: Springer, 2009.

Malphrus, Benjamin K. *The History of Radio Astronomy and the National Radio Astronomy Observatory: Evolution Toward Big Science*. Malabar, Fla.: Krieger, 1996.

Mammana, Dennis, and Donald McCarthy. *Other Suns, Other Worlds? The Search for Extrasolar Planetary Systems*. New York: St. Martin's Press, 1995.

Manuel, Oliver. *Origin of Elements in the Solar System: Implications of Post-1967 Observations*. New York: Springer, 2001.

Maran, Stephen P., ed. *The Astronomy and Astrophysics Encyclopedia*. Foreword by Carl Sagan. New York: Van Nostrand Reinhold, 1992.

Marov, Mikhail Ya, and David Grinspoon. *The Planet Venus*. New Haven, Conn.: Yale University Press, 1998.

Marschall, Laurence A. *The Supernova Story*. New York: Plenum Press, 1988.

Marten, Michael, and John Chesterman. *The Radiant Universe*. New York: Macmillan, 1980.

Mason, Brian. *Meteorites*. New York: John Wiley & Sons, 1962.

Mason, Brian, and William G. Melson. *The Lunar Rocks*. New York: Wiley-Interscience, 1970.

Meadows, A. J. *Early Solar Physics*. Elmsford, N.Y.: Pergamon Press, 1970.

Melchior, Paul. *The Earth Tides*. Oxford, England: Pergamon Press, 1966.

Melosh, H. J. *Impact Cratering: A Geologic Process*. New York: Oxford University Press, 1996.

Merrill, R. T., and M. W. McElhinney. *The Earth's Magnetic Field*. New York: Academic Press, 1983.

Meyer-Vernet, Nicole. *Basics of the Solar Wind*. Cambridge, England: Cambridge University Press, 2007.

Mezzacappa, Anthony, and George M. Fuller, eds. *Open Issues in Core Collapse Supernova Theory*. Hackensack, N.J.: World Scientific, 2005.

Michaud, Michael A. G. *Contact with Alien Civilizations: Our Hopes and Fears About Encountering Extraterrestrials*. New York: Springer, 2006.

Miller, Ron. *Extrasolar Planets*. Minneapolis, Minn.: Twenty-First Century Books, 2002.

_____. *Uranus and Neptune*. Brookfield, Conn.: Twenty-First Century Books, 2003.

Miller, S. L., and L. E. Orgel. *The Origins of Life on Earth*. Englewood Cliffs, N.J.: Prentice-Hall, 1974.

Milone, Eugene F., and William Wilson. *Solar System Astrophysics: Background Science on* the Inner Solar System. New York: Springer, 2008.

Miner, Ellis D. *Uranus: The Planet, Rings, and Satellites*. New York: Ellis Horwood, 1990.

Miner, Ellis D., and Randii R. Wessen. *Neptune: The Planet, Rings, and Satellites*. New York: Springer, 2002.

Misner, Charles W., Kip S. Thorne, and John A. Wheeler. *Gravitation*. San Francisco: W. H. Freeman, 1973.

Mitton, Simon. *Daytime Star: The Story of Our Sun*. New York: Charles Scribner's Sons, 1981.

Mobberley, Martin. *Supernovae and How to Observe Them*. New York: Springer, 2007.

Moche, Dinah L. *Astronomy: A Self-Teaching Guide*. 6th ed. New York: John Wiley & Sons, 2004.

Moldwin, Mark. *An Introduction to Space Weather*. Cambridge, England: Cambridge University Press, 2008.

Monod, Jacques. *Chance and Necessity*. Translated by Austryn Wainhouse. New York: Alfred A. Knopf, 1971.

Montesinos, Benjamín, Alvaro Giménez, and Edward F. Guinan, eds. *The Evolving Sun and Its Influence on Planetary Environments*. San Francisco: Astronomical Society of the Pacific, 2001.

Moore, Patrick. *Astronomical Telescopes and Observatories for Amateurs*. New York: W. W. Norton, 1973.

_____. *Guide to Mars*. New York: W. W. Norton, 1977.

_____. *Moore on Mercury: The Planet and the Missions*. New York: Springer, 2006.

_____. *On the Moon*. London: Cassell, 2001.

_____. *The Planet Neptune: An Historical Survey Before Voyager*. 2d ed. New York: John Wiley and Sons, 1996.

Morrison, David. *Voyages to Saturn*. NASA SP-451. Washington, D.C.: Government Printing Office, 1982.

Morrison, David, and Tobias Owen. *The Planetary System*. 3d ed. San Francisco: Pearson/Addison-Wesley, 2003.

Morrison, David, and Jane Samz. *Voyage to Jupiter*. NASA SP-439. Washington, D.C.: Government Printing Office, 1980.

Morrison, David, Sidney Wolf, and Andrew Fraknoi. *Abell's Exploration of the Universe*. 7th ed. Philadelphia: Saunders College Publishing, 1995.

Motz, Lloyd, ed. *Rediscovery of the Earth*. New York: Van Nostrand Reinhold, 1979.

Muller, Richard. *Nemesis: The Death Star*. New York: Weidenfeld & Nicolson, 1988.

Munk, W. H., and G. J. F. MacDonald. *The Rotation of Earth: A Geophysical Discussion*. New York: Cambridge University Press, 1960.

Murdin, Paul, and Lesley Murdin. *Supernova*. New York: Cambridge University Press, 1985.

Murray, Bruce, Michael C. Malin, and Ronald Greeley. *Earthlike Planets*. San Francisco: W. H. Freeman, 1981.

Mutch, Thomas A. *Geology of the Moon*. Rev. ed. Princeton, N.J.: Princeton University Press, 1972.

_____, comp. *The Geology of Mars*. Princeton, N.J.: Princeton University Press, 1976.

Nagy, B. *Carbonaceous Meteorites*. New York: Elsevier, 1975.

National Aeronautics and Space Administration. *The Case for Mars Concept Development for a Mars Research Station: Concept Development for a Mars Research Station*. San Francisco: University Press of the Pacific, 2002.

_____. *Preliminary Science Report: Apollo 11*. NASA SP-214. Washington, D.C: Government Printing Office, 1969.

_____. *Preliminary Science Report: Apollo 12*. NASA SP-235. Washington, D.C: Government Printing Office, 1970.

_____. *Preliminary Science Report: Apollo 14*. NASA SP-272. Washington, D.C.: Government Printing Office, 1971.

_____. *Preliminary Science Report: Apollo 15*. NASA SP-289. Washington, D.C.: Government Printing Office, 1972.

_____. *Preliminary Science Report: Apollo 16*. NASA SP-315. Washington, D.C.: Government Printing Office, 1972.

_____. *Preliminary Science Report: Apollo 17*. NASA SP-330. Washington, D.C.: Government Printing Office, 1973.

_____. *Viking 1, Early Results*. NASA SP-408. Springfield, Va.: National Technical Information Service, 1976.

National Aeronautics and Space Administration Advisory Council. *Earth System Science Overview*. Washington, D.C.: Government Printing Office, 1986.

National Research Council. *Changing Climate: Report of the Carbon Dioxide Assessment Committee*. Washington, D.C.: National Academy Press, 1983.

Newburn, R. L., M. Neugebauer, and Jurgen H. Rahe, eds. *Comets in the Post-Halley Era*. New York: Springer, 2007.

Nicolson, Iain. *The Sun*. New York: Rand McNally, 1982.

Nisbet, Evan G. *The Young Earth: An Introduction to Archaean Geology*. Winchester, Mass.: Unwin Hyman, 1987.

Noll, Keith S., and Harold A. Weaver, and Paul D. Feldman, eds. *The Collision of Comet Shoemaker-Levy 9 and Jupiter*. Cambridge, England: Cambridge University Press, 1996.

North, Gerald. *Observing Variable Stars, Novae, and Supernovae*. New York: Cambridge University Press, 2004.

North, John. *Cosmos: An Illustrated History of Astronomy and Cosmology*. Chicago: University of Chicago Press, 2008.

Norton, O. Richard. *The Cambridge Encyclopedia of Meteorites*. New York: Cambridge University Press, 2002.

_____. *Rocks from Space: Meteorites and Meteorite Hunters*. 2d ed. Missoula, Mont.: Mountain Press, 1998.

Norton, O. Richard, and Lawrence Chitwood. *Field Guide to Meteors and Meteorites*. London: Springer, 2008.

Novikov, Igor. *Black Holes and the Universe*. Translated by Vitaly Kisin. Cambridge, England: Cambridge University Press, 1995.

Noyes, Robert W. *The Sun, Our Star*. Cambridge, Mass.: Harvard University Press, 1982.

Ollivier, Marc, T. Encrenaz, F. Rocaves, and F. Selsis. *Planetary Systems: Detection, Formation, and Habitability of Extrasolar Planets*. New York: Springer, 2008.

Orloff, Richard W., and David M. Harland. *Apollo: The Definitive Sourcebook*. New York: Springer, 2006.

Oxlade, Chris. *Jupiter, Neptune, and Other Outer Planets*. New York: Rosen Central, 2007.

Ozima, Minoru. *The Earth: Its Birth and Growth*. Translated by J. F. Wakabayashi. New York: Cambridge University Press, 1981.

Pagel, Bernard. *Nucleosynthesis and Chemical Evolution of Galaxies*. Cambridge, England: Cambridge University Press, 1997.

Pannekoek, A. *A History of Astronomy*. London: Barnes & Noble Books, 1969.

Parker, Barry. *Einstein's Dream*. New York: Plenum Press, 1986.

Parker, Sybil P., ed. *McGraw-Hill Encyclopedia of Physics*. New York: McGraw-Hill, 1983.

Pasachoff, Jay M., and Will Tirion. *Field Guide to the Stars and Planets*. 5th ed. Boston: Houghton Mifflin, 1999.

Peek, Bertrand M. *The Planet Jupiter*. London: Macmillan, 1958.

Percy, John. *Understanding Variable Stars*. Cambridge, England: Cambridge University Press, 2007.

Peuker-Ehrenbrink, Bernhard, and Birger Schmitz, eds. *Accretion of Extraterrestrial Matter Throughout Earth's History*. New York: Kluwer Academic/Plenum, 2001.

Pickard, George L. *Descriptive Physical Oceanography: An Introduction*. 5th ed. New York: Pergamon Press, 1990.

Pomerantz, Martin A. *Cosmic Rays*. New York: Van Nostrand Reinhold, 1971.

Ponnamperuma, C., ed. *Cosmochemistry and the Origins of Life*. Dordrecht, Netherlands: Reidel, 1982.

Ponnamperuma, C., and A. G. W. Cameron. *Interstellar Communication: Scientific Perspectives*. Boston: Houghton Mifflin, 1974.

Prialnik, Dina. *An Introduction to the Theory of Stellar Structure and Evolution*. Cambridge, England: Cambridge University Press, 2000.

Pulsating Stars. 2 vols. Introductions by F. G. Smith, A. Hewish, and T. Gold. New York: Plenum Press, 1968-1969.

Rabinowitz, Avi. *Warped Spacetime, the Einstein Equations, and the Expanding Universe*. New York: Springer, 2009.

Raeburn, Paul, and Matt Golombek. *Un-*

covering the Secrets of the Red Planet. Washington, D.C.: National Geographic Society, 1998.

Raup, David M. *The Nemesis Affair*. New York: W. W. Norton, 1986.

Reid, Neil, and Suzanne Hawley. *New Light on Dark Stars: Red Dwarfs, Low-Mass Stars, Brown Stars*. 2d ed. New York: Springer Praxis, 2005.

Rey, H. A. *The Stars: A New Way to See Them*. 1952. Reprint. Boston: Houghton Mifflin, 1988.

Reynolds, Mike. *Falling Stars: A Guide to Meteors and Meteorites*. Mechanicsburg, Pa.: Stackpole Books, 2001.

Richardson, Robert S. *Exploring Mars*. New York: McGraw-Hill, 1954.

Ripley, S. Dillon. *Fire of Life: The Smithsonian Book of the Sun*. Washington, D.C.: Smithsonian Exhibition Books, 1981.

Robinson, Keith. *Spectroscopy: The Key to the Stars: Reading the Lines in Stellar Spectra*. New York: Springer, 2007.

Rosenburg, G. D., and S. K. Runcorn, eds. *Growth Rhythms and the History of the Earth's Rotation*. New York: John Wiley & Sons, 1975.

Rossi, Bruno. *Cosmic Rays*. New York: McGraw-Hill, 1964.

Rothery, David A. *Satellites of the Outer Planets: Worlds in Their Own Right*. New York: Oxford University Press, 1999.

Rucker, Rudolf. *Geometry, Relativity, and the Fourth Dimension*. New York: Dover, 1977.

Ruddiman, William F. *Earth's Climate: Past and Future*. 2d ed. New York: W. H. Freeman, 2008.

Russell, Chrisopher T. *The Cassini-Huygens Mission: Orbiter Remote Sensing Investigations*. New York: Springer, 2006.

_____. *Deep Impact Mission: Looking Beneath the Surface of a Cometary Nucleus*. New York: Springer, 2005.

Rybicki, George B., and Alan P. Lightman. *Radiative Processes in Astrophysics*. New York: Wiley, 1979.

Sagan, Carl. *Contact*. New York: Pocket, 1997.

_____. *Cosmos*. New York: Random House, 1980.

Sagan, Carl, and Ann Druyan. *Comet*. New York: Random House, 1985.

Salop, Lazarus J. *Geological Evolution of the Earth During the Precambrian*. Translated by V. P. Grudina. New York: Springer/Verlag, 1983.

Sartori, Leo. *Understanding Relativity: A Simplified Approach to Einstein's Theories*. Berkeley: University of California Press, 1996.

Savage, Candace. *Aurora: The Mysterious Northern Lights*. New York: Firefly Books, 2001.

Schaaf, Fred. *Comet of the Century: From Halley to Hale-Bopp*. New York: Copernicus, 1997.

Scharf, Caleb. *Extrasolar Planets and Astrobiology*. Herndon, Va.: University Science Books, 2008.

Schindewolf, Otto H. *Basic Questions in Paleontology: Geologic Time, Organic Evolution, and Biological Systematics*. Translated by Judith Schaefer. Chicago: University of Chicago Press, 1994.

Schlegel, Eric M. *The Restless Universe: Understanding X-ray Astronomy in the Age of Chandra and Newton*. New York: Oxford University Press, 2002.

Schmitt, Harrison J. *Return to the Moon: Exploration, Enterprise, and Energy in the Human Settlement of Space*. New York: Copernicus Books, 2006.

Schmude, Richard. *Uranus, Neptune, and Pluto and How to Observe Them*. New York: Springer, 2008.

Schneider, Peter. *Extragalactic Astronomy and Cosmology: An Introduction*. New York: Springer, 2006.

Schneider, Stephen E., and Thomas T. Arny. *Pathways to Astronomy*. 2d ed. New York: McGraw-Hill, 2008.

Schopf, J. William, ed. *Earth's Earliest Biosphere: Its Origin and Evolution*. Princeton, N.J.: Princeton University Press, 1984.

_____. *Life's Origin: The Beginnings of Biological Evolution*. Berkeley: University of California Press, 2002.

Schrunk, David, Burton Sharpe, Bonnie L. Cooper, and Madhu Thangavelu. *The Moon: Resources, Future Development, and Settlement*. New York: Springer Praxis, 2008.

Schultz, Peter H. *Moon Morphology: Interpretations Based on Lunar Orbiter Photography*. Austin: University of Texas Press, 1974.

Schwartz, Joseph, and Michael McGuinness. *Einstein for Beginners*. New York: Pantheon Books, 1979.

Schwinger, Julian. *Einstein's Legacy: The Unity of Space and Time*. New York: Scientific American Books, 1986.

Sears, D. W. *The Nature and Origin of Meteorites*. Bristol, England: Adam Hilger, 1978.

Seeds, Michael A. *Foundations of Astronomy*. 9th ed. Belmont, Calif.: Thomson Brooks/Cole, 2007.

_____. *Horizons: Exploring the Universe*. New York: Brooks/Cole, 2007.

Seibold, E., and W. Berger. *The Sea Floor*. New York: Springer, 1982.

Seielstad, George A. *At the Heart of the Web: The Inevitable Genesis of Intelligent Life*. Boston, Mass.: Harcourt Brace Jovanovich, 1989.

Sekanina, Zdenek, ed. *The Comet Halley Archive Summary Volume*. Pasadena, Calif.: Jet Propulsion Laboratory (International Halley Watch), California Institute of Technology, 1991.

Selley, Richard, Robin Cocks, and Ian Plimer, eds. *Encyclopedia of Geology*. 5 vols. Oxford, England: Elsevier Academic Press, 2005.

Serge, Brunier. *Solar System Voyage*. Translated by Storm Dunlop. New York: Cambridge University Press, 2000.

Serway, Raymond A., et al. *College Physics*. 7th ed. New York: Brooks/Cole, 2005.

Severny, A. *Solar Physics*. San Francisco: University Press of the Pacific, 2004.

Sheehan, William, and Stephen James O'Meara. *Mars: The Lure of the Red Planet*. New York: Prometheus Books, 2001.

Shklovskii, I. S., and Carl Sagan. *Intelligent Life in the Universe*. San Francisco: Holden-Day, 1966.

Short, Nicholas M. *Planetary Geology*. Englewood Cliffs, N.J.: Prentice-Hall, 1975.

Shu, Frank H. *The Physical Universe: An Introduction to Astronomy*. Mill Valley, Calif.: University Science Books, 1982.

Silk, Joseph. *The Big Bang*. Rev. ed. New York: W. H. Freeman, 1989.

Sion, Edward M., Stephane Vennes, and Harry L. Shipman. *White Dwarfs: Cosmological and Galactic Probes*. New York: Springer, 2005.

Skinner, Brian J. *The Blue Planet: An Introduction to Earth System Science*. New York: John Wiley, 1995.

Skinner, Brian J., and S. C. Porter. *The Dynamic Earth*. 5th ed. New York: John Wiley & Sons, 2006.

Slade, Suzanne. *A Look at Jupiter*. New York: PowerKids Press, 2008.

Smart, William M. *The Origin of the Earth*. 2d ed. New York: Cambridge University Press, 1953.

Smith, David G., ed. *The Cambridge Encyclopedia of Earth Sciences*. New York: Cambridge University Press, 1982.

Smith, F. G. *Pulsars*. Cambridge, England: Cambridge University Press, 1977.

Smith, Michael D. *The Origin of Stars*. London: Imperial College Press, 2004.

Snow, Theodore P. *The Dynamic Universe*. Rev. ed. St. Paul, Minn.: West, 1991.

Sobel, Dava. *The Planets*. New York: Viking, 2005.

Sonett, C. P., M. S. Giampapa, and M. S. Matthews, eds. *The Sun in Time*. Tucson: University of Arizona Press, 1991.

Soon, Willie Wei-Hock, and Steven H. Yaskell. *The Maunder Minimum and the Variable Sun-Earth Connection*. Hackensack, N.J.: World Scientific, 2003.

Spangenburg, Ray, and Kit Moser. *A Look at Mercury*. New York: Franklin Watts, 2003.

_____. *A Look at Venus*. New York: Franklin Watts, 2002.

_____. *Meteors, Meteorites, and Meteoroids*. Secaucus, N.J.: Franklin Watts, 2002.

Spencer, John R., and Jacqueline Mitton, eds. *The Great Comet Crash: The Collision of Comet Shoemaker-Levy 9 and Jupiter*. Cambridge, England: Cambridge University Press, 1995.

Spitzer, Lyman, Jr. *Physical Processes in the Interstellar Medium*. New York: Wiley, 1998.

Spudis, Paul D. *The Geology of Multi-ring Impact Basins: The Moon and Other Planets*. New York: Cambridge University Press, 1993.

Squyres, Steve. *Roving Mars: Spirit, Opportu-

nity, and the Exploration of the Red Planet. New York: Hyperion, 2006.

Stacey, F. D. *Physics of the Earth*. New York: John Wiley & Sons, 1977.

Stephenson, Bruce. *Kepler's Physical Astronomy*. Princeton, N.J.: Springer, 1994.

Stephenson, F. Richard. *Historical Eclipses and Earth's Rotation*. New York: Cambridge University Press, 2008.

Stern, Alan, and Jacqueline Mitton. *Pluto and Charon*. New York: Wiley, 1999.

Stix, Michael. *The Sun*. New York: Springer, 2004.

Stone, Robert G., Kurt W. Weiler, Melvyn L. Goldstein, and Jean-Louis Bouqueret, eds. *Radio Astronomy at Long Wavelengths*. New York: American Geophysical Union, 2000.

Strahler, Arthur N., and Alan H. Strahler. *Modern Physical Geography*. 4th ed. New York: John Wiley & Sons, 1992.

Strom, Robert G., and Ann L. Sprague. *Exploring Mercury: The Iron Planet*. New York: Springer, 2004.

Sullivan, W. T., ed. *The Early Years of Radio Astronomy*. New York: Cambridge University Press, 1984.

Sullivan, Walter. *Assault on the Unknown*. New York: McGraw-Hill, 1961.

Tabak, John. *A Look at Neptune*. London: Franklin Watts, 2003.

Taff, Laurence G. *Celestial Mechanics*. New York: Wiley-Interscience, 1985.

Tarbuck, Edward J., and Frederick K. Lutgens. *Earth: An Introduction to Physical Geology*. Illustrated by Dennis Tasa. 9th ed. Upper Saddle River, N.J.: Pearson Prentice Hall, 2008.

Tayler, Roger J. *Galaxies: Structure and Evolution*. Rev. ed. New York: Cambridge University Press, 1993.

Taylor, Edwin F., and John A. Wheeler. *Space-Time Physics*. San Francisco: W. H. Freeman, 1966.

Taylor, Stuart R. *Planetary Science: A Lunar Perspective*. Houston: Lunar and Planetary Institute, 1982.

Taylor, Stuart R., and Scott M. McLennan. *The Continental Crust: Its Composition and Evolution*. Boston: Blackwell Scientific, 1985.

Thackray, John. *The Age of the Earth*. New York: Cambridge University Press, 1989.

Thomas, Paul J., et al. *Comets and the Origin and Evolution of Life*. 2d ed. New York: Springer, 2006.

Thompson, Roy, and Frank Oldfield. *Environmental Magnetism*. London: Allen & Unwin, 1986.

Thornton, Stephen T., and Andrew Rex. *Modern Physics for Students and Engineers*. 3d ed. New York: Brooks/Cole, 2005.

Tielens, A. G. G. M. *The Physics and Chemistry of the Interstellar Medium*. Cambridge, England: Cambridge University Press, 2005.

Time-Life Books. *Comets, Asteroids, and Meteorites*. Alexandria, Va.: Author, 1990.

_____. *The Far Planets*. Alexandria, Va.: Author, 1988.

_____. *The Near Planets*. Alexandria, Va.: Author, 1989.

_____. *The New Astronomy*. Alexandria, Va.: Author, 1989.

Tipler, Paul A., and Ralph Llewellyn. *Modern Physics*. 5th ed. New York: W. H. Freeman, 2007.

Tobias, Russell R., and David G. Fisher, eds. *USA in Space*. 3d ed. Pasadena, Calif.: Salem Press, 2006.

Tocci, Salvadore. *A Look at Uranus*. New York: Franklin Watts, 2003.

Tombaugh, Clyde W., and Patrick Moore. *Out of the Darkness: The Planet Pluto*. New York: New American Library, 1981.

Trefil, James S. *The Moment of Creation: Big Bang Physics from Before the First Millisecond to the Present Universe*. New York: Charles Scribner's Sons, 1983.

_____. *Other Worlds: Images of the Cosmos from Earth and Space*. Washington, D.C.: National Geographic Society, 1999.

Trujillo, Alan, and Harold Thurman. *Essentials of Oceanography*. 9th ed. Upper Saddle River, N.J.: Prentice Hall, 2007.

Trumper, Joachin, and Gunther Hasinger. *The Universe in X Rays*. New York: Springer, 2008.

Tucker, Wallace. *The Star Splitters: The High Energy Astronomy Observatories*. NASA SP-466. Washington, D.C.: Government Printing Office, 1984.

Tucker, Wallace, and Riccardo Giacconi. *The X-ray Universe*. Cambridge, Mass.: Harvard University Press, 1985.

Tucker, Wallace, and Karen Tucker. *The Cosmic Inquirers: Modern Telescopes and Their Makers*. Cambridge, Mass.: Harvard University Press, 1986.

Tumlinson, Rick N., and Erin Medlicott, eds. *Return to the Moon*. New York: Collector's Guide, 2005.

Tyson, Neil deGrasse. *The Pluto Files*. New York: W. W. Norton, 2009.

Uchupi, E., and K. Emery. *Morphology of the Rocky Members of the Solar System*. New York: Springer, 1993.

Unsöld, Albrecht, and Bodo Baschek. *The New Cosmos: An Introduction to Astronomy and Astrophysics*. 5th ed. New York: Springer, 2001.

Van Allen, James A. *The Magnetospheres of Eight Planets and the Moon*. Oslo, Norway: Norwegian Academy of Science and Letters, 1990.

Van der Meer, Freek D., and Steven M. De Jong, eds. *Imaging Spectrometry*. New York: Kluwer Academic, 2002.

Van der Pluijm, Ben, and Stephen Marshak. *Earth's Structure*. 2d ed. New York: W. W. Norton, 2003.

Van Pelt, Michel. *Space Invaders: How Robotic Spacecraft Explore the Solar System*. New York: Springer, 2006.

Verschuur, Gerrit L. *Impact! The Threat of Comets and Asteroids*. New York: Oxford University Press, 1997.

_____. *The Invisible Universe: The Story of Radio Astronomy*. 2d ed. New York: Springer, 2006.

Vogel, Shawna. *Naked Earth: The New Geophysics*. New York: Plume, 1996.

Voit, Mark. *Hubble Space Telescope: New Views of the Universe*. New York: Harry N. Abrams, 2000.

Wagner, Jeffrey K. *Introduction to the Solar System*. Philadelphia: Saunders College Publishing, 1991.

Wall, J. V., ed. *Optics in Astronomy*. New York: Cambridge University Press, 1993.

Walter, Malcolm. *The Search for Life on Mars*. Cambridge, Mass.: Perseus Books, 1999.

Ward, Peter. *Life as We Do Not Know It: The NASA Search for (and Synthesis of) Alien Life*. New York: Penguin, 2007.

Ward, Peter, and Donald Brownlee. *Rare Earth: Why Complex Life Is Uncommon in the Universe*. New York: Springer, 2000.

Washburn, Mark. *Distant Encounters: The Exploration of Jupiter and Saturn*. New York: Harcourt Brace Jovanovich, 1982.

Wasson, John T. *Meteorites: Classification and Properties*. New York: Springer, 1974.

_____. *Meteorites: Their Record of Early Solar-System History*. New York: W. H. Freeman, 1985.

Webb, Stephen. *If the Universe Is Teeming with Aliens . . . Where Is Everybody? Fifty Solutions to Fermi's Paradox and the Problem of Extraterrestrial Life*. New York: Springer, 2002.

Wedepohl, Karl H. *Geochemistry*. New York: Holt, Rinehart and Winston, 1971.

Weeks, T. C. *Very High Energy Gamma Ray Astronomy*. New York: Taylor & Francis, 2003.

Weinberg, Steven. *Cosmology*. New York: Oxford University Press, 2008.

_____. *The First Three Minutes*. New York: Bantam Books, 1977.

Weiner, Jonathan. *Planet Earth*. New York: Bantam Books, 1986.

Weintraub, David A. *Is Pluto a Planet? A Historical Journey Through the Solar System*. Princeton, N.J.: Princeton University Press, 2006.

Wentzel, G. Donat. *The Restless Sun*. Washington, D.C.: Smithsonian Institution Press, 1989.

Wheeler, J. Craig. *Cosmic Catastrophes: Supernovae, Gamma-Ray Bursts, and Adventures in Hyperspace*. New York: Cambridge University Press, 2000.

Wheeler, John Archibald. *A Journey into Gravity and Spacetime*. New York: Scientific American Library, 1999.

Whipple, Fred L. *The Mystery of Comets*. Washington, D.C.: Smithsonian Institution Press, 1985.

White, Oran R., ed. *The Solar Output and Its Variation*. Boulder: Colorado Associated University Press, 1977.

Whitney, Charles. *The Discovery of Our Galaxy*. New York: Alfred A. Knopf, 1971.

Wicander, Reed, and James Monroe. *Historical Geology*. 5th ed. Florence, Ky.: Brooks/Cole, 2006.

Wilford, John Noble. *Mars Beckons: The Mysteries, the Challenges, the Expectations of Our Next Great Adventure in Space*. New York: Alfred Knopf, 1990.

Wilhelms, Don E. *The Geologic History of the Moon*. U.S. Geological Survey Professional Paper 1348. Washington, D.C.: Government Printing Office, 1987.

_____. *To a Rocky Moon: A Geologist's History of Lunar Exploration*. Phoenix: University of Arizona Press, 1994.

Wilkie, Tom, and Mark Rosselli. *Visions of Heaven: The Mystery of the Universe Revealed by the Hubble Space Telescope*. London: Hodder & Stoughton, 1998.

Will, Clifford. *Was Einstein Right? Putting General Relativity to the Test*. 2d ed. New York: Basic Books, 1993.

Woolfson, Michael. *The Formation of the Solar System: Theories Old and New*. London: Imperial College Press, 2007.

Wudka, Jose. *Space-Time, Relativity, and Cosmology*. New York: Cambridge University Press, 2006.

Wynn-Williams, Gareth. *The Fullness of Space: Nebulae, Stardust, and the Interstellar Medium*. Cambridge, England: Cambridge University Press, 1992.

Young, Carolynn, ed. *The Magellan Venus Explorers' Guide*. Pasadena, Calif.: Jet Propulsion Laboratory, California Institute of Technology, National Aeronautics and Space Administration, 1990.

Young, Hugh D., and Roger A. Freedman. *University Physics with Modern Physics*. 11th ed. New York: Addison Wesley, 2003.

Zeilik, Michael. *Astronomy: The Evolving Universe*. 9th ed. New York: John Wiley and Sons, 2002.

Zeilik, Michael, and Stephen A. Gregory. *Introductory Astronomy and Astrophysics*. 4th ed. Fort Worth, Tex.: Saunders College Publishing, 1998.

Zimmerman, Robert. *The Universe in a Mirror: The Saga and the Visionaries Who Built It*. Princeton, N.J.: Princeton University Press, 2008.

Zirker, J. B. *An Acre of Glass: A History and Forecast of the Telescope*. Baltimore: Johns Hopkins University Press, 2005.

_____. *Journey from the Center of the Sun*. Princeton, N.J.: Princeton University Press, 2002.

_____. *Sunquakes: Probing the Interior of the Sun*. Baltimore: Johns Hopkins University Press, 2003.

_____. *Total Eclipses of the Sun*. Expanded ed. Princeton, N.J.: Princeton University Press, 1995.

Zubrin, Robert. *Entering Space: Creating a Spacefaring Civilization*. New York: Tarcher, 2000.

Web Sites

Listed below are more than 110 authoritative Web sites current as of 2009. Although every effort has been made to ensure accuracy, Web sites are continually being updated, and there may be changes to the site address listed. A subject or keyword search through any of the major search engines will help locate a new address if the universal resource locators (URLs) listed below have been changed.

Adler Planetarium
http://www.adlerplanetarium.org

Amazing Space
http://amazing-space.stsci.edu

American Meteor Society
http://www.amsmeteors.org

Ames Research Center
http://www.nasa.gov/centers/ames

Apollo Program
http://spaceflight.nasa.gov/history/apollo

Astronomical Society of the Pacific
http://www.astrosociety.org/index.html

Astronomy Café, The
http://www.astronomycafe.net

Astronomy Magazine
http://www.astronomy.com

Astronomy Now Magazine
http://www.astronomynow.com

Astronomy Picture of the Day
http://antwrp.gsfc.nasa.gov/apod

Beijing Planetarium
http://www.bjp.org.cn/en/index.htm

British National Space Centre
http://www.bnsc.gov.uk

Cambridge Cosmology Public Home Page
http://www.damtp.cam.ac.uk/user/gr/public/
cos_home.html

Canadian Space Agency/L'Agence Spatiale
Canadienne
http://www.space.gc.ca/asc/index.html

Cassini Equinox Mission
http://saturn.jpl.nasa.gov

Cassini-Huygens Mission
http://www.nasa.gov/mission_pages/cassini/
main

Chandra X-ray Observatory Center
http://chandra.harvard.edu

Comet Shoemaker-Levy Collision with Jupiter
http://www2.jpl.nasa.gov/sl9

Compton Gamma Ray Observatory
http://cossc.gsfc.nasa.gov

Dawn Mission
http://dawn.jpl.nasa.gov

Deep Impact
http://deepimpact.jpl.nasa.gov

Discovery Program
http://discovery.nasa.gov

Earth Observations Photography, Space
Shuttle
http://earth.jsc.nasa.gov/sseop/efs

Exoplanets: The Search for Extrasolar Planets
http://exoplanets.org

Exploration System Mission Directorate
http://exploration.nasa.gov

Explorer Missions
http://nssdc.gsfc.nasa.gov/multi/
explorer.html

Galileo: Journey to Jupiter
http://www2.jpl.nasa.gov/galileo

Gamma-ray Large Area Space Telescope
http://glast.gsfc.nasa.gov

Gemini Program
http://www-pao.ksc.nasa.gov/kscpao/
history/gemini/gemini.htm

Goddard Space Flight Center
http://www.nasa.gov/centers/goddard

Gravity Probe B
http://www.gravityprobeb.com

Great Images in NASA
http://grin.hq.nasa.gov

Griffith Park Observatory
   http://www.griffithobs.org

Harvard-Smithsonian Center for Astrophysics
   http://www.cfa.harvard.edu

Hayden Planetarium
   http://www.haydenplanetarium.org

Heavens-Above Satellite Observations
   http://www.heavens-above.com

Hubblesite News Center
   http://hubblesite.org/newscenter

International Year of Astronomy: 2009
   http://www.astronomy2009.org

Jet Propulsion Laboratory (JPL)
   http://www.jpl.nasa.gov

Jodrell Bank Centre for Astrophysics
   http://www.jb.man.ac.uk

Johannesburg Planetarium
   http://www.planetarium.co.za/

Johnson Space Center
   http://www.nasa.gov/centers/johnson/home/
   index.html

JPL: Goldstone Complex
   http://deepspace.jpl.nasa.gov/dsn/gallery/
   goldstone.html

JPL: Multimedia
   http://www.jpl.nasa.gov/multimedia/
   index.cfm

JPL: The Solar System
   http://www.jpl.nasa.gov/solar-system/
   index.cfm

JPL: Solar System Dynamics
   http://ssd.jpl.nasa.gov

JPL: Stars and Galaxies
   http://www.jpl.nasa.gov/stars-galaxies/
   index.cfm

Kennedy Space Center
   http://www.nasa.gov/centers/kennedy

Lunar and Planetary Institute
   http://www.lpi.usra.edu

Lunar Prospector
   http://lunar.arc.nasa.gov

Mars Exploration Program
   http://mars.jpl.nasa.gov

Mars Exploration Rover Mission
   http://marsrovers.jpl.nasa.gov/home

Mars Global Surveyor: Mars Orbiter Camera
   http://mars.jpl.nasa.gov/mgs/msss/camera/
   images/

Mars Today
   http://www-mgcm.arc.nasa.gov

Marshall Space Flight Center
   http://www.nasa.gov/centers/marshall

Melbourne Planetarium
   http://museumvictoria.com.au/planetarium

National Aeronautics and Space
   Administration (NASA)
   http://www.nasa.gov

NASA: Images
   http://www.nasaimages.org

NASA: Multimedia
   http://www.nasa.gov/multimedia/index.html

NASA: Origins of the Universe
   http://origins.jpl.nasa.gov

NASA: Planetary Photojournal
   http://photojournal.jpl.nasa.gov

NASA: Science
   http://www.earth.nasa.gov

NASA: Solar System Mission
   http://www.nasa.gov/missions/timeline/
   current/solar-system_missions.html

NASA Space Place
   http://spaceplace.nasa.gov/en/kids

NASA Space Place: "Ask Dr. Marc"
   http://spaceplace.nasa.gov/en/kids/
   phonedrmarc

NASA: Space Science
   http://spacescience.nasa.gov

NASA: Space Science Data Center
   http://nssdc.gsfc.nasa.gov

NASA: Space Science Photo Gallery
   http://nssdc.gsfc.nasa.gov/photo_gallery

NASA: Space Science Planetary Missions
   http://nssdc.gsfc.nasa.gov/planetary/
   projects.html

NASA: Sun-Earth Connection Information Forum
http://sunearth.gsfc.nasa.gov/missions/index.php

National Astronomy and Ionosphere Center: Arecibo Observatory
http://www.naic.edu

National Geographic Astronomy
http://www.nationalgeographic.com/stars

National Geographic's Virtual Solar System
http://science.nationalgeographic.com/science/space/solar-system

National Oceanic and Atmospheric Administration
http://www.noaa.gov

National Optical Astronomy Observatory
http://www.noao.edu

National Radio Astronomy Observatory
http://www.nrao.edu

National Weather Service
http://www.nws.noaa.gov

New Horizons Pluto-Kuiper Belt Mission
http://pluto.jhuapl.edu

The Nine Planets: A Multimedia Tour of the Solar System
http://seds.lpl.arizona.edu/nineplanets/nineplanets

Pacific Science Center
http://www.pacsci.org

Phoenix Mars Lander 2007
http://phoenix.lpl.arizona.edu

Planetary Society, The
http://planetary.org

Planet Quest: Exoplanet Exploration
http://planetquest.jpl.nasa.gov

PlanetScapes
http://planetscapes.com

Russian Space Web: Chronology of Space Exploration
http://www.russianspaceweb.com/chronology.html

Satellite Information, World Data Center System
http://www.ngdc.noaa.gov/wdc

Search for Extraterrestrial Intelligence Institute
http://www.seti-inst.edu

Sky and Telescope Magazine
http://www.skyandtelescope.com

Solar and Heliospheric Observatory
http://sohowww.nascom.nasa.gov

Solar System Exploration
http://sse.jpl.nasa.gov

Space Foundation
http://www.spacefoundation.org

Space Station: Science Operations News
http://www.scipoc.msfc.nasa.gov

Space Telescope Science Institute
http://www.stsci.edu

Space Telescope Science Institute: Hubble Space Telescope
http://www.stsci.edu/hst

Space Telescope Science Institute: James Webb Space Telescope
http://www.stsci.edu/jwst

Space.com: Astronomy and Science News and Information
http://www.space.com/scienceastronomy

Spaceref.com
http://www.spaceref.com

Spaceweather.com
http://www.spaceweather.com

Spitzer Space Telescope
http://www.spitzer.caltech.edu/spitzer

Stanford Solar Center
http://solar-center.stanford.edu

Stardust Project
http://stardust.jpl.nasa.gov

Swift Gamma Ray Burst Mission
http://swift.gsfc.nasa.gov/docs/swift/swiftsc.html

The Two Micron All Sky Survey
http://pegasus.phast.umass.edu

Ulysses
  http://ulysses.jpl.nasa.gov

Ulysses: European Space Agency
  http://sci.esa.int/ulysses

U.S. Geological Survey
  http://www.usgs.gov

Views of the Solar System
  http://www.solarviews.com/eng/
    homepage.htm

W. M. Keck Observatory
  http://www.keckobservatory.org

Welcome to the Planets
  http://pds.jpl.nasa.gov/planets

Wilkinson Microwave Anisotropy Probe
  http://map.gsfc.nasa.gov

Index

Subject Index